ANTARCTIC RESEARCH SERIES

American Geophysical Union

ANTARCTIC
RESEARCH
SERIES

American Geophysical Union

Volume 1   BIOLOGY OF THE ANTARCTIC SEAS
   Milton O. Lee, *Editor*

Volume 2   ANTARCTIC SNOW AND ICE STUDIES
   Malcolm Mellor, *Editor*

Volume 3   POLYCHAETA ERRANTIA OF ANTARCTICA
   Olga Hartman

Volume 4   GEOMAGNETISM AND AERONOMY
   A. H. Waynick, *Editor*

Volume 5   BIOLOGY OF THE ANTARCTIC SEAS II
   George A. Llano, *Editor*

Volume 6   GEOLOGY AND PALEONTOLOGY OF THE ANTARCTIC
   Jarvis B. Hadley, *Editor*

Volume 7   POLYCHAETA MYZOSTOMIDAE AND SEDENTARIA OF ANTARCTICA
   Olga Hartman

Volume 8   ANTARCTIC SOILS AND SOIL FORMING PROCESSES
   J. C. F. Tedrow, *Editor*

Volume 9   STUDIES IN ANTARCTIC METEOROLOGY
   Morton J. Rubin, *Editor*

Volume 10   ENTOMOLOGY OF ANTARCTICA
   J. Linsley Gressitt, *Editor*

Volume 11   BIOLOGY OF THE ANTARCTIC SEAS III
   Waldo L. Schmitt and George A. Llano, *Editors*

Volume 12   ANTARCTIC BIRD STUDIES
   Oliver L. Austin, Jr., *Editor*

Volume 13   ANTARCTIC ASCIDIACEA
   Patricia Kott

Volume 14   ANTARCTIC CIRRIPEDIA
   William A. Newman and Arnold Ross

Volume 15   ANTARCTIC OCEANOLOGY
   Joseph L. Reid, *Editor*

Volume 16   ANTARCTIC SNOW AND ICE STUDIES II
   A. P. Crary, *Editor*

Volume 16 | ANTARCTIC RESEARCH SERIES

Antarctic Snow and Ice Studies II

A. P. Crary, *Editor*

Published with the aid of a grant from the National Science Foundation

PUBLISHER

AMERICAN GEOPHYSICAL UNION

OF THE

National Academy of Sciences—National Research Council

1971

Volume 16 | ANTARCTIC RESEARCH SERIES

ANTARCTIC SNOW AND ICE STUDIES II

A. P. Crary, *Editor*

Copyright © 1971 by the American Geophysical Union
Suite 435, 2100 Pennsylvania Avenue, N.W.
Washington, D. C. 20037

Library of Congress Catalog Card No. 64-60078
International Standard Book No. 0-87590-116-6

List Price, $24.50

Printed by
WAVERLY PRESS, INC.
Baltimore, Maryland

THE ANTARCTIC RESEARCH SERIES

The Antarctic Research Series is designed to provide a medium for presenting authoritative reports on the extensive and detailed scientific research work being carried out in Antarctica. The series has been successful in eliciting contributions from leading research scientists engaged in antarctic investigations; it seeks to maintain high scientific and publication standards. The scientific editor for each volume is chosen from among recognized authorities in the discipline or theme it represents, as are the reviewers on whom the editor relies for advice.

Beginning with the scientific investigations carried out during the International Geophysical Year, reports of research results appearing in this series represent original contributions too lengthy or otherwise inappropriate for publication in the standard journals. In some cases an entire volume is devoted to a monograph. The material published is directed not only to specialists actively engaged in the work but to graduate students, to scientists in closely related fields, and to interested laymen versed in the biological and the physical sciences. Many of the earlier volumes are cohesive collections of papers grouped around a central theme. Future volumes may concern themselves with regional as well as disciplinary aspects or with a comparison of antarctic phenomena with those of other regions of the globe. But the central theme of Antarctica will dominate.

In a sense, the series continues the tradition dating from the earliest days of geographic exploration and scientific expeditions—the tradition of the expeditionary volumes which set forth in detail everything that was seen and studied. This tradition is not necessarily outmoded, but in much of the present scientific work one expedition blends into the next, and it is no longer scientifically meaningful to separate them arbitrarily. Antarctic research has a large degree of coherence; it deserves the modern counterpart of the expeditionary volumes of past decades and centuries which the Antarctic Research Series provides.

With the aid of a grant from the National Science Foundation in 1962, the American Geophysical Union initiated the Antarctic Research Series and appointed a Board of Associate Editors to implement it. A supplemental grant received in 1966, the income from the sale of volumes in the series, and income from reprints and other sources have enabled the AGU to continue this series. The response of the scientific community and the favorable comments of reviewers cause the board to look forward with optimism to the continued success of this endeavor.

To represent the broad scientific nature of the series, the members of the Board were chosen from all fields of antarctic research. At the present time they include: Avery A. Drake, Jr., representing geology and solid earth geophysics; A. P. Crary, seismology and glaciology; George A. Llano, botany and zoology; Martin A. Pomerantz, aeronomy and geomagnetism; Morton J. Rubin, meteorology and oceanography; David L. Pawson, biology; Waldo L. Schmitt, member emeritus; and Laurence M. Gould, honorary chairman. Fred G. Alberts, secretary to the U.S. Advisory Committee on Antarctic Names, gives valuable assistance in verifying place names, locations, and maps.

Morton J. Rubin
Chairman, Board of Associate Editors
Antarctic Research Series

PREFACE

This volume of the Antarctic Research Series represents an attempt to compile in one publication the remaining results of the extensive U.S. oversnow traverses in Antarctica between 1957 and 1967. Although this attempt was not completely successful, in that some traverse results still remain to be published, I think that this volume will give the reader a fair sample of the methodology and the great effort that was required to wrest from the huge continent a few fundamental data. From the beginning of 1957, the traverses, though including an assortment of observations, had two principal objectives: measurement of the thickness of the ice and measurement of the annual accumulation of snow. Here the reader will find the old and new methods of approaching both objectives.

The discussions of subjective stratigraphic studies of the snow layers by Benson, Cameron, Koerner, Rundle, and Taylor aptly illustrate the difficulties in this method. Without the benefit of some surface stake networks at the permanent stations, the pit studies on the traverses might well have led to serious error. In retrospect, many of these reports may appear as efforts to justify the long tedious hours spent in the snow pit studies. As was said in early recruitments, to be a glaciologist one should first of all love to dig snow pits. The location of the 1954 nuclear test debris horizon and the application of the ^{240}Pb unstable isotope method by Picciotto and his colleagues have brought a new dimension to these studies.

The geophysical methods of measuring ice thickness are exemplified by the work of Bentley, Beitzel, Chang, and Jiracek. Here we can see some profitable studies that go beyond the mere identification of the echo of a sound wave from the bottom of the ice. Fortunately the radio sounding equipment saved the day in East Antarctica for the seismologists, just as the location of the 1954 nuclear test layer, plus the analysis of the radioactive materials, has saved the day for the glaciologists in that area.

The volume is completed by additional studies of a varied nature. The painstaking work by glacial geologists in the relatively sparse ice-free area of Antarctica is illustrated in the dry valley area by Calkin, who outlines a possible history of the antarctic ice sheet, as written in the glacial sediments. Hamilton and O'Kelly discuss the progress of studies in particulates, a field that should be given more emphasis in the future. Dewart writes on one of the first U.S. geophysical studies in the neighborhood of the Palmer station in the Antarctic Peninsula.

The studies included here skirt the main issues of antarctic glaciology: what is the present ice budget and what is the history of the present ice sheet? One would like to say that, with the descriptive phases of the study of the ice sheet about completed, more attention can be focused on these more topical subjects. However, this simple statement would hardly convey the extreme difficulty of tackling these problems directly, though we will certainly see many aspects attacked with renewed vigor in the future.

A. P. CRARY

CONTENTS

The Antarctic Research Series
    Morton J. Rubin... v

Preface
    A. P. Crary... vii

Geophysical Exploration in Marie Byrd Land, Antarctica
    Charles R. Bentley and Feng-Keng Chang...................... 1

Geophysical Exploration in Queen Maud Land, Antarctica
    John E. Beitzel... 39

Seismic Evidence for Moraine within the Basal Antarctic Ice Sheet
    Charles R. Bentley... 89

Seismic Anisotropy in the West Antarctic Ice Sheet
    Charles R. Bentley... 131

Gravimeter Observations on Anvers Island and Vicinity
    Gilbert Dewart... 179

Secular Increase of Gravity at South Pole Station
    Charles R. Bentley... 191

Velocity of Electromagnetic Waves in Antarctic Ice
    G. R. Jiracek and Charles R. Bentley............................. 199

Glaciological Studies on the South Pole Traverse, 1962–1963
    Lawrence D. Taylor... 209

A Stratigraphic Method of Determining the Snow Accumulation Rate at Plateau Station, Antarctica, and Application to South Pole–Queen Maud Land Traverse 2, 1965–1966
    R. M. Koerner.. 225

Snow Accumulation and Firn Stratigraphy on the East Antarctic Plateau
    Arthur S. Rundle.. 239

Accumulation on the South Pole–Queen Maud Land Traverse, 1964–1968
    E. Picciotto, G. Crozaz, and W. De Breuck..................... 257

Glaciological Studies at Byrd Station, Antarctica, 1963–1965
    Richard L. Cameron... 317

Stratigraphic Studies in the Snow at Byrd Station, Antarctica, Compared with Similar Studies in Greenland
    Carl S. Benson... 333

Investigation of Particulate Matter in Antarctic Firn
    Wayne L. Hamilton and M. E. O'Kelley.......................... 355

Glacial Geology of the Victoria Valley System, Southern Victoria Land, Antarctica
    Parker E. Calkin.. 363

GEOPHYSICAL EXPLORATION IN MARIE BYRD LAND, ANTARCTICA

Charles R. Bentley and Feng-Keng Chang

Geophysical and Polar Research Center, Department of Geology, University of Wisconsin, Madison 53705

Seismic, gravimetric, altimetric, and magnetic observations made along oversnow traverses in Marie Byrd Land and vicinity in 1959 and 1960 have provided a reconnaissance picture of this part of West Antarctica. The ice sheet surface slopes gently from a high in the region of the Executive Committee Range southwestward to the Ross ice shelf, but elsewhere exhibits a more complicated topography affected by the rugged subglacial relief. Before the formation of the ice sheet, a large island probably extended unbroken from the volcanic Executive Committee Range or Crary Mountains in the east to Edward VII Peninsula in the west, bounded on the north by open ocean and on the south by the Byrd subglacial basin. Lying off the east and northeast coast were several smaller volcanic islands. The mountains in the north-central part of this main island appear to belong to the plutonic and metamorphic province to the west. Throughout most of the region, there appears to be isostatic compensation for both the ice and the subglacial topography. Negative isostatic anomalies of -30 to -40 mgal occur near the Amundsen Sea coast, and associated with, but not superimposed upon, a subglacial trough in western Marie Byrd Land.

According to the description published before the International Geophysical Year by the *U.S. Board on Geographic Names* [1956], Marie Byrd Land is 'that portion of Antarctica lying east of Ross Ice Shelf and Ross Sea and south of the Pacific Ocean, extending approximately eastward to a line between the head of Ross Ice Shelf and Eights Coast' (about 100°W). However, 'the eastern limit of this land has been arbitrarily adopted, pending more definite mapping which may make it possible to draw boundaries along lines of natural demarcation.' No southern limit is given, since the interior of West Antarctica was nearly unknown at the time.

From geophysical observations on oversnow traverses conducted during and after the IGY, we now know that beneath a large part of the West Antarctic ice sheet the rock floor lies far below sea level. This region, which would be water-covered if the ice sheet were to melt (even after making allowance for isostatic rebound), has been named the Byrd subglacial basin. It runs from the Ross Sea south of the mountains of Marie Byrd Land as far as 100°W; eastward of this longitude it apparently forms a broad connection to the Amundsen Sea, as well as extending nearly to the Bellingshausen Sea [*Bentley*, 1964]. Even though it is not perfectly defined, the basin thus provides a natural subglacial boundary for Marie Byrd Land. We propose, then, that Marie Byrd Land be defined to be that part of Antarctica lying between the Ross Sea on the west, the Pacific Ocean on the north, and the Byrd subglacial basin on the south and east (see Figure 12). It is in this sense that the name will be used in the text of this paper. Maps, although labeled Marie Byrd Land, extend beyond its boundaries.

During 1959 and 1960, traverse parties carried out a reconnaissance examination of Marie Byrd Land. The major traverse was conducted during the 1959–1960 field season, the program including seismic reflection and refraction shooting, gravimetric and magnetic observations, and measurements of surface elevation. The traverse party left Byrd station on November 5, 1959, proceeding to the edge of the ice shelf bordering the Amundsen Sea at 73°55′S, 116°11′W via the Crary Mountains and Toney Mountain (Figure 2; see Figure 1 for index map). The same route was followed back as far as Toney Mountain (station 288 on the northward journey and station 493 returning), and re-measurements were made at all gravity, magnetic, and elevation stations. From there, the trail party traveled westward past the Usas Escarpment, Mount Petras, and the Flood Range to the Clark Mountains, thence southwestward to the Army-Navy Drive and back along this trail to Byrd station. The party reached Byrd station on February 8, 1960, after traveling a little more than 2000 km (about 1100 n. mi.).

In the preceding March, an 840-km (450-n. mi.)

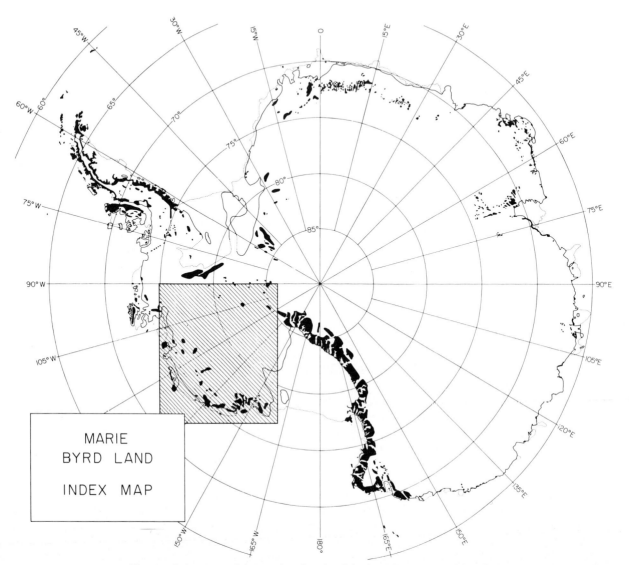

Fig. 1. Index map of Antarctica showing location of Marie Byrd Land.

trip along a triangular route was made to examine the geology of the Executive Committee Range and to cache fuel for the traverse the next summer. Surface elevation measurements were made, but time limitations and failure of equipment prevented other geophysical observations.

Included in this paper are the results of seismic soundings and elevation measurements made on the final section of the 1958–1959 Horlick Mountains traverse, from 82°08′S, 109°14′W to Byrd station, since the preceding report [*Bentley and Ostenso*, 1961] covered only observations up until the departure of the IGY party from the field on January 8, 1959. Magnetic data that were also collected have not yet been analyzed.

Members of the field party for the last part of the Horlick Mountains traverse were W. Chapman (leader), Chang, H. LeVaux, G. A. Doumani, and G. Bennett; for the Executive Committee Range traverse, J. Pirrit (leader), Chapman, Doumani, and Bennett; and for the Marie Byrd Land traverse, Pirrit (leader), Chang, P. E. Parks, LeVaux, Chapman, Doumani, Bennett, E. Boudette, K. Marks, and G. Widich.

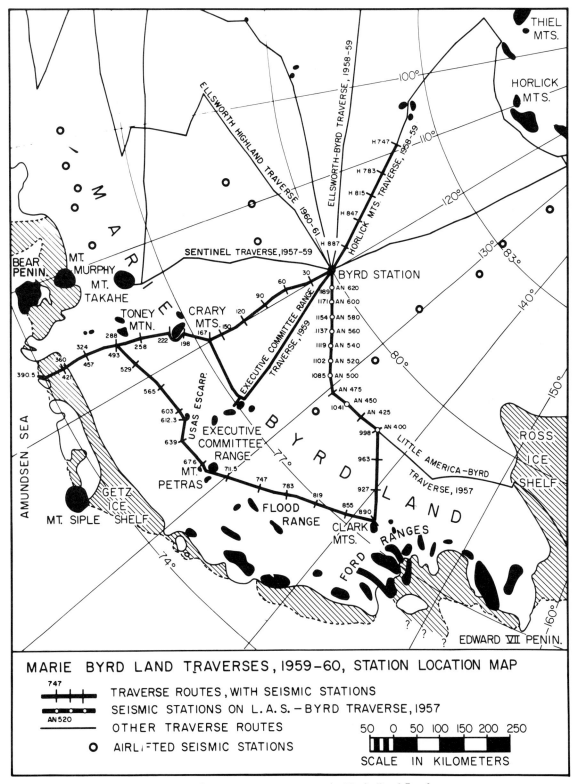

Fig. 2. Oversnow traverse routes in Marie Byrd Land.

GEOLOGIC SETTING

Discussions of the geological findings of these traverses have already been published [*Doumani and Ehlers*, 1962; *Doumani*, 1964]. The Executive Committee Range comprises five volcanic cones. Only two were visited, but according to their appearance from the air the others should be similar in nature. The rocks are exclusively basic volcanics, including basalts, andesites, and trachytes. Similar rocks are found in the Crary Mountains, which comprise four peaks, one about 25 km southeast of the other three, on Toney Mountain, which consists of two peaks about 20 km apart, and on Mount Takahe, 70 km to the east [*Anderson*, 1960]. The Usas escarpment and Mount Petras, which lie north of the Executive Committee Range, are composed, on the other hand, of rhyolitic and dacitic tuffs and flows, with relatively small amounts of granodiorite and basalt. Farther to the west, the Clark Mountains contain granodiorite, granite, and unfolded metasedimentary rocks that dip gently to the southeast. Still farther west in the Ford Ranges, beyond the limit of the traverse, highly folded geosynclinal sediments intruded by a suite of acidic batholiths were already known [*Warner*, 1945; *Passel*, 1945]; to the north in the same range, in the Fosdick Mountains, volcanic rocks are again found [*Fenner*, 1938]. The Rockefeller Mountains, in the western extremity of Marie Byrd Land, are composed of granite and metasediments [*Wade*, 1945].

FIELD EQUIPMENT AND PROCEDURES

The operating procedures in the field were similar to those of previous traverses [*Bentley and Ostenso*, 1961]. Standard seismic stations were spaced at intervals of 55 to 67 km (30 to 36 n. mi.). Besides reflection sounding, observations at seismic stations included snow-pit studies, determinations of temperature in 10-meter boreholes, gravimetric and magnetic measurements, solar observations to determine station position, and azimuth measurements on mountain peaks. Intermediate stations were made about every 5½ km (3 n. mi.), where altimetric, magnetic, gravimetric, and rammsonde data and measurements of wind speed, wind direction, and air temperature were recorded.

These procedures were followed on the Marie Byrd Land traverse up to the junction with the Army-Navy Drive. From this point on, soundings were made only at stations AN425 and AN475, where seismic measurements had not been made by the Little America–Byrd traverse party [*Bentley and Ostenso*, 1961], and gravity, magnetic, and elevation measurements were made at the stations 8 km apart that had been occupied by the earlier group. On the last part of the Horlick Mountains traverse, the standard procedures were followed, except that gravity observations were not made, since the gravimeter was not in operation.

For the seismic work, a 24-trace Texas Instruments 7000B Portable Seismograph System was used. This unit has a basic frequency range of 5 to 500 Hz and a wide range of possible filter settings. Automatic gain control and mixing are also provided, but were not generally used, as experience had shown that neither produced significant improvement in the quality of the seismograms. Furthermore, it was desirable to record true amplitudes of ground motion. Power was provided by two 12-volt heavy-duty lead-acid batteries that were charged from the vehicle generator system.

The seismic spreads normally comprised two cables, each with 12 geophones at 30-meter intervals. Three of the geophones on each cable were often placed horizontally to detect possible shear or transformed compressional-shear reflections. The cables were laid out either in line, with shots fired in the center, or in the form of an L, with shots at the corner. The charge usually consisted of a one-pound Nitramon primer or a primer with one or two pounds of Nitramon S fired in a three- or four-meter auger hole. Low-cut filtering was usually set at 60 or 90 Hz, and high-cut at 160 or 215 Hz. Failure to record a reflection occurred only at station 30 early in November, when the shot-generated noise level was still high, and at station 603, where the ice was thin. At other stations, the reflection quality was generally excellent (Figure 3).

Frost gravimeter C2-55, a temperature-controlled meter with low drift characteristics, was used for the gravity measurements on the Marie Byrd Land traverse. With a calibration constant of 0.08213 mgal/scale division, this meter had a reading range without resetting of only 125 mgal.

Before the start of the traverse, in the expectation of low ambient temperatures in the traverse vehicle, the operating temperature of the gravimeter was set to 68°F (20°C), some 50°F (28°C) below its ordinary operating temperature. This proved to be a mistake, since the temperature inside the vehicle frequently exceeded 68°F (20°C). As a result, tem-

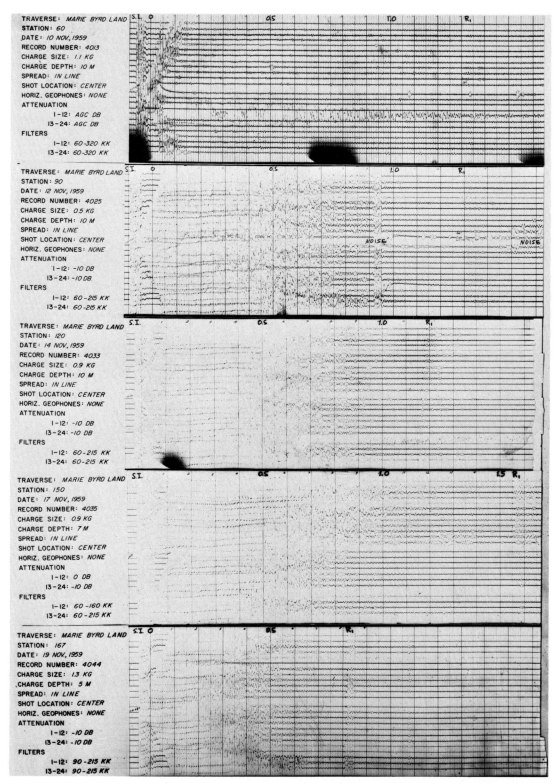

Fig. 3. Seismograms. Heavy timing lines 0.1 sec apart. *S.I.*: shot instant. R_1: reflection from ice-rock interface. R_{1P}, R_{1S}, R_{PS}: P, S, and converted P to S (and S to P) reflections from ice-water interface, respectively. R_W: reflections from water-sediment interface. R_1': reflection from sediment-rock interface. R_2: first multiple from ice-rock interface. (Continued on following pages.)

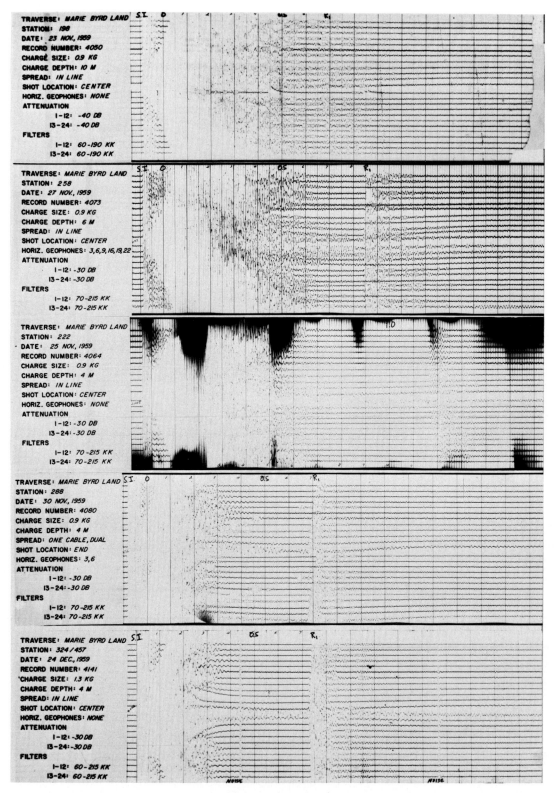

Fig. 3. (continued)

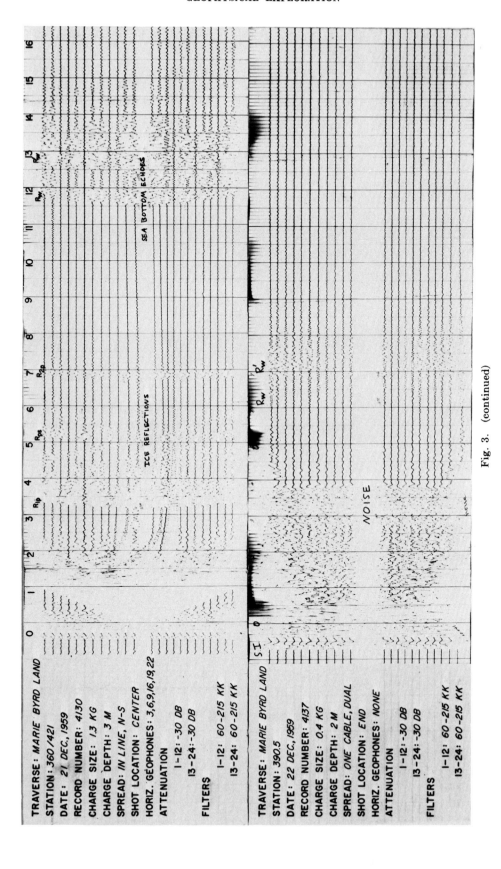

Fig. 3. (continued)

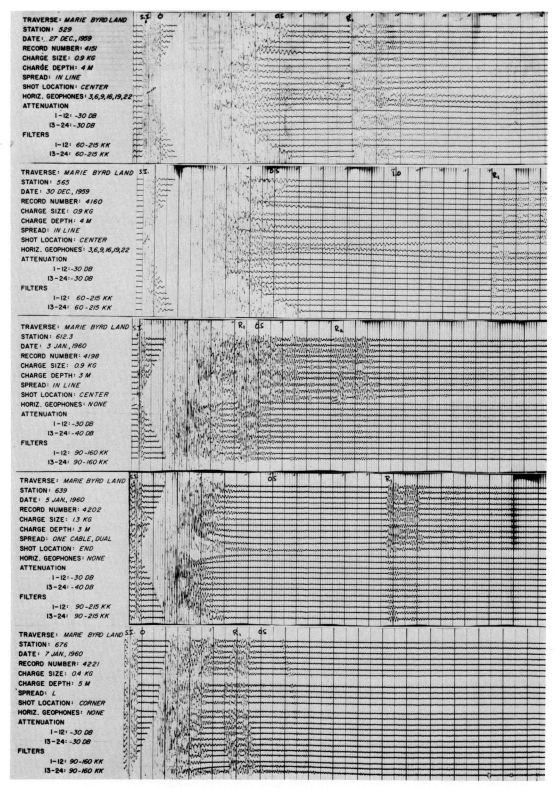

Fig. 3. (continued)

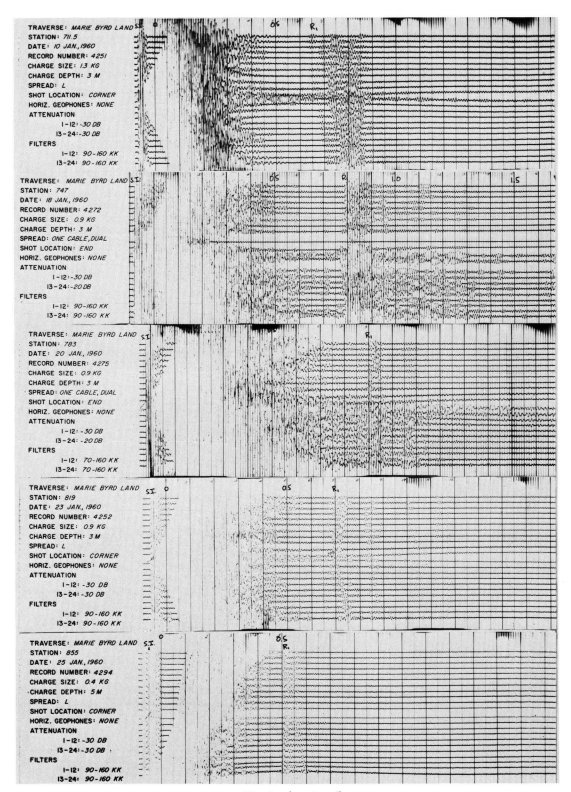

Fig. 3. (continued)

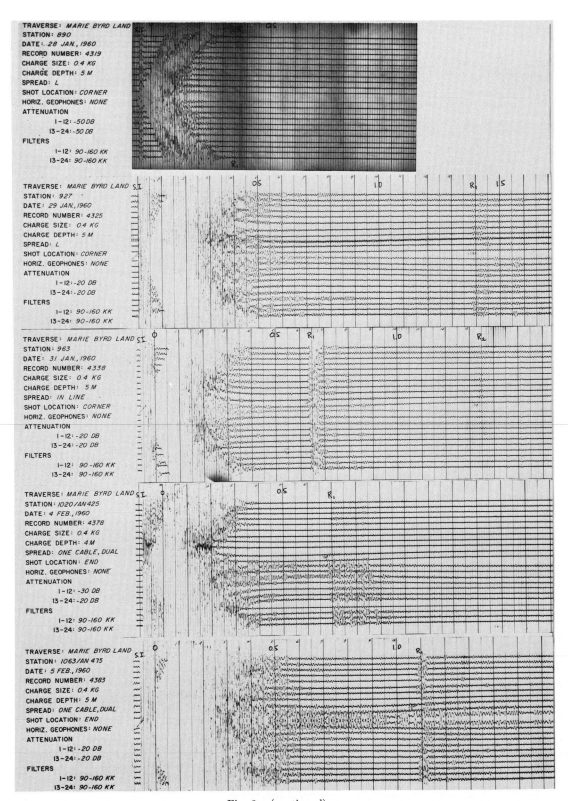

Fig. 3. (continued)

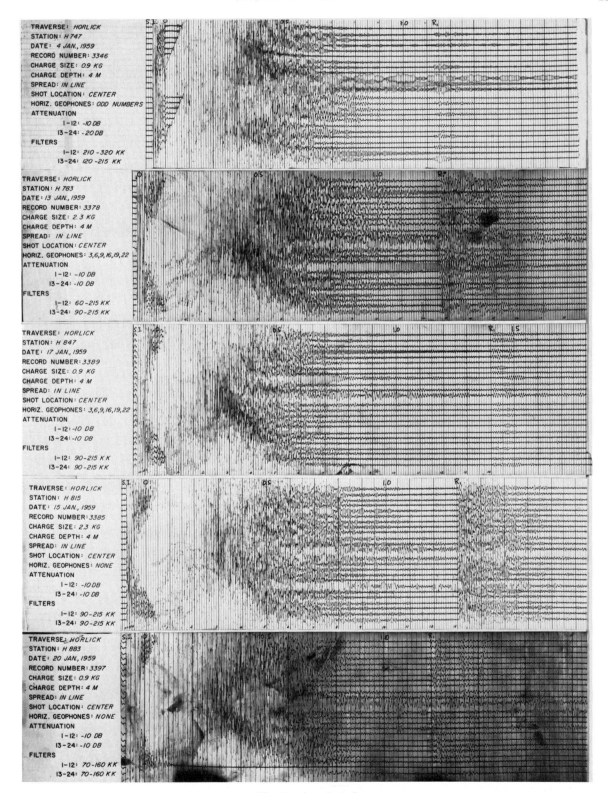

Fig. 3. (concluded)

perature regulation was lost. Rather than attempting to change the operating temperature under difficult field conditions and with time at a premium, a record of meter temperature was kept, and corrections to the observed gravity values were applied as described below.

An Arvela vertical-component magnetometer, model T7, was used for magnetic measurements. This is a wide-range null-reading instrument with temperature compensation. Two magnetic observations were made at each station, one with the magnetometer oriented toward magnetic north and the other with it oriented toward the south, in order to eliminate the effect of imperfect level of the instrument, as well as to obtain an estimate of the reading error. The calibration constant of the instrument, as determined before the traverse, was 27.3 γ/scale division. The magnetometer suffered a bad jar between stations 526 and 529, incurring a tare of about 500 γ, although there was no apparent damage to the instrument. Unfortunately, the calibration constant was not checked immediately after this accident, but at the termination of the traverse it had changed only 1% to 27.6 γ/scale division.

Elevations were measured with four Wallace and Tiernan altimeters (until one broke; see below). The altimeters were carried in two traverse vehicles traveling 5.5 km (3 n. mi.) apart, the rear vehicle occupying successive stations previously occupied by the lead vehicle. Readings were coordinated by radio. Wind speed, wind direction, and temperature measurements were taken at each station. All altimeters were read together at the beginning and end of each travel day.

Station positions (Table 1) were determined from sun lines at seismic stations and by dead reckoning at intermediate points. The observations were made by W. H. Chapman of the U.S. Geological Survey.

DATA REDUCTION

Altimetry

The altimetric data were reduced in the following way. Differences between readings of two separate pairs of altimeters at consecutive stations on the Marie Byrd Land traverse (corrected for the atmospheric temperature) were computed and averaged up to station 606, where one altimeter ceased to operate. From there on, only one set of differences was computed, the third altimeter being used as a check against reading errors. Instrument temperature corrections were applied to the indicated elevations. The index correction for altimeter pairs was calculated from the differences in indicated elevation when altimeters were read together at the beginning and end of each travel day. Any change in this index correction was distributed linearly through the day. On the Executive Committee Range traverse, only three altimeters were used, and the reduction method was the same as for the latter part of the Marie Byrd Land traverse.

Elevations determined on the two traverses along their common section of track were averaged from Byrd station to station 75, the Marie Byrd Land traverse values being used from there on. For the repeated traverse section between Toney Mountain and the Amundsen Sea coast, the values from the return journey were used, since wind measurements indicated more stable pressure conditions. Along the Army-Navy Drive, elevations computed from the two traverses were averaged.

Pressure corrections were applied according to the synoptic 700-mbar pressure maps for the Antarctic regions prepared at the International Antarctic Analysis Centre in Melbourne. The accumulated correction increased gradually from Byrd station to the Amundsen Sea, where it reached a value of +50 meters. From there it decreased slowly to a minimum value of −25 meters near the junction with the Army-Navy Drive, and returned nearly to zero upon return to Byrd station. Closure corrections of +10 meters at the Amundsen Sea (where absolute elevation was determined from optical angles measured to an iceberg) and −20 meters at Byrd station were distributed linearly along the corresponding sections of the traverse. Corrected elevation values are listed in Table 1.

The error in elevation determinations is difficult to estimate. The root-mean-square difference between values computed on the two traverses along the Army-Navy Drive was about 12 meters; between values computed along the Marie Byrd Land traverse and the Executive Committee Range traverse, about 25 meters; between those on the outgoing and return legs to the Amundsen Sea coast, about 20 meters. Other clues to elevation accuracy are provided by the closure errors and the magnitude of the pressure corrections. Because several checks were available in the course of the traverse, we estimate the error to be about ±25 meters relative to Byrd station, a value somewhat better than the average for Antarctic traverses [*Bentley*, 1964].

TABLE 1a. Station Positions and Elevations, Gravity and Seismic Results, Marie Byrd Land Traverse

Station	South Latitude	West Longitude	Surface Elevation (meters a.s.l.)	Observed Gravity (gals)	Free Air Anomaly (milligals)	Bouguer Anomaly (milligals)	Seismic Reflection Time[a] (seconds)	Shot Depth (meters)	Ice Thickness (meters)	Rock Elevation (meters a.s.l.)
0(Byrd)	79°59.2'	120°01.0'	1525	982.5960	+ 2.4				2645 b	-1120 b
3	79°56.3'	120°00.0'	1522	982.5828	- 9.7				2802	-1280
6	79°53.3'	120°00.0'	1518	982.5875	- 6.3				2725	-1207
9	79°50.3'	120°00.0'	1519	982.5754	-15.4				2841	-1322
12	79°47.3'	120°00.0'	1525	982.5795	- 7.9				2217	-1187
15	79°44.3'	120°00.0'	1535	982.5702	-12.5				2271	-1232
18	79°41.5'	120°00.0'	1543	982.5752	- 3.6				2094	-1076
21	79°38.6'	120°02.0'	1549	982.5684	- 7.0				2129	-1106
24	79°35.7'	120°02.0'	1547	982.5646	- 9.8				2171	-1126
27	79°32.8'	120°02.0'	1542	982.5556	-18.8				2900	-1238
*30	79°29.7'	120°02.0'	1538	982.5538	-20.2		N.R.		2706	-1238
33	79°26.8'	119°59.0'	1537	982.5562	-16.6				2671	-1163
36	79°23.8'	119°55.0'	1546	982.5576	-10.7				2085	-1052
39	79°20.9'	119°52.0'	1556	982.5559	- 7.7				2108	- 985
42	79°18.0'	119°49.0'	1564	982.5444	-15.1				2122	-1074
45	79°15.0'	119°46.0'	1569	982.5385	-17.9				2665	-1096
48	79°12.0'	119°44.0'	1571	982.5325	-21.6				2701	-1130
51	79°09.0'	119°42.0'	1581	982.5321	-17.2				2623	-1042
54	79°06.0'	119°40.0'	1581	982.5298	-17.9				2611	-1030
57	79°03.1'	119°37.0'	1586	982.5282	-16.3				2570	- 984
*60	79°00.2'	119°34.0'	1592	982.5234	-17.6	- 7.7	1.357	10	2573 b	- 981 b
63	78°57.4'	119°36.0'	1597	982.5242	-13.7				2530	- 933
66	78°54.3'	119°38.0'	1581	982.5271	-14.0				2530	- 949
69	78°51.1'	119°39.0'	1596	982.5148	-19.8				2642	-1046
72	78°48.3'	119°40.0'	1601	982.5204	-11.0				2526	- 925
75	78°45.3'	119°41.0'	1616	982.5269	+ 1.8				2360	- 744
78	78°42.4'	119°42.0'	1631	982.5218	+ 3.0				2369	- 738
81	78°39.4'	119°42.0'	1644	982.5026	-10.5				2595	- 951
84	78°36.3'	119°42.0'	1639	982.4967	-16.1				2690	-1051
87	78°33.3'	119°43.0'	1638	982.4993	-12.0				2633	- 995
*90	78°30.3'	119°43.0'	1655	982.4981	- 6.2	- 3.0	1.358	10	2575 b	- 920 b
93	78°27.3'	119°45.0'	1657	982.5006	- 1.3				2505	- 848
96	78°24.3'	119°46.0'	1661	982.5027	+ 3.8				2435	- 774
99	78°21.2'	119°48.0'	1675	982.4953	+ 2.6				2468	- 793
102	78°18.1'	119°49.0'	1686	982.4875	+ 0.1				2519	- 833
105	78°15.1'	119°50.5'	1690	982.4798	- 5.5				2609	- 919
108	78°12.1'	119°53.0'	1706	982.4678	- 9.8				2691	- 985
111	78°09.0'	119°55.0'	1708	982.4695	- 5.7				2633	- 925
114	78°06.0'	119°58.0'	1715	982.4751	+ 3.9				2498	- 783
117	78°03.0'	120°00.0'	1721	982.4858	+18.3				2290	- 569
*120	77°59.8'	120°01.0'	1736	982.4827	+21.7	- 6.6	1.191	10	2256 b	- 520 b
123	77°56.8'	120°01.0'	1761	982.4748	+23.5				2266	- 505
126	77°53.8'	120°01.0'	1764	982.4808	+32.2				2151	- 387
129	77°50.8'	120°01.0'	1789	982.4567	+17.7				2406	- 617
132	77°47.8'	120°00.0'	1795	982.4368	+ 1.5				2667	- 872
135	77°44.8'	120°00.0'	1802	982.4244	- 6.9				2813	-1011
138	77°41.8'	120°00.0'	1800	982.4281	- 2.2				2752	- 952
141	77°38.8'	120°00.5'	1817	982.4502	+27.3				2339	- 522
144	77°35.9'	120°01.0'	1828	982.4469	+22.9				2428	- 600
147	77°32.9'	120°01.0'	1834	982.4214	+ 1.2				2772	- 938
*150	77°29.8'	120°01.0'	1832	982.4004	-12.1	+ 0.8	1.573	7	2983 b	-1151 b
153	77°27.0'	119°58.8'	1828	982.4104	- 1.5				2656	- 828
156	77°24.0'	119°57.5'	1858	982.4247	+24.0				2139	- 281
159	77°21.2'	119°55.4'	1872	982.4189	+24.3				1985	- 113
162	77°18.4'	119°52.5'	1879	982.4053	+14.7				1972	- 93
165	77°15.4'	119°52.0'	1891	982.4018	+16.8				1788	+ 103
*167	77°13.5'	119°51.0'	1884	982.4026	+16.7	- 74.3	0.860	6	1619 b	+ 265 b
171	77°10.2'	119°40.0'	1880	982.3818	- 3.1				1912	- 32
174	77°08.1'	119°30.5'	1860	982.3790	-10.7				2006	- 146
177	77°06.0'	119°20.5'	1855	982.3773	-12.8				2033	- 178
180	77°03.0'	119°11.3'	1814	982.3812	-20.0				2100	- 286
183	77°01.6'	119°01.5'	1809	982.3729	-28.3				2218	- 409
186	76°59.7'	118°51.5'	1808	982.3780	-22.4				2128	- 320
189	76°57.5'	118°42.2'	1803	982.3746	-25.8				2175	- 372
192	76°55.3'	118°33.2'	1800	982.3902	- 9.6				1928	- 128
195	76°53.3'	118°23.2'	1813	982.3900	- 4.5				1865	- 52
*198	76°51.0'	118°14.0'	1839	982.4117	+26.7	- 74.0	0.755	10	1423 b	+ 416 b
201	76°48.1'	118°20.0'	1882	982.4037	+34.0				1380	+ 502
204	76°44.9'	118°24.0'	1894	982.3870	+26.7				1524	+ 370
207	76°41.8'	118°22.4'	1909	982.4273	+70.1				912	+ 997
210	76°38.6'	118°22.5'	1805	982.4628	+75.7				747	+1058
213	76°36.1'	118°15.9'	1667	982.4420	+14.1				1557	+ 110
216	76°35.6'	118°02.5'	1625	982.4434	+ 2.9				1706	- 81

TABLE 1a. (continued)

Station	South Latitude	West Longitude	Surface Elevation (meters a.s.l.)	Observed Gravity (gals)	Free Air Anomaly (milligals)	Bouguer Anomaly (milligals)	Seismic Reflection Time[a] (seconds)	Shot Depth (meters)	Ice Thickness (meters)	Rock Elevation (meters a.s.l.)
219	76°36.4'	117°50.0'	1567	982.4457	-13.3				1913	- 346
*222	76°37.6'	117°37.0'	1559	982.4263	-35.9	- 44.2	1.201	4	2268 b	- 709 b
225	76°34.6'	117°34.0'	1554	982.4134	-48.3				2465	- 911
228	76°31.8'	117°30.2'	1562	982.3444	-113.0				3460	-1898
231	76°29.1'	117°26.9'	1561	982.3436	-112.3				3464	-1903
234	76°26.3'	117°23.6'	1561	982.3617	-92.3				3180	-1619
237	76°23.4'	117°20.5'	1554	982.4034	-51.8				2582	-1028
240	76°20.6'	117°17.5'	1544	982.4309	-24.0				2171	- 627
243	76°17.7'	117°14.0'	1542	982.4198	-34.1				2337	- 795
246	76°15.0'	117°10.0'	1547	982.4452	- 5.3				1926	- 379
249	76°12.1'	117°06.5'	1577	982.4659	+26.7				1492	+ 85
252	76°09.4'	117°03.7'	1587	982.4405	+ 6.2				1825	- 238
255	76°06.5'	117°00.3'	1586	982.4279	+ 5.4				1852	- 266
*258	76°03.7'	116°57.0'	1588	982.4433	+13.5	- 35.0	0.929	4	1749 b	- 161 b
261	76°00.7'	116°54.2'	1587	982.4211	- 7.0				1963	- 376
264	75°58.0'	116°51.0'	1589	982.4377	+12.2				1585	+ 4
267	75°51.9'	116°48.0'	1602	982.4455	+27.4				1277	+ 325
270	75°49.0'	116°47.1'	1612	982.4613	+49.3				866	+ 746
273	75°48.9'	116°46.4'	1603	982.5004	+85.7				219	+1384
277	75°44.9'	116°48.9'	1280	982.5542	+42.7				449	+ 831
279	75°43.2'	116°44.1'	1240	982.5352	+12.7				766	+ 474
282	75°40.6'	116°38.0'	1160	982.5246	-20.7				1095	+ 65
285	75°37.9'	116°32.5'	1110	982.5126	-46.5				1339	- 229
*288/493	75°35.3'	116°27.0'	1116	982.4974	-57.7	- 78.1	0.757	4	1420 b	- 304 b
291/490	75°32.5'	116°22.1'	1069	982.5096	-56.1				1391	- 322
294/487	75°29.5'	116°21.0'	1057	982.4878	-81.4				1801	- 744
297/484	75°26.6'	116°20.0'	1103	982.5206	-32.0				1147	- 44
300/481	75°23.5'	116°17.9'	1141	982.5226	-16.1				989	+ 152
303/478	75°20.4'	116°16.9'	1128	982.5395	- 1.0				792	+ 336
306/475	75°17.4'	116°15.8'	1134	982.5540	+17.6				560	+ 574
309/472	75°14.4'	116°14.5'	1138	982.5543	+21.3				551	+ 587
312/469	75°11.3'	116°13.0'	1085	982.5304	-16.5				1117	- 32
315/466	75°08.3'	116°11.5'	1028	982.5806	+18.2				572	+ 456
318/463	75°05.3'	116°09.7'	971	982.5732	- 4.4				895	+ 76
321/460	75°02.2'	116°08.5'	894	982.5822	-16.8				1046	- 152
*324/457	74°59.4'	116°07.0'	841	982.5754	-38.0	- 33.1	0.723	4	1355 b	- 514 b
327/454	74°56.6'	116°06.0'	819	982.5697	-48.4				1482	- 663
330/451	74°53.6'	116°07.0'	764	982.5822	-50.4				1450	- 686
333/448	74°50.8'	116°09.0'	713	982.6012	-45.3				1316	- 603
336/445	74°48.0'	116°10.0'	651	982.6172	-46.2				1261	- 610
337/444	74°47.0'	116°11.8'	613	982.6329	-38.7				1107	- 494
338/443	74°46.1'	116°12.4'	575	982.6462	-38.8				1069	- 494
339/442	74°45.0'	116°11.2'	546	982.6593	-34.2				969	- 423
340/441	74°43.5'	116°13.3'	509	982.6666	-38.7				997	- 488
342/439	74°42.2'	116°12.5'	472	982.6788	-35.3				905	- 433
345/436	74°39.3'	116°14.0'	362	982.7100	-35.9				796	- 434
348/433	74°36.4'	116°14.5'	258	982.7313	-44.4				803	- 555
351/430	74°33.4'	116°14.7'	141	982.7438	-66.1				1016	- 875
352/429	74°32.4'	116°15.0'	118	982.7527	-63.8				1065	- 847
353/428	74°31.6'	116°15.8'	95	982.7685	-53.0				768	- 673
354/427	74°30.4'	116°16.0'	72	982.7800	-48.5				612 c	- 604
355/426	74°29.6'	116°17.1'	73	982.7703	-56.5				624 c	- 721
356/425	74°28.7'	116°17.5'	75	982.7607	-64.2				648 c	834
357/424	74°27.7'	116°18.0'	76	982.7438	-81.3				660 c	1089
*360/421	74°24.9'	116°18.5'	74	982.7366	-87.0	- 3.9	0.347 (ice bottom) 1.187 (sea bottom)	3	636 b	1167 b
363/418	74°21.9'	116°18.8'	64	982.7329	-91.3				514 c	-1242
366/415	74°18.9'	116°19.2'	65	982.7334	-88.2				526 c	-1207
369/412	74°15.9'	116°20.1'	63	982.7236	-96.3				502 c	-1339
372/409	74°12.9'	116°20.9'	68	982.7352	-80.8				563 c	-1124
375/406	74°10.0'	116°20.5'	60	982.7304	-85.8				465 c	-1209
378/403	74°07.0'	116°20.0'	57	982.7337	-81.0				428 c	-1152
381/400	74°04.1'	116°20.5'	55	982.7472	-65.8				404 c	- 942
384/397	74°01.5'	116°17.5'	44	982.7470	-67.3				269 c	- 976
387/394	73°58.0'	116°14.9'	39	982.7578	-55.3				208 c	- 813
*390.5	73°55.4'	116°11.0'	40	982.7738	-36.9	+ 5.7	0.655 (sea bottom)	2	220 c	- 559 b
496	75°36.3'	116°38.9'	1110	982.5413	-16.5				870	+ 240
499	75°37.0'	116°50.0'	1182	982.5183	-17.7				1033	+ 149
502	75°38.5'	117°01.5'	1232	982.5147	- 7.0				996	+ 236
505	75°39.6'	117°12.5'	1322	982.6131	+11.2				886	+ 436

TABLE 1a. (continued)

Station	South Latitude	West Longitude	Surface Elevation (meters a.s.l.)	Observed Gravity (gals)	Free Air Anomaly (milligals)	Bouguer Anomaly (milligals)	Seismic Reflection Time[a] (seconds)	Shot Depth (meters)	Ice Thickness (meters)	Rock Elevation (meters a.s.l.)
508	75°40.5'	117°23.5'	1387	982.4861	+10.7				1033	+ 354
511	75°41.5'	117°35.5'	1423	982.4781	+13.2				1105	+ 318
514	75°42.7'	117°46.5'	1475	982.4531	+ 3.4				1377	+ 98
517	75°43.7'	117°58.0'	1496	982.4536	+ 9.6				1378	+ 118
520	75°44.5'	118°09.5'	1533	982.4482	+16.1				1391	+ 142
523	75°45.5'	118°21.4'	1553	982.4386	+10.9				1563	- 10
526	75°46.6'	118°33.4'	1578	982.4374	+16.7				1574	+ 4
*529	75°47.2'	118°45.0'	1606	982.4349	+22.4	- 39.3	0.846	4	1590 b	+ 16 b
532	75°48.4'	118°56.0'	1626	982.4336	+26.4				1573	+ 53
535	75°49.6'	119°07.5'	1638	982.4261	+21.7				1680	- 42
538	75°50.7'	119°18.5'	1655	982.4126	+12.6				1854	- 199
541	75°52.2'	119°29.5'	1658	982.4111	+11.1				1902	- 244
544	75°53.5'	119°40.5'	1684	982.4153	+22.7				1777	- 93
547	75°54.2'	119°53.0'	1707	982.4302	+43.9				1505	+ 202
550	75°55.3'	120°04.5'	1712	982.3987	+13.0				1746	- 34
553	75°56.1'	120°17.0'	1714	982.4013	+15.7				1980	- 266
556	75°57.0'	120°29.0'	1766	982.3846	+14.4				2074	- 308
559	75°57.6'	120°41.0'	1769	982.3778	+12.1				2134	- 365
562	75°58.5'	120°53.5'	1808	982.3496	- 8.7				2508	- 700
*565	75°59.2'	121°05.0'	1825	982.3320	-21.5	- 25.1	1.448	4	2740 b	- 915 b
568	75°59.0'	121°16.0'	1826	982.3246	-28.5				2795	- 969
571	75°59.2'	121°28.5'	1825	982.3251	-28.4				2742	- 917
574	75°59.4'	121°40.5'	1836	982.3228	-27.4				2687	- 851
577	75°59.6'	121°53.0'	1862	982.3220	-20.4				2558	- 696
580	76°00.0'	122°04.0'	1869	982.3256	-14.9				2432	- 563
583	76°00.4'	122°17.0'	1880	982.3300	- 7.4				2279	- 399
586	76°00.7'	122°28.5'	1918	982.3402	+14.4				1940	- 22
589	76°00.9'	122°40.9'	1930	982.3341	+11.7				1941	- 11
592	76°00.9'	122°53.4'	1921	982.3582	+33.0				1967	- 46
595	76°01.0'	123°05.0'	1970	982.3650	+54.9				1232	+ 738
598	76°01.0'	123°17.5'	2031	982.3419	+50.7				1306	+ 725
601	76°00.7'	123°30.0'	2034	982.3401	+49.9				1270	+ 764
*603	76°01.0'	123°42.0'	2074	982.3550	+77.0		N.R.		852	+1222
606	76°01.0'	123°53.5'	2075	982.3412	+63.5				1005	+1070
609	76°01.9'	124°05.7'	2072	982.3745	+94.2				491	+1581
*612.3	76°00.8'	124°18.5'	2074	982.3517	+73.8	-103.4	0.414	3	763 b	+1311 b
615	76°00.0'	124°28.7'	2026	982.3440	+52.0				1069	+ 957
618	75°57.9'	124°37.9'	1981	982.3181	+13.5				1629	+ 352
621	75°55.6'	124°46.0'	1958	982.3035	- 6.5				1932	+ 26
624	75°53.4'	124°54.6'	1937	982.3247	+ 9.9				1693	+ 244
627	75°51.4'	125°03.8'	1818	982.3530	+ 2.8				1707	+ 111
630	75°49.1'	125°12.2'	1804	982.3405	-12.3				1946	- 142
633	75°47.0'	125°20.5'	1763	982.3804	+16.5				1501	+ 262
636	75°44.7'	125°29.8'	1653	982.3983	+ 2.1				1733	- 80
*639	75°42.7'	125°38.1'	1647	982.3780	-18.5	- 58.6	1.038	3	1955 b	- 308 b
642	75°43.2'	125°50.2'	1673	982.3719	-17.0				1926	- 253
645	75°44.0'	126°01.5'	1732	982.3588	-12.6				1887	- 155
648	75°44.5'	126°13.6'	1795	982.3567	+ 4.3				1665	+ 130
651	75°45.4'	126°25.0'	1826	982.3648	+22.1				1397	+ 429
654	75°45.6'	126°37.7'	1899	982.3510	+30.7				1309	+ 590
657	75°46.3'	126°49.5'	1908	982.3582	+39.5				1153	+ 755
660	75°47.0'	127°01.2'	1986	982.3352	+40.1				1190	+ 796
663	75°48.0'	127°12.5'	2023	982.3488	+64.4				831	+1192
666	75°49.0'	127°24.0'	1988	982.3605	+64.5				762	+1226
669	75°48.5'	127°36.0'	1954	982.3503	+44.2				1000	+ 954
672	75°48.0'	127°48.0'	1954	982.3555	+50.8				869	+1085
*676	75°47.5'	128°04.0'	1964	982.3615	+59.2	-111.2	0.407	5	710 b	+1254 b
678	75°47.8'	128°12.0'	2003	982.3674	+76.9				489	+1514
681	75°48.0'	128°24.3'	1892	982.4011	+80.8				326	+1566
684	75°48.3'	128°37.0'	1928	982.4036	+89.7				234	+1694
687	75°48.8'	128°43.0'	2021	982.3948	+106.5				87	+1934
690	75°50.6'	128°53.5'	1932	982.3678	+52.4				815	+1117
693	75°52.4'	129°03.5'	1894	982.3490	+19.6				1275	+ 619
696	75°53.1'	129°16.0'	1891	982.3271	- 1.6				1596	+ 295
699	75°55.0'	129°26.0'	1888	982.3331	+ 1.9				1546	+ 342
702	75°56.9'	129°37.0'	1846	982.3344	-11.1				1705	+ 141
705	75°58.6'	129°47.0'	1842	982.3348	-13.1				1737	+ 105
708	75°59.9'	130°00.0'	1838	982.3497	- 0.4				1542	+ 296
*711.5	76°02.0'	130°08.0'	1874	982.3541	+13.8	- 94.9	0.734	3	1370 b	+ 504 b
714	76°02.6'	130°18.9'	1879	982.3464	+ 7.2				1488	+ 391
717	76°05.0'	130°29.7'	1893	982.3530	+16.4				1378	+ 515
720	76°06.7'	130°40.0'	1896	982.3529	+16.0				1401	+ 495
723	76°08.4'	130°51.0'	1903	982.3254	- 9.9				1810	+ 93

TABLE 1a. (continued)

Station	South Latitude	West Longitude	Surface Elevation (meters a.s.l.)	Observed Gravity (gals)	Free Air Anomaly (milligals)	Bouguer Anomaly (milligals)	Seismic Reflection Time[a] (seconds)	Shot Depth (meters)	Ice Thickness (meters)	Rock Elevation (meters a.s.l.)
726	76°09.8'	131°01.9'	1908	982.3066	-28.2				2104	- 196
729	76°11.4'	131°12.5'	1942	982.3001	-26.1				2120	- 178
732	76°12.6'	131°24.0'	1948	982.2823	-42.7				2389	- 441
735	76°14.3'	131°34.5'	1963	982.2937	-28.0				2198	- 235
738	76°15.7'	131°45.0'	2003	982.3026	- 7.7				1947	+ 56
741	76°17.3'	131°56.0'	2048	982.2991	+ 1.5				2030	+ 18
744	76°18.8'	132°07.0'	2121	982.2882	+22.1				1646	+ 475
*747	76°20.3'	132°18.0'	2143	982.2980	+27.8	- 94.5	0.849	3	1594 b	+ 549 b
750	76°22.0'	132°28.0'	2140	982.2962	+46.2				1326	+ 814
753	76°23.6'	132°38.5'	2139	982.2981	+24.4				1663	+ 476
756	76°25.0'	132°50.0'	2159	982.3136	+45.0				1385	+ 774
759	76°26.5'	133°01.0'	2167	982.3141	+47.0				1374	+ 793
762	76°28.2'	133°12.0'	2169	982.2986	+31.0				1627	+ 542
765	76°29.7'	133°23.0'	2175	982.2956	+28.8				1666	+ 509
768	76°31.0'	133°34.0'	2174	982.2893	+21.3				1800	+ 374
771	76°32.5'	133°45.0'	2183	982.3009	+34.7				1619	+ 564
774	76°34.0'	133°56.5'	2194	982.2864	+22.5				1824	+ 370
777	76°35.5'	134°07.6'	2187	982.3025	+35.4				1634	+ 553
780	76°36.9'	134°18.7'	2177	982.3053	+34.1				1655	+ 522
*783	76°38.3'	134°30.0'	2160	982.3074	+30.1	- 85.4	0.909	3	1709 b	+ 451 b
786	76°39.6'	134°41.4'	2174	982.3003	+18.1				1882	+ 292
789	76°40.9'	134°53.5'	2129	982.3091	+20.4				1781	+ 348
792	76°42.2'	135°05.0'	2117	982.3099	+16.7				1803	+ 314
795	76°43.6'	135°16.9'	2077	982.3325	+26.0				1603	+ 474
798	76°45.0'	135°28.2'	2061	982.3465	+34.1				1444	+ 617
801	76°46.3'	135°40.0'	2033	982.3517	+29.8				1459	+ 574
804	76°47.5'	135°52.0'	2019	982.3371	+10.1				1720	+ 299
807	76°48.8'	136°04.0'	1994	982.3705	+34.8				1303	+ 691
810	76°50.3'	136°16.0'	1974	982.3754	+32.6				1295	+ 679
813	76°51.6'	136°27.5'	1926	982.3978	+39.3				1125	+ 801
816	76°52.8'	136°40.0'	1895	982.4022	+33.3				1163	+ 732
*819	76°54.0'	136°52.0'	1849	982.3967	+12.8	- 90.6	0.749	3	1403 b	+ 446 b
822	76°55.2'	137°04.0'	1846	982.3870	+ 1.9				1568	+ 278
825	76°56.4'	137°16.5'	1809	982.4333	+35.5				1032	+ 777
828	76°57.7'	137°28.5'	1775	982.4450	+36.3				991	+ 784
831	76°59.2'	137°40.0'	1749	982.4241	+ 5.9				1425	+ 324
834	77°05.0'	137°51.8'	1741	982.4304	+ 5.9				1422	+ 319
837	77°01.7'	138°04.0'	1717	982.4487	+19.0				1206	+ 511
840	77°02.9'	138°16.5'	1686	982.4819	+41.8				838	+ 848
843	77°04.1'	138°29.0'	1642	982.4949	+40.5				818	+ 824
846	77°05.2'	138°41.2'	1626	982.4796	+19.5				1122	+ 504
849	77°06.3'	138°53.5'	1628	982.4951	+34.9				898	+ 730
852	77°07.6'	139°06.0'	1572	982.5230	+44.7				699	+ 873
*855	77°09.0'	139°18.0'	1514	982.5165	+19.3	- 74.5	0.550	5	1027 b	+ 487 b
858	77°10.0'	139°30.5'	1509	982.4731	-26.2				1675	- 166
861	77°11.4'	139°43.0'	1511	982.4837	-16.0				1517	- 6
864	77°12.4'	139°55.5'	1489	982.5107	+ 3.7				1149	+ 340
867	77°13.4'	140°08.4'	1464	982.4710	-44.5				1818	- 354
870	77°14.5'	140°21.0'	1449	982.5388	+18.1				835	+ 614
873	77°15.5'	140°33.5'	1427	982.5637	+35.5				523	+ 904
876	77°16.6'	140°46.0'	1384	982.5624	+20.2				680	+ 704
879	77°17.6'	140°59.0'	1323	982.5960	+34.4				377	+ 946
882	77°18.7'	141°11.5'	1248	982.6015	+16.0				550	+ 698
885	77°19.7'	141°24.8'	1166	982.6078	- 3.7				734	+ 432
888	77°20.8'	141°37.5'	1113	982.6254	- 3.2				644	+ 469
*890	77°21.5'	141°46.0'	1107	982.6227	- 8.2	- 81.2	0.376	5	694 b	+ 413 b
894	77°24.8'	141°35.0'	1118	982.5825	-47.2				1350	- 232
897	77°27.1'	141°26.5'	1161	982.5778	-39.9				1328	- 167
900	77°29.4'	141°17.7'	1197	982.5647	-43.5				1328	- 131
903	77°31.5'	141°08.0'	1200	982.5512	-57.4				1720	- 520
906	77°33.7'	140°59.5'	1205	982.5497	-58.6				1788	- 583
909	77°36.3'	140°51.0'	1204	982.5736	-36.8				1505	- 301
912	77°38.5'	140°42.0'	1210	982.5951	-14.9				1228	- 18
915	77°41.0'	140°32.5'	1192	982.5883	-28.7				1462	- 270
918	77°43.3'	140°24.0'	1166	982.5861	-40.5				1658	- 492
921	77°45.6'	140°15.7'	1148	982.5769	-56.7				1928	- 780
924	77°48.0'	140°06.5'	1140	982.5585	-79.0				2299	-1159
*927	77°50.3'	139°57.0'	1140	982.5409	-98.1	- 34.9	1.370	5	2630 b	-1490 b
930	77°52.5'	139°47.5'	1145	982.4994	-139.4				3166	-2021
933	77°55.0'	139°38.5'	1130	982.5201	-124.9				2846	-1716
936	77°57.3'	139°30.0'	1150	982.5115	-128.7				2834	-1684
939	77°59.7'	139°21.0'	1151	982.5290	-112.4				2503	-1352
942	78°02.0'	139°12.0'	1143	982.5724	-72.9				1814	- 671

TABLE 1a. (concluded)

Station	South Latitude	West Longitude	Surface Elevation (meters a.s.l.)	Observed Gravity (gals)	Free Air Anomaly (milligals)	Bouguer Anomaly (milligals)	Seismic Reflection Time[a] (seconds)	Ice Thickness (meters)	Shot Depth (meters)	Rock Elevation (meters a.s.l.)
945	78°04.1'	139°02.2'	1131	982.6065	-43.7			1276		- 145
948	78°06.5'	138°53.5'	1112	982.6132	-44.3			1178		- 66
951	78°09.0'	138°45.0'	1064	982.6084	-65.2			1355		- 291
954	78°11.4'	138°36.0'	1059	982.5678	-109.2			1922		- 863
957	78°13.8'	138°27.5'	1050	982.5501	-131.1			2153		-1103
960	78°16.3'	138°18.6'	1062	982.5463	-132.7			2101		-1039
*963	78°18.7'	138°10.0'	1065	982.5953	-84.1	-108.4	0.687	1288 b	5	- 223 b
966	78°21.1'	138°01.8'	1063	982.6348	-46.7			847		+ 216
969	78°23.6'	137°53.0'	1020	982.6594	-36.8			778		+ 242
972	78°26.0'	137°44.5'	1036	982.6416	-51.2			1132		- 96
975	78°28.5'	137°34.0'	1037	982.6669	-27.0			891		+ 146
978	78°30.7'	137°24.4'	1039	982.6802	-14.4			827		+ 212
981	78°33.0'	137°15.0'	1013	982.6888	-15.2			935		+ 78
984	78°35.5'	137°05.5'	994	982.6943	-17.1			1066		- 72
987	78°37.8'	136°57.0'	987	982.6893	-25.5			1307		- 320
990	78°40.3'	136°46.5'	961	982.6951	-29.2			1459		- 498
993	78°42.5'	136°36.5'	915	982.6837	-56.1			1938		-1023
996	78°45.0'	136°26.5'	940	982.7003	-33.2			1742		- 802
#*998/AN400	78°45.4'	136°17.0'	946	982.7011	-30.9	- 0.6		1835 b		- 889 b
1002/AN405	78°45.0'	135°55.0'	990	982.7040	-14.7			1596		- 606
1007/AN410	78°44.6'	135°32.5'	1039	982.6787	-23.0			1690		- 651
1011/AN415	78°44.2'	135°10.0'	1041	982.6683	-34.0			1858		- 817
1015/AN420	78°43.8'	134°48.0'	1066	982.6806	-12.8			1525		- 459
*1020/AN425	78°43.4'	134°25.5'	1126	982.6694	- 5.3	- 25.1	0.763	1432 b	4	- 306 b
1024/AN430	78°42.9'	134°03.5'	1165	982.6403	-22.5			1760		- 595
1028/AN435	78°42.3'	133°41.0'	1159	982.6451	-18.7			1728		- 569
1033/AN440	78°41.7'	133°19.5'	1186	982.6422	-13.6			1710		- 524
1037/AN445	78°41.1'	132°57.0'	1216	982.6083	-37.9			2136		- 920
# 1041/AN450	78°41.5'	132°34.2'	1253	982.5934	-40.6	- 14.2		2245 b		- 992 b
1046/AN455	78°39.8'	132°13.5'	1267	982.5920	-37.7			2174		- 907
1050/AN460	78°39.0'	131°53.0'	1293	982.5910	-30.7			2054		- 761
1054/AN465	78°38.3'	131°31.7'	1314	982.5744	-39.3			2162		- 848
1059/AN470	78°37.4'	131°10.5'	1347	982.5563	-47.4			2275		- 928
* 1063/AN475	78°36.5'	130°49.5'	1388	982.5536	-36.9	- 35.1	1.121	2117 b	5	- 729 b
1067/AN480	78°35.6'	130°28.0'	1403	982.5720	-13.1			1802		- 399
1072/AN485	78°38.7'	130°09.0'	1415	982.5939	+10.8			1411		+ 4
1076/AN490	78°41.6'	129°51.0'	1407	982.5830	- 4.3			1729		- 322
1080/AN495	78°44.4'	129°32.0'	1403	982.5998	+ 8.1			1566		- 163
# 1085/AN500	78°47.4'	129°13.0'	1428	982.5872	+ 2.3	- 30.9		1705 b		- 277 b
1089/AN505	78°50.0'	128°56.5'	1436	982.5713	-11.5			1964		- 528
1093/AN510	78°52.5'	128°39.5'	1441	982.5643	-19.2			2129		- 688
1098/AN515	78°55.2'	128°22.5'	1480	982.5792	+ 6.5			1889		- 409
# 1102/AN520	78°57.8'	128°06.0'	1494	982.5992	+29.7	- 23.6		1535 b		- 41 b
1106/AN525	79°00.4'	127°49.0'	1505	982.6038	+35.7			1496		+ 9
1111/AN530	79°04.6'	127°32.0'	1501	982.6126	+41.1			1452		+ 49
1115/AN535	79°05.3'	127°14.4'	1486	982.6083	+30.1			1643		- 157
# 1119/AN540	79°07.8'	126°57.0'	1481	982.5899	+10.2	- 8.9		1975 b		- 494 b
1124/AN545	79°10.5'	126°38.0'	1479	982.5856	+ 4.3			2066		- 587
1128/AN550	79°13.2'	126°19.0'	1476	982.5845	+ 0.5			2127		- 651
1132/AN555	79°15.9'	126°01.0'	1467	982.5851	- 3.1			2177		- 710
# 1137/AN560	79°18.4'	125°42.0'	1464	982.5840	- 5.8	- 4.9		2220 b		- 756 b
J 1141/AN565	79°21.1'	125°23.0'	1451	982.5883	- 7.2			2231		- 780
1145/AN570	79°23.7'	125°03.0'	1447	982.5926	- 6.5			2219		- 772
1150/AN575	79°26.2'	124°44.0'	1449	982.5910	- 7.8			2245		- 796
# 1154/AN580	79°28.6'	124°24.0'	1453	982.5923	- 6.3	- 3.4		2230 b		- 777 b
1158/AN585	79°31.1'	124°04.5'	1439	982.5936	-11.6			2365		- 926
1163/AN590	79°33.4'	123°45.5'	1435	982.5974	-10.4			2413		- 978
1167/AN595	79°35.7'	123°26.4'	1449	982.5971	- 5.4			2423		- 974
# 1171/AN600	79°38.1'	123°06.4'	1443	982.6022	- 4.0	+ 17.4		2465 b		-1022 b
1176/AN605	79°40.4'	122°46.0'	1458	982.5979	- 6.4			2460		-1002
1180/AN610	79°42.7'	122°26.0'	1476	982.5942	- 5.5			2408		- 932
1184/AN615	79°45.0'	122°06.0'	1473	982.5950	- 6.4			2363		- 890
#1189/AN620	79°47.2'	121°46.0'	1478	982.5920	-10.5	+ 0.4		2375 b		- 897 b
1193/AN625	79°49.7'	121°25.0'	1492	982.5819	-17.5			2574		-1082
1197/AN630	79°52.1'	121°04.0'	1503	982.5790	-18.8			2685		-1182
1202/AN635	79°53.5'	120°51.8'	1510	982.5886	- 3.4			2540		-1030
1206/AN640	79°57.0'	120°22.0'	1523	982.5910	- 3.0			2627		-1104
1211/AN646 (Byrd)	79°59.2'	120°01.0'	1525	982.5960	+ 2.4	+ 28.0		2645 b		-1120 b

*Seismic station
#Seismic station, Little America-Byrd traverse, 1957

[a] Corrected to bottom of shot hole
[b] Determined from seismic reflection
[c] Determined from surface elevation

TABLE 1b. Station Positions and Elevations, Executive Committee Range Traverse, Excluding Segment Common with Marie Byrd Land Traverse

Station	South Latitude	West Longitude	Elevation (meters a.s.l.)	Station	South Latitude	West Longitude	Elevation (meters a.s.l.)
164	77°13.0'	119°51'	1883	313	77°49.6'	124°32'	1769
167	77°12.7'	120°04'	1910	316	77°52.4'	124°27'	1772
170	77°12.4'	120°18'	1916	319	77°55.1'	124°21'	1771
173	77°12.1'	120°31'	1936	322	77°57.8'	124°15'	1769
176	77°11.8'	120°44'	1951	325	78°00.5'	124°10'	1754
179	77°11.5'	120°58'	1960	328	78°03.2'	124°04'	1719
182	77°11.2'	121°11'	1966	331	78°06.0'	123°58'	1702
185	77°11.0'	121°25'	1979	334	78°08.7'	123°52'	1687
188	77°10.7'	121°38'	1990	337	78°11.4'	123°47'	1674
191	77°10.4'	121°51'	2007	340	78°14.1'	123°41'	1666
194	77°10.1'	122°05'	2020	343	78°16.8'	123°35'	1666
197	77°09.8'	122°18'	2017	346	78°19.5'	123°30'	1646
200	77°09.5'	122°31'	2025	349	78°22.2'	123°24'	1639
203	77°09.2'	122°45'	2037	352	78°25.0'	123°18'	1621
206	77°08.9'	122°58'	2042	355	78°27.7'	123°13'	1619
209	77°08.6'	123°11'	2057	358	78°30.4'	123°07'	1594
212	77°08.3'	123°25'	2054	361	78°33.1'	123°01'	1585
215	77°08.0'	123°38'	2048	364	78°35.9'	122°56'	1584
218	77°07.7'	123°51'	2050	367	78°38.6'	122°50'	1568
221	77°07.5'	124°05'	2045	370	78°41.3'	122°44'	1582
224	77°07.2'	124°18'	2079	373	78°44.0'	122°38'	1569
227	77°06.9'	124°32'	2083	376	78°46.7'	122°33'	1565
230	77°06.6'	124°45'	2098	379	78°49.4'	122°27'	1567
233	77°06.3'	124°58'	2080	382	78°52.2'	122°21'	1555
236	77°06.0'	125°12'	2107	385	78°54.9'	122°16'	1548
239	77°05.7'	124°25'	2076	388	78°57.6'	122°10'	1538
242	77°02.9'	124°31'	2175	391	79°00.3'	122°04'	1532
245	77°00.2'	124°36'	2126	394	79°03.0'	121°59'	1530
248	76°57.4'	125°42'	2256	397	79°05.7'	121°53'	1526
251	76°54.6'	125°48'	2297	400	79°08.5'	121°47'	1524
254	76°57.4'	125°42'	2256	403	79°11.2'	121°42'	1527
257	77°00.2'	125°36'	2126	406	79°13.9'	121°36'	1519
260	77°02.9'	125°31'	2175	409	79°16.6'	121°30'	1516
263	77°05.7'	125°25'	2076	412	79°19.3'	121°24'	1512
266	77°07.7'	125°34'	2035	415	79°22.1'	121°19'	1500
269	77°09.6'	125°43'	1973	418	79°24.8'	121°13'	1493
271	77°11.6'	125°52'	1943	421	79°27.5'	121°07'	1491
274	77°14.3'	125°46'	1938	424	79°30.2'	121°02'	1486
277	77°17.0'	125°41'	1927	427	79°32.9'	120°56'	1490
280	77°19.8'	125°35'	1931	430	79°35.6'	120°50'	1500
283	77°22.5'	125°29'	1920	433	79°38.4'	120°45'	1500
286	77°25.2'	125°23'	1916	436	79°41.1'	120°39'	1500
289	77°27.9'	125°18'	1891	439	79°43.8'	120°33'	1490
292	77°30.6'	125°12'	1877	442	79°46.5'	120°28'	1503
295	77°33.3'	125°06'	1824	445	79°49.2'	120°22'	1514
298	77°36.1'	125°01'	1822	448	79°52.0'	120°16'	1509
301	77°38.8'	124°55'	1809	451	79°54.7'	120°10'	1525
304	77°41.5'	124°49'	1815	454	79°57.4'	120°05'	1521
307	77°44.2'	124°44'	1790	Byrd	79°59.2'	120°01'	1525
310	77°46.9'	124°38'	1800				

TABLE 1c. Station Positions, Elevations, and Seismic Results, Last Part of Horlick Mountains Traverse

Station	South Latitude	West Longitude	Surface Elevation (meters a.s.l.)	Seismic Reflection Time (seconds)	Ice Thickness (meters)	Rock Elevation (meters a.s.l.)
747	82°08.6'	109°14'	1808	1.185	2230	- 422
750	82°06.6'	109°28'	1783			
753	82°04.6'	109°43'	1787			
756	82°02.6'	109°57'	1774			
759	82°00.6'	110°11'	1753			
762	81°58.6'	110°26'	1738			
765	81°56.6'	110°40'	1747			
768	81°54.5'	110°54'	1752			
771	81°52.5'	111°09'	1738			
774	81°50.5'	111°23'	1717			
777	81°48.5'	111°37'	1703			
780	81°46.5'	111°52'	1690			
783	81°44.5'	112°06'	1684	1.274	2405	- 721
787	81°41.6'	112°22'	1678			
791	81°38.7'	112°39'	1655			
795	81°35.8'	112°56'	1640			
799	81°32.9'	113°12'	1612			
803	81°29.9'	113°28'	1595			
807	81°27.0'	113°45'	1600			
811	81°24.1'	114°02'	1600			
815	81°21.2'	114°18'	1608	1.390	2625	-1017
819	81°18.1'	114°34'	1574			
823	81°15.0'	114°51'	1554			
827	81°11.9'	115°08'	1522			
831	81°08.8'	115°24'	1512			
833	81°07.2'	115°32'	1521			
835	81°05.6'	115°40'	1517			
835.5	81°06.0'	115°42'	1516			
839	81°02.5'	115°57'	1510			
843	80°59.4'	116°14'	1492			
847	80°56.3'	116°30'	1490	1.490	2815	-1335
851	80°53.3'	116°41'	1489			
855	80°50.3'	116°53'	1503			
859	80°47.2'	117°04'	1497			
863	80°44.2'	117°16'	1513			
867	80°41.2'	117°27'	1505			
871	80°38.2'	117°38'	1526			
875	80°35.2'	117°50'	1530			
879	80°32.1'	118°01'	1519			
883	80°29.1'	118°13'	1524	1.266	2390	- 885
887	80°26.1'	118°24'	1505			
891	80°22.7'	118°36'	1486			
895	80°19.4'	118°48'	1483			
899	80°16.0'	118°01'	1478			
903	80°12.7'	118°13'	1492			
907	80°09.3'	118°25'	1494			
911	80°05.9'	118°38'	1501			
915	80°02.6'	118°50'	1523			
Byrd	79°59.2'	120°02'	1525		2645	-1120

Gravity

Considerable difficulty was encountered in reducing the gravity data. The first problem was the correction for the gravimeter temperature variations, which covered the range 68–72°F (20–22°C). To determine the effect of temperature on instrument reading, Parks conducted controlled experiments during the winter of 1960 at Byrd station. Gravimeter dial readings and thermometer indications from field stations where there was more than one reading were also compared. The latter data were in good agreement with the coefficient of 5.3 mgal/°F (9.5 mgal/°C) indicated by the observations at Byrd station (Figure 4). Field measure-

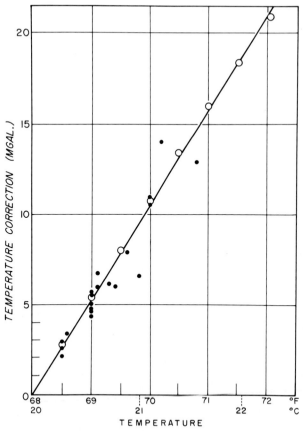

Fig. 4. Temperature calibration of gravimeter. Dots indicate field observations; open circles indicate observations at Byrd station.

differences from the end of the traverse to station 189, and from the beginning to station 186, a much more modest change of 3.4 mgal was obtained. This correction greatly improved the agreement between the values for the difference in ice thickness between station 168 and station 198 as computed from seismic soundings and free-air anomalies, respectively.

An airlifted tie from Byrd station to station 783 was made with a Worden gravimeter by E. C. Thiel on January 21. From this point to the junction of the traverse with the Army-Navy Drive, there was a misclosure of 9.2 mgal. Correction for this was applied linearly along this section of the traverse. Gravity values were adjusted to fit previously measured values (*Bentley and Ostenso* [1961], adjusted for the most recent value of 982.5960 given for Byrd station by *Woollard and Rose* [1963]) at the junction with Army-Navy Drive (station AN400) and at Byrd station, but not elsewhere along this trail. Because of the instrumental difficulties encountered on the Marie Byrd Land traverse, the adjusted gravity values from the Little America–Byrd traverse are believed to be more reliable, and are listed in Table 1 with a few minor corrections.

Free-air gravity anomalies were calculated for all

ments were complicated by a lag between temperature changes at the thermometer and the corresponding internal changes affecting instrument readings. This is illustrated in Figure 5, which shows a typical variation of indicated gravity reading with time, the indicated temperature being written at points along the curve. The error in observed gravity resulting from temperature instability (and from a thermometer reading error of about $\pm 0.1°F$) was estimated from Figure 5 to be less than ± 5 mgal.

Another problem resulted from irregular drift. Two tares, probably caused by temperature shocks to the instrument, occurred while at a station and could thus be corrected for: a decrease of 9 mgal at station 258, and an increase of 5 mgal at station 855. Between stations 186 and 189, an unreasonably large apparent change of 60 mgal was attributed to a tare, since it occurred in a region of otherwise moderate gravity gradients. By computing gravity

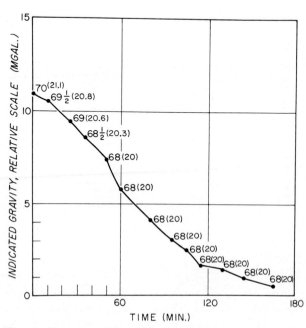

Fig. 5. Typical variation of indicated gravity reading with the time and temperature. Temperature in °F is noted next to each observed point.

stations. At those stations where the ice thickness was measured, 'ice-corrected' Bouguer anomalies were computed, i.e. correction was made for the entire ice column, using a density of 0.9 g/cm^3, and then for the remaining surplus or deficit of rock mass relative to sea level, assuming a density of 2.67 g/cm^3. This is the same as a simple Bouguer correction applied for the equivalent elevation of the rock, i.e. the elevation of a land surface that would have the same load above a given level as that of the actual rock and superimposed ice combined.

It is clear from the previous discussion that the observed gravity values could be in error by 10 mgal or more along some sections of the traverse. The error in gravity anomalies could reach 20 mgal, although the relative error between adjacent stations would be substantially less.

Ice Thickness

The ice thickness at each seismic station (Table 1) was computed assuming an average velocity of 3820 m/sec for vertically traveling waves beneath the near-surface low-velocity firn. A constant negative correction of 30 meters was applied to the computed ice thickness values to allow for the near-surface low velocities. The average velocity beneath these layers was computed by *Bentley and Ostenso* [1961] assuming a velocity of 3850 m/sec through 90% of the ice, and 3600 m/sec in the lowest 10%. A maximum error of ±40 meters in total ice thickness, caused chiefly by uncertainty in the thickness of the basal low velocity layer, was estimated. The velocity of 3840 ± 25 m/sec obtained in the ice near Toney Mountain [*Chang*, 1964] is in good agreement with the earlier measurements.

Free-air gravity anomalies were used to interpolate rock elevation between seismic reflection stations. To determine the best conversion factor to use, a plot of the difference in rock elevation between seismic stations versus the difference in free-air gravity anomaly between the same stations was prepared (Figure 6). Omitting extreme values, a least-squares regression line with slope 16.2 m/mgal was fitted to the majority of the points (solid points in Figure 6), which were considered to be representative of the area as a whole. The slope was rounded off to 15 m/mgal; it is clear from the scatter of the points in Figure 6 that this is not significantly different from the computed value. The points represented by open circles were omitted from the determination of slope as having too great an effect on the least-squares computation compared with their actual importance in the analysis. Since very large elevation differences are almost surely connected with relative maxima and minima in rock height, the associated difference in gravity anomaly would probably be unrealistically small, owing to the effect of the surrounding subglacial terrain. The points indicated by crosses are simply extreme deviations from the norm. The root-mean-square deviation of elevation differences from the 15-m/mgal relation is 230 meters excluding these points and 440 meters including them.

On the assumption that a 1-mgal anomaly represents a 15-meter change in rock level, the change in rock elevation equivalent to the change in free-air anomaly relative to the nearest preceding seismic station was calculated for each gravity station. The difference between the rock elevation computed from the free-air anomaly and that computed from seismic soundings at the next seismic station was distributed linearly among the gravity stations. In regions of rough rock topography, the probable error in this process is clearly several hundred meters.

An empirical relation between surface elevation and thickness of the floating ice shelf was determined from the differences between the elevation and ice thickness (from seismic reflections) at station 360, and the elevation (43 meters) and ice thickness (257 meters) at Little America station [*Crary et al.*, 1962]. The resulting factor—change in ice thickness equals 11 times change in elevation—was used between station 354 and the Amundsen Sea. Ocean bottom depths were determined from seismic reflections at stations 360 and 390.5, assuming a wave velocity in the sea water of 1.44 km/sec. Depths elsewhere were determined from the free-air anomalies assuming a specific gravity of 1.03 for sea water and 2.67 for rock.

Magnetometry

Since the small change in calibration constant of the magnetometer between departure from and return to Byrd station was not significant in view of the over-all uncertainties in the magnetic observations, a mean factor of 27.5 γ/scale division was used for the whole magnetic profile. The profile was computed from the readings with the magnetometer ori-

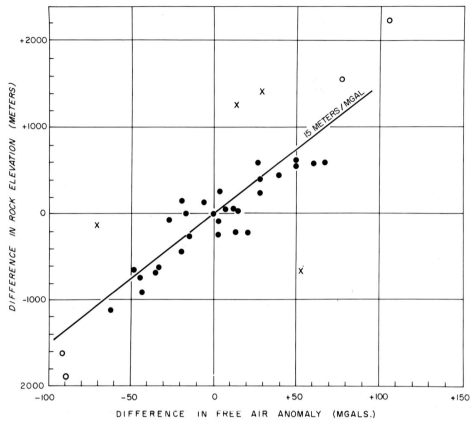

Fig. 6. Comparison of differences in free-air anomalies and rock elevation between seismic stations. Open circles indicate points not used in regression slope determination; crosses indicate points separately considered in error estimation.

ented southward, since this was the orientation for the repeated observations at seismic stations.

An estimate of the accuracy of the magnetic observations was derived from the readings obtained with northward and southward orientation of the magnetometer. The mean difference between readings was 550 γ (attributable to an error of about 1° in the level bubble) with a standard error of ±75 γ. The corresponding error in an individual reading is thus $75/\sqrt{2} \simeq 50$ γ. (A reduction in the error by another factor of $\sqrt{2}$ could have been gained by computing the profile using the mean of the two readings; in view of other errors, this insignificant improvement was clearly not worth the computational effort.)

No corrections were applied for temporal magnetic variations, previous experience having shown the futility of doing so with control stations hundreds of kilometers away [*Ostenso and Bentley*, 1959]. The raw readings, multiplied by the calibration factor and adjusted to the station value at Byrd station at the time of departure (58,560 γ at 1935 GMT on November 5, 1959 [*U.S. Coast and Geodetic Survey*, 1962]), are shown in Figure 7. A closure error of approximately 1000 γ was immediately apparent. Correction for the jump of 500 γ between stations 526 and 529 (975 km along the traverse) reduced this error by half. It was further found that a regional gradient could much more satisfactorily be fitted to the observed profile if another instrumental jump of 500 γ was assumed to have occurred between stations 87 and 96 (160–180 km along the traverse), the closure error at the same time being eliminated. Although there was no field evidence for such a tare, the assumption was adopted as the most satisfactory explanation of the observations. The resulting regional magnetic curve, shown by the dashed line in Figure 7, was used to construct the regional map of vertical intensity (Figure 8).

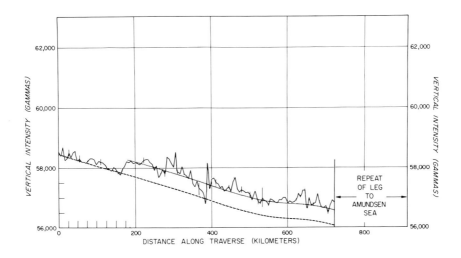
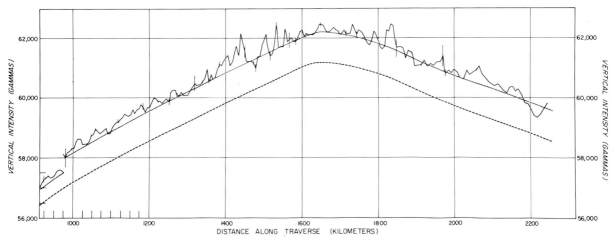

Fig. 7. Observed vertical magnetic intensity. Upper and lower dotted lines indicate regional curves fitted before and after tare corrections, respectively. Vertical line segments show the range of observed values at seismic stations.

The regional field has been removed from the observed vertical intensity values to produce the anomaly profile shown in Figure 9. To provide a guide in interpreting the variations in vertical intensity, we have superimposed on the profile a series of lines indicating the level of activity at Byrd station during the times of measurement. The height of each line segment above the zero line gives the difference, to the nearest 50 γ, between the maximum and minimum value of vertical intensity recorded at Byrd station during the period starting three hours before the initial field observation and ending three hours after the last field observation for each continuous interval of travel. Thus anomalies that substantially exceed the corresponding activity level in amplitude can be regarded with confidence as resulting from geologic sources, whereas those of lesser amplitude may very well result from ionospheric activity alone.

For the repeated section of the traverse from mile 288 to mile 390.5, the Z values from the outgoing leg have been used, since the activity level for the return trip was much higher, 450 to 650 γ. Nevertheless, the agreement between repeated observations was within ±200 γ at all stations, reproducing in gross the 500 γ anomaly observed along this section of the route.

RESULTS

The profile of the ice sheet along the Marie Byrd Land traverse route, as deduced from the seismic, gravity, and altimetry observations, is shown in

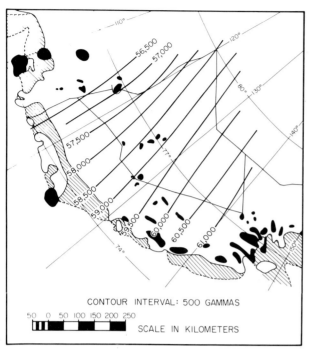

Fig. 8. Map of regional magnetic intensity.

Figure 10. As we shall see in a later section, there is good reason to believe that the land has been depressed by the load of the ice to produce approximate isostatic equilibrium. The amount of this depression has been estimated on the basis of Archimedes' principle, assuming that mantle rock of density 3.3 g/cm³ has been displaced by ice of density 0.9 g/cm³. Ice thickness values were averaged over a distance of 110 km along the traverse route to introduce a degree of 'regionality' into the isostatic computation; the crustal depression was then taken to be 0.27 times the mean ice thickness, reduced, in the appropriate region, by the effect of the superimposed water load that is deduced to have existed before glacierization.

The result is shown in Figure 10, the negative ordinate of the curve relative to sea level indicating the amount of depression. In this way, the same curve indicates 'adjusted sea level,' i.e. the height of the sea relative to the rock surface in the absence of ice, with no allowance for eustatic change due to melting of the ice. Any global change in sea level could easily be superimposed; it would in any event be small compared with the inferred crustal warping.

Contour maps of ice surface and subglacial topography and ice thickness (Figures 11, 12, and 13) have been prepared using the results from this and adjacent traverses [*Bentley and Ostenso*, 1961; *Thiel*, 1961; *Behrendt et al.*, 1962]. On the map of subglacial topography (Figure 12), the estimated 'adjusted shoreline,' i.e. the adjusted sea-level contour, has been drawn, thus approximating the preglacierization coastline. Allowance for a higher stand of the sea would shift this coastline only slightly in most places because of the steepness of the rock surface.

Surface Topography

In the Executive Committee Range (76.5°S, 127°W), the ice sheet surface (Figure 11) reaches an elevation of more than 2300 meters, its highest in Marie Byrd Land. The slope southwestward toward the Ross ice shelf is gentle and regular across the low-lying rock surface. A surface ridge runs westward from the Executive Committee Range into the Flood Range, reaching surprisingly near the coast at 135°W longitude. The ridge also runs eastward and then southeastward, with slowly decreasing elevation, to merge with the divide between the Amundsen Sea and Ross Sea drainage systems. Damming of the ice behind both the Crary Mountains and Toney Mountain is indicated by the sharp drop in surface elevation northward across each mountain group (Figure 10). A slight but definite decrease in elevation southward from Toney Mountain toward the Crary Mountains may be a reflection of the deep underlying subglacial trough.

On the southwestern flank of the central ridge, the mean surface slope is 4 m/km. If we assume that the average basal shear stress, $\bar{\tau}$, is given by $\bar{\tau} = \rho g \bar{\alpha} \bar{h}$, where ρ is the density of ice, g is the acceleration of gravity, $\bar{\alpha}$ is the mean ice surface slope, and $\bar{h}$ is the mean ice thickness [*Nye*, 1952], then $\bar{\tau}$ in this region ($\bar{h}$ = 1500 meters) is roughly 0.5 bar.

Northward from the central ridge, the slope is much steeper. The mean slope is 15 m/km for the first 50 km inland from the ice shelf north of Toney Mountain, and just as great between Mount Petras and the coast, where the elevation drops 2000 meters in 130 km. Along the former section, the average ice thickness is about 1000 meters, leading to a computed $\bar{\tau}$ of 1.3 bars, twice the value on the other side of the ice divide. Although the ice thickness is not known north of Mount Petras and elsewhere between the traverse route and the coast, the basal shear stress must surely reach this same average

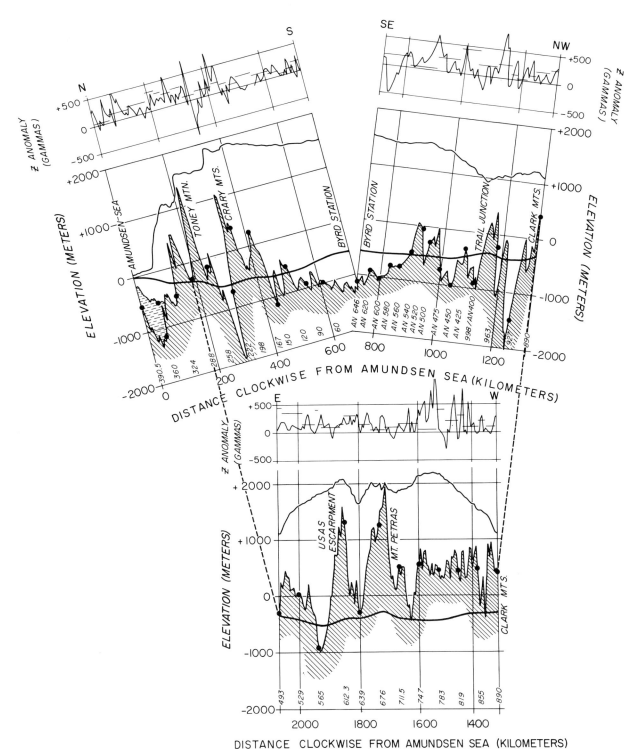

Fig. 9. Profile of ice sheet and residual vertical magnetic intensities. Byrd station is in center of upper diagram. Solid line through rock topography indicates inferred crustal depression, or 'adjusted sea level.' Horizontal line segments on magnetic profile indicate concurrent activity range at Byrd station.

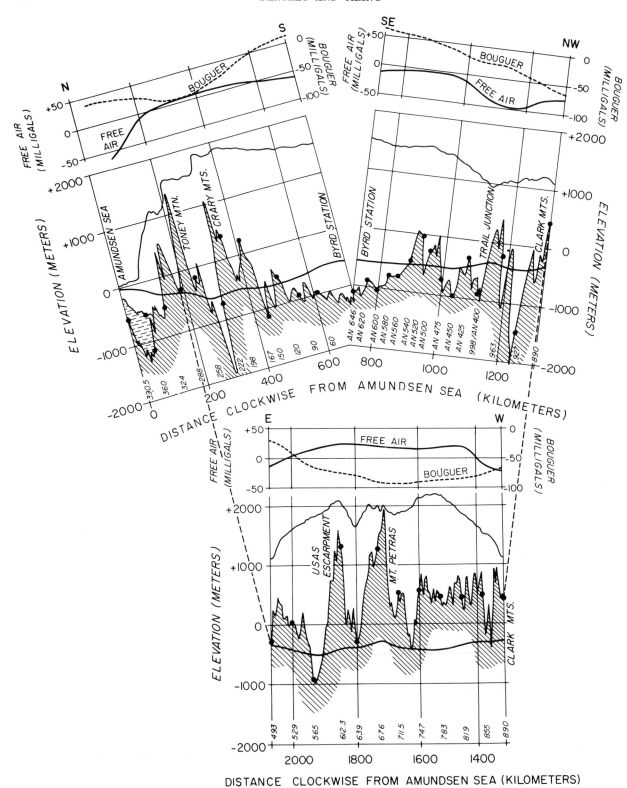

Fig. 10. Profile of ice sheet and mean gravity anomalies. Byrd station is in center of upper diagram. Scales for free-air and Bouguer anomalies are, respectively, at left and right end of each profile section. Solid line through rock topography indicates inferred crustal depression, or 'adjusted sea level.'

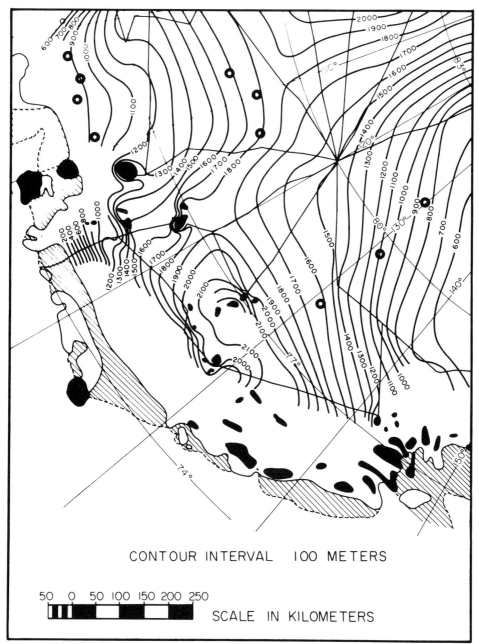

Fig. 11. Ice surface topography in Marie Byrd Land.

value over extensive regions. It appears, therefore, that the basal shear stress is roughly twice as great in the coastal areas as in the more central parts of West Antarctica.

At least part of this difference is probably due to the greater accumulation and lesser ice thickness in the coastal regions. We can obtain an estimate of the magnitude of this effect by a simple calculation. Although the mechanism of the differential horizontal motion in the ice is not certain, it can be safely assumed that most of the shearing strain takes place at or near the base of the ice [*Nye*, 1959].

It is quite sufficient for our purposes to assume that the outward velocity of the ice is uniform with depth, and that it depends upon the basal shear stress raised to some power m:

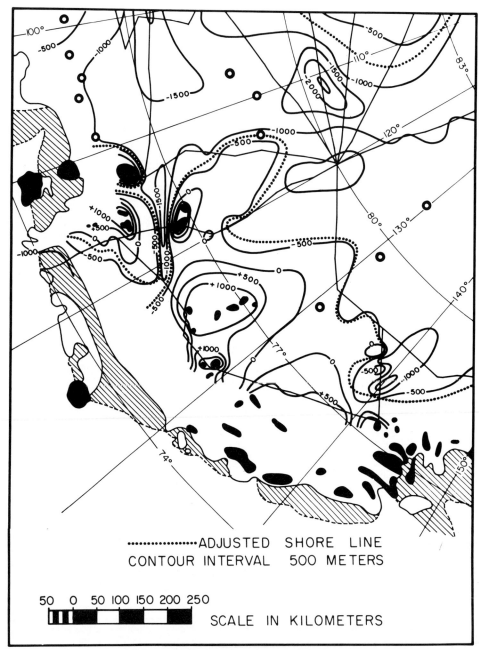

Fig. 12. Subglacial topography in Marie Byrd Land. Dotted line shows estimated position of pre-glacierization shore line.

$$v = \bar{\tau}^m$$

If steady-state flow conditions exist, i.e. the ice sheet is neither growing nor shrinking, then

$$ax = hv$$

where x is the distance from the center of the ice sheet, h is the ice thickness, and a is the ice accumulation rate, assumed to be constant in one region. Combining the two equations, we find

$$\bar{\tau} = (ax/h)^{1/m}$$

For similar values of x in the two regions, h is about twice as great in the central regions as near the coast, and a is about twice as great near the coast

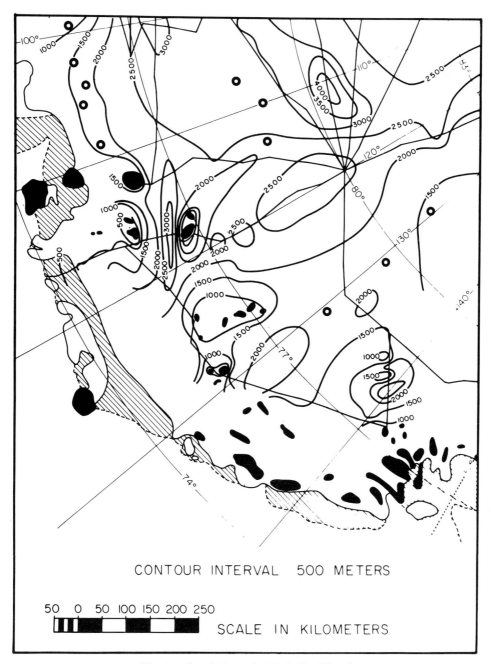

Fig. 13. Ice thickness in Marie Byrd Land.

[*Giovinetto*, 1964]; we thus find that the shear stress ratio should be $4^{1/m}$. Calculations and experiments indicate that m in all probability lies between 2 and 3 [*Glen*, 1955; *Weertman*, 1957; *Nye*, 1959], so that the stress ratio would be between 1.6 and 2, which is about the observed value. Thus these factors alone may be enough to account for the asymmetry of the Marie Byrd Land ridge.

Another factor that may be important is a difference in temperature at the base of the ice. Unfortunately, it is not possible to obtain a meaningful quantitative estimate of this factor, since both the value of the basal temperature and its effect upon the shear stress depend to a large degree upon the mechanism of differential motion, which is uncertain. Whatever the mechanism, however, we can say that a colder temperature would tend to increase the surface slope, other things being equal. In the absence of important frictional heating in the bottom ice layers, the basal temperatures in the coastal regions would be lower, so that we can at least conclude that a temperature difference could contribute to the observed asymmetry in surface profile.

Rock Topography

The rock topography of Marie Byrd Land is extremely rugged, particularly in the eastern part. There are a number of peaks that rise several thousand meters above the level of the surrounding land. A deep trench cuts to nearly 2000 meters below sea level between the Crary Mountains and Toney Mountain, becoming shallower to the northwest and broadening into the Byrd subglacial basin to the southeast. The rock surface between the Executive Committee Range and the Crary Mountains on the south side of the trench is assumed, for the sake of mapping, to lie between present and adjusted sea level. The reason for this is that most of the rock surface beneath the traverse tracks lies above adjusted sea level, even at distances as large as 100 km from the nearest exposed rock; examination of Figures 10 and 12 shows that between the Byrd subglacial basin and the edge of the continent rock elevations below adjusted sea level are found only in the deep trench. Nevertheless, it is entirely possible that the Executive Committee Range and the Crary Mountains were separated by open water before glacierization. There is also the possibility, of course, that a high subglacial ridge connects these peaks. This is unlikely, however, because of the volcanic nature of the mountains [*Doumani and Ehlers*, 1962] and the low rock elevations elsewhere in the vicinity.

The symmetry of Mount Takahe, coupled with the rapid southward drop in subglacial rock elevation found on the 1957–1958 traverse [*Bentley and Ostenso*, 1961], makes it probable that the rock surface falls below adjusted sea level between Mount Takahe and Toney Mountain, and between Mount Takahe and Mount Murphy to the north.

A northwestward extension of the trench between Toney Mountain and the Crary Mountains is found with shallower depth beneath the north leg of the traverse. Farther to the west, subglacial spurs of the Usas escarpment and Mount Petras exhibit a relief of some 2000 meters, being separated by a valley whose floor is close to adjusted sea level.

West of Mount Petras is another valley reaching about to adjusted sea level. Still farther west, the subglacial terrain is considerably less rough, maintaining an elevation averaging 700 meters above adjusted sea level to the Clark Mountains. This relative smoothness disappears immediately to the south, where the subglacial slopes of the mountains fall to a valley cutting 1600 meters below adjusted sea level. The rough terrain continues for about 200 km to the southeast, with an average elevation near adjusted sea level before reaching the relatively smooth floor of the Byrd subglacial basin.

From the actual soundings, no matter what interpretation is applied in contouring, it is apparent that there are substantial areas of true land in Marie Byrd Land. Since there is no reason to believe that the traverse route did not cross a representative section of central Marie Byrd Land, it seems probable that before the formation of the ice sheet there was a large island extending from the Executive Committee Range or even farther east to the Edward VII Peninsula, bounded by the Byrd subglacial basin on the south. Lying off the coast of this island to the east and northeast were several smaller islands including Mount Murphy and Mount Siple, Bear Peninsula, and possibly Mount Takahe, each of which was probably a single volcanic cone. It is in the realm of speculation whether the Crary Mountains were a part of the main island or lay offshore; in either case, the associated land extended, continuously or discontinuously, for some 250 km to the southeast. Similarly, the land associated with Toney Mountain extended 100 km or so to the north and east, and might have included Mount Takahe.

The map of ice thickness (Figure 9) largely reflects the subglacial topography, since the subglacial relief is so much greater than that of the ice sheet surface. A maximum thickness of more than 3000 meters occurs in the trough between the Crary Mountains and Toney Mountain. Assuming that the traverse route represents an average sampling of the region, the mean ice thickness in Marie Byrd Land is 1770 meters.

Magnetics

The regional magnetic map (Figure 8) shows a regular increase in the vertical intensity (Z) from geographic east to west, i.e., toward the magnetic pole, as one would expect. The gradient appears slightly steeper in the eastern than in the western part of Marie Byrd Land. The reality of this distinction is uncertain, however, in view of the errors in observation, in regional curve fitting, and in drawing the map with no more than two points on any one isomagnetic line.

With station spacing of 5½ km, it is apparent that no anomaly shapes in the profile of residual vertical intensity (Figure 12) will be well enough determined to permit valid source depth calculations, even if the subglacial and ionospheric contributions could be satisfactorily distinguished. A comparison of large anomalies with subglacial topographic variations, however, can provide a few clues to the nature of the buried geology.

The most striking anomaly is that associated with the Crary Mountains, having an apparent peak-to-peak amplitude of nearly 1400 γ. Since the feature is narrow, the apparent maxima and minima having been observed on consecutive stations, the true amplitude may be considerably larger. The large susceptibility of the subglacial rock indicated by this anomaly is supported by the analyses of *Doumani and Ehlers* [1962], who found a magnetite content of 4% to 15% in six samples of andesites, trachytes, and basalts from the Crary Mountains. A magnetite content of 10% implies a susceptibility (k) of the order of 0.05 cgs, a very high value. With this susceptibility, for example, a vertical dike 1 km high and 400 meters wide, buried under 1 km of ice, would give rise to an anomaly 1200 γ in amplitude. Thus the measured magnetite content is more than sufficient to explain the large amplitude of the anomalies associated with the Crary Mountains. The large size of the negative part of the anomaly, however, suggests an important degree of remanent magnetization with a direction of magnetization substantially different from that of the present field, which is nearly vertical (about 75° dip). Southward from the Crary Mountains, there appears to be a good qualitative correlation between the subglacial peaks near miles 167 and 120 and corresponding Z anomaly maxima, suggesting that these buried peaks are petrologically similar to the Crary Mountains.

In view of the strong magnetic effect of the Crary Mountains, it is surprising to observe the absence of any significant anomaly associated with Toney Mountain, where the exposed rocks are similar [*Doumani and Ehlers*, 1962]. Since the subglacial spur of this peak which lies beneath the traverse track rises some 1500 meters above the rock level on either side, reaching within a few hundred meters of the ice surface, the susceptibility of the rock must be much lower than that found in the Crary Mountains. Order-of-magnitude calculations indicate that the absence of an anomaly greater than a few hundred gammas implies a magnetite content of no more than a few tenths of a percent. Although *Doumani and Ehlers* [1962] reported no measurable amount of magnetite in the single sample (a trachyandesite) from Toney Mountain, the probability that a mass of rock the size of this buried spur could consist of the same basic volcanics as those of the Crary Mountains, yet be almost lacking in magnetite, is very small. The implication is strong that the subglacial peak comprises more acidic rocks than were collected from Toney Mountain, perhaps similar to the rhyolites and dacites found to the west in the Usas escarpment and Mount Petras, or perhaps a young acidic differentiate of the Toney Mountain magma. The assumption of volcanics of some kind is justified in view of the nearness of Toney Mountain and the lack of evidence that any other rock type exists in eastern Marie Byrd Land.

The buried peak with a relief of roughly 1 km north of Toney Mountain (between miles 288 and 324) also shows no associated magnetic anomaly, suggesting a composition similar to that of the Toney Mountain spur. Still farther north, there appears to be a renewed association between magnetic highs and peaks in the bedrock topography.

There are no large anomalies associated with the spurs of the Usas escarpment and Mount Petras on the northern leg of the traverse. In 18 of 19 rock samples obtained from these mountain groups, *Doumani and Ehlers* [1962] found no more than a trace of magnetite. The one exception, containing 17% magnetite, is taken from olivine basalt capping the western exposed peak of the Usas escarpment. Apparently this basalt is not present in large quantity beneath the traverse track. The rest of the rocks are rhyolitic to dacitic tuffs and flows and granodiorite.

Between Mount Petras and the Clark Mountains, the anomaly amplitudes are once again large. Some of them are associated with the subglacial topography, such as the three consecutive peaks between km 1400 and 1500, whereas some are not, for example the very large anomaly just east of station 783. On the basis of geographic proximity plus a mean rock elevation nearly 1 km above adjusted sea level, this region can be considered a southern extension of the Flood Range. The relatively low topographic relief and the variable magnetic character suggest that this region belongs to the plutonic and metamorphic province of western Marie Byrd Land rather than to the Cenozoic volcanic province to the east.

The anomalies diminish again west of km 1400 and are low in the vicinity of the Clark Mountains, which are made up largely of granites and metasediments [*Doumani and Ehlers*, 1962] without significant magnetite content.

The profile from the Clark Mountains to Byrd station shows large anomalies that lack good correlation with the subglacial topography. The anomaly amplitudes suggest the presence of highly susceptible rocks, such as those of the Crary Mountains, but the lack of topographic correlation implies that such rocks are not the major constituents of the buried peaks. Unfortunately, there are no nearby outcrops for comparison.

The deep, broad negative anomaly near Byrd station is a peculiar feature, one that suggests a temporal change in the magnetic field rather than a geologic effect. The Byrd station record does show a decrease in vertical intensity, but it amounts to only 150 γ and occurs two hours after the minimum was recorded on the traverse profile. The anomaly, therefore, must truly reflect the buried rock. Because there is no prominent positive anomaly, despite the nearly vertical magnetic inclination, it is difficult to explain the negative anomaly without resort to remanent magnetization in a direction having a large component opposite the current field direction. This suggests the presence of a basaltic flow with reversed polarity.

Gravity

As the gravity anomaly at any individual station depends primarily on the subglacial rock elevation, some method of eliminating the effect of the topography is necessary before any other conclusions can be reached from the gravity observations. Where seismic soundings provide the ice thickness, a Bouguer anomaly can be computed, but there is no information on which to base corrections for the surrounding subglacial terrain; furthermore, the stations are very widely spaced. At most stations, no independent measurement of ice thickness at all is available, and only a free-air anomaly can be calculated. In both cases, however, averaging over many stations will tend to eliminate the effect of the terrain, since the buried terrain effect is as likely to be negative as positive. (This, of course, is not true for surface terrain effects.) Although individual gravity anomalies cannot be used for any local geological interpretation, we can hope to find meaningful regional variations from the mean anomalies.

Regional averages have been computed as part of a study of the gravity variations of West Antarctica [*Bentley*, 1968]. In this study, the map of West Antarctica has been divided into squares 111 km (1° of latitude) on a side, and the average free-air anomaly has been computed for each square. Means for squares 222 km (2°) on a side were taken as the arithmetic average of the means of the smaller squares, regardless of the number of actual observations in each 1° square, to prevent bias toward the areas with greater concentrations of gravity measurements. The total number of observed values ranged from 10 to 130 in a single 1° square, and from 40 to 200 in a single 2° square. Since the traverse routes were chosen largely without regard to the subglacial topography, we can expect this procedure to provide a reasonable mean of topographic variations too short to be isostatically compensated. One should keep in mind, however, that in some cases, notably along the northern leg of the Marie Byrd Land traverse, the 2° means comprise merely linear averages along sections of a single traverse route. It is thus possible that they may differ significantly from true areal means.

The sampling problem is much more serious for Bouguer anomalies than for free-air anomalies. The number of seismic soundings ranges from 1 to 4 per 1° square, and from 3 to 16 per 2° square. In places of irregular subglacial topography, these few soundings may be quite unrepresentative of the area as a whole. We have therefore used the additional information about ice thickness furnished by the gravity measurements by computing the mean equivalent elevation (see section *Gravity* above) from all data for each square, and then calculating the mean Bouguer anomalies from these means and the mean free-air anomalies. Although this process is in part redundant, it is justified by the fact that rock elevations are computed from relative rather than absolute free-air anomalies. The mean Bouguer anomaly in each square is, in effect, calculated from a number of *independent* measurements equal to the number of seismic soundings, and then corrected for deviations from the topographic mean.

Mean free-air anomalies in the southern part of

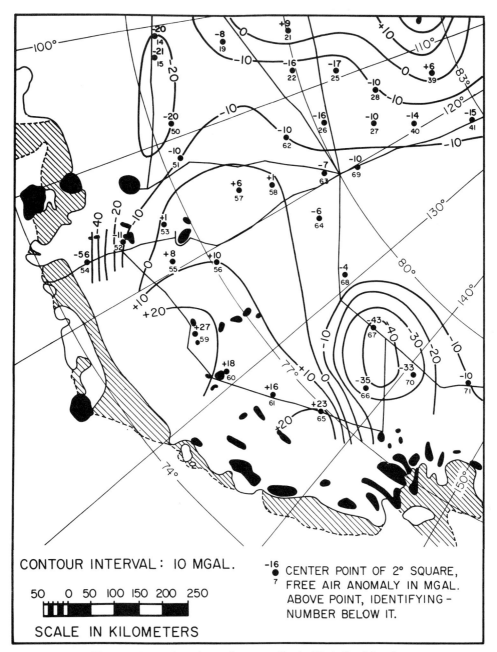

Fig. 14. Mean free-air gravity anomalies in Marie Byrd Land.

Marie Byrd Land and vicinity (Figure 14) are small. This region is bordered in eastern and western Marie Byrd Land by strong negative anomalies reaching −56 and −43 mgal, respectively, and in the north by positive anomalies up to +27 mgal. The profile along the traverse route (Figure 10) shows that these anomalies are generally associated with mean topographic variations, but that the western minimum is offset by 100 km from the associated subglacial trough. The Bouguer anomalies (Figure 15) decrease in a more or less regular fashion from +10 mgal near Byrd station to −90 mgal in the vicinity of Mount Petras and the Flood Range, with no striking anomalies.

It is well known that mean free-air anomalies over very large regions are a good approximation to mean isostatic anomalies, since the average gravitational field value depends much more on the total

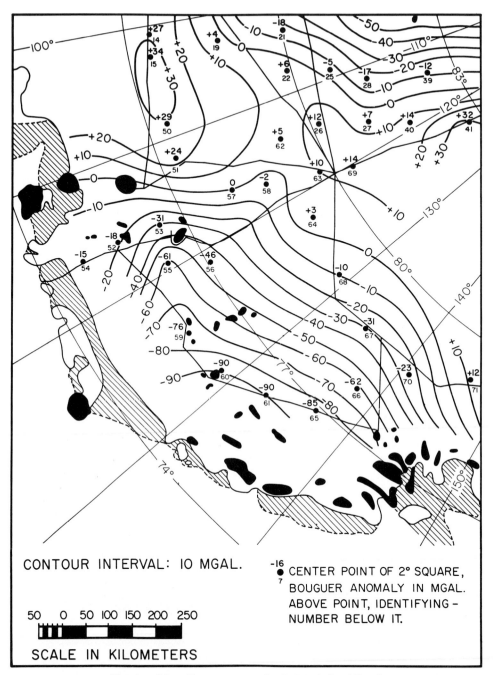

Fig. 15. Mean Bouguer anomalies in Marie Byrd Land.

mass beneath a region than on the vertical distribution of that mass. For areas 200 km in a side, however, we expect to find a correlation between mean free-air anomaly and topographic elevation, assuming isostatic balance, resulting from the finite size of the topographic and compensating masses. The gravity difference to be expected for a given topographic difference depends not only on the depth of compensation, but also on the distribution of masses in the surrounding areas. We do not, therefore, expect to find a closely defined linear regression of anomalies upon elevation, but we can expect to find a consistent trend throughout a region.

This is clearly shown in Figure 16, where we have

TABLE 2. Mean Free-Air and Bouguer Anomalies and Equivalent
Elevations, Marie Byrd Land and Vicinity, Averaged over 2° Squares

2° Square Index No.	No. of 1° Squares	No. of Gravity Stations	No. of Seismic Stations	Mean Equiv. Elev., meters	Mean Free-Air Anom., mgal	Mean Bouguer Anom., mgal
14	4	114	13	−420	−20	+27
15	3	99	11	−490	−20	+34
19	3	60	8	−110	−8	+4
21	3	79	9	+240	+9	−18
22	3	85	10	−200	−16	+6
25	4	109	11	−110	−17	−5
26	4	232	16	−250	−16	+12
27	4	253	13	−180	−10	+7
28	3	121	8	+10	−10	−17
40	3	114	7	−300	−14	+14
41	2	45	3	−420	−15	+32
50	3	62	6	−440	−20	+29
51	3	63	6	−300	−10	+24
52	3	96	7	+60	−11	−18
53	3	89	8	+290	+1	−31
54	2	41	3	−370	−56	−15
55	3	95	8	+620	+8	−61
56	3	63	7	+500	+10	−46
57	3	64	6	+50	+6	0
58	4	102	10	+30	+1	−2
59	2	45	4	+920	+27	−76
60	2	46	4	+940	+15	−90
61	2	41	4	+940	+15	−90
62	3	93	10	−130	−10	+5
63	4	211	14	−150	−7	+10
64	3	60	9	−80	−6	+3
65	2	43	3	+960	+23	−85
66	3	78	6	+240	−35	−62
67	2	52	7	−110	−43	−31
68	2	30	7	+50	−4	−10
69	3	175	11	−210	−10	+14
70	3	71	6	−90	−33	−23
71	2	30	3	−200	−10	+12

plotted mean free-air anomalies for Marie Byrd Land against mean equivalent elevations. A regression line with a slope of 28 mgal/km has been fitted by least squares to values from all squares except 54, 66, 67, and 70, which are clearly anomalous. This slope is large compared with those found elsewhere [*Uotila*, 1960; *Strange and Woollard*, 1964], but is in accord with data from the rest of West Antarctica (Bentley, unpublished data). We can conclude that the observations are consistent with a state of isostatic balance within most of Marie Byrd Land.

It can further be concluded that there is over-all isostatic compensation for the ice, which produces a mean attraction of 64 mgal. The intercept of the regression line at zero equivalent elevation is −6 mgal. Although from the standpoint of internal consistency this intercept may appear to be significant, it cannot be considered as such, since the over-all mean free-air anomaly could be in error by this amount, or the mean surface elevation for the region could be in error by 20 meters.

The anomalously negative gravity values apparent in Figure 16 occur in the two areas previously noted in Figure 14. Square 54, in which the mean anomaly falls 40 mgal below the regression line,

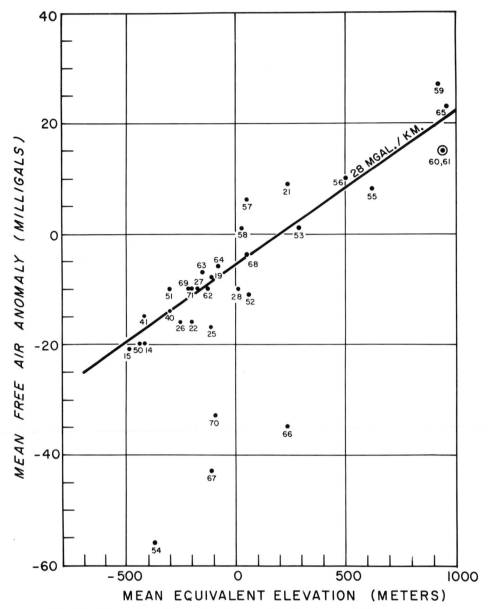

Fig. 16. Comparison of mean free-air anomaly with mean equivalent elevation.

adjoins the Amundsen Sea coast in eastern Marie Byrd Land. As a possible explanation for this low value, let us consider the effect on the gravity of a recent retreat of the ice, which would produce a large change in ice thickness near the coast.

Sample computations based on an equilibrium profile given by *Nye* [1959, p. 498], taking the central ice thickness as 2000 meters and the ice sheet half-width as decreasing from 400 to 300 km, indicate a decrease in ice thickness at the present edge of the grounded ice that would be about 1000 meters greater than that 150 km inland. This would correspond to an anomaly difference of 40 mgal, just that observed. Changing assumptions about the maximum thickness and the half-width do not substantially alter this value.

In order to explain the whole anomaly in this way, however, it would be necessary to assume that there has been essentially no isostatic rebound after load removal. This is unrealistic. The amount of crustal depression should decrease by a factor of exp $(-t/t_0)$ in t years after load removal; from measurements in Fennoscandia, t_0 appears to be on the order of 5000 years [*Heiskanen and Vening Meinesz*, 1958, part 10B]. Even after only 2000 years, the gravity anomaly should have diminished by a third, and after 5500 years by two thirds. Observations by *Black and Berg* [1964] and *Bennett* [1964] suggest that the ice retreated in the Ross Sea area contemporaneously with the end of the Wisconsin (Wurm) glaciation. It is unlikely that a major retreat would have occurred much more recently in the present case.

The other abnormally low gravity values (squares 66, 67, and 70) are all found in western Marie Byrd Land. The minimum value is about 36 mgal below the regression line (equivalent, for example, to an uncompensated lowering of the rock surface by 320 meters). Since an anomaly of this numerical size and limited areal extent (Figure 14) obviously cannot be explained in terms of changing ice thickness, it appears that there are current or recently relaxed tectonic forces maintaining isostatic disequilibrium. Unfortunately, it is not possible from the current traverse coverage to define the shape of the area of deficient gravity very well. A comparison of Figures 9 and 11 does indicate, however, that the anomaly pattern may be of different orientation, as well as offset, from the associated trough.

SUMMARY

The ice surface in Marie Byrd Land is highest in the vicinity of the Executive Committee Range, whence it slopes gently toward the Ross ice shelf, and more steeply and erratically toward the Amundsen Sea. This asymmetry can be explained by the difference in accumulation rate and ice thickness in the two regions, although a variation in basal temperature may also have an effect.

The subglacial topography of the region is extremely rugged, the total relief of the rock surface exceeding 5000 meters. Before the formation of the ice sheet, there were substantial areas of true land, particularly in the west. A large island probably extended unbroken from the Executive Committee Range or Crary Mountains in the east to Edward VII Peninsula in the west, bounded on the north by open ocean and on the south by the Byrd subglacial basin. Lying off the east and northeast coast were several smaller volcanic islands.

The magnetic profiles suggest that a subglacial spur of Toney Mountain must be different in composition from the highly susceptible rock of the Crary Mountains, perhaps a young acidic differentiate of the alkaline magma which produced the exposed mountain. The Flood Range appears to belong to the plutonic and metamorphic province of the Ford Ranges rather than to the Cenozoic volcanic province to the east. Strong negative magnetic anomalies in the Crary Mountains and near Byrd station suggest the presence of remanent magnetization in a direction quite different from that of the current magnetic field.

Marie Byrd Land exhibits general isostatic compensation for the load of the ice sheet. Negative free-air anomalies near the Amundsen Sea coast are too large to be caused by a recent retreat of the ice, except in the unlikely event that the ice front has receded on the order of 100 km in the last thousand years. A region of negative isostatic anomalies in western Marie Byrd Land is associated with, but not superimposed upon, a subglacial trough cutting 2000 meters below sea level.

Acknowledgment. The authors gratefully acknowledge the assistance of the members of the field parties in collecting the field data. Particular thanks go to P. E. Parks, who not only conducted a major part of the field program, but also aided in the reduction of the gravity observations.

Contribution 190 from the Geophysical and Polar Research Center, Department of Geology, University of Wisconsin.

REFERENCES

Anderson, V. H., The petrography of some rocks from Marie Byrd Land, Antarctica, *Ohio State Univ. Res. Found. Rept. 825-2,* part 8, 27 pp., 1960.

Behrendt, J. C., T. S. Laudon, and R. J. Wold, Results of a geophysical and geological traverse from Mt. Murphy to the Hudson Mts., Antarctica, *J. Geophys. Res., 67,* 3973–3980, 1962.

Bennett, H. F., A gravity and magnetic survey of the Ross ice shelf area, Antarctica, *Univ. Wis. Geophys. and Polar Res. Ctr. Rep. 64-3,* 97 pp., 1964.

Bentley, C. R., The structure of Antarctica and its ice cover, *Research in Geophysics,* Vol. 2, *Solid Earth and Interface Phenomena,* pp. 335–389, M.I.T. Press, Cambridge, Mass., 1964.

Bentley, C. R., Gravity maps, in *Magnetic and Gravity Maps of the Antarctic, Antarctic Map Folio 9,* edited by J. C. Behrendt and C. R. Bentley, American Geographical Society, New York, 1968.

Bentley, C. R., and N. A. Ostenso, Glacial and subglacial topography of West Antarctica, *J. Glaciol., 3,* 882–911, 1961.

Black, R. F., and T. E. Berg, Glacier fluctuations recorded by patterned ground, Victoria Land, in *Antarctic Ge-*

ology, edited by R. J. Adie, pp. 107–122, Interscience, New York, 1964.

Chang, F. K., Report of seismic wave studies in northwest Marie Byrd Land, Antarctica, *Bull. Seismol. Soc. Amer. 54,* 51–65, 1964.

Crary, A. P., E. S. Robinson, H. F. Bennett, and W. W. Boyd, Jr., Glacial studies of the Ross ice shelf, Antarctica, *IGY Glaciol. Rep. 6,* 193 pp., American Geographical Society, New York, 1962.

Doumani, G. A., Volcanoes of the Executive Committee Range, Byrd Land, in *Antarctic Geology,* edited by R. J. Adie, pp. 666–675, Interscience, New York, 1964.

Doumani, G. A., and E. G. Ehlers, Petrography of rocks from mountains in Marie Byrd Land, West Antarctica, *Bull. Geol. Soc. Amer., 73,* 877–882, 1962.

Fenner, C. N., Olivine fourchites from Raymond Fosdick Mountains, Antarctica, *Bull. Geol. Soc. Amer., 49,* 367–400, 1938.

Giovinetto, M. B., The drainage systems of Antarctica: Accumulation, in *Antarctic Snow and Ice Studies, Antarctic Res. Ser.,* vol. 2, pp. 127–155, AGU, Washington, D.C., 1964.

Glen, J. W., The creep of polycrystalline ice, *Proc. Roy. Soc. London, A, 228,* 519–528, 1955.

Heiskanen, W. A., and F. A. Vening Meinesz, *The Earth and its Gravity Field,* New York, 470 pp., 1958.

Nye, J. F., A method of calculating the thickness of the ice sheets, *Nature, 169,* 529–530, 1952.

Nye, J. F., The motion of ice sheets and glaciers, *J. Glaciol., 3,* 493–507, 1959.

Ostenso, N. A., and C. R. Bentley, The problem of elevation control in Antarctica, and elevations on the Marie Byrd Land traverses, 1957–58, *IGY Glaciol. Rep. 2,* IV-1 to IV-26, American Geographical Society, New York, 1959.

Passel, C. F., Sedimentary rocks of the southern Edsel Ford Ranges, Marie Byrd Land, Antarctica, *Proc. Amer. Phil. Soc., 89,* 123–131, 1945.

Strange, W. E., and G. P. Woollard, The use of geologic and geophysical parameters in the evaluation, interpolation, and prediction of gravity, *Hawaii Inst. Geophys. Rep. HIG-64-17,* 1964.

Thiel, E. C., Antarctica, one continent or two?, *Polar Record, 10,* 335–348, 1961.

U.S. Board on Geographic Names, *Geographic Names of Antarctica,* Department of Interior, Washington, D.C., 1956.

U.S. Coast and Geodetic Survey, *Magnetograms and Hourly Values, Byrd Station, Antarctica, 1959,* Department of Commerce, Washington, D.C., 1962.

U.S. Coast and Geodetic Survey, *Magnetograms and Hourly Values, Byrd Station, Antarctica, 1960,* Department of Commerce, Washington, D.C., 1963.

Uotila, V. A., Investigations on the gravity field and shape of the earth, *Publ. Inst. Geod. Phot. Cart., 10,* Columbus, 92 pp., 1960.

Wade, F. A., The geology of the Rockefeller Mountains, King Edward VII Land, Antarctica, *Proc. Amer. Phil. Soc., 89,* 67–77, 1945.

Warner, L. A., Structure and petrography of the southern Edsel Ford Ranges, Antarctica, *Proc. Amer. Phil. Soc., 89,* 78–122, 1945.

Weertman, J., On the sliding of glaciers, *J. Glaciol., 3,* 33–38, 1957.

Woollard, G. P., and J. C. Rose, *International Gravity Measurements,* Society of Exploration Geophysicists, Tulsa, Okla., 518 pp. 1963.

GEOPHYSICAL EXPLORATION IN QUEEN MAUD LAND, ANTARCTICA

John E. Beitzel

Geophysical and Polar Research Center, University of Wisconsin, 6118 University Avenue, Middleton 53562

Two reconnaissance traverses, South Pole–Queen Maud Land traverses 1 and 2, provided ice thickness and surface elevation data, as well as some information on the properties of the subglacial terrain. Electromagnetic soundings were used for the first time on a major Antarctic traverse, augmenting the seismic and gravimetric ice thickness determinations. The experimental use of distributed charges and arrays of multiple receivers successfully improved the signal-to-noise ratios of seismic reflection records. The ice surface, at an average elevation of 2950 meters above sea level in the traverse area, slopes toward the Weddell Sea at a grade of about 0.2% and is incised by a large ice stream flowing from the center of the ice sheet toward the Recovery glacier. Much of the ice surface topography was found to relate to the subglacial relief, in accordance with the mechanism described by Robin, wherein the effects of longitudinal stress variations are taken into account. The subglacial surface, with a total relief of about 2.5 km, displays a strong, roughly north-south grain and is dominated by a large high at about 30°E and a sub-sea level valley to the west. Magnetic measurements suggest that the rocks beneath the ice are probably but not necessarily part of a crystalline complex. The ice sheet and subglacial relief together were found to be in isostatic equilibrium throughout most of the area.

The first two parts of the South Pole–Queen Maud Land traverse were carried out during the 1964–1965 and 1965–1966 summer seasons. The first segment, called SPQMLT 1, began at South Pole station on December 4, 1964, and concluded at the abandoned Russian Pole of Inaccessibility station (82°07′S, 55°02′E) on January 28, 1965. SPQMLT 2, the second part of the traverse, began on December 15, 1965, at Pole of Inaccessibility, and ended at Plateau station (79°15.1′S, 40°31′E) on January 29, 1966. The location of Queen Maud Land and routes of the traverses are shown in Figure 1.

Major stations, which were 50 to 80 km apart, were located by multiple sun lines determined with a Kern DKM-2 theodolite. Borehole temperature and density measurements and seismic reflection measurements were made at the major stations. Minor stations, about 9 km apart, were established en route for the purpose of making gravity readings and glaciological and meteorological observations. The track direction was maintained by sun compass when feasible and by magnetic compass when necessary, and positions between major stations were determined by the Sno-Cat odometers, which were calibrated against tellurometer lines. The navigation was carried out by N. Peddie during SPQMLT 1 and D. Elvers during SPQMLT 2, both of the Coast and Geodetic Survey.

The traverse vehicles were two Tucker 843 Sno-Cats, one 742 Sno-Cat with a ground auger mounted on the back, and several sleds and trailers. The caravan carried all supplies, equipment, and much of the fuel necessary to support the nine to eleven men in the field party.

A summary of relevant statistics is presented in Table 1.

This report includes a description of the field methods and data reduction techniques, and the results obtained from the gravity, magnetic, seismic, electromagnetic sounding, and altimetry observations.

ALTIMETRY

The altimetry was conducted in the same way during both traverses. Twelve altimeters were separated into six pairs, matched on the basis of similar drift characteristics. Six of the altimeters were carried in one Sno-Cat and the matching six were carried in another. The vehicles traveled one station interval apart (6 to 10 km), and both sets of altim-

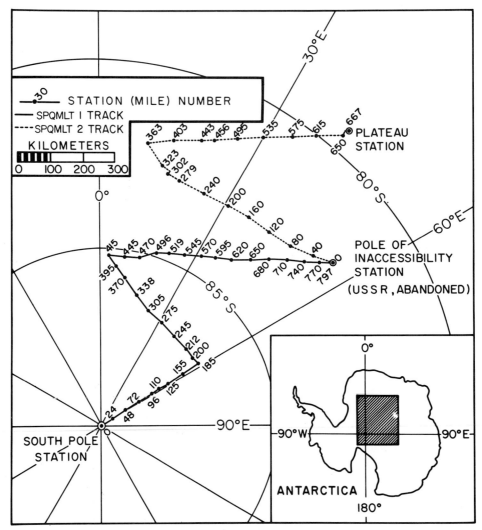

Fig. 1. Traverse location map, SPQMLT 1 and 2.

eters were read simultaneously at the respective stations. Thus measured elevation differences from station to station were free of temporal pressure effects. In addition, one altimeter in each Sno-Cat was read every quarter mile. The instruments were all read together at the beginning and end of each travel day to determine the reference correction, i.e., the difference in readings of each pair at the same elevation. Any change in the reference correction was distributed linearly with time.

The altimeter results were corrected for instrument temperature and air temperature effects, and the final elevation differences were then determined by using the means of the various differences between pairs of altimeters. Questionable and obviously spurious values were rejected. For this reason, many of the elevation determinations were based on data from five altimeter pairs, rather than six.

A root-mean-square standard deviation was computed for each mean elevation difference; the average of the standard deviations for all the station pairs, s, was 1.6 meters for each of the traverses. The standard error in the sample mean was estimated by using

$$\sigma_{\bar{x}} = \frac{\bar{s}}{N}\left(\frac{i}{\sqrt{m}} + \frac{j}{\sqrt{n}} + \cdots\right)$$

where N is the total number of final mean elevation differences, and $i, j, \cdots$, are the numbers of mean values determined using $m, n, \cdots$, instrument pairs, respectively. For each of the two traverses $\sigma_{\bar{x}} =$

TABLE 1. Description of South Pole–Queen Maud Land Traverses 1 and 2

	SPQMLT 1	SPQMLT 2
Departure date	Dec. 4, 1964	Dec. 15, 1965
Termination date	Jan. 28, 1965	Jan. 29, 1966
Elapsed time	54 days	46 days
Departure point	90°S	82°07′S, 55°06′E
Termination point	87°07′S, 55°06′E	79°14.8′S, 40°30′E
Turning point (s)	86°45.0′S, 58.6°E (Dec. 16); 85°10.2′S, 1.6°E (Jan. 6)	82°00′S, 9°55′E (Jan. 10)
Total distance	821 n. mi. (1520 km)	725 n. mi. (1341 km)
Average elevation	2780 meters	3090 meters
Average ice thickness	2740 meters	2770 meters
Maximum elevation	3718 meters (mi. 797)	3718 meters (mi. 0)
Minimum elevation	2628 meters (mi. 415)	2512 meters (mi. 363)
Maximum ice thickness	3580 meters (mi. 680)	3580 meters (mi. 279)
Minimum ice thickness	1805 meters (mi. 240)	1565 meters (mi. 225)
Number of personnel	10–9	11
Number of vehicles	Two 843 Sno-Cats, one 742 Sno-Cat, 2 rollitrailers, 4 sleds	Two 843 Sno-Cats, one 742 Sno-Cat, 2 rollitrailers, 6 sleds
Number of gravity stations	180	136
Number of seismic stations:		
Vertical reflection	29	19
Wide-angle reflection	3	3
Short refraction	3	3
Long refraction	2	4

0.70 meter, implying an uncertainty of the order of 1 meter for each station interval.

The SPQMLT 1 elevations are based on values of 2800 meters above mean sea level for South Pole (mile 0), obtained on a traverse originating from McMurdo station [*Crary*, 1963], and 3718 meters above mean sea level [*Kapitsa and Sorokhtin*, 1965] for the Pole of Relative Inaccessibility (mile 797), obtained on a traverse originating from Mirnyy station. The difference in elevations between these two stations as obtained on the SPQMLT 1 traverse was 867 meters instead of 918 meters. Despite the probable errors inherent throughout the complete 3000-mile McMurdo, South Pole, Pole of Inaccessibility, Mirnyy loop, this misclosure of 51 meters has been adjusted linearly between the South Pole and Pole of Inaccessibility. For SPQMLT 2 elevations, no closures are available at this time. Recent Japanese traverses to Plateau station, however, are expected to result in a three-way check (McMurdo, Mirnyy, Shōwa) on the elevations of this high plateau area of Antarctica.

The elevation data are presented graphically in the profiles of Figure 2 and are tabulated in Appendix 1.

SURFACE SLOPES

In addition to the altimeter data recorded every ¼ mile (about 0.46 km), ice surface slope information was obtained from slope shots made at each 5-mile altimeter station where visibility permitted. The procedure for taking slope shots involved scanning the horizon with a transit until the highest and lowest points were found. The azimuths and vertical angles to these points were then recorded.

The field method is quite sensitive to local topog-

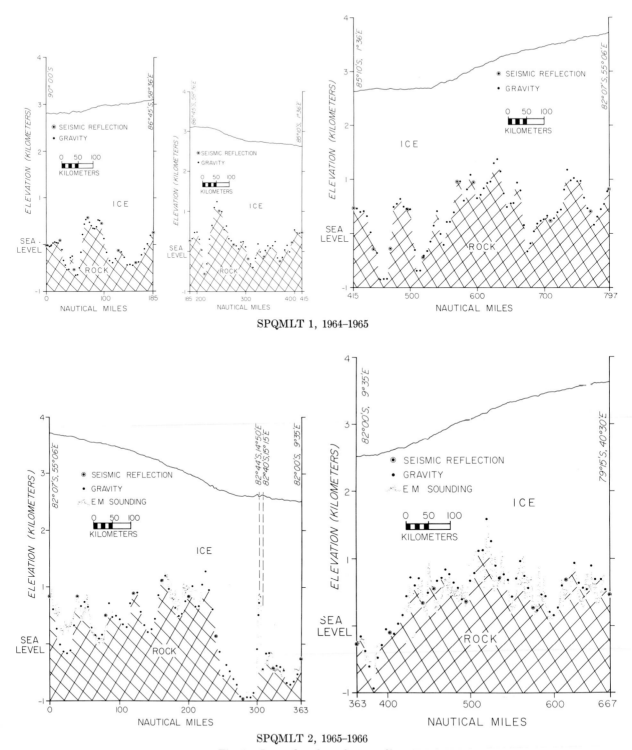

Fig. 2. Ice and rock surface profiles.

raphy and is therefore likely to yield regional slope information only when the data are taken in statistical quantities or are smoothed. Smoothed data are presented in Figure 3 in the form of frequency distributions of the azimuths of highs and lows for each of the two traverses. The smoothing was ac-

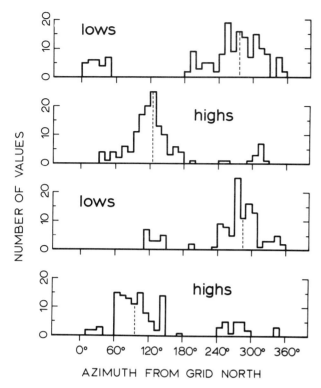

Fig. 3. Frequency distributions of ice surface azimuths, from transit observations. The two upper graphs show the distribution of azimuths of highs and lows, respectively, from SPQMLT 1; the lower two graphs are for SPQMLT 2.

complished by combining data from five consecutive stations and finding the azimuth of the bisector of a 30° sector that included the largest number of slope-shot azimuths (30° was selected, because it was found to be the smallest sector that would consistently enclose two or more azimuths). The running modes of this type have mean down-slope azimuths of 290° and 280° for SPQMLT 1 and SPQMLT 2, respectively.

Local slope directions as indicated by slope shots were used in constructing the ice surface contour map (Figure 4). Only those slope shots with highs and lows consistent within ±20° were used. These data indicate mean down-slope grid azimuths of 280° ± 40° and 280° ± 60° for SPQMLT 1 and SPQMLT 2, respectively, and mean vertical angles of 12 ± 11 min (slope of 0.0036 radians) and 10 ± 11 min (slope of 0.0030 radians). The data appear in Figure 4 in the form of vectors attached to the traverse track at the appropriate locations and pointing down slope. The length of each vector is the horizontal distance required for the surface elevation to drop one contour interval (100 meters), and, therefore, ideally, the length of the vector should equal the local distance between contours. The contours shown in Figure 4 are based on the altimetry data and the illustrated slope-shot vectors. A broader view of the ice surface topography is provided by Figure 5, which shows the SPQMLT 1 and SPQMLT 2 data augmented by contours modified from Bentley et al. [1964].

Description of the Ice Surface

The traverse tracks lie along the west slope of the East Antarctic ice sheet, approaching the crest in their most easterly excursions. The highest part of the ice surface appears to lie in the vicinity of 81°S, 75°E. The traversed area displays a fairly uniform gradient sloping downward toward the Weddell Sea, dropping 1000 meters in about 500 km of lateral extent (2 m/km), and also displays a curvature of the ice surface contours that suggests the existence

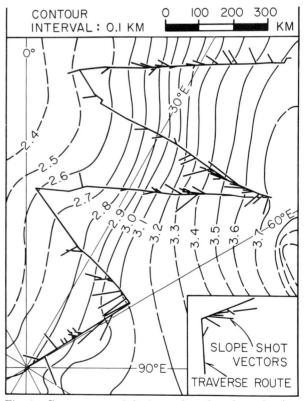

Fig. 4. Contour map of the ice surface elevations, showing the traverse route and slope shot vectors.

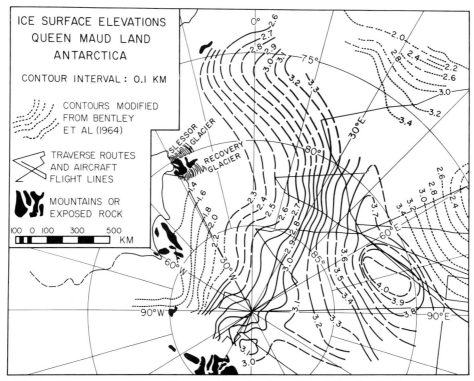

Fig. 5. Contour map of ice surface elevations for Queen Maud Land and adjacent areas.

of a large stream flowing from the center of the ice sheet toward the Recovery glacier (Figure 5).

Detailed profiles reveal topography of the order of 10 to 30 meters in amplitude with half-wavelengths of 10 to 30 km. Superimposed on these are features generally 2 to 4 km in extent and with up to 6 or 8 meters of relief. The ice surface features and their relation to the subglacial surface and to glacial mechanics are considered below.

Relation of Ice Thickness and Surface Slope

Profiles of the ice and rock surfaces (Figure 6) show maximum ice surface slopes occurring directly above peaks in the rock surface, as has been previously reported by, for example, *Robin* [1958] in northern Queen Maud Land and *Bourgoin* [1956] in Greenland. Because visual inspection suggests the existence of a relationship between ice surface slopes and subglacial topography, and because older theory, which neglects longitudinal stress gradients [e.g., *Nye*, 1957], is inadequate, the method of *Robin* [1967] was applied to the present data. The assumptions required in applying Robin's method are that the slope of the bed β is small (i.e., $\sin \beta \simeq \beta$) and that the shear stress component τ_{xy} is small compared with the normal stress difference when averaged throughout the ice thickness [*Collins*, 1968].

Electromagnetic soundings and altimetry at approximately half-kilometer spacing along miles 0 to 80, 298 to 335, and 420 to 640 of SPQMLT 2 provided data suitable for testing the method. The traverse route in these locations lay approximately along flow lines (Figure 4) except between miles 303 and 308, where a detour was made around a large crevasse field. The last area is of special interest, however, because of the particularly steep relief of the subglacial surface. The ice thickness and measured surface slope values were smoothed by taking running means over 3.7 km, which is of similar magnitude to the ice thickness. In order to remove the regional ice surface slopes, a second-degree least-squares polynomial was fitted to, and subtracted from, the surface slope values, yielding residual surface slope values.

Robin's equation 3 is

$$\alpha = \frac{\tau}{\rho g h} - \frac{1}{\rho g} \frac{d\bar{\sigma}_x^0}{dx} \qquad (1)$$

where α is ice surface, τ is the bottom shear stress,

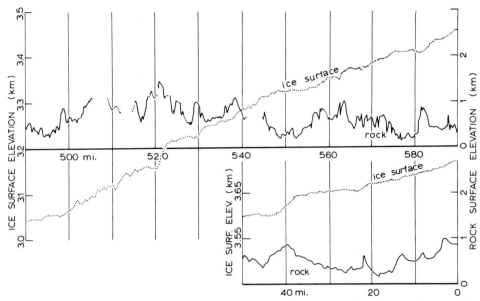

Fig. 6. Detailed ice and rock surface profiles from SPQMLT 2 showing relationships of rock peaks and ice surface slope maximums. Breaks in the rock surface curves are at locations of no data or intermittent data. Ice elevation scale is ten times rock elevation scale.

$\bar{\sigma}_x^0$ is the departure of the longitudinal stress from the weight of the overlying ice, h is the ice thickness, ρ is the ice density, and g is the value of gravity. Since the slope over large distances (100 km) depends principally on the first term on the right-hand side of equation 1, the regional slope values provide estimates of the shear stress. The interval from mile 298 to 335 is too short to be reliable, but the average slope for mile 0 to 80 yields a shear stress of 0.3 bar and miles 420 to 640 give shear stress values ranging from 0.3 to 0.6 bar.

The residual surface slope should be dependent on the second term in the right-hand side of (1). This term has been computed for the several data sets, following Robin's technique: estimating the longitudinal strain rates in the ice from outward velocities computed on the basis of the steady-state assumption (that the ice thickness does not change with time). The departure $\bar{\sigma}_x^0$ can then be determined with the aid of a flow law of the form

$$\sigma = b\dot{\epsilon}^{1/n} \qquad (2)$$

where σ is the uniaxial stress, $\dot{\epsilon}$ is the strain rate, b is the temperature-dependent 'viscosity' coefficient, and n is a constant equal to 1 for a Newtonian fluid and equal to infinity for a plastic material. Two flow laws were used in the computations, one with $n = 1$ and the other with $n = 4$. Laboratory determinations [Budd, 1968] suggest that $n = 1$ is appropriate for areas of low temperatures and low strain rates, while $n = 4$ is nearer the value conventionally used, as by Glen [1955]. The results are shown in Figures 7, 8, and 9, along with the residual measured surface slopes and the rock surface profiles. The values of b used in the flow laws were chosen to provide reasonable agreement between the computed and measured slope values and are given below:

	For $n = 1$	For $n = 4$
mile 0–80	10^5 bar yr	10 bar yr$^{1/4}$
298–335	10^4	10
420–640	10^4	10

Generally, good qualitative agreement between the computed and measured curves is found throughout the profiles. The near coincidence of the curves makes them difficult to distinguish in many areas. Between miles 424 and 650, the linear flow law provides a better fit than the power law at the upstream end of the profile, where the strain rates are lower, whereas the opposite is true downstream (see further discussion by Beitzel [1970]). The power law appears to fit better on miles 0 to 80, but if the linear b value is reduced to 3×10^4 bar yr, making it closer to the other linear b values, the best agreement is obtained. On the short segment, miles

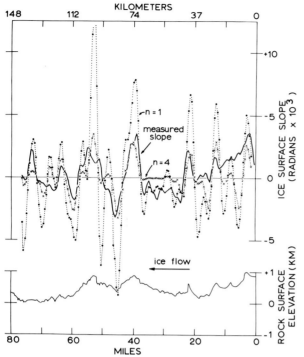

Fig. 7. Comparison of computed and measured ice surface slopes for miles 0 to 80 of SPQMLT 2. The upper solid curve shows measured ice surface slope values and the lower solid curve is the rock surface profile. Computed surface slope values are shown by dots for $n = 1$ and by crosses for $n = 4$.

298 to 335, the power law fits better, as might be expected for an area of high strain rates.

SEISMIC STUDIES

Standard seismic reflection methods were used to measure the ice thickness, and some long-distance refraction shooting was done in an attempt to determine properties of the rocks beneath the ice. The velocities required for interpretation of the reflection data were obtained from short refraction and wide-angle reflection profiles. All of the seismic records were made on a photographically recording 24-channel Houston Technical Laboratories 7000-B seismograph, which provides for various filter and gain combinations. Refraction shooting was done with the filters open, but the high passband of 210–320 Hertz and high gains were found to yield the best reflection signals. Shooting procedures and spread geometries are described in the sections below.

Velocities

Because the wave velocities in the ice sheet reach a maximum in the upper few hundred meters, the velocity functions for the entire thickness of ice cannot be determined directly. Instead, short-range refraction measurements are used to find the near-surface velocity structure, where the greatest variation in velocity occurs, and the wide-angle reflection method is used to obtain an average velocity for the total ice thickness.

Short refraction. A brief description of the six short refraction profiles appears in Table 2. Two-meter geophone spacing was used throughout, and some additional measurements were made with half-meter spacing at close range, to detail the top few meters. Compressional (P) waves were generated by firing caps beneath a horizontal metal plate at the surface of the snow, and shear (S) waves were generated by striking a metal plate thrust vertically into the snow.

Times on the short refraction records were read to the nearest 0.1 msec, and the times are probably accurate to 0.2 msec. Travel-time plots were con-

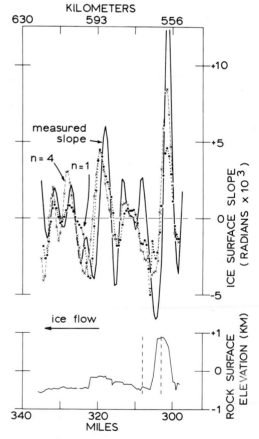

Fig. 8. Comparison of computed and measured ice surface slopes for miles 298 to 335 of SPQMLT 2. Symbols are the same as in Figure 7.

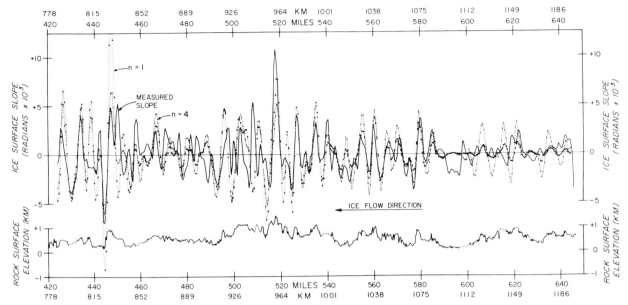

Fig. 9. Comparison of computed and measured ice surface slopes for miles 420 to 650 of SPQMLT 2. Symbols are the same as in Figure 7.

TABLE 2. Short Refraction Profiles

Mile No.	Position	Length of Profile, meters	Type of Waves Measured
220, SPQMLT 1	86°47.3′S, 47.7°E	300	P and S
496, SPQMLT 1	84°58.1′S, 17.9°E	300	P and S
220, SPQMLT 2	82°54′S, 27°15′E	494	P
279, SPQMLT 2	82°54.5′S, 18°11′E	348	S
363, SPQMLT 2	82°01.0′S, 9°28′E	344	P
456, SPQMLT 2	81°29′S, 21°55′E	444	P
		394	S

structed for each profile; velocities were measured from tangents to the travel-time plots in the curving parts of the curves, and least-squares analysis was used to determine velocities along the more nearly linear parts of the curves. The resulting velocities were plotted versus distance and were fitted with smooth curves (Figure 10). From these curves, distance and velocity values were picked which, by means of the Wiechert-Herglotz-Bateman integral, were integrated to yield velocity versus depth curves (Figure 11). The integrations were performed graphically, using a planimeter.

The Wiechert-Herglotz-Bateman integral is applied with the attendant assumption that the velocity increases with depth. This assumption is probably not violated on a large scale; borehole measurements of the physical properties of ice indicate that the seismic velocity does indeed increase smoothly downward in the near-surface part of the ice sheet.

Where both P- and S-wave velocities are available, Poisson's ratio σ can be determined using

$$\sigma = \frac{1 - 2Vs^2/Vp^2}{2(1 - Vs^2/Vp^2)}$$

where Vs and Vp are the S-wave and P-wave velocities, respectively. Poisson's ratio was computed at three stations where both P- and S-wave velocities were available and from P-wave velocities determined at mile 220 and S-wave velocities determined

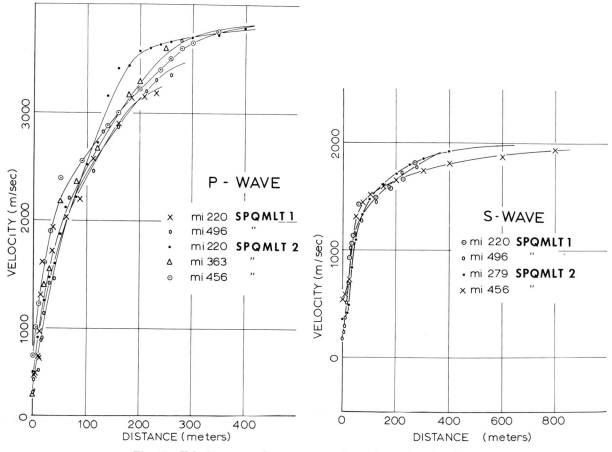

Fig. 10. Velocity versus distance curves from short refraction data.

at mile 279 on SPQMLT 2. The four curves of Poisson's ratio versus depth (Figure 12) show marked near-surface differences, then converge at depths of 30 meters and greater toward values between 0.30 and 0.35.

Wide-angle reflection profiles. During each of the two seasons, three wide-angle profiles were attempted. Four of the profiles provided useable data, the other two being unsuccessful because of poor reflections and very low signal-to-noise ratios. Each of the profiles extended to approximately six kilometers with five to seven shot locations and one instrument layout of twenty-four geophones covering 700 meters. Times were picked from the records to the nearest millisecond but probably have uncertainties of 2 or 3 msec. The distances from shot to spread were measured by Tellurometer to a probable accuracy of a decimeter. Reflection times were corrected for shot depths, bottom topography, and near-surface effects. Mean P-wave velocities (Table 3) were obtained by least-squares fitting of regression lines to standard $t^2 - x^2$ plots for each profile (Figure 13).

Though four values are too few to indicate any trends, previous data are available from East Antarctica, providing a broader sampling. *Kapitsa* [1965] has cited velocities of '3800 m/sec in the central regions of Antarctica where ice thickness exceeds 2000 m, and 3700–3750 m/sec in the coastal regions.' *Sorokhtin et al.* [1960] previously reported a 3750 ± 70 m/sec velocity for Komsomol'skaya, where the ice thickness exceeds 3000 meters. *Crary* [1963] obtained a velocity of 3728 m/sec from a good-quality reflection at 87°7.3′S, 156.4°E, and *Robinson* [1964], from wide-angle reflection profiles in the vicinity of the South Pole, found velocities of 3780 and 3860 m/sec at 89°6.7′S, 114.9°W and 88°32.3′S, 48.9°W, respectively. An estimate of 3820 m/sec for East Antarctica was presented by *Bentley* [1964], based on a review of the available data.

The velocity values from Komsomol'skaya and the South Pole traverse were combined with the

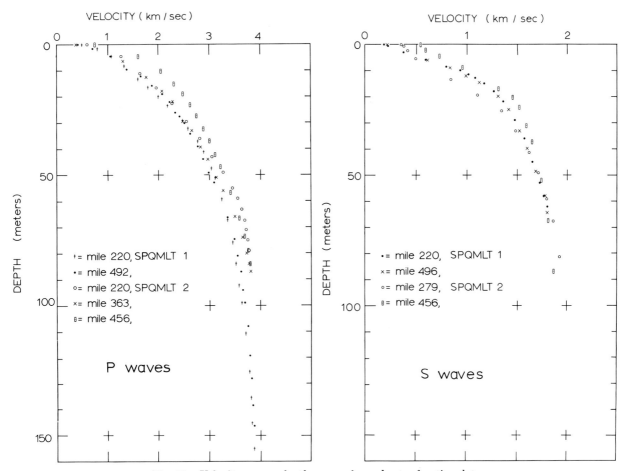

Fig. 11. Velocity versus depth curves from short refraction data.

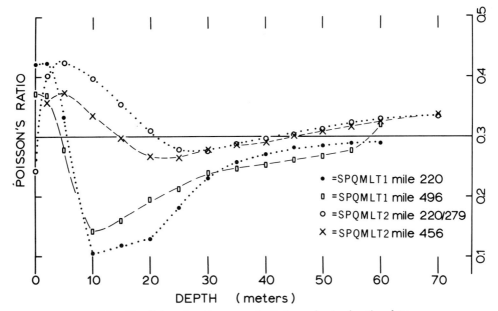

Fig. 12. Poisson's ratio versus depth from short refraction data.

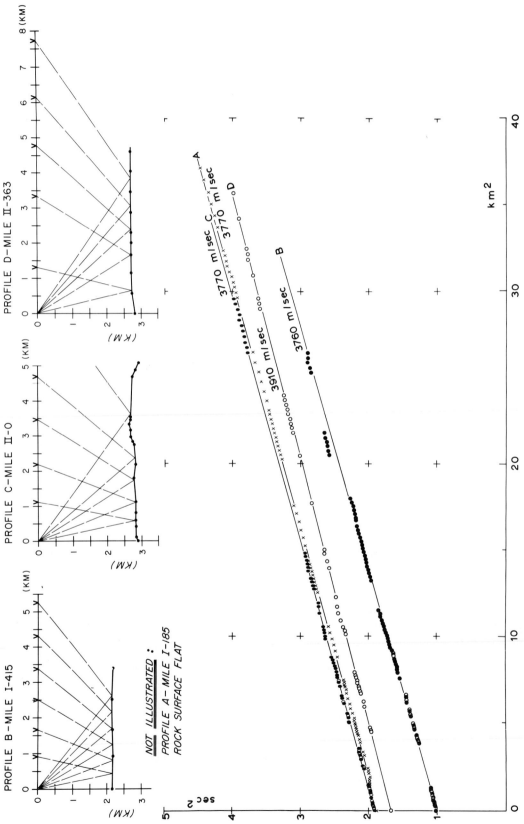

Fig. 13. Seismic wide-angle reflection profiles. Curves A–D are least-squares-fitted t^2 versus x^2 lines for the corresponding profiles shown above. The profiles indicate shot locations and ice thicknesses. Ice thickness at the beginning of each profile was determined by seismic reflection; the other thickness values were determined by gravity in profiles A and B and by electromagnetic sounding in profiles C and D.

SPQMLT values in a search for some consistent trend. The velocities plotted against ice surface elevation (Figure 14) show no apparent trend, although the highest velocities occur at the lowest surface elevations, which is surprising, since the velocity is expected to vary inversely with temperature [e.g., *Robin*, 1958, p. 87]. There is, however, a group of consistent values clustered about 3770 m/sec standing in contrast to the two highest velocities. The computed velocity shown in Figure 14 is a mean value calculated from a velocity profile for a 3000-meter-thick ice sheet, the velocity profile in turn having been determined from temperatures found by *Robin's* [1955] method. The distribution of measured velocity values about the computed value is probably attributable to anisotropy in the ice, as in West Antarctica [*Bentley*, this volume].

Maximum P-wave velocity values also were determined from wide-angle reflection profiles and from long refraction data, using first arrivals and multiple P-wave arrivals. Velocities determined in this way from long-range wide-angle shots are listed in Table 4. Long refraction results (discussed below) yielded the ice velocities and Poisson ratios listed in Table 5.

Seismic Reflection Measurements

Ice thicknesses were determined by vertical reflections at 25 stations during SPQMLT 1 and at 18 stations during SPQMLT 2. The usual spread configuration was an L with the shot hole at the corner, although occasionally an in-line spread with the shot point in the middle was used. Electrotech 4-, 7-, 20-, and 30-Hz geophones were used, both singly and in multiple arrays. The multiple arrays are discussed below.

The shot holes were machine augered to a depth of approximately forty meters, the relatively deep holes being used to help suppress the surface noise, which is a shot-generated phenomenon, and to generate higher-frequency energy. DuPont Nitramon S in charges of one to three pounds was used as the energy source. Distributed charges composed of 8-foot cardboard tubes wrapped with Primacord and designed to propagate energy at the seismic velocity were tested at several stations in 1964–1965, and commercially produced delay units (DuPont Elcord) were used during the 1965–1966 season. The results of their use are discussed in the next section.

Sources of error in seismic reflection soundings

TABLE 3. Average Seismic Velocities in the Ice (t^2 versus x^2)

Mile No.	Avg. Velocity	No. of Picks
185, SPQMLT 1	3770 ± 10	97
415, SPQMLT 1	3760 ± 5	122
0, SPQMLT 2	3760 ± 20	85
363, SPQMLT 2	3910 ± 20	46

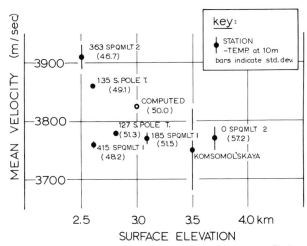

Fig. 14. Average seismic velocity values (t^2 versus x^2) for the East Antarctic ice sheet and their relation to surface elevation. Data from traverses are labeled with mile numbers. The numbers in parentheses are 10-meter temperatures in minus degrees Celsius.

include errors in picking, poorly defined shot instant, and velocity uncertainties. *Bentley* [1964] has suggested ±10 meters as a representative value for the relative accuracy between ice thickness values determined in a particular region, including all normal errors except velocity uncertainty.

For the sake of uniformity and because the variations of velocity throughout the traversed area are not known, one formula was used for computing all the vertical reflection depths. An average velocity of 3770 m/sec was selected as representative for the total thickness of ice, excluding a 250-meter surface layer. The velocity versus depth data from the short refraction profiles were used to correct the wide angle reflection data for near surface effects. The resulting formula for seismically determined ice thicknesses is

TABLE 4. Ice Velocities at Medium Ranges
(From Wide-Angle Reflection Profiles)

Mile No.	Record	Velocity, m/sec	Range, meters
185, SPQMLT 1			
	7049	3910	2150–2850
	7048	3970	2800–3500
	7055	3905	3800–4500
	7055	3935*	
	7056	3940	4835–5540
	7056	3930*	
	7057	3932*	5856–6558
415, SPQMLT 1			
	7125	3885	2165–2870
	7124	3890	3030–3730
	7124	3959*	
	7114	3880	3970–4670
	7114	3993*	
	7121	3920	4985–5690
	7121	3945*	
0, SPQMLT 2			
	8012	3980	2420–3195
	8013	3960	3375–4150
	8013	3980*	
	8015	3945*	5081–5857
363, SPQMLT 2			
	8094	3920	2300–3050
	8095	3950	4550–5300
	8095	3920*	
	8101	3920	5830–6580
	8101	3910*	
	8106	3910*	3446–4195

*Maximum velocities, obtained from refracted P multiples.

$$d = [(t_{1/2} - 0.079)\, 3770 + 250] \text{ meters}$$

where $t_{1/2}$ is the one-way time in seconds, and d is ice thickness. The vertical travel time through the top 250 meters of ice is 0.079 sec. Correction for the shot depth is applied by adding the up-hole time, recorded with a shot-point geophone, to the two-way travel time. The resulting ice thicknesses are listed in Appendix 1, and the seismic reflection travel times and other relevant data are listed in Appendix 2. Sample seismic reflection (vertical) records appear in Appendix 3, along with a sample of typical wide-angle reflection records. Because we had many requests for ice thickness data before the final velocity values were computed, thicknesses were determined with an assumed average velocity of 3810 m/sec (exclusive of the near surface), and these values were used in constructing the profiles of Figure 2. Because the corrections required to bring the profiles into agreement with the final velocity data were small, the drawings were not re-done. Correct ice thickness values are given in Appendix 1.

Surface Noise and Signal Enhancement

During times of low temperatures, shot-generated surface noise poses a major problem to successful

TABLE 5. Ice Velocities at Long Ranges (From Refraction Profiles)

Mile No.	Range, km	Velocity, km/sec		Poisson Ratio
		P	S	
275, SPQMLT 1	16.7	3.95	1.93	0.344
620, SPQMLT 1	11.7	3.90		
0, SPQMLT 2	12.5	3.92		
323, SPQMLT 2	19.5	3.90	1.97	0.329

seismic shooting on the East Antarctic plateau. The recommended procedures for overcoming the problem of surface noise include the use of deep shot holes, the employment of multiple arrays of geophones, the selection of a pass band of the highest frequencies possible, and the discontinuation of shooting during the coldest parts of the season. Each of the first three procedures was followed on the SPQMLT, the second receiving special attention.

The use of multiple geophone arrays to reduce surface noise depends upon the incoherence of the noise across short distances. Experiments by *Nakaya* [1959] suggest that elements of the order of decimeters in dimensions may be responsible for the generation of the surface noise [*Bentley*, 1964]. This being the case, an array of geophones connected in parallel and spaced a few decimeters apart should help reduce the incoherent noise by destructive interference, while reinforcing the coherent reflected signal. Therefore various patterns of paralleled geophones were used during the traverses.

To facilitate the application of multiple arrays, a set of paralleling cables was provided for the 1964–1965 traverse, permitting convenient attachment of many geophones to a single take-out. Generally, ten or twelve geophones of the same frequency were set in a rectangular grid, with a spacing of about a half meter. To enhance the signal amplitudes, arrays of five or six series pairs, all connected in parallel, were sometimes used. Also, combinations of two or four geophones variously connected were tried.

At South Pole station, where the arrays were first tested, signal and noise amplitudes were measured from five records, yielding mean signal-to-noise ratios (snr) of 2.8 ± 1.2 for the single geophones and 4.9 ± 2.5 for the arrays. Arrays of four geophones, composed of two series pairs connected in parallel, were also used, and at mile 445, for example, yielded a mean snr of 2.8 ± 1.3 as compared with 3.8 ± 2.2 for arrays of twelve geophones, composed of six series pairs connected in parallel. Similar results were obtained during SPQMLT 2. Records from six SPQMLT 2 stations shot during the colder parts of the season (when surface noise is generally worse) using arrays of ten geophones each, as well as single geophones and pairs, gave mean snr of 2.9 ± 1.1 for the arrays and 2.2 ± 1.2 for the others.

Various array geometries were used, ranging from rows of geophones at two-meter spacing to square grid arrangements with half-meter spacing. The use of half-meter spacing, as opposed to greater spacing, did not significantly reduce the snr enhancement, supporting the notion of decimeter-size noise sources. Because of this, and because the most coherent signal could presumably be obtained with the smallest geophone spacing, the half-meter grid arrays were put into general use. The exact geometry of the arrays does not appear to be critical, but the minimum number of geophones is most important, eight to ten generally being the number required to produce measurable improvement.

Another technique that was applied in the hope of reducing surface noise was the use of distributed charges. If the ignition of the charge occurs in the down-hole direction at a rate equal to the velocity of seismic waves, there will ideally be destructive interference in the upward-traveling energy, reducing the surface noise due to the shot. The Primacord devices described above were difficult to couple and inconvenient to prepare, and they failed to produce enough energy for a good reflection signal except in the thinnest ice. Their successors, the Elcord units, were easy to couple, could be adjusted for various velocities of propagation, and could be assembled with interspersed Nitramon charges. Again adequate energy was a problem, because the use of Nitramon units to augment the charge tended to annul the beneficial effect of the distributed charge. Therefore the Elcord units, although more promising than the homemade Primacord devices, were also generally limited in application to thinner ice. Where their use was feasible, however, they were fairly successful, yielding improved signal-to-noise ratios.

Long Refraction Studies

Six long refraction profiles were attempted in order to determine seismic velocities of the upper crustal layers. Four of the attempts yielded useable records, but only three provided apparent velocities in the subglacial material. Failures were due variously to inadequate radio communication, insufficient energy, and a misfire. Independent control on the ice thickness was provided by a vertical seismic reflection measurement at mile 275 of SPQMLT 1 and by gravity measurements at mile 620. At mile 0 of SPQMLT 2, seismic, gravity, and EMS ice thickness measurements were made.

The distances from shots to receivers were measured by Tellurometer and Sno-Cat odometer. Tellurometer distances are presumed to be accurate to ±0.1 meter. Odometer measurements, which were used to extend beyond the useable range of the tellurometers, were repeatable to within 3%.

The records were taken at 25-cm/sec paper speed, and times were picked to the nearest millisecond. The picks are judged accurate ±3 msec. The time breaks were monitored by radio, the tone appearing on trace twelve of the twenty-four channel records. The combination of low signal and high noise levels reduced the accuracy of the time reference on some of the records, contributing an uncertainty of up to 5 msec.

The composition of the receiving spread varied but generally included twelve vertical and twelve horizontal geophones or arrays, the horizontal ones aligned alternately in line with and perpendicular to the cable. There were generally twelve arrays consisting of eight or ten geophones connected in paral-

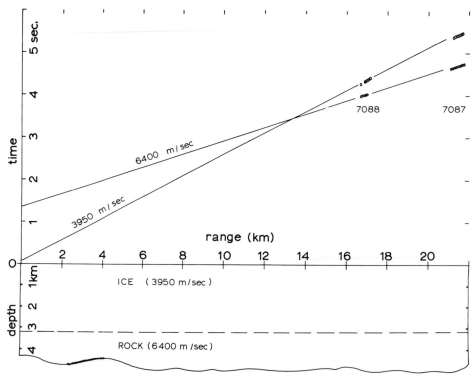

Fig. 15. Seismic long refraction results for mile 275, SPQMLT 1. Upper plot is travel-time plot; lower plot is the model derived from the data. The line indicating the top of the second layer (rock) in the model is dashed to indicate low confidence.

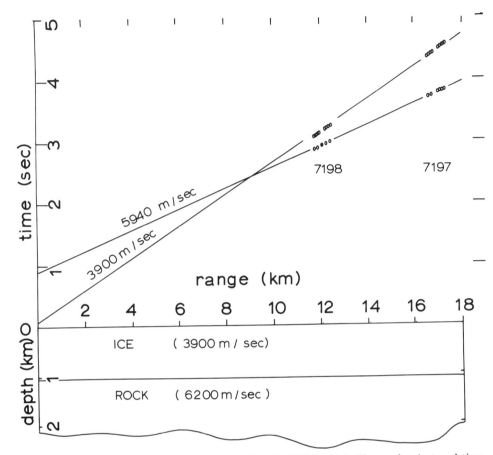

Fig. 16. Seismic long refraction results for mile 620, SPQMLT 1. Upper plot is travel-time plot; lower plot is the model derived from the data.

lel, some arrays in each of the three orientations; the other twelve receivers were either single geophones or pairs connected in series.

Standard refraction formulations for flat-lying beds were used to determine depths. As for the reflection calculations, the average velocity of 3770 m/sec was used for the ice layer after suitable corrections to 250 meters below the surface.

SPQMLT 1, mile 275. The first long refraction profile (Figure 15) was shot at mile 275 of SPQMLT 1 (86°38′S, 30.6°E) on December 26, 1964. The receivers were parallel to the track at mile 275, and the two 250-lb shots were beyond mile 275, 16.7 and 21.1 ± 0.1 km from the nearest receiver of the 702-meter spread. Velocities of 3950 and 6400 m/sec were determined for the ice and subglacial material, respectively; the ice velocity was well determined by multiple P arrivals. The ice layer thickness computed from the intercept time is 3200 meters for the nearer shot and 3050 meters for the farther, while the thickness determined by seismic reflection is 2667 meters, suggesting the possible existence of an undetected layer or that the rock velocity should be nearer 6000 m/sec.

SPQMLT 1, mile 620. A refraction profile was shot at mile 620 of SPQMLT 1 (84°10′S, 37.6°E) on January 20, 1965. The 702-meter recording spread was fixed at the station, and shots of 250 lb and 350 lb were placed along the track 16.6 and 11.7 km beyond the station, respectively. The ice velocity of 3900 m/sec (Figure 16) was determined from arrival times on the two records. The higher-velocity arrivals could be picked on only five or six traces on each of the two records, yielding apparent velocities of 5900 ± 100 m/sec, with commensurate uncertainty in the intercept times. Taking the two records together, a subglacial apparent velocity V_2 of 5940 m/sec was obtained. Correcting for the dip of the rock surface from shot to receiver, which according to gravity data is 2° (0.035 radians), gives a rock velocity of 6200 m/sec and an ice thickness of 2240 meters, in fair agreement with the ice thickness of 2134 meters determined from gravity data.

Pole of Inaccessibility. The long refraction profile at the Pole of Inaccessibility (82°07'S, 55°06'E), shot on December 10, 1965, was laid out with the 766-meter receiving spread at mile 0 of SPQMLT 2 and with two 300-lb shots located between the two traverse tracks at distances of 12.5 and 15.8 km from mile 0. Electromagnetic soundings provided ice thickness data to 5.5 km, and thicknesses at greater distances were estimated by projection of measurements taken along nearby traverse tracks.

Interpretation of the data was hampered by the absence of timing lines and of the time break on the 15.8-km record, which displayed the only high-velocity arrivals. Two approaches were used to circumvent the problem. In the first, comparison of the frequencies of noise on both records and uniformity of paper speed after the first second were assumed to permit picking of velocity data. The second approach made use of the ice layer velocity of 3920 m/sec (Figure 17) determined from P_n multiples on record 8007 (12.5-km range) to find a time scale for record 8008, and the scale was then used to determine the velocities of the other arrivals. The first approach yielded velocities of 5030 and 6370 km/sec. While the subglacial velocities are dubious at best, intercept-time determination of the ice thickness using 5030 m/sec gives 3080 meters, compared with 2895 meters measured at the receivers and 3080 meters found by interpolation at the 12.5-km shot point. Using the above velocities, the first subglacial layer would have a thickness of 1070 meters.

Following the second approach, velocities of 4900 and 6260 m/sec were determined, yielding an ice thickness of 2900 meters and a second layer thickness of 1350 meters.

ELECTROMAGNETIC SOUNDER

During SPQMLT 2, electromagnetic soundings were carried out along 530 miles, or about 80% of the route. The equipment, loaned by the U.S. Army Electronics Command, comprised a 30-MHz pulsed radar transmitter and receiver, as well as a special camera. The travel times of the vertically reflected pulses were measured on an oscilloscope by comparison with a graticule. The time base was calibrated against an accurate oscillator. Travel times were read visually and recorded every 0.2 n. mi. (0.37 km); the signals were photographed approximately every nautical mile. Figure 18 shows sample recordings.

The electronic equipment was mounted in a Sno-Cat, and transmission lines were extended back along the train of sleds to a specially constructed antenna sled at the rear of the caravan. The antennas, which were mounted parallel to each other, were rigid folded dipoles (5 meters long), each with a single reflector element 2 meters above the antenna. The reflectors were found to enhance the signal by 3 to 5 db.

The electromagnetic wave velocity in the ice was determined by comparison of travel times with seismic travel times. Figure 19 is a comparison of the electromagnetic travel times and seismic travel times at 15 stations, with a regression line fitted to the points. The slope of the line is 22.8 ± 1.0 μsec/sec, which means that for a seismic velocity of 3820 m/sec the electromagnetic wave velocity would be

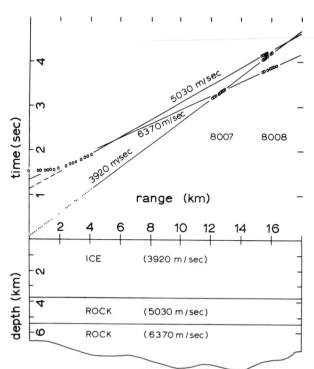

Fig. 17. Seismic long refraction results for mile 0, SPQMLT 2. Upper plot is travel-time plot. The data points between 0 and 4 were obtained from a wide-angle profile. The lower plot is the model derived from the data.

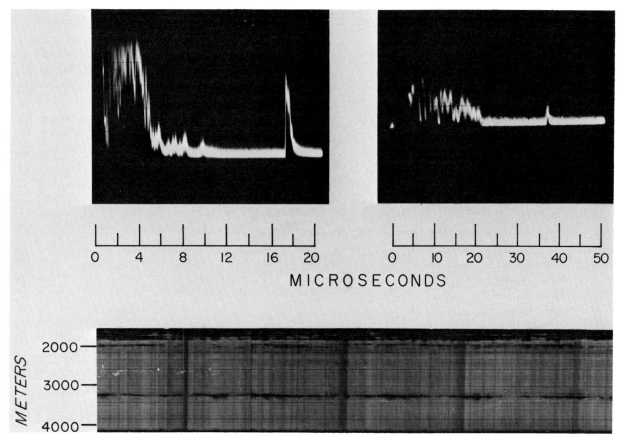

Fig. 18. Electromagnetic recordings, photographed from the oscilloscope screen with a 35-mm camera (upper figures) and taken with a sequence camera (lower figure). Distance covered in the lower figure is about 0.25 km; bottom of ice is indicated by the dark band near 3200 meters.

168 m/μsec (more recently, successful wide-angle electromagnetic reflection profiles have been obtained in Queen Maud Land, indicating a mean velocity of 171 m/μsec [Clough and Bentley, 1970]). The EMS results shown in Figure 2 were adjusted to the seismic reflection ice thickness values, as determined at the time of plotting.

ROCK SURFACE ELEVATIONS

Rock surface elevation values, determined from ice surface elevations and ice thicknesses, are shown as profiles in Figures 2 and 20, as a contour map in Figure 21, and are tabulated in Appendix 1. The rock surface (Figure 21) displays a striking grain which trends north-northeast, and, though the contours are based on widely separated lines of data, the reality of this trend is supported by the persistence of the gross topographic features from line to line, as shown in Figure 20.

The dominant feature is a massive central block 250 to 300 km wide, flanked on the Greenwich meridian side by a 90-km-wide valley below sea level. More than 2 km of relief was found between miles 220 and 280 of SPQMLT 2, on the west flank of the central high; this is comparable to the relief in the western Basin and Range province of the United States. Electromagnetic sounding data indicate that the surface of the central block is dissected by relief of the order of hundreds of meters occurring over a few kilometers.

MAGNETICS

During the two traverses, Varian M-49 proton-precession magnetometers were used for total magnetic-field measurements. In the 1964–1965 season, two instruments, one in each of two Sno-Cats, provided almost continuous data along the traverse route. The two magnetometers were monitored according to the system described above for the altimeters, with simultaneous readings at each station

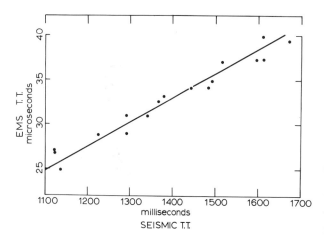

Fig. 19. Comparison of electromagnetic and seismic reflection travel times for SPQMLT 2 showing least-squares-fitted line.

and nonsynchronous readings at quarter-mile intervals between. Extensive instrument failures limited the 1965–1966 data to readings from one instrument taken at five-mile intervals. During both seasons, a U.S. Coast and Geodetic Survey observer determined declination, inclination, and field components at each major station, using a C.A.R.L. magnetometer.

The total field readings for SPQMLT 1 were plotted against distance and against time in an attempt to distinguish temporal and spatial variations. It had been presumed that simultaneous readings at different locations would provide difference values between stations that would be free from temporal effects. However, the two profiles show little or no systematic variations with time, while displaying excellent correlation with distance (Figure 22). The anomalies agree between the two sets of data to generally within the order of 20 γ. The differences are not systematic with time and are probably attributable to reading errors or perhaps the proximity of a vehicle, except at miles 600 to 640. Here the anomaly peaks are offset in the correct direction to be temporal in origin and are separated by the vehicle interval. Comparison of the readings from the two instruments read at the same time and in the same place indicate that the magnetometers were mutually consistent, certainly within the reading accuracy of the reed meters (10 to 20 γ with a weak signal).

Temporal changes were apparent at the overnight stations (e.g., mile 110; see Figure 22), where readings were taken at two or more times, hours apart. Therefore, the several readings at each major station were averaged and the result was taken to be the correct absolute total field value. The data in the succeeding interval were shifted by a constant to tie with the mean station value, and the closure at the next station was linearly distributed with distance back along the interval. In this manner a single profile was prepared, using the more complete data set (the lower curve of Figure 22) for values between major stations. Gaps in the data were filled with values from the other set. Two complete profiles were not deemed necessary, because anomalies of less than 100 γ were not considered useful for analysis, and anomalies of the order of hundreds of gammas in magnitude agree well between the two sets of data. Furthermore, exact correlation of the quarter-mile readings was hampered by noncoincidence of the reading locations.

The resulting profiles are shown in Figure 23 with a linear regional gradient removed. The regional trends were fitted by eye to the three track segments and equal -4.3 γ/mile from South Pole to mile 185, -10.2 γ/mile from mile 185 to mile 415, and $+1.5$ γ/mile from mile 415 to the Pole of Inaccessibility. During SPQMLT 2, the magnetic readings were taken only at 5-mile stations and thus appear as discrete points in Figure 23. The SPQMLT 2 regionals were -6.6 γ/mile from mile 0 to mile 360 and $+2.3$ γ/mile from mile 360 to mile 667.

Depth Estimates

The total-field curve displays several types of anomalies in terms of width and amplitude. Broad undulations of the order of 100 to 200 km in wavelength, with maximum average amplitudes of 100 γ, were delineated by fitting a spline to the total field curves. These anomalies are shown in Figure 24 along with a second class (intermediate width) of anomalies that range from about 40 to 100 km in wavelength and average approximately 140 γ in maximum amplitude. The anomalies of the second group were graphically fitted with smooth curves after the removal of the longer trends. The remaining anomalies, composing a third class, are generally 4 to 15 km in width and up to 200 γ in ampli-

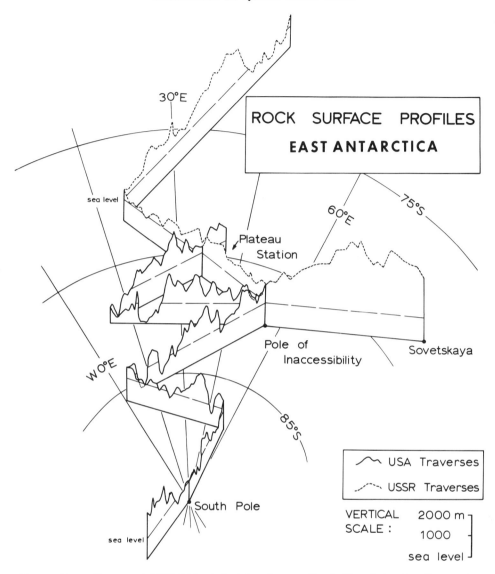

Fig. 20. Fence diagram of SPQMLT subglacial surface profiles in East Antarctica. Intersecting USSR traverse profiles are shown as dashed lines. Sea level is represented by long dashes.

tude. The sharpness and width of these narrowest residual anomalies, as well as their locations, suggest that their sources are at, or very near, the rock surface.

Peter's [1949] half-slope depth estimates were made for several representative anomalies in the two largest size classes, and the results are given in Table 6. Estimated depths computed from fourteen of the narrow anomalies are shown in Table 7. The results must be viewed with caution, because few of the anomalies are due to a single geometrically simple source, and because the assumptions of the Peter's half-slope method are often violated; nonetheless, the computed depths may give useful estimates of the maximum depths of origin.

Though the broad anomalies indicate deep sources, the narrow anomalies appear generally to have sources near the rock surface. Seven narrow anomalies gave depths that differ from the ice thickness by no more than 100 meters, and the mean difference between the ice thickness and the estimated depth to source for all fourteen anomalies is

260 meters. Furthermore, while the topography of the rock surface is not known in detail because the data are at five-mile intervals, eight of the shallow anomalies do display a general correlation with the rock surface features (Figure 23). Thus some of the anomalies are almost certainly associated with sources near the rock surface.

To aid in investigating the relationship of magnetic anomalies to the rock topography, a total field curve was computed using as a model the rock surface elevation data (at about 9-km spacing) from the third leg of SPQMLT 1. The set of data includes many relatively large magnetic anomalies and considerable subglacial topographic relief, with magnetic highs that in general appear to correspond with the peaks in the rock surface. The model profile was computed by the method of *Talwani and Heirtzler* [1964], which is applicable to two-dimensional models. Thus the rock features are assumed to be of infinite extent normal to the plane of the profile. The model total-field values were computed for the appropriate magnetic inclination of 73°, and the rocks in the model were considered to have uniform magnetic properties across the entire profile. The assumed susceptibility was 0.005 cgs, a reasonable value for rocks of average composition similar to granodiorite, for example, and no remanent component was included.

The resulting total-field curve is shown in Figure 25 with the measured total field values, the latter modified by the removal of a linear regional curve. The rock surface profile is shown beneath. General agreement in character and amplitude between the two total-field curves prevails across most of the profile with the notable exception of miles 650–700, where an inverse relationship exists. The correspondence of the character of the rock surface and the magnetic profile (Figure 23) in this area, though inverse, is quite consistent, suggesting locally reversed remanent magnetization.

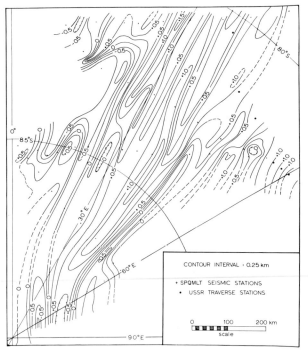

Fig. 21. Contour map of subglacial surface, East Antarctica.

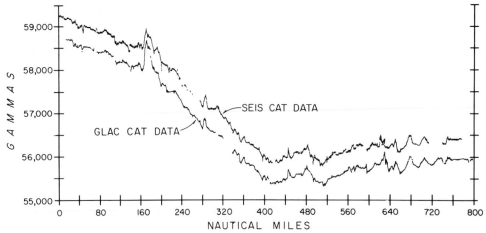

Fig. 22. Magnetic total field data, SPQMLT 1 (the seis cat data have been offset +400 γ to facilitate comparison). The readings are uncorrected for temporal variations.

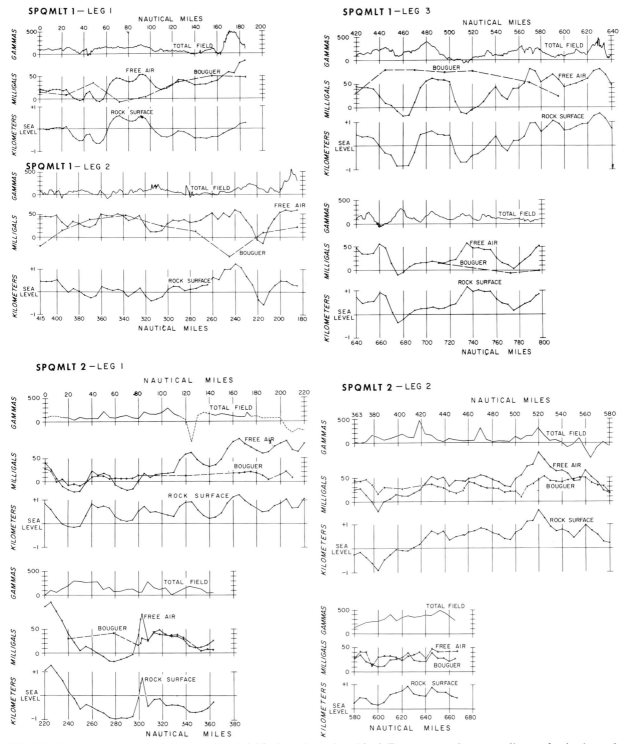

Fig. 23. Traverse geophysical data: residual field, free-air and residual Bouguer gravity anomalies, and seismic- and gravity-determined rock surface elevation profiles. Reference levels of the gravity curves have been shifted relative to the scale shown. Zero-mgal levels above correspond to these actual values (absolute values appear in Appendix 1):

Traverse	Miles	Free-Air	Bouguer		Traverse	Miles	Free-Air	Bouguer
SPQMLT 1	0–185	−50	−150	no regional re-	SPQMLT 2	0–220	−10	−10
	185–415	−25	−125	moval		220–363	−30	−30
						363–580	−30	−30
	415–797	−25	0			580–667	−40	−40

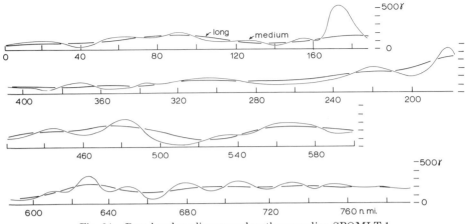

Fig. 24. Broad and medium wavelength anomalies, SPQMLT 1.

TABLE 6. Computed Source Depths for Broad and Intermediate Width Anomalies: SPQMLT 1 Magnetics

Mile No.	Width, km	Depth, km
42–66	44	10.2
80–104	33	12.6
163–183	37	6.0
215–235	37	10.2
472–490	32	12.5
660–695	33	10.4
456–500	31	12.6
538–600	89	29.2
610–660	70	29.6
50–120	100	35.0

Correlation of Magnetic Anomalies

The SPQMLT 1 residual total magnetic field and rock surface elevation profiles are shown in position along the route in Figure 26. Among the most prominent features is the pair of anomalies adjacent to the first turning point. Their proximity and similarity of character and magnitude suggest that they are correlative features. A third, very similar anomaly occurs near the center of the last (uppermost in Figure 26) leg of the traverse. The three anomalies are shown together in Figure 27. The similarity in shape of the anomaly at miles 185–200 to the anomaly at miles 620–640 is striking and suggests that the two may be correlative, indicating a trend approximately parallel to the 20°E meridian. The angles of this trend to the various profiles are listed in Figure 27. The greater width of the anomaly at miles 160–185 is consistent with the relatively small angle between the track of leg one and the trend.

The possible trend cannot be extended northward by correlation of magnetic anomalies, because the total intensity data from SPQMLT 2 are too widely spaced. However, the anomalies under discussion are associated with rock surface highs, and the strong, short-wavelength components of the anomalies indicate sources at or near the bottom of the ice. Thus it may be possible to use rock surface features to test the validity of northward extrapolation of the trend. The residual (narrow-width anomalies) total field curve for miles 620–640, SPQMLT 1, is shown in Figure 28 (bottom curve), and the associated rock surface profile is displayed immediately above. The corresponding features from SPQMLT 2, as indicated by the north-northeast trend, are shown by the upper two curves. The rock surface profile centered on mile 160 of SPQMLT 2 was computed partly from the ice surface slopes and tied to available electromagnetic values, whereas the uppermost curve is based entirely on electromagnetic measurements. The relationship of the several magnetic and rock surface curves supports the validity of correlation from line to line along the proposed trend.

GRAVITY

LaCoste-Romberg geodetic gravity meter 5 was read at 3- to 5-mile intervals (each minor and major station) along the traverse route. The gravity data were initially used to interpolate ice thickness

TABLE 7. Computed Source Depths for Narrow Anomalies: SPQMLT 1 Magnetics

Mile No.	Max. Amp., γ	Depth, meters	Ice Thickness, meters	Quality
20–31	80	2680	2720–3150	Poor
36–43	160	2800	3080–3350	Fair
204–207	110	2320	2750–2920	Fair
246–252	60	2500	2000–2010	Poor
280–284	300	2660	2540–2750	Good
322–326	60	2110	2550–2780	Fair
330–336	100	2660	2760–2920	Fair
339–347	120	3300	2580–2700	Fair
360–364	135	2780	2470–2860	Fair
399–402	150	1600	2120–2190	Fair
403–408	80	2080	2120–2180	Poor
530–537	100	2900	2320–2840	Poor
666–687	150	3700	3230–3730	Poor
746–752	85	3030	2550–2640	Good

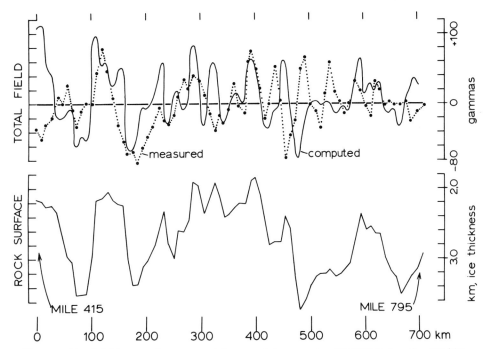

Fig. 25. Total-field anomaly computed from rock topography, SPQMLT 1, miles 415–795.

values between seismic stations, but, with the advent of virtually continuous electromagnetic soundings, meaningful Bouguer anomalies could be computed.

The gravity meter used was temperature controlled and was found to be virtually drift-free both in previous field work and from long-term observations at McMurdo and South Pole stations. The relative errors in observed gravity probably do not exceed 1 or 2 mgal. The absolute value of gravity

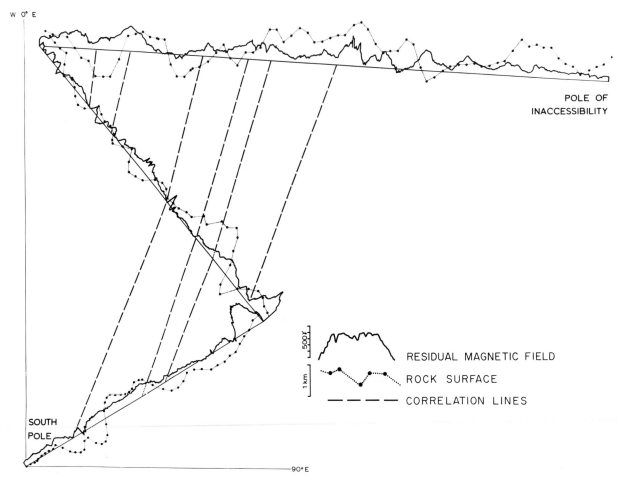

Fig. 26. Residual total field profiles plotted along route of SPQMLT 1; several anomaly correlations are indicated.

anomalies, however, is subject to much greater error, owing to uncertainty in elevations. A likely value for the uncertainty in elevations obtained by altimetry is ±50 meters, corresponding to an error of ±15 mgal in the free-air anomaly. A total error of ±20 mgal has been cited [Bentley, 1964] as a conservative value for both free-air and Bouguer anomalies.

Ice Thickness

In computing ice thicknesses from gravity observations, relative free-air anomalies are used to determine changes in the subglacial elevation. Changes in gravity from station to station across the 60- to 75-km interval between seismic reflection stations are used, and any subglacial elevation closure is distributed linearly along the interval. Thus the effects of absolute errors in elevation are eliminated, except for random, nonlinear errors across the interval. The latter are unlikely to produce errors in free-air gravity greater than 2 mgal. Additional sources of error in ice thickness determination include lateral density variations in subglacial material and terrain effects due to the subglacial surface topography.

The factor used to relate free-air gravity changes to rock surface elevation changes was 20 m/mgal, a value that has been found to be applicable throughout the continent [Bentley, 1964, pp. 354–355]. Changes in free-air anomaly versus changes in rock surface elevation for the South Pole–Queen Maud Land seismic stations are plotted in Figure 29, and the resulting distribution shows that the above factor is generally applicable in the traversed area. The standard deviation from the 20-m/mgal line through the origin is 19.0 mgal. The gravity-determined rock surface elevations are indicated on the traverse profiles of Figure 2 by solid circles, and the

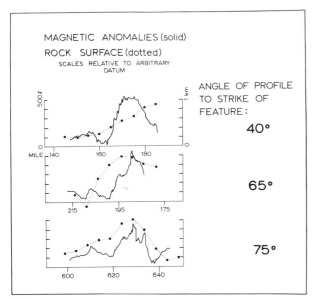

Fig. 27. Curves showing correlation of large magnetic anomalies, SPQMLT 1.

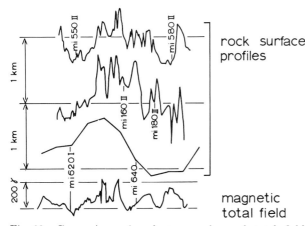

Fig. 28. Comparison of rock topography and total field curves, SPQMLT 1 and 2. The top curve is the electromagnetic rock surface profile from a part of leg 2 of SPQMLT 2, and the second curve is a part of the rock surface profile from leg 1 of SPQMLT 2. The third curve shows rock surface topography (determined by gravity and seismics) from leg 3 of SPQMLT 1 and the associated residual total field (curve below).

seismic reflection depths to which they are tied are shown as double circles.

Electromagnetic-sounding depth measurements made on the second phase of the SPQMLT provide an opportunity to evaluate the quality of gravity-determined rock elevations. Because velocities of electromagnetic waves in ice are not yet accurately known, the electromagnetic measurements were re-duced using a velocity that brought the depths into agreement with the seismic values. Thus both the gravity and the electromagnetic sounding are tied to the seismic measurements. It is preferable, therefore, to compare the changes in electromagnetic travel time between successive stations with the corresponding changes in free-air anomalies. Figure 30 is a plot showing this comparison, with a regression line through the origin fitted to the distribution. The slope is -0.227 μsec/mgal with an estimated variance in the slope of 0.018. Applying the appropriate electromagnetic velocity of 171 m/μsec, the equation of the fitted line indicated a rock surface elevation change of 19.4 m/mgal, in good agreement with the seismic–free air comparison.

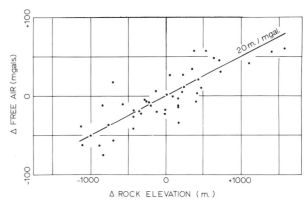

Fig. 29. Relationship of free-air anomaly changes and seismically determined rock surface elevation changes, SPQMLT 1 and 2.

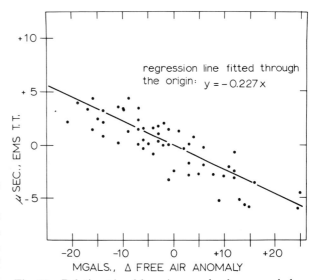

Fig. 30. Relationship of free-air anomaly changes and electromagnetic travel-time changes.

Bouguer Anomalies

Values of observed gravity (relative to South Pole station) free-air anomalies, and Bouguer anomalies are listed in Appendix 1, and the last two are plotted in Figure 23. In order to keep Figure 23 compact, to facilitate comparison of curves, the gravity values have been plotted against various datum levels, as indicated in the caption, and the Bouguer anomalies are residual values, obtained by removing the linear regionals listed below.

SPQMLT 1, mile	0–185	0
	185–415	0
	415–797	−0.39 mgal/mile
SPQMLT 2, mile	0–363	+0.37
	363–667	−0.38

In areas where no electromagnetic ice thickness data were available, the Bouguer anomalies were computed at each station with seismic control, using densities of 0.90 g/cm^3 for ice and 2.67 g/cm^3 for rock. Where continuous ice profiling was accomplished, terrain corrections were computed for the subglacial surface using Talwani's adaptation of Hubbert's integral method [*Talwani et al.*, 1959], which determines the gravity effect across a body in, say, the x direction, where x is perpendicular to the strike of the body, and where the cross section of the body can be represented by a set of polygons in the xz plane, assuming infinite extent in the y direction.

Comparison of the simple Bouguer anomalies determined by seismic soundings with the complete Bouguer anomalies computed from the electromagnetic-sounding profile data, indicates that the terrain corrections are less than 10 mgal.

TABLE 8. Mean Free-Air Anomalies along SPQMLT 2 Track

Interval, miles	Mean Free-Air Anomaly, mgal
0–100	−9
100–200	+42
200–363	+6
363–500	0
500–667	+6

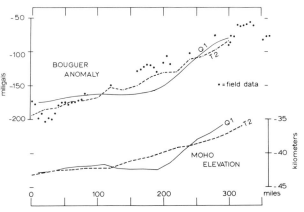

Fig. 31. Bouguer anomalies computed from crustal models $Q1$ and $T2$. The model profiles are along leg 1 of SPQMLT 2, with the zero of the horizontal axis at mile 360 and traverse mile numbers decreasing to the right. Data points are measured Bouguer anomaly values. The lower curves are the Moho elevations used in the models to generate the computed Bouguer curves (above), labeled $Q1$ and $T2$.

Regional Bouguer Anomalies

The Talwani gravity program was applied to simple crustal models to generate regional Bouguer gravity curves, which in turn were compared with the Bouguer anomaly curve from field data. The models were computed using surface elevations and an assumed standard-sea-level crustal column, with the concomitant assumption of isostatic balance. The latter assumption appears justified by the fact that average free-air anomalies are near zero throughout the traverse. As is indicated in Table 8, miles 100–200 are the major exception. Here the free-air anomaly is continuously strongly positive.

Broad-scale Bouguer gravity variations were investigated using a sea-level reference section modified from *Robinson* [1964]: a 5-km layer of density 2.67 (sea level to −5 km), a 27-km layer of density 2.91 (−5 to −32 km), and a 28-km layer of density 3.31 (−32 to −60 km). Using this reference section and varying only the Moho elevation, hydrostatically balanced crustal models were computed for miles 0–360 of SPQMLT 2, and Bouguer anomalies were computed for the models. To account for crustal rigidity, the effect of the excess mass was distributed by averaging model crustal thicknesses over intervals of 100 miles (185 km) in model $Q1$ (Figure 31) and 200 miles (370 km) in model $T2$ (Figure 32). Model $T2$ was extended to 1500 miles (2500 km) in length by incorporating data from Soviet traverses from Mirnyy to the Pole of Inaccessibility [Sorokhtin et al., 1960].

For the first leg of SPQMLT 2, the over-all regional trend of the Bouguer anomaly is very nearly linear, as is evident in Figure 31. The model-generated curves fit the same linear trend fairly well, except, in Q1, for a −25-mgal bulge centered on about mile 180 and corresponding to the large positive free-air anomaly of miles 150 to 230. If the difference in Bouguer gravity between Q1 and the field determinations is presumed to be due to the absence of part of the root required by the assumption of isostasy, a simple slab-effect calculation will serve to estimate the amount of excess root used in the model. With a density contrast of $3.31 - 2.91 = 0.40$ g/cm^3 and a maximum gravity difference of 25 mgal, we apply the formula

$$\Delta g = 2\pi \Delta \rho \gamma h$$

where $2\pi\gamma = 0.04185$ for h the slab thickness in meters, Δg is the gravity effect in milligals, and $\Delta \rho$ is the density contrast in g/cm^3, to obtain

$$h = \frac{\Delta g}{0.04185 \Delta \rho} = 1493 \text{ meters}$$

Eliminating 1.5 km, at mile 180, from the second layer of the crustal model (Q1) effectively eliminates the downward bulge in the model Moho, reducing Moho elevation profile to a rather uniform sublinear slope. Without the bottom 1.5 km of the root, the surface topography in the vicinity of mile 180 (including the ice) is 85% compensated with the 50-mile averaging radius.

The apparent spuriousness of the bulge in the Q1 mantle suggests that a broader averaging radius might be more appropriate. The model T2 radius is 100 miles (185 km), and the resulting Bouguer anomaly is a fairly good approximation to the field data. Miles 1140 to 1500 comprise the SPQMLT 2 part of the East Antarctic profile. Comparing the model Moho configurations of T2 and Q1, it is seen that the bulge does not appear on the former, and hence that the corresponding Bouguer curve approximates a linear trend. The largest remaining discrepancy is that associated with the region of negative free-air anomalies at miles 10–85 of SPQMLT 2 and 785–670 of SPQMLT 1. The amplitude of departure of the Bouguer anomaly from the regional (and from the model results) is 25 mgal at mile 20.

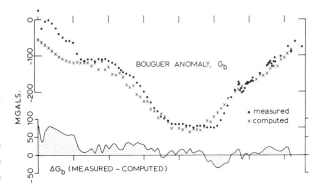

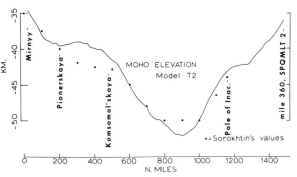

Fig. 32. Bouguer anomaly computed from crustal model T2 for profile across East Antarctica. Data points shown as solid circles are measured Bouguer anomaly values; crosses are computed points, determined from crustal model T1 with Moho elevations shown by the lowest curve. The ΔG_b curve shows the difference, in milligals, obtained by subtracting the computed Bouguer values from the measured ones.

The free-air lows are coincident with a pair of topographic basins about 600 meters deep and from 35 to 75 km wide.

The T2 Bouguer values display two notable departures from the field data. The first is in the vicinity of Mirnyy (Figure 32, mile 0 to 200), where, on the basis of high free-air values, Sorokhtin et al. [1960] have postulated an uncompensated uplift. The other major disagreement occurs at mile 950–1100, along one flank of the Gamburtsev Mountains root, where the computed values are algebraically larger than the Soviet field results. Compensation for the Gamburtsev Mountains may involve a larger part of the crust than is assumed in the model or, alternately, lateral density changes may occur. A third explanation is that isostatic compensation is not complete.

Isostasy

East Antarctica is generally considered to be isostatically balanced as a whole [*Ushakov,* 1965; *Bentley,* 1964], on the basis of free-air anomalies, although large structures such as the Golitsin and Gamburtsev Mountains are reflected in high free-air anomalies, and an extensive zone of strongly negative free-air anomalies exists in Victoria Land. As was indicated above, the SPQMLT data support this conclusion. The relation of the regional Bouguer anomaly trend to average elevation changes and the results of crustal models developed on the basis of hydrostatic equilibrium of the crust indicate broad isostatic compensation. The direct relation of Bouguer anomalies to local topography suggests that features of the order of tens of kilometers in width are not locally compensated. The elevated central part (SPQMLT 2, mile 150–240) of the rock surface in the traversed area was found to be incompletely compensated for a crustal model ($Q1$) using an averaging radius of 93 km. Extending the radius over which compensation is distributed to 185 km (model $T2$) effectively eliminated a local bulge in the model Moho, but at the expense of overcompensation in adjacent areas. The approximately 165-km-wide zone (in which rock elevations in excess of +1 km occur) therefore appears to be an incompletely compensated uplifted block.

GEOLOGICAL SETTING AND INTERPRETATION

The major part of East Antarctica has generally been considered a shield. Precambrian crystalline basement composed of migmatized and granitized gneisses with widespread charnockites is exposed in block mountains along the East Antarctica coast from the Greenwich meridian to 145°E [*Klimov et al.,* 1963]. The Precambrian complex is the principal lithology exposed along the coast of Queen Maud Land, north of the traversed region. To the west of the traversed region, the Paleozoic to Mesozoic sedimentary and intrusive rocks of the Transantarctic Mountains lie on a presumably Precambrian and Cambrian basement. According to *Robinson* [1964], the mountain system, which is block-faulted, is bounded on its West Antarctic side by a fault, but no evidence for a fault has been found on the East Antarctic side. The rocks of the Transantarctic Mountains may extend under East Antarctic ice, but limited refraction data suggest that the sediments are much thinner, if present at all.

The 'Gamburtsev Subglacial Mountains,' east of the traversed area, were envisioned by Soviet scientists [*Kapitsa and Sorokhtin,* 1965] as reaching elevations in excess of 2.5 km and extending at elevations of 2 km for approximately 1100 km along an arcuate trend. The eastern part of the mountains is mapped as trending in a direction parallel to a line through Komsomol'skaya and the Pole of Inaccessibility, and the western segment parallels the Greenwich Meridian. *Ushakov* [1965] stated that the 'Gamburtsev Subglacial Mountains' are unlikely to be the result of folding, being located near the center of the platform, and are probably, instead, an activated sector of the post-Proterozoic platform. On the basis of the great crustal thickness and the degree of erosion, Ushakov postulated activization during late Mesozoic-early Cenozoic time.

The classification of East Antarctica as a shield is not without question. Before glaciation, much of East Antarctica was a plateau, standing at an average elevation of the order of a kilometer, as opposed to an average elevation of hundreds of meters for a typical shield area. On the basis of group velocity data, the East Antarctic crust is also significantly thicker than the Canadian Shield and has lower sub-Moho velocities than the latter [*Dewart and Toksöz,* 1965]. The seismic velocities thus determined for East Antarctica agree with those of the Basin and Range province of the United States. The picture of East Antarctica as a shield is also marred by the Victoria Land isostatic anomaly, which was found to be inadequately explained by glacial fluctuations [*Robinson,* 1964]. The alternative hypothesis of tectonic forces suggests that the region of Victoria Land paralleling the Transantarctic Mountains is not part of a stable shield. The isostatic anomalies are associated with, though not coincident with, a zone of depressions in the rock surface reaching more than 0.5 km below sea level. A narrower but equally deep depression was found in the western part of the traversed area with less pronounced associated low free-air anomalies. One tenuous possibility is that the anomalous zone extends into Queen Maud Land on a reduced scale.

The nature of the rocks underlying the ice sheet of central Queen Maud Land remains undetermined at present. The surrounding geology that is observable and the traverse magnetic data point toward a normal continental crystalline complex. The magnetic total field profiles show a general parallelism with the rock surface features along most of SPQMLT 1 (Figures 2 and 26), and the magnetic model of Figure 25 demonstrates that, with granitic to intermediate rock types, the rock topography is a generally adequate source for most of the observed magnetic anomalies. Many of the broader anomalies over elevated parts of the rock surface could be explained as the summation of many smaller anomalies due to the collective topographic relief. Furthermore, the magnetic depth estimates (Table 8) largely support the notion of magnetic sources at the bottom of the ice. In addition, the limited refraction seismic studies suggest the absence of major sedimentary sections, although some sediments ($\leqslant 1$ km) may lie beneath the ice, at least locally. The subglacial velocities measured are appropriate for rocks of a crystalline complex, composed of metasediments or granitic material or, more likely, both. At 88°39'S, 157.5°W, *Robinson* [1964] found a velocity of 4.1 km/sec (typical of sediments) beneath the ice, although at 87°55.0'S, 126.2°W and 87°8.8'S, 112.1°W. Robinson gives subglacial velocities of 5.4 and 5.7 km/sec, respectively, with no sedimentary section apparent until the Transantarctic Mountains are reached. The sparsely distributed information is consistent with a model of sediments in the Transantarctic Mountains thinning toward the interior, with little or no sedimentary cover farther south.

Whatever the nature of the lithology, some of the magnetic anomalies cannot be readily explained by rock surface topography alone. For example, the large persistent anomaly that correlates from mile 170 to 190 to 630 of SPQMLT 1 (Figures 2 and 26) is out of proportion to the topography, suggesting that the anomaly is at least due to a deeper source or to lithographic variations. A series of other large anomalies extends to the east of it on track 3 of SPQMLT 1 from mile 640 to 710, similar in size to the anomaly at mile 180, but atypical of the remainder of the profile. Large anomalies of similar or greater amplitude and great linear extent have been investigated in the vicinity of the Queen Mary Coast [*Glebovskiy*, 1959a, b] and were there attributed to basic intrusions localized in the bounding fractures associated with block faulting. The large SPQMLT 1 anomalies may reflect a zone of similar intrusives, or they may be associated with variations in the basement.

Another feature of the traversed area that is worthy of consideration is the apparent tectonic grain. The correlation of the gross topographic features from line to line (Figure 20) and of the magnetic anomalies (Figure 26) support the contoured interpretation of the rock surface map (Figure 21) and also suggest geologic control. The pervasive grain of the relief might be attributable to strong joint control, faulting, or holomorphic folding.

One argument in favor of faulting is the presence of relatively steep-sided blocks such as the features that appear near mile 500 of SPQMLT 1 and mile 300 of SPQMLT 2 (Figure 2). In addition, the magnetic anomaly associated with mile 500 of SPQMLT 1 can be explained by the topography alone, requiring no major lithologic differences between the peak and the surrounding terrain, such as might be expected if intrusives, for example, were involved. The notion of vertical movement, furthermore, is consistent with the extensive block faulting thought to exist in the nearby Transantarctic Mountains and with the isostatically unbalanced block uplifts of the 'Golitsin Subglacial Mountains' and along the George V Coast [*Ushakov*, 1965].

Acknowledgments. This work was supported by National Science Foundation grants GA-149, GA-215, and GA-1126, administered by the Office of Antarctic Programs. The author is grateful to Dr. C. R. Bentley for guidance and numerous suggestions and wishes to thank J. W. Clough for helpful discussions and other members of the South Pole-Queen Maud Land traverses for assistance in the field. Traverse logistic support was provided by the U.S. Navy.

Contribution 224, Geophysical and Polar Research Center, University of Wisconsin.

REFERENCES

Beitzel, J. E., The relationship of ice thicknesses and surface slopes in Dronning Maud Land, in *International Symposium on Antarctic Glaciological Exploration (ISAGE)*, Publ. 86, International Association of Scientific Hydrology and Scientific Committee on Antarctic Research, Cambridge, England, 1970.

Bentley, C. R., The structure of Antarctica and its ice cover, *Research in Geophysics*, vol. 2, chap. 14, MIT Press, Cambridge, Mass., 1964.

Bentley, C. R., Seismic anisotropy of the West Antarctic ice sheet, this volume.

Bentley, C. R., R. L. Cameron, C. Bull, K. Kojima, and A. J. Gow, Physical characteristics of the Antarctic ice sheet,

Antarctic Map Folio Series, folio 2, American Geographical Society, New York, 1964.

Bourgoin, Jean-Paul, Quelques caracteres analytiques de la surface et du socle de l'inlandsis Groenlandais, *Ann. Geophys.*, *12*(1), 75–83, 1956.

Budd, W. F., The dynamics of ice masses, Ph.D. thesis, University of Melbourne, Melbourne, Australia, 1968.

Clough, J. W., and C. R. Bentley. Electromagnetic wave velocities in the East Antarctic ice sheet, in *International Symposium on Antarctic Glaciological Exploration (ISAGE)*, Publ. 86, International Association of Scientific Hydrology and Scientific Committee on Antarctic Research, Cambridge, England, 1970.

Collins, I. F., On the use of equilibrium equations and flow law in relating the surface and bed topography of glaciers and ice sheets, *J. Glaciol.*, *7*(50), 199–204, 1968.

Crary, A. P., Results of United States traverses in East Antarctica, *IGY Glaciological Report 7*, American Geographical Society, New York, 1963.

Dewart, G., and M. N. Toksoz, Crustal structure in East Antarctica from surface wave dispersion, *Geophys. J.*, *10*(2), 127–139, 1965.

Glen, John W., The creep of polycrystalline ice, *Proc. Royal Soc. London, A*, *228*, 519–538, 1955.

Glebovskiy, Yu. S., The existence of a subglacial ridge in the Pionerskaya area, *Sov. Antarctic Exped. Bull.*, *1*, 285–288, 1959a.

Glebovskiy, Yu. S., Subglacial Mt. Brown–Gaussberg ridge, *Sov. Antarctic Exped. Bull.*, *1*, 381–385, 1959b.

Kapitsa, A. P., The 1964 Vostok–Pole of Inaccessibility traverse, *Inf. Bull. Sov. Antarctic Exped.*, No. 51, 13–18, 1965.

Kapitsa, A. P., and O. G. Sorokhtin, Ice thickness measurements on the Vostok–Molodezhnaya traverse, *Inf. Bull. Sov. Antarctic Exped.*, No. 51, 19–22, 1965.

Klimov, L. V., M. G. Ravich, and D. S. Solov'ev, Geological structure of the Antarctic platform, First Int. Symposium on Ant. Geology, in *Antarctica: Commission Reports 1963*, pp. 6–19, National Science Foundation, Washington, D.C., 1963.

Nakaya, U., Visco-elastic properties of processed snow, *Res. Rep. 58*, Snow, Ice, Permafrost Res. Estab., Wilmette, Illinois, 22 pp., 1959.

Nye, J. F., The distribution of stress and velocity in glaciers and ice sheets, *Proc. Royal Soc. London, A*, *239*, 113–133, 1957.

Peters, L. J., The direct approach to magnetic interpretation and its practical applications, *Geophysics*, *14*, 290–320, 1949.

Robin, G. de Q., Ice movement and temperature distribution in glaciers and ice-sheets, *J. Glaciol.*, *2*, 523–532, 1955.

Robin, G. de Q., Seismic shooting and related investigations, in *Norwegian-British-Swedish Antarctic Expedition, 1949–52, Sci. Results 5, Glaciology 3*, Norsk Polarinstitutt, Oslo University Press, Oslo, 1958.

Robin, G. de Q., Surface topography of ice sheets, *Nature*, *215*, 1029–1032, 1967.

Robinson, E. S., Geological structure of the Transantarctic Mountains and adjacent ice-covered areas, Antarctica, Ph.D. dissertation, University of Wisconsin, 1964.

Sorokhtin, O. G., O. K. Kondratyev, and Yu. N. Avsyuk, Methods and principal results of seismic and gravimetric studies of the structure of the Eastern Antarctic, *Bull. (Izv.) Acad. Sci. USSR, Geophys. Ser.*, No. 3, 265–268, 1960.

Talwani, M., and J. R. Heirtzler, Computation of magnetic anomalies caused by two-dimensional structures of arbitrary shape, in *Computers in the Mineral Industries*, edited by G. A. Parks, pp. 469–480, Stanford University Press, Palo Alto, Calif., 1964.

Talwani, M., J. L. Worzel, and M. Landisman, Rapid computations for two-dimensional bodies with application to the Mendocino submarine fracture zone, *J. Geophys. Res.*, *64*, 49–59, 1959.

Ushakov, S. A., Geophysical studies of the crustal structure in the East Antarctica, Acad. of Sci. USSR Geophys. Comm., transl. by Aeronautical Chart and Information Center, St. Louis, Mo., 1965.

APPENDIX 1. GEOPHYSICAL DATA FROM SPQMLT 1 AND 2

Mile	Latitude	Longitude	Surface Elevation, meters	Bed Rock Elevation, meters	Observed Gravity,† mgal	Free-Air Anomaly, mgal	Bouguer Anomaly, mgal	Total Field Intensity, γ
			SPQMLT 1					
0*	90°00′		2800	+22	0	−29	−135	58720
3	89°57′	60.0°	2807	+20	−2.86	−30		717
6	54	.0	2798	−13	−2.37	−32		714
8	52	.0	2798	+1	−2.14	−32		745
9	51	.0	2796	+21	−0.74	−31		702
12	48	.0	2799	+61	−0.34	−30		677
15	45	.0	2801	+81	+0.48	−29		681
18	42	.0	2792	+72	+1.22	−30		688
21	39	.0	2804	+56	−3.96	−32		630
24*	89°36′	60.0°	2821	+122	−7.09	−29	−142	58672
27	32	.1	2819	−76	−14.73	−38		669
30	29	.1	2818	−305	−24.30	−47		617
33	26	.2	2814	−414	−26.94	−51		559
36	23	.2	2812	−511	−29.62	−54		491
39	20	.3	2799	−494	−23.15	−52		573
42	17	.4	2814	−243	−13.82	−38		605
45	14	.4	2852	−208	−22.24	−34		477
48*	89°10′	58.5°	2854	−493	−35.88	−47	−115	58519
51	07	.4	2846	−619	−41.37	−55		538
54	04	.4	2844	−613	−42.39	−56		562
57	01	.3	2848	−468	−38.17	−51		546
60	88°58′	.3	2836	−137	−19.68	−36		534
63	54	.2	2843	+248	−4.53	−18		523
66	51	.1	2866	+467	−2.50	−09		502
69	48	.1	2870	+569	−0.37	−05		506
72*	88°45′	58.0°	2893	+602	−7.47	−05	−157	58465
75	42	.1	2892	+505	−10.64	−08		449
78	38	.2	2924	+426	−23.26	−11		437
81	35	.3	2931	+377	−26.47	−12		441
84	32	.4	2930	+368	−25.27	−11		429
87	29	.6	2941	+403	−25.53	−07		448
90	26	.7	2963	+553	−23.68	+02		452
93	22	.8	2964	+542	−23.30	+03		435
96*	88°19′	58.9°	2973	+480	−27.87	+01	−145	58389
100	15	.9	2985	+157	−45.48	−12		361
105	10	59.0°	2983	−136	−56.91	−24		249
110	88°05′	59.0°	2973	−319	−60.15	−30		58272
115	87°60′	.1	2967	−354	−56.21	−28		212
120	54	.2	2987	−270	−55.60	−21		221
125*	87°48′	59.2°	2989	−92	−44.70	−09	−113	58156
130	43	.3	3006	−122	−50.84	−09		157
135	38	.4	3003	−240	−55.35	−14		098
140	32	.5	3007	−402	−64.22	−21		039
145	27	.7	3015	−387	−65.32	−19		072
150	22	.8	3027	−406	−69.50	−19		069
155*	87°17′	58.9°	3034	−345	−68.08	−14	−100	58072
160	12	.8	3044	−341	−68.94	−11		57982
165	06	.8	3057	−189	−63.36	−01		58022
170	01	.7	3070	−59	−58.83	+08		397
175	86°56′	.7	3076	+45	−53.60	+06		373
180	50	.6	3090	+232	−46.72	+29		198

Mile	Latitude	Longitude	Surface Elevation, meters	Bed Rock Elevation, meters	Observed Gravity,† mgal	Free-Air Anomaly, mgal	Bouguer Anomaly, mgal	Total Field Intensity, γ
185*	86°45'	58.6°	3106	+294	−46.60	+34	−103	57958
190	46	57.0°	3102	+353	−47.79	+32		58169
195	47	55.5	3115	+505	−49.55	+34		57839
200	48	53.9	3112	+517	−53.31	+29		57545
205	86°48'	52.3°	3107	+333	−66.59	+14		57539
210	48	50.8	3104	+66	−84.58	−05		452
215*	86°48'	49.2°	3099	−529	−118.27	−40	−115	57491
220	47	47.7	3106	−297	−112.03	−31		479
225	47	46.1	3092	+342	−78.81	−03		489
230	46	44.6	3082	+694	−61.23	+12		399
235	46	43.1	3068	+1103	−39.59	+29		289
240	45	41.5	3048	+1265	−28.34	+35		210
245*	86°44'	40.0°	3022	+1030	−35.23	+20	−169	57112
250	43	38.4	2994	+1016	−18.64	+28		082
255	42	36.9	2946	+564	−17.68	+14		56951
260	41	35.3	2935	+659	−0.77	+28		921
265	40	33.7	2896	+339	+3.93	+21		896
270	39	32.2	2882	+296	−0.66	+12		835
275*	86°38'	30.6°	2857	+190	+3.10	+08	−112	56775
280	36	29.2	2841	+158	+5.98	+07		651
285	35	27.8	2815	+93	+20.82	+13		657
290	34	26.4	2804	+258	+21.73	+10		648
295	33	25.0	2784	+275	+28.26	+12		610
300	31	23.6	2768	+145	+16.29	+05		591
305*	86°30'	22.2°	2756	−159	+14.34	−10	−100	56562
310	27	21.1	2745	−285	+14.55	−13		516
315	24	20.0	2749	−366	+12.20	−13		451
320	20	18.8	2748	−142	+26.70	+01		375
325	17	17.7	2740	+215	+50.07	+23		380
330	14	16.6	2742	+13	+42.39	+16		234
336	11	15.2	2731	−163	+40.64	+12		129
338*	86°09'	14.8°	2726	−104	+46.28	+16	−77	091
340	08	14.4	2722	+53	+54.18	+23		122
345	03	13.5	2718	+164	+58.12	+27		076
350	86°00'	12.5°	2704	+109	+56.88	+22		56025
355	85°57'	11.6°	2708	+169	+55.87	+23		55845
360	53	10.6	2702	+265	+59.58	+26		954
365	49	09.7	2692	−140	+39.70	+03		814
370*	85°46'	8.7	2695	−242	+30.81	−03	−86	55808
375	42	7.9	2692	−128	+32.98	−01		805
380	38	7.1	2682	+55	+41.87	+06		746
385	34	6.3	2679	+177	+45.50	+10		668
390	30	5.5	2676	+3	+34.40	−02		590
395*	85°26'	4.7°	2672	+247	+44.43	+08	−109	55412
400	22	3.9	2659	+562	+63.17	+24		537
405	18	3.2	2650	+489	+61.21	+20		492
410	14	2.4	2639	+482	+63.25	+20		361
415*	85°10'	1.6°	2628	+491	+65.99	+20	−144	55386
420	11	2.6	2639	+486	+62.65	+20		350
425	11	3.6	2645	+412	+57.61	+17		407
430	12	4.6	2656	+440	+56.00	+18		444
435	12	5.7	2664	+350	+49.46	+14		516
440	13	6.7	2674	+49	+31.72	−01		498
445*	85°13'	7.7°	2679	−265	+14.85	−16	−95	55575
450	13	8.8	2685	−338	+9.46	−20		491
455	12	9.9	2672	−882	−4.93	−33		432

Mile	Latitude	Longitude	Surface Elevation, meters	Bed Rock Elevation, meters	Observed Gravity,† mgal	Free-Air Anomaly, mgal	Bouguer Anomaly, mgal	Total Field Intensity, γ
460	12	11.0	2669	−813	−9.05	−43		493
465	11	12.1	2669	−805	−8.47	−42		529
470*	85°11′	13.2°	2688	−250	+13.49	−14	−95	55535
475	08	14.1	2674	+516	+57.37	+26		651
480	06	15.0	2671	+572	+62.42	+31		742
485	04	15.9	2678	+662	+66.08	+37		668
490	01	16.8	2669	+519	+63.50	+33		574
495	84°59′	17.7	2673	+479	+61.01	+32		495
496*	84°58′	17.9°	2676	+462	+59.50	+32	−101	55460
500	57	18.6	2681	+469	+54.92	+29		439
505	55	19.4	2690	−282	+10.23	−13		400
510	54	20.3	2687	−665	−12.25	−36		417
515	52	21.1	2679	−556	−13.93	−39		383
519*	84°51′	21.8°	2683	−419	−6.78	−31	−99	55427
520	50	21.9	2687	−386	−6.05	−29		442
525	49	22.8	2688	−301	−1.26	−23		485
530	47	23.6	2682	−128	+10.27	−13		529
535	45	24.4	2718	+171	+15.47	+04		603
540	44	25.2	2742	+440	+21.59	+18		566
545	84°42′	26.0°	2792	+13	−16.13	−01		55560
550	40	26.8	2785	−190	−22.78	−09		594
555	38	27.6	2816	+238	−9.75	+14		663
560	36	28.5	2834	+237	−14.39	+15		732
565	34	29.3	2849	+404	−9.60	+25		696
570*	84°32′	30.1°	2872	+978	+12.95	+56	−123	55760
575	30	30.9	2930	+986	−9.37	+52		748
580	28	31.6	2958	+622	−41.08	+29		703
585	26	32.4	2968	+835	−38.42	+36		644
590	24	33.1	3001	+1104	−40.12	+45		587
595*	84°22′	33.9°	3046	+966	−65.78	+33	−152	55650
600	20	34.6	3068	+671	−88.50	+18		633
605	18	35.4	3087	+731	−92.81	+21		676
610	15	36.1	3108	+854	−94.24	+26		780
615	13	36.9	3144	+955	−101.81	+31		703
620	10	37.6	3172	+976	−105.96	+36		686
625	08	38.2	3189	+1318	−95.59	+52		911
630	05	38.8	3226	+1397	−104.47	+56		861
635	02	39.3	3263	+1173	−128.58	+44		750
640	83°59′	39.9	3283	+762	−156.87	+23		55700
645	56	40.5	3289	+509	−175.52	+10		750
650	83°53′	41.3°	3314	+574	−180.85	+10		55880
655	49	41.8	3322	+588	−182.34	+12		760
660	46	42.4	3335	+964	−169.25	+31		570
665	42	42.9	3361	+823	−186.09	+23		660
670	39	43.4	3385	+180	−227.24	−10		720
675	36	44.0	3394	−309	−256.26	−35		910
680	83°32′	44.5°	3397	−158	−251.22	−28		55960
685	28	45.0	3405	+66	−244.36	−17		802
690	25	45.5	3428	+212	−245.99	−10		785
695	21	46.0	3447	+270	−251.76	−09		727
700	18	46.5	3465	+289	−257.12	−08		869
705	14	47.0	3482	+359	−260.81	−05		972
710*	83°10′	47.5°	3497	+276	−271.28	−09	−159	55874
715	07	48.0	3500	+310	−272.71	−09		848
720	04	48.5	3518	+421	−274.83	−04		808
725	00	49.0	3526	+479	−276.65	−02		847

74　　　　　　　　　　　　　　　　JOHN E. BEITZEL

Mile	Latitude	Longitude	Surface Elevation, meters	Bed Rock Elevation, meters	Observed Gravity,† mgal	Free-Air Anomaly, mgal	Bouguer Anomaly, mgal	Total Field Intensity, γ
730	82°57′	49.5	3528	+843	−261.24	+15		936
735	54	50.0	3559	+1211	−254.59	+33		911
740	82°50′	50.5°	3577	+1010	−272.29	+22		55855
745	46	50.9	3578	+1063	−272.37	+23		828
750	42	51.3	3606	+995	−286.87	+19		962
755	39	51.7	3611	+1004	−290.33	+19		935
760	35	52.1	3640	+695	−317.15	+02		889
765	31	52.5	3638	+502	−328.58	−09		902
770*	82°27′	52.9°	3659	+440	−340.55	−13	−181	55894
775	23	53.3	3666	+191	−352.53	−21		899
780	20	53.7	3663	+337	−343.37	−11		918
785	16	54.1	3681	+479	−339.97	−01		858
790	12	54.5	3692	+567	−337.16	+07		897
795	08	54.8	3708	+807	−328.42	+23		917
797*	82°07′	55.0°	3718	+865	−328.44	+27	−175	56010
				SPQMLT 2				
0*	82°07′	55°06′	3718	+865	−328.4	+27	−173	55850
5	10	54 35	3700	+632	−335.6	+13	−177	854
10	12	54 05	3688	+298	−349.7	−06	−198	828
15	15	53 35	3676	+17	−361.2	−22	−191	761
20	18	53 05	3669	−108	−366.1	−30	−203	720
25	20	52 35	3655	−141	−364.9	−34	−198	644
30	22	52 05	3653	−85	−362.6	−33	−200	667
35	24	51 35	3645	+351	−339.1	−14	−191	600
40*	82°27′	51°05′	3620	+859	−307.5	+10	−179	55574
45	29	50 30	3600	+698	−309.2	+01	−175	533
50	31	49 55	3598	+766	−305.1	+04	−174	651
55	33	49 20	3566	+685	−299.1	−01	−178	594
60	34	48 45	3551	+313	−313.3	−20	−175	452
65	36	48 10	3538	+181	−316.0	−27	−174	454
70	38	47 35	3536	+125	−318.0	−31	−172	442
75	40	47 00	3524	+188	−311.0	−28	−170	350
80*	82°42′	46°25′	3520	+533	−292.5	−12	−162	55303
85	43	45 45	3497	+747	−269.5	+04		406
90	44	45 05	3482	+492	−272.7	−04		324
95	44	44 25	3474	+517	−264.0	+02		292
100	45	43 45	3456	+455	−256.3	+03		300
105	46	43 05	3438	+369	−250.0	+04		348
110	46	42 25	3425	+318	−243.6	+06		196
115	47	41 45	3410	+744	−212.7	+32		114
120*	82°48′	41°05′	3391	+913	−193.1	+45	−149	55017
125	48	40 25	3356	+926	−179.5	+48		54517
130	49	39 45	3333	+588	−187.2	+33		55022
135	50	39 05	3319	+352	−192.5	+23		062
140	50	38 25	3290	+228	−187.6	+19		002
145	50	37 45	3273	+282	−177.5	+23		54922
150	51	37 05	3256	+458	−161.4	+34		932
155	52	36 25	3244	+875	−134.7	+57		876
160*	82°52′	35°45′	3205	+1138	−107.3	+72	−128	54816
165	52	35 04	3155	+1228	−86.4	+78	−126	769
170	52	34 24	3133	+994	−90.2	+67	−122	742
175	52	33 43	3114	+830	−91.4	+60	−119	705
180	52	33 03	3090	+674	−90.8	+53	−115	693
185	53	32 21	3070	+532	−90.8	+47	−123	602
190	53	31 41	3064	+577	−85.5	+50	−130	575

Mile	Latitude	Longitude	Surface Elevation, meters	Bed Rock Elevation, meters	Observed Gravity,† mgal	Free-Air Anomaly, mgal	Bouguer Anomaly, mgal	Total Field Intensity, γ
195	53	31 01	3017	+804	−58.6	+62		
200*	82°53′	30°20′	2992	+897	−45.2	+68	−116	54516
205	53	29 34	2970	+1075	−31.8	+74	−106	308
210	54	28 48	2939	+706	−42.9	+54	−118	217
215	54	28 02	2916	+709	−38.1	+51		195
220	54	27 15	2889	+1088	−13.1	+68		124
225	54	26 29	2840	+1312	+11.0	+77		192
230	54	25 43	2792	+987	+7.3	+58		115
235	55	24 56	2795	+638	−13.3	+38		159
240*	82°55′	24°10′	2751	+164	−25.7	+12	−102	54182
245	82°55′	23°24′	2707	−114	−32.0	−08		249
250	54	22 37	2706	−539	−48.9	−25		206
255	54	21 51	2681	−342	−32.3	−16		153
260	54	21 05	2651	−454	−29.5	−22		130
265	53	20 19	2637	−593	−33.1	−30		107
270	53	19 33	2619	−691	−35.8	−38		53904
275	82°52′	18 47	2611	−889	−41.7	−46		53930
279*	82°52′	18°10′	2609	−938	−44.3	−49	−75	808
285	50	17 21	2605	−907	−23.1	−43		734
290	49	16 40	2599	−914	−33.7	−40		702
295	47	15 59	2606	−842	−29.4	−33		640
300	46	15 18	2644	−69	+0.5	+09	−90	498
302	45	15 02	2626	+531	+37.4	+41	−86	474
303	44	14 50	2593	+741	+58.6	+52	−76	471
308	40	15 15	2627	−433	−8.8	−03	−77	620
313	38	14 34	2606	−149	+14.6	+14	−60	423
318	36	13 53	2592	−145	+21.7	+18	−62	306
323*	82°34′	13°20′	2567	−420	+18.6	+08	−61	53274
328	30	12 51	2565	−374	+17.2	+05	−59	066
333	26	12 24	2555	−396	+14.9	+04	−56	048
338	21	11 55	2541	−457	+11.8	−02	−61	053
343	17	11 27	2544	−666	−3.8	−15		023
348	13	10 59	2532	−699	−6.1	−19		52935
353	08	10 31	2528	−636	−6.1	−18	−82	837
358	04	10 03	2516	−504	−0.2	−14	−77	680
363*	82°00′	9°35′	2512	−236	+10.2	−04	−76	52612
368	81°58′	10 13	2519	−128	+9.5	−02	−79	600
373	56	10 51	2541	−344	−11.9	−16	−83	638
378	55	11 29	2530	−579	−24.1	−30	−91	786
383	53	12 07	2534	−900	−45.3	−50	−112	748
388	51	12 45	2534	−448	−26.5	−30	−100	701
393	49	13 24	2569	−252	−31.4	−24		744
398	48	14 01	2580	−23	−27.1	−15		
403*	81°46′	14°40′	2597	−69	−30.7	−18	−109	52865
408	44	15 17	2617	−92	−38.9	−19		810
413	43	15 54	2640	+61	−39.1	−12		804
418	81°42′	16°30′	2643	+186	−34.6	−06		53189
423	40	17 07	2652	+466	−24.2	+08		52903
428	38	17 44	2691	+746	−23.0	+22	−108	888
433	37	18 21	2722	+625	−39.4	+16	−116	807
438	36	18 58	2735	+728	−39.1	+21	−119	777
443*	81°34′	19°35′	2760	+365	−65.8	+02	−129	52861
449	32	20 16	2771	+512	−58.4	+14	−136	840
455	29	20 57	2836	+579	−81.0	+12	−132	
456*	81°29′	21°04′	2837	+576	−81.7	+12	−128	52839
460	27	21 31	2852	+691	−81.7	+18	−115	852

Mile	Latitude	Longitude	Surface Elevation, meters	Bed Rock Elevation, meters	Observed Gravity,† mgal	Free-Air Anomaly, mgal	Bouguer Anomaly, mgal	Total Field Intensity, γ
465	25	22 05	2887	+740	−91.5	+20	−121	876
470	22	22 39	2919	+862	−96.7	+25	−129	53145
475	20	23 13	2954	+796	−112.7	+22	−134	52910
480	17	23 47	2987	+662	−130.7	+15	−136	879
485	15	24 22	3009	+594	−142.2	+11	−139	903
490	12	24 56	3041	+421	−162.2	+02	−149	907
495*	81°10′	25°30′	3055	+381	−170.0	0	−148	52906
500	07	26 03	3067	+690	−157.6	+18	−151	53020
505	04	26 35	3095	+806	−159.6	+25	−164	52965
510	02	27 08	3120	+1143	−149.6	+44	−144	53038
515	80 59	27 40	3152	+1150	−158.4	+47	−140	048
520	56	28 13	3156	+1611	−135.8	+72	−133	207
525	54	28 45	3219	+1282	−171.0	+57	−126	056
530	51	29 18	3227	+919	−190.7	+41	−138	52941
535*	80°48′	29°50′	3259	+732	−209.2	+34	−142	52960
540	45	30 19	3284	+795	−217.6	+35	−145	910
545	42	30 49	3312	+703	−234.8	+28	−141	810
550	39	31 18	3323	+454	−254.5	+12	−141	870
555	36	31 47	3327	+722	−246.2	+24	−139	995
560	33	32 17	3352	+991	−244.3	+35	−141	785
565	80 30′	32°46′	3374	+764	−266.4	+21	−154	52585
570	27	33 16	3386	+592	−282.6	+10	−162	52805
575*	80°24′	33°45′	3407	+305	−307.4	−07	−165	52915
580	21	34 11	3412	+258	−316.4	−13	−169	839
585	17	34 36	3435	+492	−316.9	−05	−169	883
590	14	35 02	3455	+471	−329.2	−09	−171	938
595	10	35 28	3473	+231	−352.0	−24	−201	952
600	07	35 53	3471	+190	−358.6	−30	−184	961
605	04	36 19	3483	+299	−361.9	−28	−184	53000
610	00	36 45	3495	+623	−354.6	−15	−185	095
615*	79°57′	37°10′	3512	+708	−360.8	−14	−190	52969
620	53	37 36	3522	+773	−362.3	−10	−198	53038
625	50	38 01	3539	+959	−359.8	−01	−190	072
630	46	38 27	3557	+658	−382.0	−16	−189	041
635	42	38 53	3563	+589	−389.1	−19	−185	050
640	39	39 18	3575	+566	−384.7	−19	−199	084
645	35	39 44	3589	+922	−383.5	−01	−181	078
650	31	40 10	3605	+721	−400.3	−11	−189	187
657	23	40 15	3611	+712	−406.5	−11		126
662	18	40 10	3617	+565	−418.1	−18		070
667*	79°15′	40°30′	3624	+488	−425.4	−22	−192	52971

*Seismic reflection station. †Relative to South Pole.

Longitude values are given to tenths of a degree rather than in minutes because longitude can be determined less accurately than latitude in polar regions, owing to the convergence of the meridians.

APPENDIX 2. SEISMIC REFLECTION RESULTS FROM SPQMLT 1 AND 2

Mile	Record	Reflection Time, sec	Hole Depth, meters	Ice Thickness, meters	Rock Elevation above msl, meters	Gains 1–12	Gains 13–24	Filters Low	Filters High	No. of Records	Signal Quality
					SPQMLT 1						
0	7012	1.495	20	2778	22	9	8	210, 120	out	5	+
24	7020	1.460	32	2699	122	9	9	210	out	1	0
48	7022	1.805	38	3347	-493	10	10	210	out	1	0
72	7033	1.284	30	2291	602	10	9	210	out	5	-
96	7036	1.345	36	2493	480	10	9	210, 120	out	2	+
125	7039	1.656	37	3081	-92	10	10	210	out	2	0
155	7041	1.816	38	3379	-345	10	9	210, 120	out	2	+
185	7044	1.518	38	2812	294	9	9	210	out	4	+
215	7062	1.941	40	3628	-529	9	9	120, 210	320	2	0
245	7083	1.078	40	1992	1030	8	8	210	out	2	0
275	7086	1.441	41	2667	190	9	9	210	out	1	-
305	7089	1.574	40	2915	-159	9	9	210	out	3	+
338	7094	1.530	16	2830	-104	9	9	210	out	3	-
370	7097	1.572	40	2917	-242	9	9	210	out	4	-
395	7109	1.305	38	2425	247	9	9	210	out	2	+
415	7117	1.112	40	2137	491	8	8	210	out	5	+
445	7139	1.588	32	2944	-265	9	9	210	out	5	+
470	7142	1.581	40	2938	-250	10	10	210	320	2	+
496	7148	1.205	38	2214	462	9	9	210	out	4	0
519	7174	1.670	40	3102	-419	9	9	210	out	3	+
570	7186	1.025	36	1874	978	8	8	210	out	1	+
595	7192	1.132	39	2080	966	8	8	210	out	1	0
710	7230	1.703	40	3221	276	10	10	210	out	2	-
770	7239	1.656	39	3219	440	10	10	210	out	3	+
797	7242	1.561	40	2853	865	9	9	210	out, 320	2	-
					SPQMLT 2						
0	8004	1.539	37	2853	865	7	7	210	out	8	+
40	8017	1.490	40	2761	859	9	9	210	320	1	0
80	8023	1.610	40	2987	533	8	9	210	320	2	0
120	8030	1.340	41	2478	913	8	8	210	320	2	-
160	8036	1.122	40	2067	1138	9	9	210	320	2	0
200	8044	1.137	40	2095	897	9	8	210	320	1	0
240	8065	1.398	43	2587	164	10	10	210	320	2	0
279	8078	1.907	47	3547	-938	10	10	210	320	4	+
323	8085	1.610	44	2987	-420	10	10	210	320	4	+
363	8092	1.483	43	2748	-236	9	10	210	320	2	+
403	8118	1.440	44	2666	-69	10	9	210	320	2	0
443	8120	1.296	48	2395	365	9	9	210	320	1	-
456	8150	1.225	35	2261	576	9	8	210	320	3	-
495	8158	1.444	30	2674	381	9	9	210	320	5	0
535	8161	1.366	40	2527	732	10	10	210	320	2	+
575	8169	1.671	35	3102	305	10	9	210	320	3	+
615	8172	1.513	41	2804	708	10	10	210	320	6	0
667	8177	1.689	47	3136	488	8	8	210	320	4	+

APPENDIX 3. SAMPLE SEISMIC RECORDS: WIDE-ANGLE REFLECTION AND VERTICAL REFLECTION RECORDS

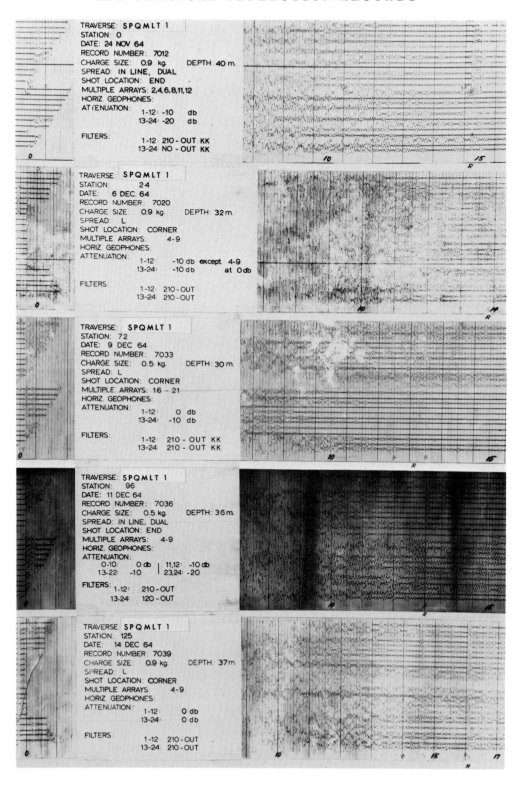

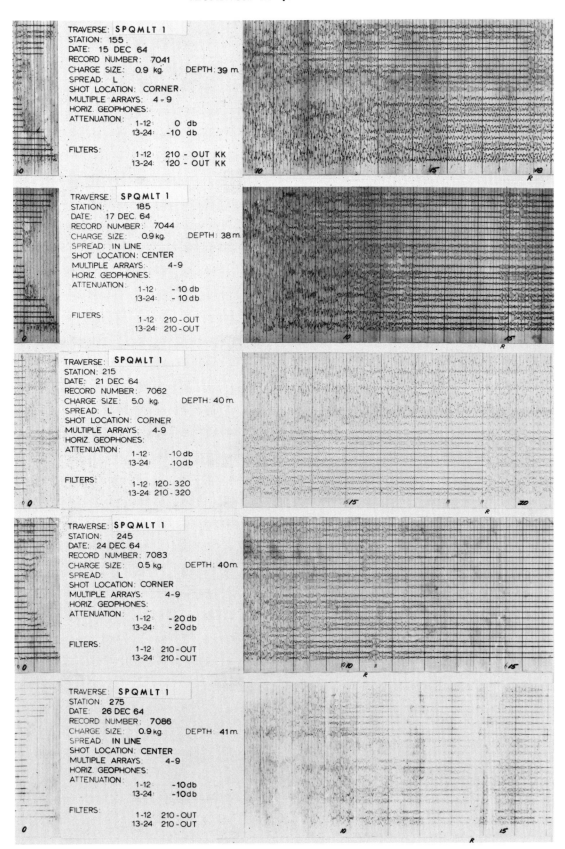

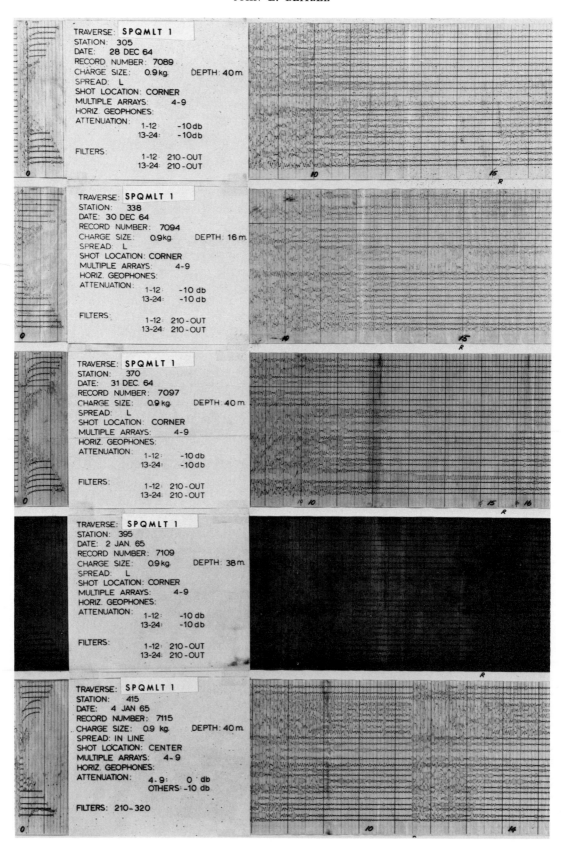

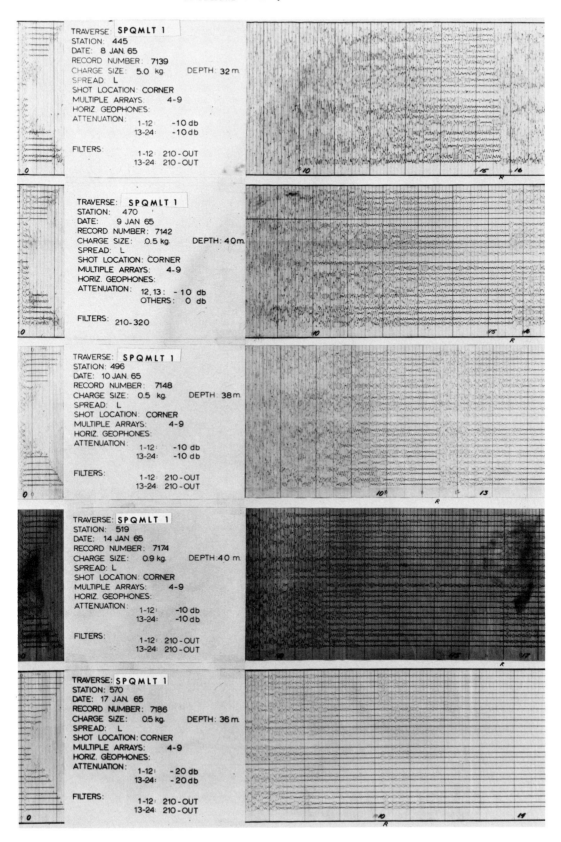

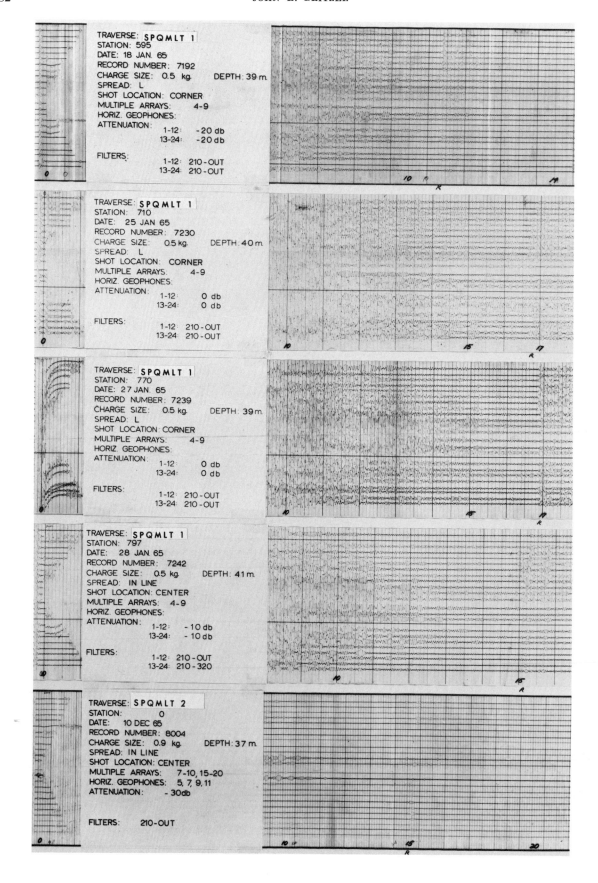

GEOPHYSICS IN QUEEN MAUD LAND

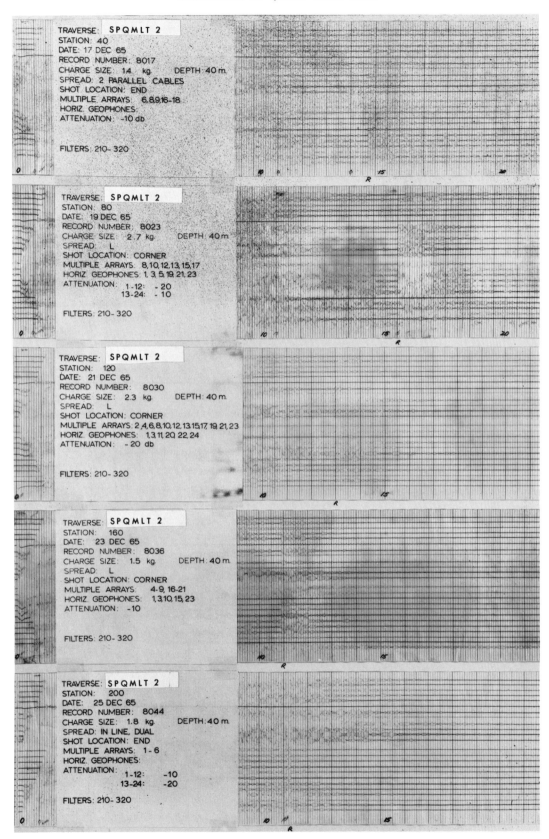

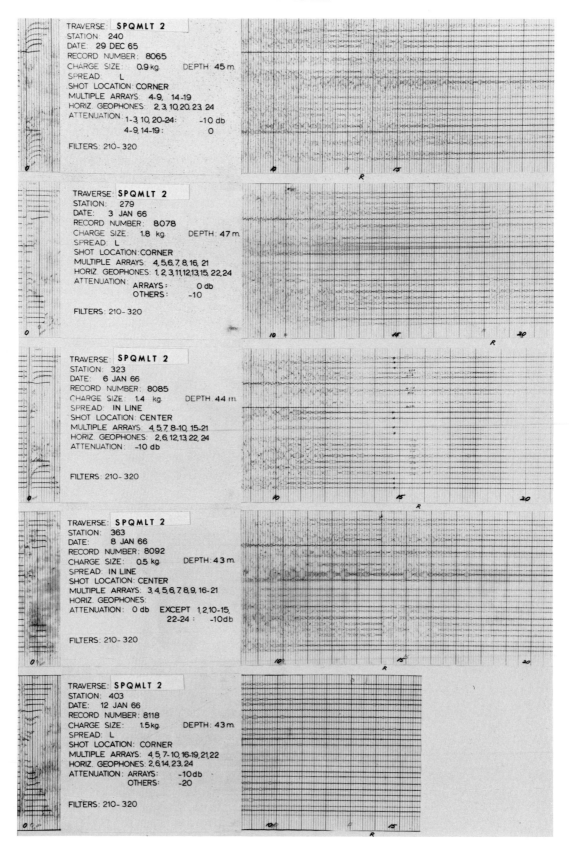

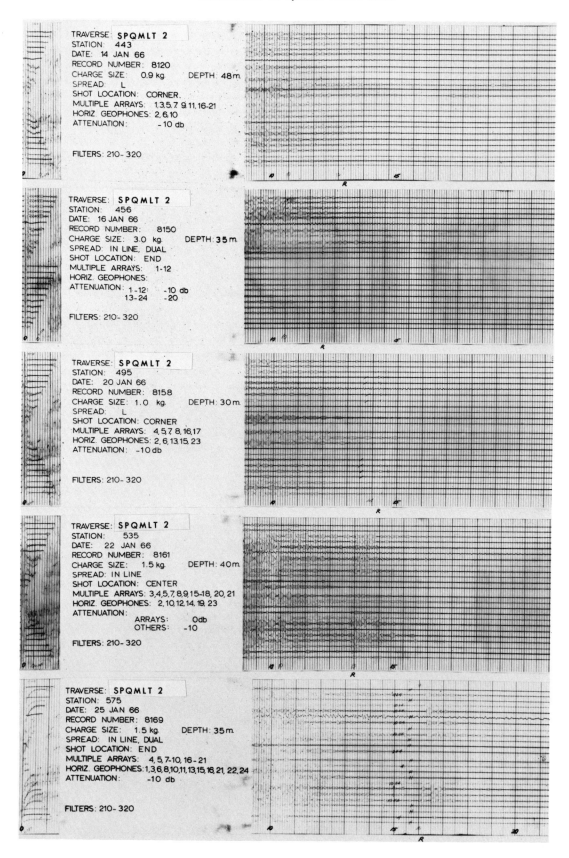

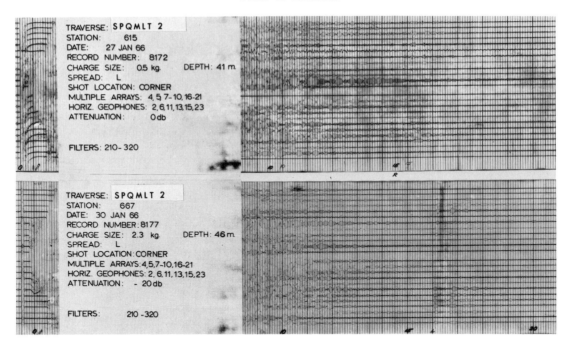

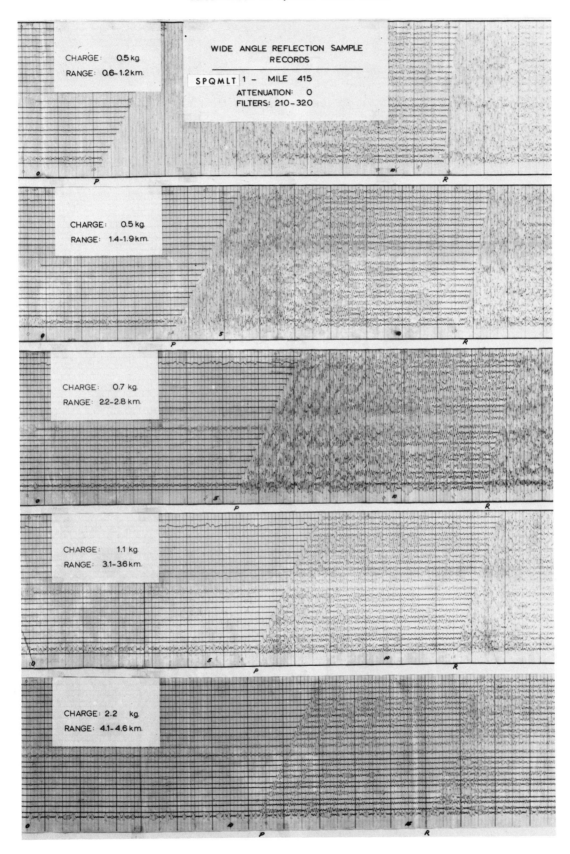

SEISMIC EVIDENCE FOR MORAINE WITHIN THE BASAL ANTARCTIC ICE SHEET

Charles R. Bentley

University of Wisconsin, Department of Geology and Geophysics
Geophysical and Polar Research Center, Middleton 53562

A low-amplitude seismic echo reflecting from a horizon a few hundred meters above the bottom of the ice sheet has been widely observed in West Antarctica. The echo varies widely from place to place in amplitude, duration, and coherence, and in the indicated reflector dip. The weakness of the reflected signal, together with wave velocity information, indicates that the basal layer beneath the reflector is composed primarily of ice. The reverberative, incoherent, and variable characteristics of the reflections leads to the conclusion that they probably arise in most places from one or several bands of morainal debris within the ice. In a few instances, however, the source may be the upper surface of a zone of ice at the melting point.

In the course of seismic reflection measurements carried out during the austral summer of 1957–1958 in central West Antarctica, a low-amplitude echo (R_e) arriving a fraction of a second before the main bottom echo (R_1) was discovered [*Bentley and Ostenso*, 1961]. Because of its small amplitude, about one-tenth that of R_1, it is possible to observe R_e only where the signal level on reflection seismograms is very low for a few tenths of a second immediately before R_1. On the Antarctic ice sheet, the signal generated from a shot (and recorded locally) diminishes at a rate that depends strongly upon the temperature in the surface layers of the firn. Where the temperature is relatively high, the signal will decrease below the ambient noise level in half a second or so; in colder regions the same decay may require several seconds. The ice surface is comparatively low and warm in central and northern West Antarctica, yet the ice is generally thick, so that this region is well suited for the observation of R_e, and it has been widely recorded there. Failure to record it elsewhere in Antarctica can almost always be attributed to insufficiently quiet records and thus sheds no light on the existence of the corresponding reflecting surface at depth.

As might be expected, R_e has been particularly well observed on traverses crossing the thick ice of the Byrd subglacial basin. The best and most numerous samples of this echo were recorded on the Sentinel traverse (ST) in 1957–1958 [*Bentley and Ostenso*, 1961], and on the Ellsworth Highland traverse (EHT) in 1960–1961. Only from the latter traverses has any significant evidence for the absence of R_e been obtained. Other examples of R_e were recorded on the Little America–Byrd traverse (LAB) in 1957 [*Bentley and Ostenso*, 1961], the Marie Byrd Land traverse (MBLT) in 1959–1960 [*Bentley and Chang*, this volume], the Antarctic Peninsula traverse (APT) in 1961–1962 [*Behrendt*, 1963], at Byrd Station in 1959 by E. S. Robinson, and at an airlifted station in 1960 by J. C. Behrendt. The location of stations and the orientation of reflection profiles from which data were obtained are shown in Figure 1.

REFLECTION CHARACTER

R_e typically occurs anywhere from 0.03 to 0.3 sec before the main bottom echo (Appendix). Filter settings permitting, the frequency of the signal is almost always between 100 and 150 Hz, although occasionally (EHT 492 and 612, APT 432) it is around 200 Hz. The duration of one wave group ranges from about 0.02 sec (which is also about the duration of the outgoing signal from the shot) to 0.12 sec; in addition, there is sometimes more than one wave group. The echo amplitudes, both in comparison to that of the main echo and in absolute value, vary widely from place to place and sometimes from trace to trace on the same seismogram. The dip of the reflecting surface, usually quite small, reaches several degrees in a few locations. The R_e surface is not necessarily parallel to the base of the ice, the dip of the former being sometimes more and sometimes

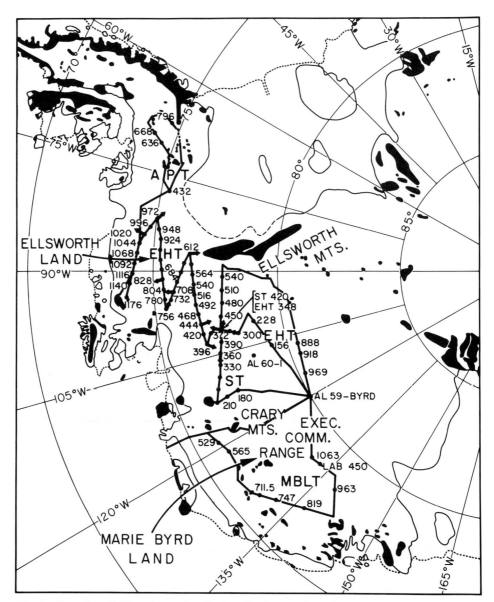

Fig. 1. Location of West Antarctic seismic reflection stations. R_e was observed at numbered stations, and was significantly absent at stations marked by unnumbered dots. Short arrows show the orientation of reflection profiles. APT, Antarctic Peninsula traverse; EHT, Ellsworth Highland traverse; MBLT, Marie Byrd Land traverse; AL 59, airlifted survey, 1959; AL 60, airlifted survey, 1960.

less than that of the latter. The maximum recorded dip of the R_e surface is about 19° (ST 390).

The weakness of R_e indicates (see equation 4 below) that the corresponding acoustic impedance contrast is small (acoustic impedance is the product of wave velocity and density). Because of the complicating effect of anisotropy in the main body of the ice sheet, the compressional wave velocity (V_p) in the basal layer cannot be determined accurately. Nevertheless, as is shown below, we can say that V_p lies within the range 3.3–4.3 km/sec, so that the density must be, correspondingly, between 1.1 and 0.8 g/cm³. The basal layer must therefore be composed primarily of ice, since no other geologically reasonable material possesses nearly this peculiar combination of high velocity and low density. The problem to be resolved, then, is in what way the ice below the reflector is modified from that above.

There are two general possibilities for producing an acoustic impedance contrast that can be recognized: the ice in the layer might be clean and thus of the same density as the ice above, the acoustic impedance contrast being the result of a velocity change alone, or the ice in the layer might be contaminated with other material, so that both the density and the wave velocity may differ from that in the ice above. We will henceforth refer to these as 'pure-ice models' and 'impure-ice models,' respectively. Two explanations have been put forward.

1. The ice in the basal layer is at the melting point. In this case, V_p in the layer could be expected to be between 3500 and 3700 m/sec on the basis of measurements on temperate glaciers [*Robin*, 1958, Table 29, p. 75], whereas the velocity in the overlying ice, just a few tenths of a degree below melting, would be close to 3800 m/sec. A very rapid change of velocity with temperature occurs between $-0.2°C$ and $-0.1°C$, presumably owing to the effect of melting at crystal boundaries. If we assume an ice sheet 3000 meters thick with surface at $-30°C$ and base at the melting point, the minimum temperature gradient to be expected near the base is $30 \div 3000$, or $0.1°C$ in 10 meters. In comparison, a P wave of frequency 120 Hz would have a wavelength of about 30 meters. It follows that there would be a significant velocity change over a depth interval of only a fraction of a wavelength, and that energy would therefore be reflected.

It is worth noting two likely characteristics of this model. First, in the basal layer V_p would be within 300 m/sec of the velocity above the layer, so that the acoustic impedance ratio would be no less than 0.9. Second, there would be a single discrete reflecting surface that would produce an impulsive echo. This would necessarily be true unless it were considered possible to have an oscillatory temperature-depth function.

2. The ice in the basal layer is contaminated. If the contaminant is morainal material [*Bentley and Ostenso*, 1961; *Behrendt*, 1963], the reflecting surfaces would probably be rough, and several surfaces could exist, depending upon the distribution of contaminating material, so that the reflected signal could well be reverberatory. The mean P-wave velocity in the layer would not necessarily be much different from that above, even without consideration of other effects, since the reflecting bands might make up only a small proportion of the total thickness.

If the contaminant is volcanic ash, we would expect that R_e would be particularly strong along flow lines leading from known sources of volcanism, and that the age of the reflecting surface would be the same, if not everywhere, at least over wide regions.

Another 'pure-ice' possibility that must be considered, especially in the light of recent observations by *Gow* [1968] on the deep ice core from Byrd station, is an abrupt change in ice fabric across the interface. At Byrd station, strongly oriented ice, with c axes concentrated around the vertical direction, was recovered from depths roughly between 1000 and 1800 meters. The bottom 300 meters of the core, however, extending to the base of the ice sheet, comprised ice of roughly random crystal orientation. The change from anisotropic to isotropic ice occurs largely in a distance of less than, and possibly much less than, 50 meters; a more precise description must await further analysis on the core.

The velocity difference (for vertically propagating P waves) that would be expected at the boundary between ice with 100% vertical alignment of the c axes and ice with random crystal orientation is about 200 m/sec [*Bennett*, 1968]. The characteristics of this model as the cause of R_e would be similar to those of the temperature model, both in the maximum velocity contrast that could exist and in the singularity of the reflecting surface. Although the mechanism that produces the phenomenon is not known, it is reasonable to assume that the ice flow vectors at the depth of the boundary would lie in the boundary, hence that the dip of that surface would not exceed the bedrock slope.

Results from the deep hole at Byrd station have not yet, at least, shed much light on the other suggested explanations for the early reflections. Morainal material was found only within a few meters of the bottom of the hole. Thin ash layers were observed between 450 and 850 meters of the bottom, but without concentrations of ash great enough that seismic reflections would be expected. The base of the ice sheet is at the pressure-melting temperature, as is indicated by the presence of water; temperatures in the bottom few hundred meters of the ice have not been measured, but reasonable interpolation does preclude the existence of a constant-temperature zone more than a few tens of meters thick [Gow, 1968]. It should be pointed out that R_e was not observed under moderately favorable conditions in reflection shooting about 2 km away from the drill site, although it was well recorded about 12 km away, near old Byrd station.

In order to choose the most likely explanation for R_e, we will now consider its characteristics and distribution in more detail.

RECORDING EQUIPMENT AND PROCEDURES

A Texas Instruments 7000B Portable Seismograph System was used for all the reflection measurements. This system has a basic frequency range of 5 to 500 Hz (between points 6 db down), a recording oscillograph employing 500-Hz galvanometers, and a wide range of possible filter settings. On most shots, high-cut filters were set at either 160 or 215 Hz, and low-cut filters either were not used at all or were set at 60 or 90 Hz. Occasionally a higher range, 210–320 Hz, was used, but only at one station (EHT 156; see Appendix), where the noise level in the usual recording band was unusually high, was R_e better recorded at the higher frequencies.

Amplifier gains, set to desired maximum levels by potentiometer adjustment, could then be varied over a range of 100 db in 10-db steps by attenuator switches. Automatic gain control was also available on the equipment but was never used, since amplitude information was more important than events in the early parts of the records. Mixing was also not used.

The geophones were of the Electro-Tech EVS type, some with 7-Hz, some with 20-Hz, and some with 30-Hz natural frequency. The calibration constants were essentially the same over the frequency band of interest: 0.2 volt/cm sec^{-1}. On many of the shots, several geophones were oriented horizontally to record shear waves: this should be noted in the examination of the sample seismograms (Appendix), since the corresponding traces do not show R_e. The geophone interval was 31 meters for all shots.

Shot holes were drilled to a depth of 3–5 meters with a hand auger. Charge sizes ranged from 0.4 to 5.0 kg, the usual charge being 0.9 kg (1-lb Nitramon with a primer of 0.9-lb equivalent weight). 'Sprung' holes (ones in which a charge has already been fired) were often used to take advantage of improved energy coupling from shot to firn.

AMPLIFIER CALIBRATION

Amplifier gain levels were determined in two ways. A number of test records were run using the test oscillator built into the equipment. Unfortunately, there were not enough of these test records on all traverses, and the amplitude of the test signal was itself uncertain. These records were supplemented, therefore, by measurements of amplifier noise levels on records from all stations at which R_e was recorded.

Calibration data were the best on the Sentinel traverse. A mean value of 0.37 μv/mm for stations ST 240 to ST 969, as shown by the test records, was adopted as the standard. The calibration for ST 180 and ST 210 (see Table 1 for all calibration constants) was taken according to the noise ratio relative to the standard. For EHT, the oscillator tests and noise ratios were in good agreement, the latter indicating an abrupt change in gain levels between EHT 156 and EHT 300, and between EHT 420 and EHT 444. Using the calibration constants determined in this way, the mean normalized amplitudes of R_e were essentially the same for the Sentinel and Ellsworth Highland traverses.

For the Marie Byrd Land traverse, oscillator test records and noise level checks were not in good agreement, giving calibration constants of 0.06 and 0.033 μv/mm, respectively. An intermediate value of 0.05 μv/mm, which resulted in a mean normalized amplitude of R_e the same as that for the Sentinel and Ellsworth Highland traverses, was used.

For the Antarctic Peninsula traverse, no test signal records were available. It was assumed that the calibration was unchanged from the end of the

TABLE 1. Amplifier Calibration Constants

Station Intervals	Calibr. Constants, μv/mm
ST 180 and ST 210	0.16
ST 240 to ST 969	0.37
MBLT (all)	0.05
EHT 156 to EHT 228	0.033
EHT 300 to EHT 420	0.10
EHT 420 to EHT 1176	0.20
APT (all)	0.20
LAB 450	0.24
AL59 Byrd	0.45
AL60 1	0.07

preceding season (EHT), this being consistent with the noise level ratio. Test oscillator records were used for LAB 450; noise levels were used for the recordings made at Byrd station and the airlifted site.

The oscillator tests on the Sentinel traverse showed a calibration variation of 30% (standard deviation) from the mean. Amplifier noise levels vary by about the same amount between stations on any traverse at which the calibration constant is nominally the same. From this evidence, and from the similarity in mean reflection amplitude from traverse to traverse, we will assume that the calibration constants are in error by no more than 50%.

TRAVEL TIMES

Reflection times were read to the nearest millisecond wherever possible. Most R_1 times are accurate within a very few milliseconds. The travel times for R_e are considerably less certain; the reflections are often sharp in onset and can be read with an error of only a few milliseconds, whereas weak arrivals or noisy background can cause errors of several hundredths of a second, so that the basal layer thickness could be underestimated by as much as 100 meters.

All travel times were converted to thicknesses using a mean velocity of 3840 m/sec. This velocity is in some doubt, owing to the uncertain effects of anisotropy; the mean velocity used could sometimes be in error by as much as 100 m/sec [Bentley, this volume], and the velocity applicable in the basal layer could differ by even more. Even this uncertainty, however, amounts to an error of only a few percent, which is not significant for the purposes of this paper. Depths below the surface were corrected for the effect of the near-surface low-velocity layers. Travel times and layer thickness are listed in Table 2.

REFLECTION AMPLITUDES

The maximum peak-to-peak deflection in R_e was measured on each seismogram trace, and the mean amplitude (half peak-to-peak) was found for each seismogram. Separate means were recorded for different seismograms at the same station in order to retain an indication of the amplitude variability. Mean trace amplitudes were reduced to ground velocity amplitudes by using the calibration constants for amplifiers and geophones discussed above and the known factor for the amplifier attenuation setting.

Adjustment for variation in charge size from the standard of 0.9 kg was made assuming that the energy generated was proportional to the charge weight. A study of seismograms from the Sentinel traverse and the 1958–1959 Horlick traverse showed that this assumption was justified on the average, but that there was a wide variation with individual shots, the standard deviation being ±60%.

A correction for efficiency of coupling from shot to firn, in accordance with the degree of 'springing,' was also made. On the basis of the study of Sentinel and Horlick traverse seismograms, a correction factor of 0.5, relative to an unsprung hole, was applied to the observed amplitudes when the springing charge was the same size as the recorded shot; factors of 0.7 and 0.3 were used for smaller and larger springing charges, respectively. When more than one previous charge had been fired in the same shot hole, the springing charge was taken to be the sum of the individual charges. The standard deviation found from the Sentinel-Horlick study for the effect of springing was ±35%. As a check on the correction factors applied for hole springing, mean corrected R_e amplitudes for all sprung and unsprung holes were compared; no significant difference was found.

A normalizing factor was applied to correct for the effect of geometrical spreading from the shot, which was assumed to result in an amplitude decrease proportional to the distance traveled. The factor used was simply the depth in kilometers to the R_e reflecting surface.

It is to be expected that there also will be energy loss due to signal absorption in the ice, and consequently that the signal will undergo attenuation varying exponentially with distance. If we denote as α the energy absorption constant, then the amplitude absorption factor for R_e will be $\exp(-\alpha h_e)$, where h_e is the depth to the reflector. To see whether the absorption was measurable, amplitudes normalized for spreading were plotted logarithmically against h_e (Figure 2). It is clear from inspection that there is no statistically significant correlation between the two, and hence that the absorption constant is not measurable from these data. We should note, however, that the data are biased in that weak reflections from large depths would not be observed at all. If we consider only the strongest reflections, there is a suggestion of an amplitude variation consistent with $\alpha = 0.2$ km^{-1}, as is indicated by the solid line in Figure 2. Robin [1958] finds $\alpha = 0.65$ km^{-1} (dashed line), which is apparently too large to fit the present data. It is possible, as Robin points out, that reflection amplitudes in his study were

TABLE 2a. Travel Times and Layer Thicknesses for Single Stations

Station	Lat. S	Long. W	Record	Filters	Mean Travel Time, sec		Ice Thickness, km	
					R_e	R_1	Layer	Total
ST 180	77°28'	112°22'	1146	0–160	1.02	1.15	0.25	2.17
			1147	60–160		1.15		2.17
ST 210	77°00'	112°47'	1168	60–160	1.02	1.33	0.60	2.52
			1168	60–160	1.16	1.33	0.33	2.52
			1171	0–160	1.02	1.33	0.60	2.52
			1171	0–160	1.16	1.33	0.33	2.52
ST 270	76°44'	111°22'	1232	0–215		1.33		2.48
ST 300	76°58'	109°07'	1243	0–215		1.26		2.34
ST 330	77°07'	107°01'	1278	0–215	1.11	1.29	0.35	2.44
ST 390	77°21'	102°41'	1322	0–215	1.43	1.62	0.37	3.06
ST 420	77°28'	100°26'	1352	0–215	1.58	1.78	0.38	3.38
			1353	0–320	1.58	1.78	0.38	3.38
ST 450	77°32'	98°11'	1383	0–320	1.67	1.80	0.24	3.41
ST 480	77°34'	95°56'	1413	0–215	1.75	1.90	0.28	3.60
ST 510	77°36'	93°44'	1444	60–215	1.33	1.47	0.26	2.78
ST 540	77°41'	91°28'	1476	60–215	1.16	1.27	0.20	2.40
ST 888	80°27'	107°01'	1849	60–215	1.45	1.69	0.46	3.21
ST 918	80°23'	109°51'	1871	0–215		2.28		4.34
			1872	60–215	1.94	2.28	0.64	4.34
ST 969	80°14'	114°40'	1913	0–160	1.43	1.57	0.27	3.00
MBLT 529	75°47'	118°45'	4151	60–215	0.17	0.85	0.26	1.59
			4151	60–215	0.79	0.85	0.12	1.59
MBLT 565	75°59'	121°05'	4162	60–215	1.16	1.45	0.56	2.74
MBLT 711.5	76°02'	130°08'	4250	90–160	0.68	0.72	0.08	1.37
MBLT 747	76°19'	132°18'	4272	90–160	0.78	0.85	0.13	1.59
MBLT 819	76°54'	136°52'	4284	90–160	0.70	0.76	0.11	1.40
MBLT 963	78°19'	138°10'	4339	90–160	0.60	0.69	0.18	1.29
MBLT 1063	78°36'	130°47'	4383	90–160	1.00	1.14	0.26	2.12
EHT 156	79°20'	106°07'	6022	210–320	1.62	1.78	0.31	3.36
			6023	210–320		1.78		3.36
EHT 228	78°50'	100°11'	6053	210–320	1.22	1.40	0.33	2.64
EHT 300	78°12'	101°50'	6077	60–215	1.18	1.32	0.27	2.50
			6082	210–320	1.17	1.32	0.29	2.50
EHT 324	77°49'	101°10'	6097	60–160	1.27	1.46	0.36	2.76
			6098	90–160	1.26	1.46	0.37	2.76
EHT 348	77°28'	100°27'	6101	0–215		1.80		3.41
			6102	0–215	1.61	1.80	0.36	3.41
EHT 372	77°08'	101°07'	6157	0–320	1.43	1.58	0.29	2.98
			6158	0–215	1.43	1.58	0.28	2.98
			6159	0–160	1.44	1.58	0.27	2.98
EHT 396	76°53'	102°35'	6163	0–215		1.34		2.56
EHT 420	76°52'	100°47'	6176	0–215		1.21		2.29
			6178	0–215	1.11	1.21	0.18	2.29
EHT 444	76°48'	99°02'	6180	0–320		1.12		2.11
			6181	0–215	0.98	1.12	0.26	2.11
EHT 468	76°47'	97°16'	6191	0–215	0.88	1.05	0.34	1.98
			6192	0–320	0.87	1.05	0.35	1.98
			6193	0–215	0.87	1.05	0.35	1.98

TABLE 2a. (continued)

Station	Lat. S	Long. W	Record	Filters	Mean Travel Time, sec		Ice Thickness, km	
					R_e	R_1	Layer	Total
EHT 492	76°43'	95°31'	6194	0–215		0.90		1.70
			6196	0–215	0.76	0.90	0.28	1.70
			6197	0–215	0.74	0.90	0.31	1.70
EHT 516	76°41'	93°41'	6204	0–215	0.84	1.02	0.33	1.92
			6204	0–215	0.90	1.02	0.23	1.92
EHT 540	76°39'	91°55'	6207	60–160	0.94	1.12	0.34	2.12
EHT 564	76°36'	90°07'	6213	60–215	1.04	1.18	0.27	2.23
EHT 588	76°32'	88°25'	6217	0–215		1.31		2.47
			6218	0–215		1.31		2.47
EHT 612	76°26'	86°41'	6223	0–320	1.18	1.27	0.17	2.42
EHT 684	75°58'	91°28'	6246	0–215	0.93	1.02	0.18	1.92
EHT 708	75°49'	93°00'	6254	0–215	0.68	0.79	0.22	1.49
			6255	0–160	0.68	0.79	0.23	1.49
			6257	0–215	0.68	0.79	0.22	1.49
			6258	0–160	0.68	0.79	0.22	1.49
EHT 732	75°38'	94°30'	6270	0–215		0.99		1.90
			6271	60–160		0.99		1.90
EHT 756	75°27'	95°56'	6275	0–160	0.96	1.04	0.15	1.97
EHT 780	75°25'	94°19'	6310	0–215	0.98	1.04	0.13	1.97
EHT 804	75°23'	92°43'	6317	0–160		1.07		2.05
EHT 828	75°21'	91°06'	6322	0–215		1.10		2.09
EHT 900	75°17'	86°16'	6343	0–160		1.06		2.02
EHT 924	75°14'	84°40'	6361	0–215	1.06	1.13	0.14	2.13
			6362	0–215	0.93	1.13	0.37	2.13
			6362	0–215	1.06	1.13	0.14	2.13
EHT 948	75°08'	83°09'	6365	0–215	0.82	0.94	0.22	1.76
			6366	0–215	0.83	0.94	0.20	1.76
			6367	0–215	0.82	0.94	0.22	1.76
EHT 972	74°58'	81°41'	6388	60–215	0.62	0.68	0.11	1.25
EHT 996	74°45'	83°01'	6401	0–215		0.84		1.61
EHT 1020	74°31'	84°16'	6407	0–160		0.79		1.34
EHT 1044	74°16'	85°32'	6420	60–160	0.58	0.68	0.18	1.26
EHT 1068	74°12'	87°03'	6424	0–215	0.93	1.01	0.15	1.90
EHT 1092	74°06'	88°29'	6429	0–215		0.63		1.20
EHT 1116	74°00'	89°58'	6433	0–215		0.77		1.48
EHT 1140	73°58'	91°27'	6442	90–215		0.50		0.95
EHT 1176	73°51'	93°37'	6445	0–215		0.90		1.70
			6446	0–215	0.78	0.90	0.24	1.70
APT 432	75°14'	77°09'	2	0–215	0.93	1.15	0.42	2.17
APT 636	74°16'	70°10'	1	90–215	0.77	1.07	0.59	2.02
APT 668	73°54'	69°26'	1	90–215	0.62	0.85	0.44	1.60
			2	90–215	0.62	0.85	0.44	1.60
APT 796	74°27'	67°08'	6	90–320	0.47	0.52	0.09	0.97
LAB 450	78°40'	132°35'	140	0–160	0.95	1.19	0.46	2.24
AL59 Byrd	79°59'	120°01'	8	210–320	1.07	1.31	0.46	2.48
AL60 1	78°42'	107°07'	11	90–320	1.46	1.66	0.38	3.15

TABLE 2b. Travel Times and Layer Thicknesses for Profiles

Station	Distance along Profile	Record	Filters	Mean Travel Time, sec		Ice Thickness, km	
				R_e	R_1	Layer	Total
ST 360	0	1310	0–215		1.56		2.98
	6.5	1316	0–160	1.38	1.55	0.31	2.95
	7.9	1317	0–160	1.38	1.54	0.31	2.93
	10.1	1318	0–160	1.34	1.53	0.36	2.91
	13.7	1319	0–160	1.31	1.49	0.35	2.86
	13.7	1320	0–160	1.31	1.49	0.35	2.86
EHT 300	0	6077	60–215	1.18	1.32	0.27	2.50
	3.3	6082	210–320	1.17	1.32	0.29	2.50
	3.3	6091	60–160	1.20	1.35	0.29	2.57
	2.8	6092	60–160	1.21	1.36	0.29	2.59
	2.0	6093	90–215	1.20	1.33	0.25	2.53
	1.3	6094	60–160	1.19	1.35	0.31	2.57
EHT 348	0	6101	0–215		1.80		3.41
	7.8	6102	0–215	1.61	1.80	0.36	3.41
EHT 348E	7.8	6144	0–215	1.50	1.76	0.50	3.35
	12.0	6145	0–215	1.64	1.81	0.33	3.45
	10.0	6146	0–215	1.54	1.78	0.46	3.39
	4.0	6147	0–215	1.56	1.78	0.42	3.39
	2.0	6148	0–215	1.64	1.80	0.31	3.43
EHT 348N	6.9	6152	90–160	1.50	1.68	0.35	3.21
EHT 396	0	6163	0–215		1.34		2.56
	2.1	6174	0–215	1.24	1.44	0.44	2.74
EHT 444	0	6180	0–320		1.12		2.11
	0	6181	0–215	0.98	1.12	0.26	2.11
	2.8	6188	60–215	1.01	1.14	0.25	2.16
	2.1	6189	0–215	1.02	1.15	0.25	2.18
	1.1	6190	0–215	1.00	1.17	0.33	2.22
EHT 612	0	6223	0–320	1.18	1.27	0.17	2.42
	0.9	6233	60–215	1.19	1.29	0.19	2.45
	1.5	6234	60–215	1.20	1.30	0.19	2.47
	2.2	6235	60–215	1.21	1.32	0.21	2.51
	2.8	6236	60–215	1.21	1.33	0.23	2.53
EHT 708	0	6254	0–215	0.68	0.79	0.22	1.49
	0	6255	0–160	0.68	0.79	0.23	1.49
	0	6257	0–215	0.68	0.79	0.22	1.49
	0	6258	0–160	0.68	0.79	0.22	1.49
	0.9	6264	60–215	0.65	0.74	0.17	1.40
	1.6	6265	60–215	0.66	0.74	0.15	1.40
	1.3	6266	60–215	0.66	0.74	0.15	1.40
	0.6	6267	60–215	0.65	0.75	0.19	1.42
EHT 756	0	6275	0–160	0.96	1.04	0.15	1.97
	3.7	6295	30–160	0.93	1.04	0.22	1.98
EHT 828	0	6322	0–160		1.10		2.09
	1.0	6333	0–160	0.94	1.06	0.23	2.01
	2.0	6335	0–160	0.96	1.06	0.18	2.01
EHT 972	0	6388	60–215	0.62	0.68	0.11	1.25
	0.9	6386	60–215	0.66	0.72	0.12	1.37
	1.6	6387	0–215	0.63	0.68	0.10	1.28
	2.2	6388	60–215	0.62	0.68	0.12	1.28
	2.9	6389	0–215	0.62	0.67	0.11	1.27

TABLE 2b. (continued)

Station	Distance along Profile	Record	Filters	Mean Travel Time, sec		Ice Thickness, km	
				R_e	R_1	Layer	Total
EHT 972	3.9	6390	0–160	0.56	0.65	0.17	1.23
	4.9	6391	48–160	0.57	0.64	0.13	1.20
	5.9	6392	out	0.58	0.60	0.05	1.14
	6.9	6393	60–215	0.53	0.56	0.06	1.05
	12.9	6397	60–215	1.08	1.33	0.48	2.53
EHT 1020	0	6407	60–160		0.79		1.34
	0.9	6415	0–215	0.72	0.77	0.10	1.45
	1.6	6416	60–160	0.73	0.79	0.11	1.49
	2.2	6417	60–160	0.75	0.85	0.19	1.61

affected by AGC, particularly where the ice was less than 1 km thick, thus increasing the apparent attenuation.

For lack of any better value, we will thus take $\alpha = 0.2$ km^{-1}. Since amplitudes were normalized for spreading loss to $h_e = 1$ km, we will use the same standard depth for absorption corrections. All amplitudes, consequently, have been multiplied by $\exp 0.2(h_e - 1)$ to produce the final, fully normalized amplitudes of ground velocity. Since the absorption correction varies by less than a factor of 2 over the whole observed range of h_e, the error introduced by improper assessment of the absorption should be relatively unimportant.

Combining the estimated errors in calibration, charge size, and springing factors as if they were true standard errors, and allowing for the uncertainty in absorption, we conclude that the absolute normalized velocity amplitudes are probably reliable within a factor of 2, i.e a value of 4×10^{-6} cm sec^{-1} can be taken to mean $(2-8) \times 10^{-6}$ cm sec^{-1}.

Comparison of ground velocity amplitudes $(\dot{u}_z)$ calculated from 43 seismograms recorded at the 19 stations at which more than one recording of R_e was obtained suggests that this error estimate is conservative. A standard error was calculated from these observations as follows. First, a standard error $(s)_j$ at each station was calculated from the relation

$$(s)_j = \left[\sum_{i=1}^{n} (\delta \dot{u}_z)_i^2 / (n-1) \right]^{1/2}$$

where $(\delta \dot{u}_z)_i$ is the geometric deviation of the individual velocity amplitude $(\dot{u}_z)_i$ from the mean $\langle \dot{u}_z \rangle$ at that station:

$$(\delta \dot{u}_z)_i = \frac{\langle \dot{u}_z \rangle}{(\dot{u}_z)_i} - 1 \qquad (\dot{u}_z)_i < \langle \dot{u}_z \rangle$$

$$= \frac{(\dot{u}_z)_i}{\langle \dot{u}_z \rangle} - 1 \qquad (\dot{u}_z)_i > \langle \dot{u}_z \rangle$$

and n is the number of observations at the station. (Usually n was only 2; needless to say, no significance can be placed on the individual $(s)_j$.) The over-all standard error was then simply taken to be

$$s = \left[\sum_{j=1}^{m} (s)_j^2 / (m-1) \right]^{1/2}$$

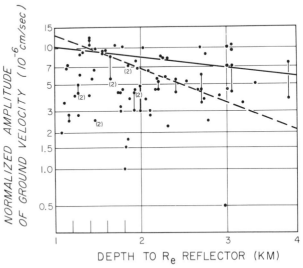

Fig. 2. Amplitude of R_e corrected for spreading loss versus depth to basal layer. Solid line represents factor assumed for absorption corrections. Dashed line represents Robin's [1958] absorption factor.

where m is the number of stations. The standard error so obtained was ±30%, which should serve as a reasonable estimate of the expected relative error for observations not too far apart on the same traverse. Inclusion of the calibration error leads to a standard error for absolute amplitudes of about a factor of 1.6, which can be compared with the factor of 2 obtained above.

At stations where R_e was not observed, the trace amplitudes a few tenths of a second before R_1 were examined. In the great majority of cases, R_e could have been present with a significant amplitude without being detected. At several stations, however, it was possible to set a meaningful upper limit on the size of R_e, on the assumption that a signal-to-noise ratio greater than 1 would be observable (and using the total ice thickness as the normalizing factor for energy spreading). At some fifteen locations, the upper limit was smaller than all but a very few observed values.

Mean amplitudes and normalizing factors for all seismograms are listed in Table 3.

REFLECTION COEFFICIENTS

There are two methods that can be used to estimate the reflectivity of the R_e surface: analysis of the absolute amplitudes on the basis of the expected energy generated by the shot, and analysis of relative amplitudes compared to those for R_1, which eliminates errors due to calibration uncertainties but introduces new errors resulting from variations in the reflection coefficient at the base of the ice.

Absolute Amplitudes

The energy E per unit area of wave front in one wavelength of a reflection from depth h_e in a uniform medium is given by

$$E = I_0 \frac{r_e^2 \exp(-2\alpha h_e)}{(2h_e)^2}$$

where I_0 is the energy per steradian in one wavelength generated at the source, and r_e is the amplitude reflectivity of the R_e surface. This must be modified to allow for the effect of the inhomogeneity of the firn around the shot point, becoming [Robin, 1958]

$$E = \frac{I_0 r_e^2 \exp(-2\alpha h_e)}{(2h_e)^2} \left(\frac{V_0}{V_m}\right)^2 \quad (1)$$

where V_0 and V_m are the compressional wave velocities at the shot point and at depth, respectively.

At the same time, we can say for a simple harmonic plane wave

$$E = \tfrac{1}{2}\rho\lambda\dot{a}^2$$

where ρ is the density, λ is the wavelength, and $\dot{a}$ is the particle velocity amplitude. For a compressional wave impinging vertically on the free upper surface of the ice sheet, the vertical velocity of the surface, $\dot{u}_z$, is the sum of the particle velocities associated with the incident and reflected waves:

$$\dot{u}_z = 2\dot{a}$$

whence

$$E = \tfrac{1}{8}\rho_0\lambda_0\dot{u}_z^2 \quad (2)$$

where ρ_0 and λ_0 are the density and the wavelength near the surface, respectively. Combining equations 1 and 2 and solving for r_e, we obtain

$$r_e = (\rho_0\lambda_0/2I_0)^{1/2}(V_m/V_0)h_e \exp(\alpha h_e)\dot{u}_z \quad (3)$$

Because of the steep velocity gradient near the surface of the ice sheet, ρ_0 and λ_0 are not clearly defined, varying greatly within one wavelength of the surface. We will assume values appropriate within approximately a quarter wavelength of the surface; since r_e depends only upon the square roots of these quantities, the uncertainty involved is negligible compared to calibration errors. Assuming a frequency for R_e of 120 Hz and taking variations of velocity and density with depth typical for West Antarctica, we estimate $\rho_0 = 0.4$ g/cm³ and $\lambda_0 = 10$ meters. We also find, for a shot depth of 4 meters, $V_m/V_0 = 2$.

To estimate the value of I_0, we refer again to the study by Robin [1958]. Examination of his Table 26 (p. 64) supports his suggestion that in estimating I_0, as in calculating absorption, the results for relatively thin ice are affected by AGC action. If we consider only stations where the ice thickness was greater than 900 meters, and also neglect those stations for which Robin considers the results abnormal, we find that all values of energy generation lie between 1 and 2×10^5 ergs/ster in 1 wavelength for a charge of 360 grams. For the standard West Antarctic charge of 0.9 kg, we will accordingly assume $I_0 = 3.5 \times 10^5$ ergs.

TABLE 3a. Amplitudes and Reflectivities for Single Stations

Station	Record	Amplitude of R_e Trace, mm	Amplitude of R_e Ground Vel., $\times 10^{-6}$ cm sec^{-1}	Normalizing Factors Chg.	Normalizing Factors Spring.	Normalizing Factors Depth	Normalizing Factors Absorp.	Norm. Ground Vel., $\times 10^{-6}$ cm sec^{-1}	r_e	Ampl. Ratio R_e/R_1
ST 180	1146	5.8	2.3	1	1	1.92	1.2	5.3	0.032	0.16
	1147	1.7	2.2	1	1	1.92	1.2	5.0	0.030	0.12
ST 210	1168	1.7	2.2	1	1	1.92	1.2	5.0	0.030	0.14
	1168	1.8	2.3	1	1	2.19	1.3	6.4	0.038	0.14
	1171	4.9	6.3	0.6	0.5	1.92	1.2	4.3	0.026	
	1171	6.6	8.4	0.6	0.5	2.19	1.3	7.0	0.042	
ST 240	1204	<2	<1.8	1	1	2.26	1.3	<5.4	<0.032	
ST 270	1232	<1.5	<1.4	1	1	2.48	1.4	<4.6	<0.028	
ST 300	1243	<1	<0.9	1	1	2.34	1.3	<2.8	<0.017	
ST 330	1278	6.8	6.3	1	0.5	2.09	1.3	8.2	0.049	0.26
ST 360	1310	<1	<0.9	1	0.5	2.98	1.5	<2.0	<0.012	
ST 390	1322	4.0	3.7	1	1	2.69	1.4	13.9	0.083	0.14
ST 420	1352	3.7	3.4	1	1	3.00	1.5	15.2	0.091	0.20
	1353	6.0	5.5	0.6	0.7	3.00	1.5	10.4	0.062	
ST 450	1383	2.7	2.5	0.6	0.7	3.17	1.5	5.1	0.031	0.19
ST 480	1413	3.8	3.5	0.6	0.7	3.32	1.6	7.8	0.047	0.07
ST 510	1444	5.2	4.8	0.6	0.7	2.52	1.4	6.9	0.041	0.07
ST 540	1476	2.6	2.4	1	1	2.20	1.3	6.7	0.040	0.09
ST 888	1849	1.1	3.3	1	1	2.75	1.4	12.7	0.076	0.13
ST 918	1871	2.2	2.0	1	1	3.70	1.7	12.9	0.077	0.17
	1872	2.5	2.3	0.6	0.7	3.70	1.7	6.1	0.037	0.11
ST 969	1913	2.6	7.7	0.6	0.3	2.73	1.4	5.3	0.032	0.13
MBLT 529	4151	2.4	9.6	1	0.7	1.33	1.1	9.6	0.058	0.10
	4151	2.3	9.2	1	0.7	1.47	1.1	10.4	0.062	0.08
MBLT 565	4162	1.3	5.2	0.8	0.5	2.18	1.3	4.6	0.028	0.09
MBLT 711.5	4250	5.3	6.6	1	0.7	1.29	1.1	6.4	0.038	0.22
MBLT 747	4272	6.9	8.6	1	0.5	1.46	1.1	6.9	0.041	
MBLT 819	4284	2.9	11.6	0.8	0.7	1.29	1.1	8.9	0.053	0.43
MBLT 963	4339	3.6	4.5	1	0.7	1.11	1.0	3.6	0.022	0.09
MBLT 1063	4383	2.0	2.5	1.5	0.5	1.86	1.2	4.1	0.025	0.19
EHT 156	6022	8.1	2.1	1	1	3.05	1.5	9.7	0.058	
	6023	1.7	1.4	1	1	3.05	1.5	6.4	0.038	
EHT 228	6053	3.6	0.9	1	1	2.31	1.3	2.8	0.017	
EHT 300	6077	4.8	3.8	1	1	2.23	1.2	9.9	0.059	0.13
	6082	1.7	1.4	0.6	0.3	2.23	1.2	0.6	0.004	0.10
EHT 324	6097	4.2	3.4	1	0.7	2.40	1.3	7.4	0.044	0.12
	6098	5.3	4.2	0.6	0.7	2.40	1.3	5.6	0.034	0.12
EHT 348	6101	2.6	2.1	1.5	1	3.05	1.5	14.3	0.086	0.23
	6102	6.0	4.8	1	0.7	3.05	1.5	15.4	0.092	0.23
EHT 372	6157	2.8	2.2	1	1	2.69	1.4	8.4	0.050	0.21
	6158	5.1	4.1	0.6	0.7	2.69	1.4	6.4	0.038	0.32
	6159	12.4	3.1	0.6	0.7	2.69	1.4	4.8	0.029	
EHT 396	6163	<1	<0.8	0.6	0.7	2.56	1.4	<1.2	<0.007	
EHT 420	6176	1.7	1.4	1	1	2.11	1.3	3.6	0.022	0.03
	6178	10.0	2.5	1	0.5	2.11	1.3	3.3	0.020	
EHT 444	6180	0.5	2.5	1.5	1	1.85	1.2	8.2	0.049	0.04
	6181	10.6	5.3	1	0.7	1.85	1.2	8.2	0.049	
EHT 468	6191	1.4	2.2	1.5	1	1.63	1.1	6.2	0.037	0.06
	6192	2.1	3.4	1	1	1.63	1.1	6.2	0.037	
	6193	14.5	7.2	1	0.7	1.63	1.1	9.4	0.056	

TABLE 3a. (continued)

Station	Record	Amplitude of R_e		Normalizing Factors				Norm. Ground Vel., $\times 10^{-6}$ cm sec^{-1}	r_e	Ampl. Ratio R_e/R_1
		Trace, mm	Ground Vel., $\times 10^{-6}$ cm sec^{-1}	Chg.	Spring.	Depth	Absorp.			
EHT 492	6194	1.0	5.0	1.5	1	1.40	1.1	11.4	0.068	0.09
	6196	7.5	12.0	1	0.7	1.40	1.1	19.1	0.115	
	6197	12.6	20.2	0.6	0.7	1.40	1.1	19.2	0.115	
EHT 516	6204	3.7	5.9	1	1	1.59	1.1	10.7	0.064	0.08
	6204	3.9	6.2	1	1	1.69	1.2	12.1	0.073	0.09
EHT 540	6207	5.1	8.2	1	0.7	1.78	1.2	11.9	0.071	0.18
EHT 564	6213	3.0	4.8	1	0.7	1.96	1.2	8.0	0.048	0.14
EHT 588	6217	<1	<1.6	1	0.7	2.47	1.3	<3.7	<0.022	
	6218	<2.5	<1.3	1	1	2.47	1.3	<4.1	<0.025	
EHT 612	6223	1.5	2.4	1.5	1	2.25	1.3	10.4	0.062	0.04
	6224	8.1	4.1	0.6	0.7	2.25	1.3	4.9	0.029	
EHT 636	6238	<2	<3.2	1	0.7	2.07	1.2	<6.6	<0.040	
EHT 660	6242	<0.8	<4.0	1	0.7	2.07	1.2	8.2	<0.049	
EHT 684	6246	2.1	3.4	1	0.7	1.74	1.2	4.7	0.028	0.06
	6247	4.4	7.0	0.6	0.7	1.74	1.2	4.2	0.025	
EHT 708	6254	2.9	4.6	1	0.7	1.27	1.1	4.4	0.026	0.07
	6255	5.6	9.0	0.6	0.7	1.27	1.1	5.1	0.031	
	6257	2.6	4.2	1	0.5	1.27	1.1	2.8	0.017	
	6258	5.0	8.0	0.6	0.7	1.27	1.1	4.6	0.028	
EHT 732	6270	<0.5	<0.8	1	0.7	1.90	1.2	<1.3	<0.008	
	6271	<2	<1.0	0.6	0.7	1.90	1.2	<1.0	<0.006	
EHT 756	6275	4.2	2.1	0.6	0.7	1.82	1.2	1.9	0.011	0.04
EHT 780	6310	0.9	1.4	1	0.7	1.84	1.2	2.2	0.013	0.05
EHT 804	6317	<2	<1.0	0.6	0.7	2.05	1.2	<1.1	<0.007	
EHT 828	6322	<1.5	<0.8	0.6	0.7	2.09	1.2	<0.8	<0.005	
EHT 852	6337	<2	<3.2	1	1		1.1	<5.8	<0.035	
EHT 876	6341	<2	<3.2	1	1.6	1.87	1.2	<7.1	<0.043	
EHT 900	6343	<2.5	<4.0	0.6	0.7	2.02	1.2	<4.2	<0.025	
EHT 924	6361	1.7	2.7	1	1	1.99	1.2	6.6	0.040	0.08
	6362	2.7	4.3	0.6	0.7	1.76	1.2	3.7	0.022	0.06
	6362	2.2	3.5	0.6	0.7	1.99	1.2	3.6	0.022	0.05
EHT 948	6366	3.6	5.8	1	1	1.55	1.1	10.0	0.060	
	6367	8.6	13.8	0.6	0.7	1.55	1.1	10.0	0.060	
EHT 972	6370	<0.6	<1.0	1	1	1.26	1.1	<1.3	<0.008	
EHT 996	6401	<0.5	<0.8	1.4	0.5	1.61	1.1	<1.0	<0.006	
EHT 1020	6407	<2	<3.2	1	1	1.34	1.1	<4.6	<0.028	
EHT 1044	6420	5.1	2.6	1.4	0.5	1.08	1.0	2.0	0.012	
EHT 1068	6424	6.9	3.4	1.4	0.5	1.75	1.2	4.9	0.029	0.08
EHT 1092	6429	<3	<1.5	1.4	0.5	1.20	1.0	<1.3	<0.008	
EHT 1116	6433	<2	<1.0	1.4	0.5	1.48	1.1	<1.1	<0.007	
EHT 1140	6442	<1	<1.6	1	0.7	0.95	1.0	<1.1	<0.007	
EHT 1176	6445	1.1	1.8	1	1	1.46	1.1	2.8	0.017	0.05
	6446	6.4	3.2	1	0.5	1.46	1.1	2.6	0.016	0.03
APT 432	432-2	1.4	2.2	1.5	0.5	1.75	1.2	3.4	0.020	0.06
	432-11	1.6	2.6	1.5	0.3	1.63	1.1	2.1	0.013	
APT 636	636-1	1.6	2.6	1.5	1	1.43	1.1	6.0	0.036	0.06
APT 668	668-1	1.0	1.6	1.5	1	1.16	1.0	2.9	0.017	0.04
	668-2	3.3	5.3	0.6	0.7	1.16	1.0	2.6	0.016	
APT 796	796-6	6.6	10.6	1.5	0.3	0.88	1.0	4.1	0.025	
LB 450	140	2.2	4.2	0.6	1	1.78	1.2	5.3	0.032	0.20
AL59 Byrd	8	4.6	16.6	1	0.3	2.02	1.2	12.4	0.074	
AL60 1	11	5.5	3.1	1	0.3	2.77	1.4	3.6	0.022	

TABLE 3b. Amplitudes and Reflectivities for Profiles

Station	Dist. along Profile	Record	Amplitude of R_e		Normalizing Factors				Norm. Ground Vel., $\times 10^{-6}$ cm sec^{-1}	r_e	Ampl. Ratio R_e/R_1
			Trace, mm	Ground Vel., $\times 10^{-6}$ cm sec^{-1}	Chg.	Spring.	Depth	Absorp.			
ST 360	0	1310	<1	<0.9	1	0.5	2.98	1.5	<2.0	<0.012	
	6.5	1316	1.9	1.7	1	1	2.64	1.4	6.4	0.038	0.12
	7.9	1317	1.3	1.2	1	1	2.62	1.4	4.3	0.026	0.08
	10.1	1318	1.0	0.9	1	1	2.55	1.4	3.0	0.018	0.09
	13.7	1319	1.9	1.7	1	1	2.51	1.4	5.9	0.035	0.06
	13.7	1320	4.0	3.7	0.6	0.7	2.51	1.4	5.2	0.031	
EHT 300	0	6077	4.8	3.8	1	1	2.23	1.2	9.9	0.059	0.13
	0	6082	1.7	1.4	0.6	0.3	2.23	1.2	0.6		0.10
	3.3	6091	2.4	1.9	1.5	1	2.30	1.3	8.6	0.052	0.20
	2.8	6092	2.9	2.3	1.5	1	2.32	1.3	10.5	0.063	
	2.0	6093	1.2	1.0	1.5	1	2.30	1.3	4.5	0.027	0.11
	1.3	6094	2.3	1.8	1.5	1	2.28	1.3	8.1	0.049	
EHT 348	0	6101	2.6	2.1	1.5	1	3.05	1.5	14.3	0.086	0.23
	0	6102	6.0	4.8	1	0.7	3.05	1.5	15.4	0.092	0.23
EHT 348E	5.8	6143	1.8	0.4	1	1	3.10	1.5	2.1	0.013	0.17
	7.8	6144	2.8	0.7	1	1	2.88	1.5	2.9	0.017	0.08
	12.0	6145	1.9	0.5	1	1	3.15	1.5	2.3	0.014	0.18
	10.0	6146	0.7	0.2	0.6	1	2.95	1.5	0.5	0.003	
	4.0	6147	4.4	1.1	0.6	1	3.00	1.5	2.9	0.017	0.10
	2.0	6148	4.1	1.0	0.6	1	3.15	1.5	3.0	0.018	
EHT 348N	12.9	6149	<3	<2.4	0.6	1	3.4	1.6	<7.9	<0.047	
	9.9	6150	noise		0.6	1	3.4	1.6			
	6.9	6151	<2.5	<2.0	0.6	1	3.4	1.6	<6.6	<0.040	
	6.9	6152	1.5	1.2	1	0.3	2.86	1.6	2.0	0.012	0.08
	5.9	6153	<3	<2.4	0.6	1	3.4	1.6	<7.9	<0.047	
	3.9	6154	<3	<2.4	0.6	1	3.4	1.6	<7.9	<0.047	
	1.9	6155	<1.5	<1.2	0.6	1	3.4	1.6	<4.0	<0.024	
EHT 396	0	6163	<1	<0.8	0.6	0.7	2.56	1.4	<1.2	<0.007	
	3.2	6173	<1.6	<1.3	1	1	2.66	1.4	<4.7	<0.028	
	2.1	6174	0.4	0.3	1	1	2.36	1.2	0.9	0.005	0.04
	1.0	6175	<0.9	<0.7	1	1	2.67	1.4	<2.7	<0.016	
EHT 444	0	6180	0.5	2.5	1.5	1	1.85	1.2	8.2	0.049	0.04
	0	6181	10.6	5.3	1	0.7	1.85	1.2	8.2	0.049	
	2.8	6188	8.6	4.3	1	1	1.94	1.2	10.1	0.061	
	2.1	6189	2.6	4.2	1	1	1.96	1.2	9.9	0.059	
	1.1	6190	4.8	7.7	1	0.7	1.92	1.2	12.4	0.074	
EHT 612	0	6223	1.5	2.4	1.5	1	2.25	1.3	10.4	0.062	0.04
	0	6224	8.1	4.1	0.6	0.7	2.25	1.3	4.9	0.029	
	0.9	6233	1.3	2.1	1	1	2.27	1.3	6.1	0.037	
	1.5	6234	3.6	5.8	1	0.5	2.28	1.3	8.5	0.051	
	2.2	6235	1.0	1.6	1	0.5	2.30	1.3	2.4	0.014	0.03
	2.8	6236	1.4	2.2	1	0.5	2.30	1.3	3.4	0.020	0.03
EHT 708	0	6254	2.9	4.6	1	0.7	1.27	1.06	4.4	0.026	0.07
	0	6255	5.6	9.0	0.6	0.7	1.27	1.06	5.1	0.031	
	0	6257	2.6	4.2	1	0.5	1.27	1.06	2.8	0.017	
	0	6258	5.0	8.0	0.6	0.7	1.27	1.06	4.6	0.028	
	0.9	6264	3.1	5.0	1	0.5	1.25	1.1	3.3	0.020	0.06
	1.6	6265	2.5	4.0	1	0.5	1.28	1.1	2.7	0.016	0.17
	1.3	6266	2.6	4.2	1	1	1.27	1.1	5.6	0.034	0.07
	0.6	6267	2.2	3.5	1	1	1.23	1.1	4.5	0.027	0.07

TABLE 3b. (continued)

| Station | Dist. along Profile | Record | Amplitude of R_e | | Normalizing Factors | | | | Norm. Ground Vel., $\times 10^{-6}$ cm sec^{-1} | r_e | Ampl. Ratio R_e/R_1 |
			Trace, mm	Ground Vel., $\times 10^{-6}$ cm sec^{-1}	Chg.	Spring.	Depth	Absorp.			
EHT 756	0	6275	4.2	2.1	0.6	0.7	1.82	1.2	1.9	0.011	0.04
	2.7	6285	<1.7	<0.9	0.8	0.5	1.99	1.2	<0.8	<0.005	
	1.7	6286	<1.7	<0.9	1	0.5	1.99	1.2	<1.0	<0.006	
	4.4	6294	<1.2	<0.6	1	1	1.97	1.2	<1.4	<0.008	
	3.7	6295	1.8	0.9	1	0.3	1.76	1.2	0.6	0.003	0.03
EHT 828	0	6322	<1.5	<0.8	0.6	0.7	2.09	1.2	<0.8	<0.005	
	1.0	6333	1.5	0.8	1	0.5	1.78	1.2	0.8	0.005	0.02
	1.7	6334	<1.4	<0.7	1	0.5	2.01	1.2	<0.9	<0.006	
	2.0	6335	2.4	1.2	1	0.5	1.83	1.2	1.3	0.008	
	3.0	6336	<1.5	<0.8	1	0.5	2.03	1.2	<0.9	<0.006	
EHT 972	0	6370	<0.6	<1.0	1	1	1.26	1.1	<1.3	<0.008	
	0.9	6386	2.2	3.5	1.5	0.5	1.28	1.1	3.6	0.022	0.05
	1.6	6387	3.4	17.0	1	0.5	1.21	1.0	10.7	0.064	
	2.2	6388	8.4	13.4	1	0.5	1.19	1.0	8.3	0.050	
	2.9	6389	2.2	35.2	1	0.5	1.18	1.0	21.5	0.129	0.13
	3.9	6390	11.2	17.9	1	0.5	1.08	1.0	9.9	0.059	0.13
	4.9	6391	3.9	6.2	1	0.3	1.10	1.0	2.1	0.013	0.04
	5.9	6392	0.7	1.1	1	1	1.09	1.0	1.2	0.007	
	6.9	6393	2.9	4.6	1.5	1	0.99	1.0	6.9	0.041	
	7.9	6394	<1.9	<3.0	1.5	1	1.07	1.0	<4.9	<0.029	
	9.9	6395	<0.7	<1.1	1.5	1	2.18	1.3	<4.6	<0.028	
	10.9	6396	<0.4	<0.6	1.5	1	2.35	1.3	<3.0	<0.018	
	12.9	6397	0.9	1.4	1	1	2.05	1.2	3.6	0.022	0.01
EHT 1020	0	6407	<2.1	<3.4	1	1	1.34	1.1	<4.8	<0.029	
	0.9	6415	0.6	3.0	1	0.5	1.35	1.1	2.1	0.013	0.04
	1.6	6416	4.1	6.6	1	0.5	1.40	1.1	5.0	0.030	0.05
	2.2	6417	3.1	5.0	1	0.5	1.45	1.1	3.9	0.023	

Since ground velocities have been normalized to the standard charge and to $h_e = 1$, we can now express the reflection coefficient directly in terms of the normalized velocities, $\dot{u}_z'$, by substitution into equation 3, obtaining, closely enough,

$$r_e = 0.006 \, \dot{u}_z'$$

where $\dot{u}_z'$ is expressed in units of 10^{-6} cm/sec. Values of r_e calculated from this equation are listed in Table 3.

Observed velocities of ground motion (Table 3) range from 0.5×10^{-6} to 22×10^{-6} cm/sec, and the geometric mean is about 6×10^{-6} cm/sec. Correspondingly, we have $0.003 < r_e < 0.13$ and a mean reflection coefficient $\bar{r}_e$ of about 0.035.

Relative Amplitudes

In considering the relative amplitudes of R_e and R_1, we shall concern ourselves here only with the mean for all stations, since we have no independent way of estimating the variations in reflectivity at the base of the ice. The over-all geometric mean amplitude ratio is 0.1 after spreading corrections are applied. The problem then reduces simply to estimating a mean reflection coefficient, $\bar{r}_1$, at the base of the ice.

If R_1 were typically impulsive, we could adopt a representative value for the reflection coefficient at an ice-rock interface. *Robin* [1958, Table 22, p. 55] gives a coefficient for energy of 0.31 for sedimentary rock and 0.48 for gabbroic rock; we might then reasonably take $\bar{r}_1 = 0.6$, yielding $\bar{r}_e = 0.06$. Since R_1 is not, in reality, a simple reflected pulse, we can assume instead that $\bar{r}_e \leq 0.06$.

It is also possible to estimate a minimum acceptable value for $\bar{r}_e$. Examination of the reflection records reveals that the bottom echo typically lasts for two or three tenths of a second, or ten to fifteen times as long as the shot-generated pulse. Let us assume, conservatively, that the whole reflected signal contains one-third of the incident energy, the rest having been lost by scattering and transmission into the deepest bottom layer. The mean energy in the reflection is thus about 0.03 times that in the incident wave. From the seismograms it is clear that the maximum amplitude in R_1 is at least twice the minimum, hence that the maximum energy is at least 1½ times the mean. It follows that $\bar{r}_1^2$ is at least 0.04, and thus that $\bar{r}_e \geq 0.02$.

The estimate of $\bar{r}_e$ obtained above from consideration of absolute amplitudes falls well within the limits set by the analysis of amplitude ratios. We conclude, therefore, that the calibration constants and correction factors determined above are not seriously in error, and that individual reflection coefficients can be considered reliable within the quoted uncertainty.

The reflection coefficient for specular reflection at the boundary between homogeneous mediums is related to the wave velocities V_e below the boundary and V_i above it, and the corresponding densities ρ_e and ρ_i, by the expression

$$r_e = \frac{\rho_e V_e - \rho_i V_i}{\rho_e V_e + \rho_i V_i} \qquad (4)$$

The quantity r_e will be positive or negative depending upon whether the acoustic impedance ρV increases or decreases across the boundary from above. It is not possible, however, to assign a sign to the observed values of r_e. We should then write, instead of equation 4,

$$r_e = \frac{|\rho_e V_e - \rho_i V_i|}{\rho_e V_e + \rho_i V_i}$$

For a pure ice boundary, this obviously reduces to

$$r_e = \frac{|V_e - V_i|}{V_e + V_i} \qquad (5)$$

If the material below the reflecting surface comprises ice contaminated by rock fragments, both the density and the P-wave velocity in the material will depend upon the proportional content of rock, designated by q. However, *Röthlisberger* [1971] has shown, using equations of *Bruggeman* [1937], that the change in P-wave velocity is slight, assuming isolated spheres of granite imbedded in the ice up to a 20% rock content by volume. Thus for a contaminated-ice boundary we have, nearly enough,

$$|r_e| = \frac{|\rho_e - \rho_i|}{\rho_e + \rho_i}$$

Taking the rock:ice density ratio to be 3, we then have

$$\rho_e = (1 + 2q)\rho_i$$

whence

$$r_e = q/(1 + q)$$

or ignoring second-order quantities in q, simply

$$q = r_e$$

Using this simple approximation, we find that the observed reflectivities (Table 3) would correspond to rock contents in the reflecting layer of from 0.3 to 13%, with a mean of about 4%. If the model of the reflector as a zone of dirty ice is correct, however, the assumption of specular reflection from the surface of a homogeneous layer is probably not justified, and we would expect energy losses due not only to transmission through the boundary, but also to scattering from irregularities in the surface. On the other hand, the degree of coherence between traces on individual records is great enough to justify the assumption the R_e does not arise by reflection from widely spaced, individual scattering centers. This also accords with glaciological expectations, since the linear dimensions of included morainal particles would surely be very small compared with the 30-meter wavelength of the seismic wave. We are thus concerned with scattering due to the roughness, on a scale comparable to a wavelength, of a continuous surface.

Since little is known of the method of implantation of debris into the ice, the roughness is difficult to guess. If transport along shear planes is a necessary mechanism (as is discussed below), we might well expect that irregularities of the surface would be small compared to a wavelength, and that scattering losses would not be large. Nevertheless, we should keep in mind that at least some of the R_e amplitudes may be significantly reduced from this cause, and that estimated percentages of included rock should consequently be taken as minimum values.

Another factor that affects r_e is the sharpness of the boundary. A gradual (on the scale of a wavelength) increase in rock content with depth would produce no echo. Here again, the calculated percentage of morainal material would be a lower limit on the actual amount. Observations at the edge of the ice sheet, where a vertical section through the ice can be observed, do indicate a gradual increase of debris content with depth. From a study by

Evteev [1959] at the Bunger Hills on the Knox Coast, the percentage of morainal material in a 40-meter section of dirty ice increased from 0 to about 3% over the upper 30 meters, by another 1% on the next 5 meters, and then by some 8% in the bottom 5 meters. With this kind of a distribution, the value of q calculated from an observed reflection would be substantially less than the total of 12%, probably more in the neighborhood of 8%.

Part of the observed variations in r_e undoubtedly result from these factors. Nevertheless, differences in scattering losses and sharpness of gradational boundaries could hardly explain the wide range of observed r_e values or the distinct geographical patterns (see below). It remains probable that reflectivity variations primarily reflect differences in the mean rock content of a debris-laden band of ice.

For pure-ice models, we find, corresponding to the observed range of r_e, velocity differences between 80 and 800 m/sec, and a mean of about 300 m/sec. Since the maximum differences attributable to temperature and anisotropy are 300 and 200 m/sec, respectively, we conclude that a density contrast occurs at least at stations where R_e is stronger than average, and probably at most stations. Even a combination of maximum temperature and anisotropic effects does not appear sufficient to produce the largest observed values of r_e in the absence of a density contrast.

WAVE VELOCITIES IN THE BASAL LAYER

In earlier studies [*Bentley and Ostenso*, 1961; *Behrendt*, 1963], wave velocities in the basal layer were estimated from mean velocities for reflected waves traveling through the whole ice sheet, assuming the ice above the R_e reflector to be isotropic and everywhere at a temperature below the melting point. It was P-wave velocities around 3600 m/sec thus obtained that first suggested the possibility of a zone of ice at the pressure-melting point; in later analysis, P- and S-wave velocity estimates were combined in an attempt to determine a mean morainal content on the assumption of a zone of dirty ice.

It has since become apparent that the assumption of isotropy for the major part of the ice sheet is grossly unjustified. A high degree of anisotropy

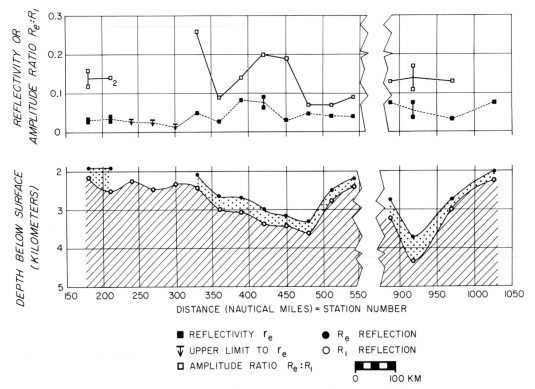

Fig. 3. Basal layer reflectivity, reflection amplitude ratios, and basal layer thickness, Sentinel traverse.

throughout much of the ice column has been observed directly in the deep drill hole at Byrd station [*Gow*, 1968] and by seismic measurements at many stations in West Antarctica [*Bentley*, this volume]. The effect of anisotropy on the mean wave velocities is large enough to mask a velocity difference of a few hundred meters per second in the basal layer. Thus the conclusions as to the nature of the layer that were based on reflection velocities are invalid.

We can, however, put rough limits on V_p in the layer. Examining the reflection profiles given by *Bentley* [this volume], which assumed no velocity differential in the basal layer, we see from the goodness of fit of the anisotropic models that an unaccounted-for velocity effect greater than, say, 50 m/sec could hardly be present. Taking the basal layer thickness as one-tenth that of the entire ice sheet, we find that the velocity in the layer surely cannot differ from the mean velocity by more than ±0.5 km/sec.

AREAL DISTRIBUTION OF REFLECTIVITY

Significant variations in r_e appear along the traverse routes with a fairly regular pattern. On the Sentinel traverse (Figure 3), r_e is relatively great over the deeper parts of the Byrd subglacial basin, i.e. where the ice is more than 2.5 km thick. On the Ellsworth Highland traverse, similarly large reflectivities are found near the crossing with the Sentinel traverse (EHT 348 and ST 420), but a more pronounced maximum in r_e appears over a relative high in the subglacial topography (Figure 4), (stations EHT 420 to 588). Farther to the north, r_e is generally smaller, although high values again appear over rough subglacial topography between stations EHT 924 and EHT 1068 (Figure 5).

The amplitude ratio of R_e and R_1 shows generally similar changes, but with some differences. Along the Sentinel traverse and on the adjacent segments of EHT, the amplitude ratio is mostly largest where r_e

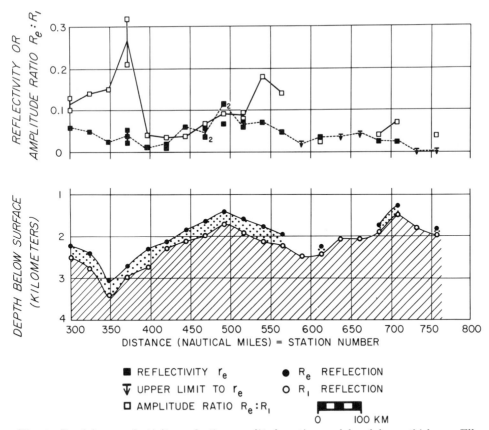

Fig. 4. Basal layer reflectivity, reflection amplitude ratios, and basal layer thickness, Ellsworth Highland traverse, first section.

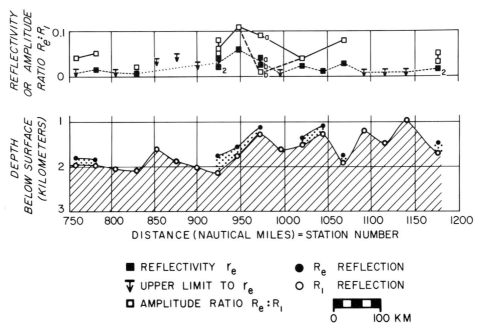

Fig. 5. Basal layer reflectivity, reflection amplitude ratios, and basal layer thickness, Ellsworth Highland traverse, second section.

is largest. It also is relatively great in comparison to r_e, thus indicating a poor reflecting surface at the base of the ice. Where r_e shows a maximum between EHT 420 and EHT 516, on the other hand, the amplitude ratio, while showing a similar relative maximum, is scarcely larger than r_e, indicating a very good basal reflector. Although a calibration error is suggested by the fact that the curves actually cross (implying $r_1 > 1$), there is clearly an abrupt change in the amplitude ratio relative to stations both to the east and to the west that is not reflected in r_e itself. In the subglacially rough northeastern region, an intermediate basal reflectivity is indicated.

In Figure 6, the observed values of r_e (split into octave ranges) are superimposed on a map of bedrock topography. Also shown in the figure is the direction of regional surface slope, which is assumed to be indicative of the direction of ice movement. Upper limits of reflectivity are shown where the evidence for the absence of R_e was significant.

R_e is absent or poorly developed at stations near ice divides and, at least usually, wherever the bedrock surface stands higher than at any other point upstream. It is generally strongest downstream from regions of high subglacial relief, such as the Ellsworth Mountains and the mountains of Marie Byrd Land. The contrast is particularly striking in comparing the reflectivities in the central subglacial basin, where $\bar{r}_e = 0.051$ and few values are less than 0.04, with those in western Ellsworth Land, where $\bar{r}_e = 0.023$ (including stations where only the upper limit on r_e was known), and not a single value exceeds 0.04. The boundary between these reflectivity regions is sharp, and coincides with that between ice flowing northwestward from the Ellsworth Mountains and ice flowing outward from a local divide in Ellsworth Land. Note that the boundary occurs between successive stations on the same traverse, so that calibration errors are minimal.

A necessary (and perhaps sufficient) condition for the development of R_e in a particular place thus appears to be the existence somewhere upstream of a higher bedrock elevation. Also, there are significant variations between adjacent stations that can reasonably be explained in terms of varying rock concentrations, but that would not be expected from a pure-ice boundary. There is thus a clear inference that R_e results primarily from morainal material plucked off topographic irregularities and carried along in the ice. Local sources of debris cannot be ruled out on the basis of the evidence presented here, but it is difficult to imagine any mechanism for raising the material locally several hundred meters above the base of the ice that would be sufficiently widespread to explain the extensive occurrence of the phenomenon.

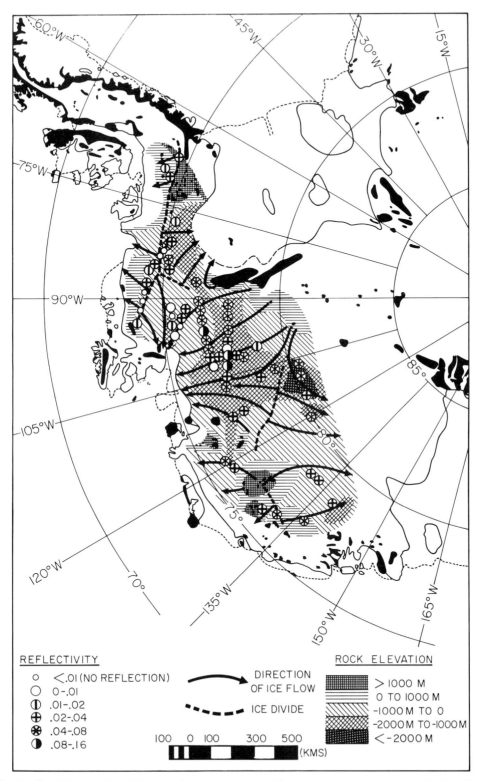

Fig. 6. Basal layer reflectivity superimposed on map of bedrock topography and glacial flow lines.

In some places that are up the 'regional' bedrock slope from the ice divide, the occurrence of R_e could be attributed to relatively local topography. Such is the case, for example, at Byrd station. Gravity measurements indicate a bedrock hump several hundred meters high to the east, even though the regional slope is downward. This interpretation is supported by the observation in 1967 that R_e at new Byrd station, if it occurs at all, has an amplitude less than one-third that at old Byrd station, despite a station separation of only a few miles.

Reflectivities plotted along the wide-angle reflection profiles (Figure 7) mostly do not display significant variations, indicating that changes over distances of a few kilometers are usually small. There are, however, several exceptions. At EHT 348 (and at the common station ST 420) r_e is very large, yet it is consistently small along the entire adjoining profile. Just the reverse is true at ST 360; r_e is too small to be observed at the zero point of the profile, yet is fairly large at the other reflection sites. These two profiles show that r_e can differ markedly over distances of the order of the ice thickness, and, further, that the differences are not dependent upon bedrock topography, since both profiles are in the region of the smallest subglacial relief found anywhere under the grounded ice of West Antarctica. The large variations in r_e also occur at EHT 972; here, however, they are associated with a very rough bedrock surface, one that drops off more than a kilometer between points only 2 km apart along the profile.

Evidence against the pure-ice models is provided by abrupt changes of reflectivity, especially over a smooth bottom. It would seem more difficult to produce a significant change in the vertical tempera-

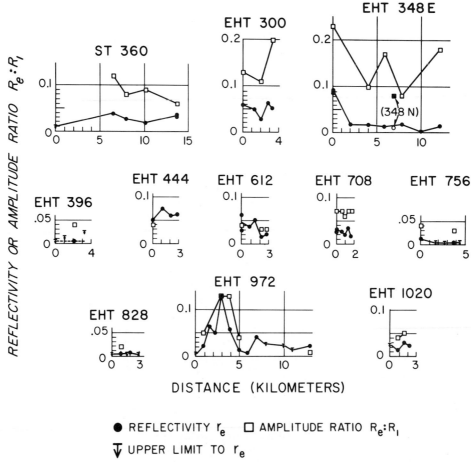

Fig. 7. Basal layer reflectivity and reflection amplitude ratios along wide-angle reflection profiles.

ture gradient or the mean crystal orientation over a horizontal distance of a few kilometers where the subglacial topography is slight than to produce variations in moraine content, which would depend largely upon the source areas.

BASAL LAYER THICKNESS

There is a considerable degree of regularity also in the thickness of the basal layer. On both the traverse sections (Figures 3–5) and the smaller-scale profiles (Figure 8) the near parallelism between the upper and lower surfaces of the layer is obvious. There is a tendency for the layer thickness to be large where the total ice thickness is large, especially when the central Byrd subglacial basin region is compared with western Ellsworth Land to the north (Figure 9). Correlations between these quantities within regions, however, are not statistically significant, nor is there any consistent trend between neighboring stations (connected by arrows in Figure 9).

These general patterns show clearly on a map of layer thickness (Figure 10). The map shows that, as in the case of the reflectivity, there is an abrupt boundary between the central basin region, where the mean layer thickness is 430 meters, and western Ellsworth Land, where it is only 180 meters. The boundary again just coincides with that between ice flowing from different divides. It is further quite reasonable to suspect, on the basis of the properties of R_e, that the northern boundary of the Ellsworth Mountains drainage system lies south of the traverse track near 75.5°S, 97°W rather than intersecting it. The position of the flow line drawn through this area could easily, because of uncertainties in the surface topography, be in error by the required amount.

The components of dip along the wide-angle profiles corresponding to R_e and R_1 are similar in most places, the most striking parallelism being at EHT 972 (Figure 8). On the other hand, at those stations, exclusively in the central Byrd subglacial basin, where an L-shaped geophone spread permitted strike and dip to be calculated, there appears to be no relation between the dips of the R_e and R_1 surfaces, in either magnitude or direction (Table 4). The L spreads and the wide-angle profiles were not at the same stations, however, and there is a weak suggestion that parallelism is poorly developed on those profiles closest to the central basin (in particular, ST 360 and EHT 300). This might be taken as

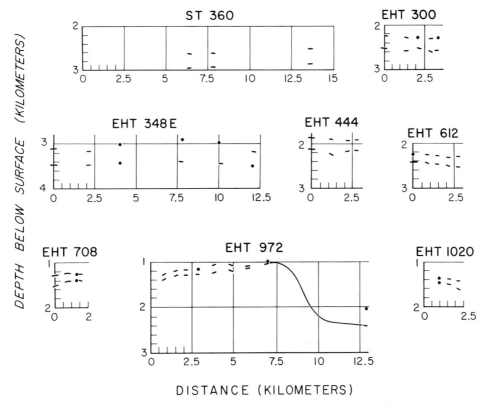

Fig. 8. Basal layer thickness along wide-angle reflection profiles.

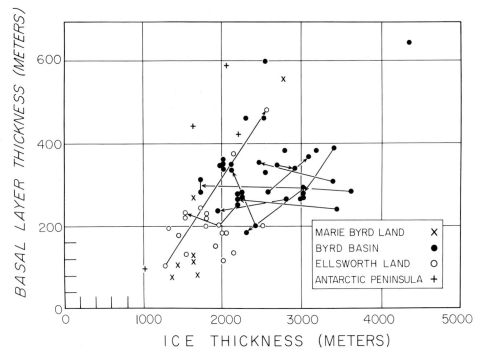

Fig. 9. Basal layer thickness versus total ice thickness. Arrows connect neighboring stations.

an indication that the R_e and R_1 surfaces are more closely related in the northern region than in the central Byrd subglacial basin, but the data sample is much too small to justify any definite conclusions.

There is no apparent relationship between the slope of the ice surface and that of the R_e reflector. The R_e dips on the L spreads are in widely differing directions relative to the surface slope, and R_e on those profiles, at EHT 444 and 1020, shot most nearly up or down slope shows no characteristics that differ in any discernible way from those on the other profiles.

The dip characteristics are particularly inconsistent with the hypothesis of a temperature boundary as the source of R_e. In the smooth central regions of the Byrd subglacial basin, the temperature isotherms in the ice would surely be closely parallel to the base of the ice; in fact, this is just where the dips of R_e and R_1 seem to differ the most.

PURE-ICE BOUNDARIES

An interesting exception to the general pattern is found at EHT 612. On all the reflection seismograms at this station and on the associated wide-angle profile, R_e is short and highly coherent between traces. In the best example (see Appendix), the appearance on each trace is of two similar pulses separated by 0.01 sec. The time interval between pulses is about twice what it would take an impulse to travel from the shot point to the upper surface of the ice. The second impulse may thus represent the wave which has reflected once from the ice surface before propagating downward.

Whether or not this explanation of the detailed characteristics of R_e is valid, the high degree of regularity and coherence of R_e here compared with most other stations (e.g., ST 210, Appendix) is striking. This suggests the possibility of a pure-ice boundary at EHT 612. Further evidence consistent with a pure-ice boundary is found in the near constancy of basal layer thickness and the small dips of the layer surface along the wide-angle reflection profile (Figure 8).

Less favorable to this hypothesis is the apparent change in reflectivity along the profile (Figure 7). On the other hand, the mean value for r_e, 0.034, corresponds, from equation 5, to a velocity difference of 250 m/sec, which is in reasonable agreement with that deduced for either the temperature or the ice-fabric effect. It is possible, considering the calibration errors, that the apparent changes in r_e along the profile are not real, or at least are exaggerated. We can thus conclude that at EHT 612 there is a distinct possibility that R_e is produced by a layer of wet or homogeneous ice at the base of the ice sheet.

Because of this conclusion, it is interesting to see

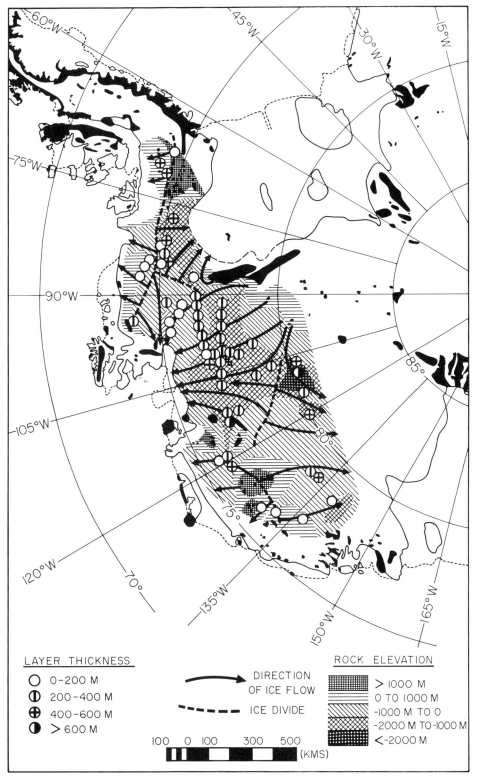

Fig. 10. Basal layer thickness superimposed on map of bedrock topography and glacial flow lines.

TABLE 4. Attitudes of Layer and Basal Surfaces

Station	R_e Dip	R_e Strike	R_1 Dip	R_1 Strike	Difference Dip	Difference Strike
ST 210	3°	170°	6°	250°	−3°	80°
	5	180	6	250	−1	70
ST 330	5	060	11	070	−6	10
ST 390	19	200	2	030	+17	170
ST 420	2	070	3	160	−1	90
ST 450	10	220	4	000	+6	140
ST 480	8	120	0	180	+8	60
ST 510	2	110	1	250	+1	140
ST 540	4	350	5	040	−1	50
ST 969	0		10	160	−10	
MBLT 819	10	?	15	?	−5	20
EHT 924	0		6	150	−6	
EHT 948	0		2	340	−2	

whether the mechanism for producing a basal layer of wet ice suggested by *Lliboutry* [1966], which requires an upward component of ice particle velocity relative to the bottom, could be operative. We first assume that flow lines are parallel in the horizontal plane, originating at the ice divide to the northwest. For stations EHT 612 and 636, *Shimizu* [1964] gives accumulation rates of 34 and 40 g/cm² yr; we will adopt 40 g/cm² yr as a round figure. The distance to the divide is 60 km, and the mean ice thickness is 2.2 km. The resultant steady-state velocity of ice movement would be 10 m/yr at EHT 612, with a mean to the divide of 5 m/yr. The mean gradient of ice thickness in the same direction is no more than 0.01. It follows from Lliboutry's equation 13 that there will be a downward relative motion of the ice at the surface of 0.3 m/yr. Upward motion would only be obtained for velocities greater than 40 m/yr, and it would require a mean velocity from the ice divide of some 80 m/yr to produce a temperate layer 200 meters thick.

As Lliboutry points out, however, the layer can be produced over steeper local slopes and carried downstream without rapid thinning. Halfway from EHT 612 to the divide, gravity anomalies indicate a gradient of ice thickness of 0.025 over a distance of 15 km. The velocity required to produce the observed layer would be around 30 m/yr, still 6 times that estimated above.

The velocity may indeed be this high. Although there are no surface elevation data between EHT 612 and the north end of the Ellsworth Mountains, comparison with Sentinel traverse data shows that a northward drop of several hundred meters must exist. Any reasonable interpolation of the surface contours necessitates a strongly convergent flow around the north end of the mountains, the convergence probably being effective at EHT 612. Measurements on the interpolated contours indicate a mean convergence (the mean depends little upon how the contours are drawn) of about sixfold. If this applied also over the region of steep thickness gradient discussed above, the velocity would be raised from 5 m/yr to the required value of 30 m/yr.

Because of the converging flow, it is very difficult to ascertain just which direction is upslope from EHT 612. The section along the traverse route may, in fact, deviate significantly from the flow line, so that the calculations above are not specifically relevant. We have shown, however, that the high velocity resulting from strongly convergent flow could act, together with a local slope of a steepness and extent known to exist in the vicinity, to effect the Lliboutry mechanism.

Short, coherent-pulse, R_e echoes, which are perhaps similar but are less clearly displayed, were also recorded at MBLT 565 and 819, EHT 372, 708, 756, and 780, and APT 796. The signal-to-noise ratios at MBLT 565 and APT 796 are very poor, so that more extended signals of slightly smaller amplitude could exist. Furthermore, these stations are in mountainous regions where there is good reason to expect moraine.

At MBLT 819, R_e was well recorded. The dips of the R_e and R_1 surfaces are 10 and 15°, respectively, differing in direction by only 20°; considering the indicated irregularity in bottom topography, these differences are probably not too large to be consistent with pure ice boundary. On the other hand, the reflectivity is rather high (0.053, corresponding, for no density contrast, to a velocity difference of 400 m/sec), and the location is downstream from a good potential source of moraine. The situation at EHT 372 is similar: the short pulse characteristics are good and the reflectivity is only slightly high (0.040, corresponding to a velocity difference of 300 m/sec), but the station is surrounded by others at which the proper echo characteristics definitely do not exist. At both MBLT 819 and EHT 372, therefore, the morainal model is preferred, the short pulse form then arising simply from a single well-defined dirty ice layer.

Echoes at EHT 708 more closely fit the pure-ice models. The signal-to-noise ratio is good on several records, coherence is good, $\bar{r}_e = 0.025$ and r_e differs exceptionally little along the associated reflection profile, and the dips on R_e are similar to, but always less than, those on R_1. Taking this station by itself, all the evidence supports a pure-ice model. Furthermore, the location may well be in the region of ice drainage from a local divide, not from the Ellsworth Mountains.

Supporting evidence from neighboring stations is not so good, however. At EHT 684, R_e varies widely in amplitude from trace to trace and shows a double pulse. At all the remaining stations nearby (EHT 732 to 828), the echoes are very weak or unobservable, so that R_e echo characteristics cannot be well determined. Indicated velocity contrasts at these stations range from less than 50 to about 100 m/sec. The reflectivities are thus too small for a sharp velocity boundary, although they could theoretically correspond to velocity changes spread out over a distance comparable to one wavelength. It follows that, whereas the properties of R_e at EHT 708 can be nicely explained in terms of a velocity boundary, it is necessary to rely upon a model that differs significantly at adjacent stations where the velocity gradient must be greatly reduced.

In terms of the Lliboutry temperate-ice mechanism, in particular, the conditions at EHT 708 are not favorable, since, as far as can be determined from the traverse coverage, the ice thickens upstream, both locally and regionally.

At APT 432, the station for which Lliboutry suggests a temperature boundary, the echo duration on two records is 6 to 8 cycles, too long to fit the hypothesis unless the outgoing pulses were atypically drawn out.

CONCLUSIONS

From the examination of the characteristics and distribution of R_e in West Antarctica, several conclusions about the nature of the basal reflecting layer in the ice sheet can be drawn.

1. In the great preponderance of examples, the evidence convincingly demonstrates that R_e must arise from a layer of ice contaminated by foreign material, rather than from a layer of pure ice exhibiting a different wave velocity from that in the ice above, produced either by temperature or crystal fabric effects. The reason is that the pure-ice layer would necessarily have a discrete, continuous upper surface, resulting in an impulsive, coherent reflection, whereas the introduction of contaminants could easily result in multiple, discontinuous 'dirty' layers, giving rise to reverberative, incoherent reflections. The pertinent observations are as follows:

a. In most examples, the duration of the echo is several times that of the outgoing pulse from the shot, and R_e often exhibits more than one amplitude maximum, separated by time intervals as much as ten times the incident pulse length. Furthermore, the echo amplitude sometimes varies greatly from trace to trace on a single record. These basic characteristics of R_e provide the strongest evidence in favor of an impure-ice model.

b. Reflectivities associated with R_e vary widely on a regional basis, the observed range being from 0.003 or less to 0.13, almost 2 orders of magnitude. The maximum reflectivity to be expected from a velocity contrast in pure ice is only about 0.04. Any lesser value could result, depending on the steepness of the velocity gradient constituting the boundary, but there would be no explanation for the higher values, which comprise about half the observations.

c. At several places, abrupt changes in the reflectivity occur over distances of only 1 or 2 km. At two of these places, in particular, where the subglacial relief is very slight, it seems improbable that a mechanism would exist to produce the necessary rapid horizontal change in vertical temperature gradient or fabric modification.

d. In many places there appears to be no relation between the attitudes of the upper and lower surfaces of the basal layer, the upper surface often having the greater dip. Both flow planes and, in the absence of strong, localized heat sources just beneath the ice, isothermal surfaces in the ice would be smoother and flatter than the bedrock surface.

e. There is a striking contrast between the mean reflectivity and mean layer thickness for the central region of the Byrd subglacial basin and the corresponding means for western Ellsworth Land, both being more than twice as great in the Byrd subglacial basin. The boundary between these zones coincides, within the limit of error in estimating flow lines, with the boundary between ice draining the rugged subglacial (and exposed) terrain of the Ellsworth Mountains, and ice flowing outward from local divides over relatively subdued subglacial topography. This observation, however, is not necessarily incompatible with the pure-ice models, which also depend, in different ways, upon the flow history of the ice.

2. It seems highly probable that the contaminating material is morainal debris. This conclusion is supported by the association of thick basal layers and strong reflectivities with areas where there is a

region of high, rugged subglacial topography upstream, and is consistent with all the observations mentioned above. The only other contaminant that might be added to the ice in significant quantities is volcanic ash. Ash layers can be discounted as the primary source of R_e for several reasons.

   a. The only region containing exposed, recent volcanism is Marie Byrd Land, yet the strongest reflections are found in ice that, when at the surface, was somewhere in southeastern West Antarctica. Without tracing flow lines in detail back to the paleo-surface of the ice sheet, however, it is not possible to discount entirely the possibility that the distribution of R_e reflects patterns of atmospheric circulation.

   b. More surely significant are the abrupt local changes in reflectivity; these would hardly be expected for ash layers far from the source.

   c. Steep dips would not be found on ash layers, especially where the base of the ice was level. Parallelism between the R_e and R_1 surfaces would be pronounced, in contrast to the observations.

   d. Since ash layers far from the source would surely be contemporaneous over wide areas, the ice at the top of the basal layer should be regionally the same age. One would then expect a much better correlation between total ice thickness and layer thickness than is found, particularly between neighboring stations on the same flow line (arrows in Figure 9).

   Volcanic eruptions could, on the other hand, be responsible for R_e in a few individual cases. In particular, the unusually high reflector at ST 210, found in ice that may well have been at the surface near the Crary Mountains, might well have such an origin. The double echo at MBLT 529, downstream from the Executive Committee Range, might be another example.

   3. Reflections probably arise from one or several bands of debris, not necessarily thicker than about ten meters, separated by relatively clean ice. Although some lengthening of R_e will result from surface roughness, the longer reverberations require several surfaces of acoustic impedance contrast.

   4. At one station there is a good possibility that a temperature-controlled velocity boundary is the source of R_e. The echoes at EHT 612 and on the adjacent reflection profile are unusually short and regular. Because of strongly convergent ice flow around the north end of the Ellsworth Mountains, flow velocities may be great enough to affect *Lliboutry's* [1966] mechanism even though the station is close to the ice divide. At other stations, where R_e appears to be impulsive and coherent, supporting evidence generally does not favor the temperature hypothesis.

   5. The means of introducing moraine into the ice and then carrying it to levels several hundred meters above the base are not known. As *Lliboutry* [1966] points out, vertical divergence of flow in thickening ice is far insufficient as a mechanism, and it seems necessary to rely upon transport along shear planes. Although there is as yet no other evidence for such shear planes, it at least is possible that they could develop where the subglacial topography is rugged.

   Acknowledgment. Contribution 227 from the Geophysical and Polar Research Center, University of Wisconsin.

REFERENCES

Behrendt, J. C., Seismic measurements on the ice sheet of the Antarctic Peninsula, *J. Geophys. Res., 68,* 5973–5990, 1963.

Bennett, H. F., An investigation into velocity anisotropy through measurements of ultrasonic wave velocities in snow and ice cores from Greenland and Antarctica, Ph.D. thesis, University of Wisconsin, Madison, 1968.

Bentley, C. R., Anisotropy in the West Antarctic ice sheet, this volume.

Bentley, C. R., and F. K. Chang, Geophysical exploration in Marie Byrd Land, Antarctica, this volume.

Bentley, C. R., and N. A. Ostenso, Glacial and subglacial topography of West Antarctica, *J. Glaciol., 3,* 882–911, 1961.

Bruggman, D. A. G., Berechnung verschiedener physikalisher Konstanten von heterogenen Substanzen, III. Die elastischen Konstanten der quasi-isotropen Mischkörper aus isotropen Substanzen, *Ann. Phys. 5. F. 29,* 160–178, 1937.

Evteev, S. A., Determination of the amount of morainic material carried by glaciers of the East Antarctic coast, *Soviet Antarctic Expedition, Info. Bull. 2* (English Edition), 7–9, 1959.

Gow, A. J., H. T. Ueda, and D. E. Garfield, Antarctic ice sheet: Preliminary results of first core hole to bedrock, *Science, 161,* 1011–1013, 1968.

Lliboutry, L., Bottom temperatures and basal low velocity layer in an ice sheet, *J. Geophys. Res., 71,* 2535–2543, 1966.

Robin, G. de Q., Seismic shooting and related investigations, in *Norwegian-British-Swedish Antarctic Expedition, 1949–52, Sci. Results 5, Glaciology 3,* Norsk Polarinstitutt, Oslo University Press, Oslo, 1958.

Röthlisberger, H., Seismic exploration in cold regions, *Cold Regions Sci. Eng. Monograph II–A2a,* Cold Regions Res. Eng. Lab., Hanover, N.H., in press, 1971.

Shimizu, H., Glaciological studies in West Antarctica 1960–62, in *Antarctic Snow and Ice Studies, Antarctic Res. Series,* vol. 2, pp. 37–64, AGU, Washington, D.C., 1964.

Wood, A. B., *A Textbook of Sound,* G. Bell & Sons, London, 1941.

APPENDIX: REFLECTION SEISMOGRAMS SHOWING EARLY REFLECTION (R_e)

(Scale factors are factors by which R_e amplitudes should be multiplied to reduce all records to a common level with respect to amplifier gain, charge size, and springing of shot holes.)

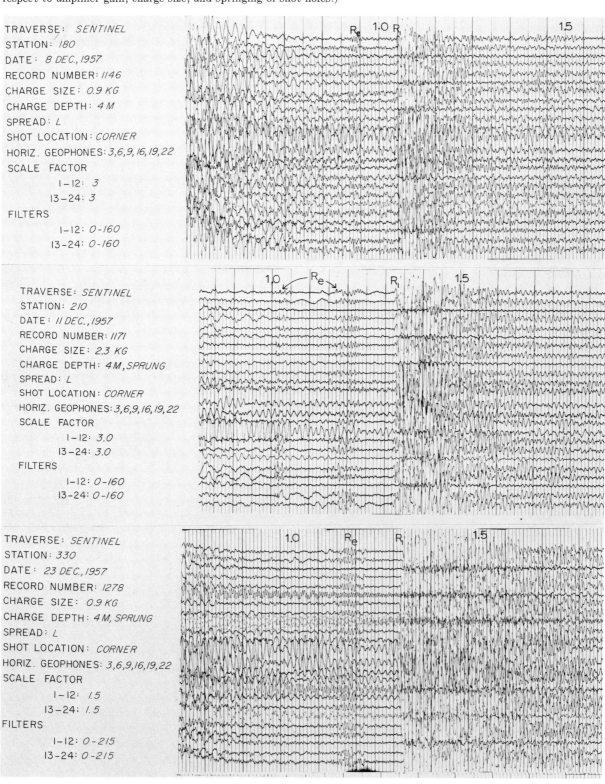

TRAVERSE: SENTINEL
STATION: 180
DATE: 8 DEC., 1957
RECORD NUMBER: 1146
CHARGE SIZE: 0.9 KG
CHARGE DEPTH: 4 M
SPREAD: L
SHOT LOCATION: CORNER
HORIZ. GEOPHONES: 3,6,9,16,19,22
SCALE FACTOR
    1–12: 3
    13–24: 3
FILTERS
    1–12: 0-160
    13–24: 0-160

TRAVERSE: SENTINEL
STATION: 210
DATE: 11 DEC., 1957
RECORD NUMBER: 1171
CHARGE SIZE: 2.3 KG
CHARGE DEPTH: 4M, SPRUNG
SPREAD: L
SHOT LOCATION: CORNER
HORIZ. GEOPHONES: 3,6,9,16,19,22
SCALE FACTOR
    1–12: 3.0
    13–24: 3.0
FILTERS
    1–12: 0-160
    13–24: 0-160

TRAVERSE: SENTINEL
STATION: 330
DATE: 23 DEC., 1957
RECORD NUMBER: 1278
CHARGE SIZE: 0.9 KG
CHARGE DEPTH: 4M, SPRUNG
SPREAD: L
SHOT LOCATION: CORNER
HORIZ. GEOPHONES: 3,6,9,16,19,22
SCALE FACTOR
    1–12: 1.5
    13–24: 1.5
FILTERS
    1–12: 0-215
    13–24: 0-215

TRAVERSE: *SENTINEL*
STATION: *360*
DATE: *27 DEC., 1957*
RECORD NUMBER: *1316*
CHARGE SIZE: *0.9 KG*
CHARGE DEPTH: *3 M*
SPREAD: *ONE CABLE, DUAL*
SHOT LOCATION: *NEAR 12*
HORIZ. GEOPHONES: *3,6,9*
SCALE FACTOR
    1-12: *3*
    13-24: *10*
FILTERS
    1-12: *0-160*
    13-24: *OUT*

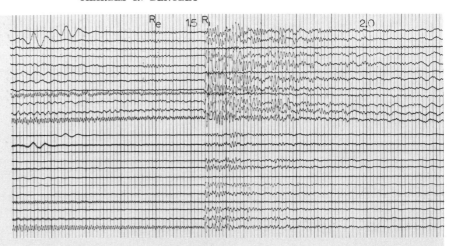

TRAVERSE: *SENTINEL*
STATION: *390*
DATE: *29 DEC., 1957*
RECORD NUMBER: *1322*
CHARGE SIZE: *0.9 KG*
CHARGE DEPTH: *4 M*
SPREAD: *L*
SHOT LOCATION: *CORNER*
HORIZ. GEOPHONES: *3,6,9,16,19,22*
SCALE FACTOR
    1-12: *3*
    13-24: *3*
FILTERS
    1-12: *0-215*
    13-24: *0-215*

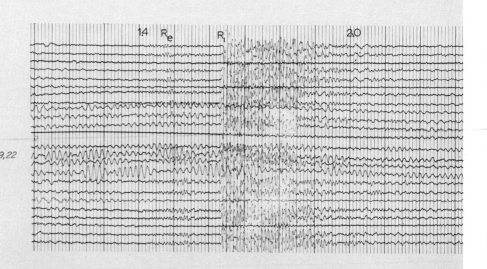

TRAVERSE: *SENTINEL*
STATION: *420*
DATE: *31 DEC., 1957*
RECORD NUMBER: *1353*
CHARGE SIZE: *0.9 KG*
CHARGE DEPTH: *4 M*
SPREAD: *L*
SHOT LOCATION: *CORNER*
HORIZ. GEOPHONES: *3,6,9,16,19,22*
SCALE FACTOR
    1-12: *3*
    13-24: *3*
FILTERS
    1-12: *0-215*
    13-24: *0-215*

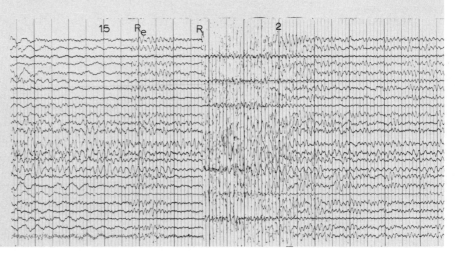

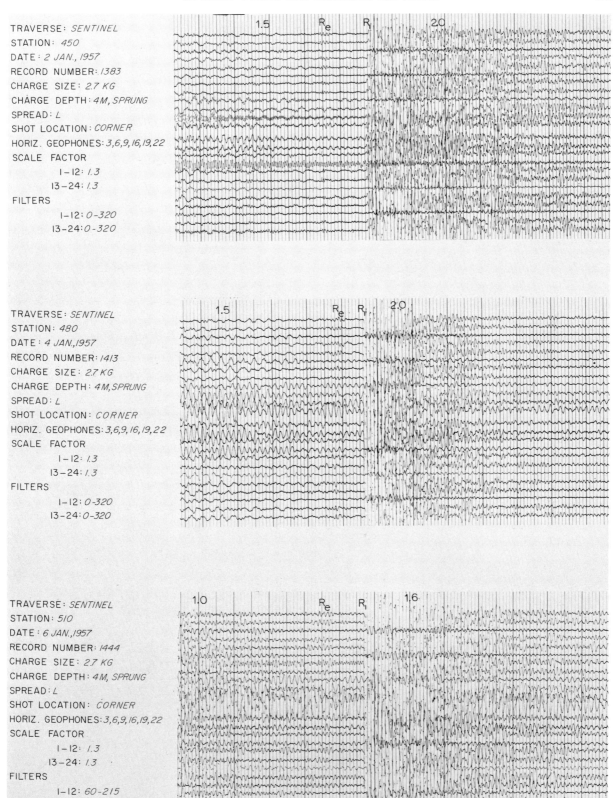

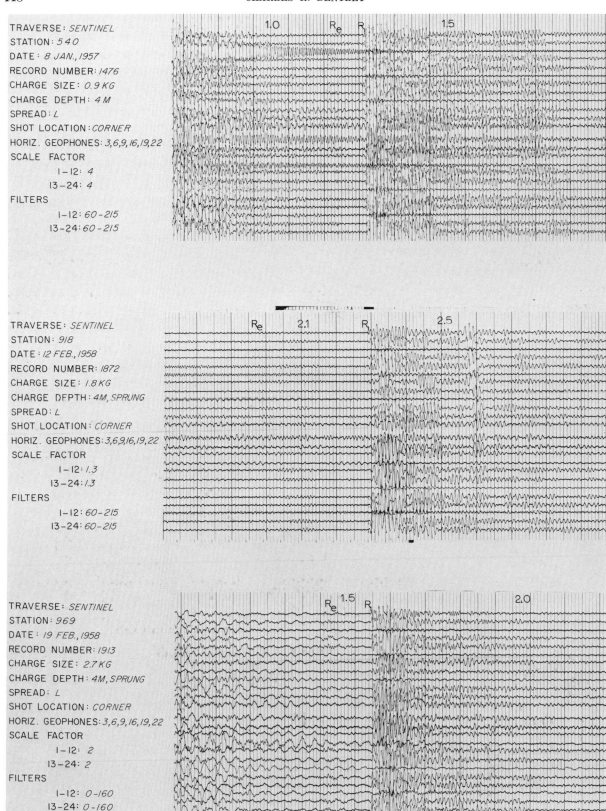

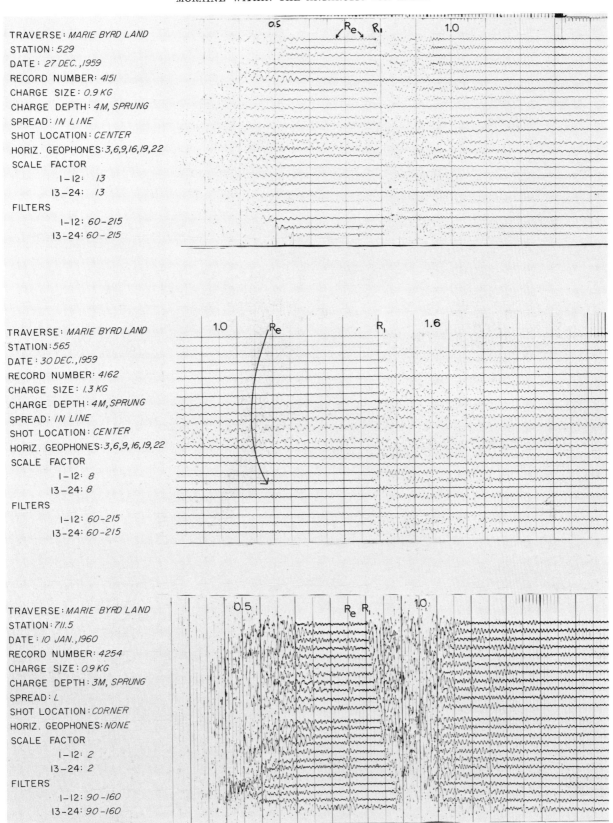

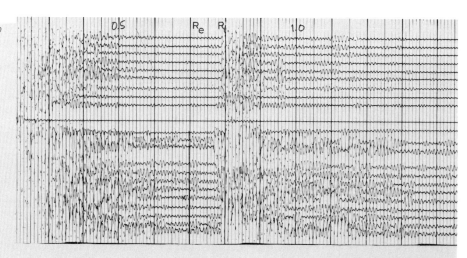

TRAVERSE: MARIE BYRD LAND
STATION: 747
DATE: 18 JAN., 1960
RECORD NUMBER: 4272
CHARGE SIZE: 0.9 KG
CHARGE DEPTH: 3M, SPRUNG
SPREAD: ONE CABLE, DUAL
SHOT LOCATION: NEAR 12
HORIZ. GEOPHONES: NONE
SCALE FACTOR
    1–12: 10
    13–24: 3
FILTERS
    1–12: 90–160
    13–24: 90–160

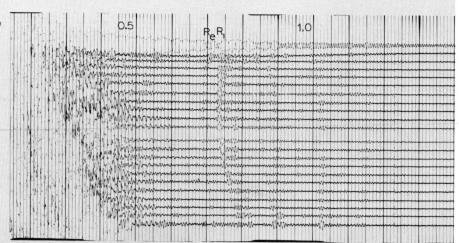

TRAVERSE: MARIE BYRD LAND
STATION: 819
DATE: 23 JAN., 1960
RECORD NUMBER: 4284
CHARGE SIZE: 1.3 KG
CHARGE DEPTH: 3M, SPRUNG
SPREAD: L
SHOT LOCATION: CORNER
HORIZ. GEOPHONES: NONE
SCALE FACTOR
    1–12: 10
    13–24: 10
FILTERS
    1–12: 90–160
    13–24: 90–160

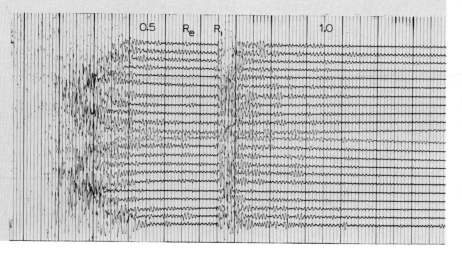

TRAVERSE: MARIE BYRD LAND
STATION: 963
DATE: 31 JAN., 1960
RECORD NUMBER: 4339
CHARGE SIZE: 0.9 KG
CHARGE DEPTH: 5M, SPRUNG
SPREAD: IN LINE
SHOT LOCATION: CENTER
HORIZ. GEOPHONES: NONE
SCALE FACTOR
    1–12: 4
    13–24: 4
FILTERS
    1–12: 90–160
    13–24: 90–160

TRAVERSE: MARIE BYRD LAND
STATION: 1063
DATE: 5 FEB., 1960
RECORD NUMBER: 4383
CHARGE SIZE: 0.4 KG
CHARGE DEPTH: 5M, SPRUNG
SPREAD: ONE CABLE, DUAL
SHOT LOCATION: NEAR 12
HORIZ. GEOPHONES: NONE
SCALE FACTOR
    1-12: 4
    13-24: 4
FILTERS
    1-12: 90-160
    13-24: 90-160

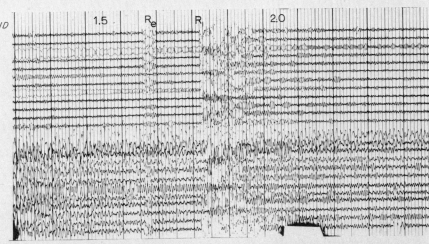

TRAVERSE: ELLSWORTH HIGHLAND
STATION: 156
DATE: 24 NOV., 1960
RECORD NUMBER: 6022
CHARGE SIZE: 0.9 KG
CHARGE DEPTH: 4M
SPREAD: IN LINE
SHOT LOCATION: CENTER
HORIZ. GEOPHONES: 3,6,9,16,19,22
SCALE FACTOR
    1-12: 2
    13-24: 2
FILTERS
    1-12: 210-320
    13-24: 90-215

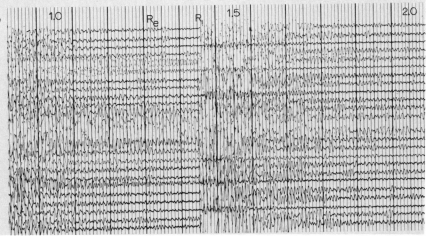

TRAVERSE: ELLSWORTH HIGHLAND
STATION: 300
DATE: 2 DEC., 1960
RECORD NUMBER: 6077
CHARGE SIZE: 0.9 KG
CHARGE DEPTH: 4M
SPREAD: IN LINE
SHOT LOCATION: CENTER
HORIZ. GEOPHONES: 3,6,9,16,19,22
SCALE FACTOR
    1-12: 2
    13-24: 2
FILTERS
    1-12: 60-215
    13-24: 60-215

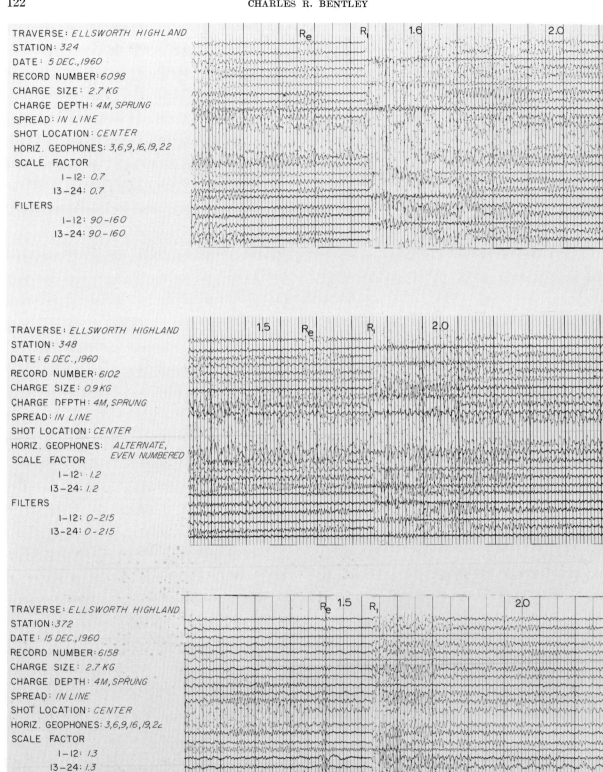

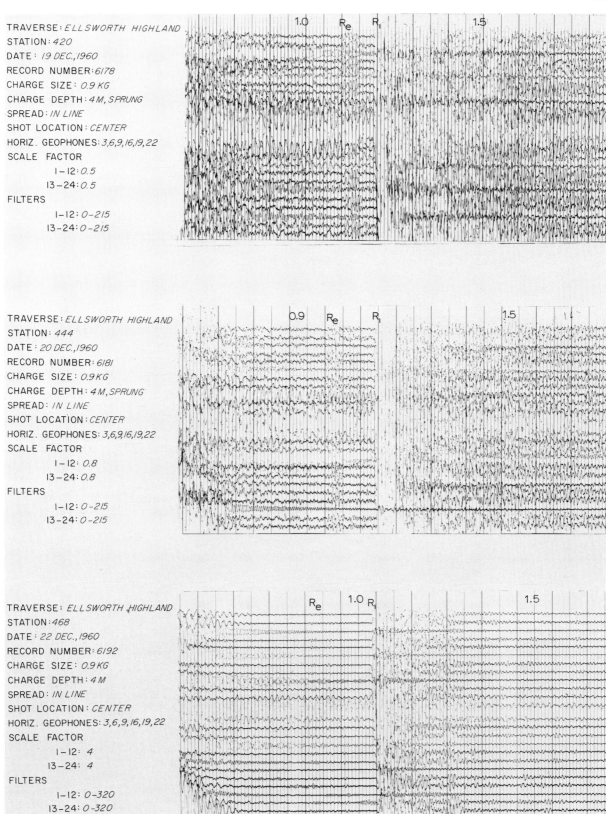

TRAVERSE: *ELLSWORTH HIGHLAND*
STATION: *492*
DATE: *24 DEC., 1960*
RECORD NUMBER: *6196*
CHARGE SIZE: *0.9 KG*
CHARGE DEPTH: *4M, SPRUNG*
SPREAD: *IN LINE*
SHOT LOCATION: *CENTER*
HORIZ. GEOPHONES: *3,6,9,16,19,22*
SCALE FACTOR
  1-12: *3*
  13-24: *3*
FILTERS
  1-12: *0-215*
  13-24: *0-215*

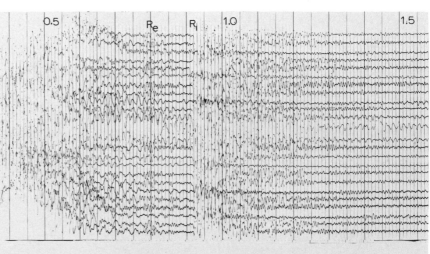

TRAVERSE: *ELLSWORTH HIGHLAND*
STATION: *516*
DATE: *25 DEC., 1960*
RECORD NUMBER: *6204*
CHARGE SIZE: *0.9 KG*
CHARGE DEPTH: *4M*
SPREAD: *IN LINE*
SHOT LOCATION: *CENTER*
HORIZ. GEOPHONES: *3,6,9,16,19,22*
SCALE FACTOR
  1-12: *5*
  13-24: *5*
FILTERS
  1-12: *0-215*
  13-24: *0-215*

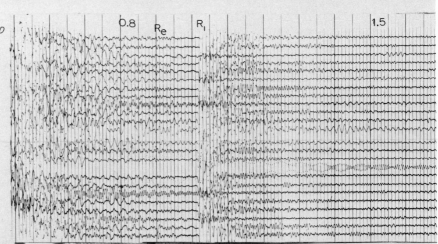

TRAVERSE: *ELLSWORTH HIGHLAND*
STATION: *540*
DATE: *26 DEC., 1960*
RECORD NUMBER: *6207*
CHARGE SIZE: *0.9 KG*
CHARGE DEPTH: *4M, SPRUNG*
SPREAD: *IN LINE*
SHOT LOCATION: *CENTER*
HORIZ. GEOPHONES: *3,6,9,16,19,22*
SCALE FACTOR
  1-12: *4*
  13-24: *4*
FILTERS
  1-12: *0-215*
  13-24: *0-215*

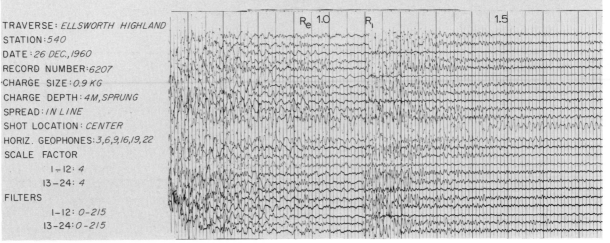

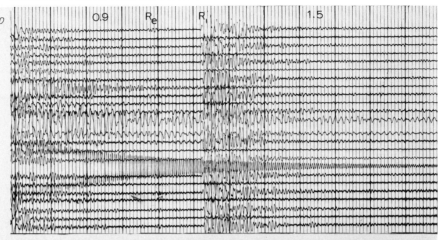

```
TRAVERSE: ELLSWORTH HIGHLAND
STATION: 564
DATE: 28 DEC., 1960
RECORD NUMBER: 6213
CHARGE SIZE: 0.9 KG
CHARGE DEPTH: 4M, SPRUNG
SPREAD: IN LINE
SHOT LOCATION: CENTER
HORIZ. GEOPHONES: 2,4,6,8,10,15
SCALE FACTOR      17,19,21,23
    1-12: 4
    13-24: 4
FILTERS
    1-12: 60-215
    13-24: 60-215
```

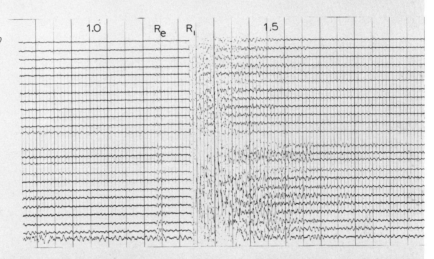

```
TRAVERSE: ELLSWORTH HIGHLAND
STATION: 612
DATE: 31 DEC., 1960
RECORD NUMBER: 6234
CHARGE SIZE: 2.7 KG
CHARGE DEPTH: 4M, SPRUNG
SPREAD: ONE CABLE, DUAL
SHOT LOCATION: NEAR 12
HORIZ. GEOPHONES: NONE
SCALE FACTOR
    1-12: 8
    13-24: 3
FILTERS
    1-12: 0-320
    13-24: 60-215
```

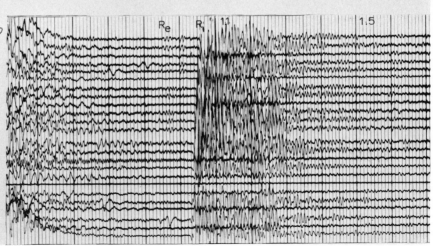

```
TRAVERSE: ELLSWORTH HIGHLAND
STATION: 684
DATE: 6 JAN., 1961
RECORD NUMBER: 6247
CHARGE SIZE: 2.7 KG
CHARGE DEPTH: 4M, SPRUNG
SPREAD: IN LINE
SHOT LOCATION: CENTER
HORIZ. GEOPHONES: 3,6,9,15,17
SCALE FACTOR      19,21,23
    1-12: 4
    13-24: 4
FILTERS
    1-12: 0-215
    13-24: 0-215
```

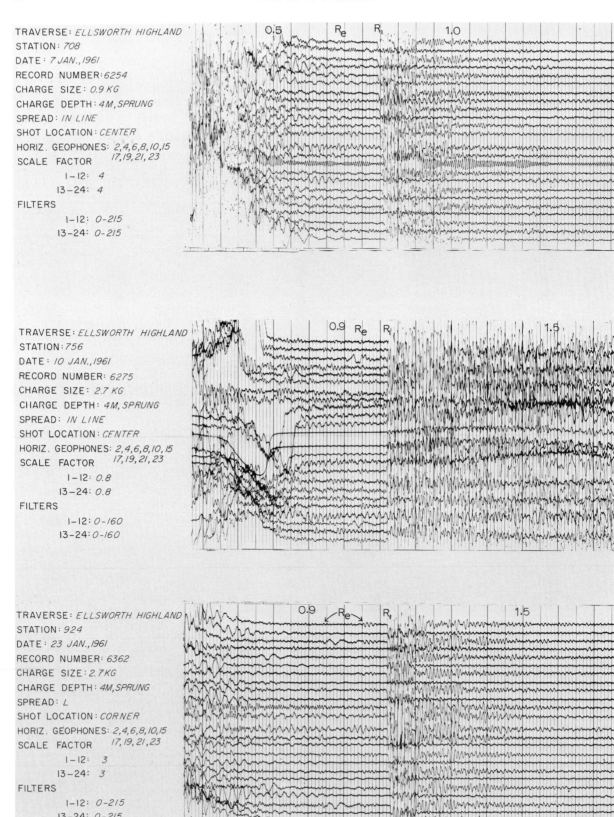

TRAVERSE: ELLSWORTH HIGHLAND
STATION: 948
DATE: 24 JAN., 1961
RECORD NUMBER: 6367
CHARGE SIZE: 2.7 KG
CHARGE DEPTH: 4M, SPRUNG
SPREAD: L
SHOT LOCATION: CORNER
HORIZ. GEOPHONES: 2,4,6,8,10,15
                  17,19,21,23
SCALE FACTOR
    1-12: 3
    13-24: 3
FILTERS
    1-12: 0-215
    13-24: 0-215

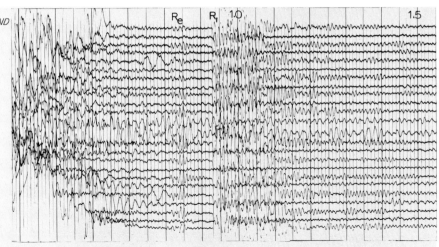

TRAVERSE: ELLSWORTH HIGHLAND
STATION: 972
DATE: 28 JAN., 1961
RECORD NUMBER: 6388
CHARGE SIZE: 0.9 KG
CHARGE DEPTH: 3M, SPRUNG
SPREAD: ONE CABLE, DUAL
SHOT LOCATION: NEAR 12
HORIZ. GEOPHONES: NONE
SCALE FACTOR
    1-12: 10
    13-24: 3.2
FILTERS
    1-12: 0-215
    13-24: 60-215

TRAVERSE: ELLSWORTH HIGHLAND
STATION: 1020
DATE: 31 JAN., 1961
RECORD NUMBER: 6416
CHARGE SIZE: 0.9 KG
CHARGE DEPTH: 3M, SPRUNG
SPREAD: ONE CABLE, DUAL
SHOT LOCATION: NEAR 12
HORIZ. GEOPHONES: NONE
SCALE FACTOR
    1-12: 3.2
    13-24: 10
FILTERS
    1-12: 60-215
    13-24: 0-215

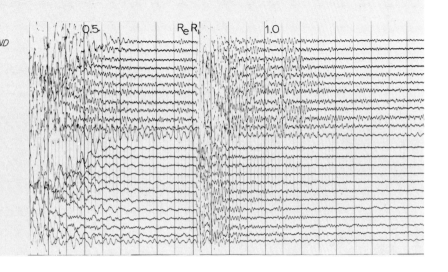

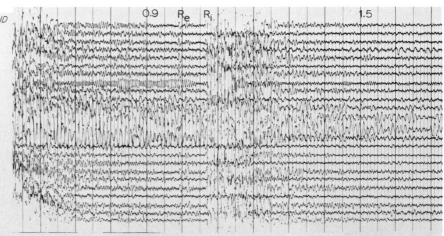

TRAVERSE: ELLSWORTH HIGHLAND
STATION: 1068
DATE: 2 FEB., 1961
RECORD NUMBER: 6424
CHARGE SIZE: 0.5 KG
CHARGE DEPTH: 4M, SPRUNG
SPREAD: IN LINE
SHOT LOCATION: CENTER
HORIZ. GEOPHONES: 2,4,6,8,10,15
SCALE FACTOR 17,19,21,23
  1-12: 1.4
  13-24: 1.4
FILTERS
  1-12: 0-215
  13-24: 0-215

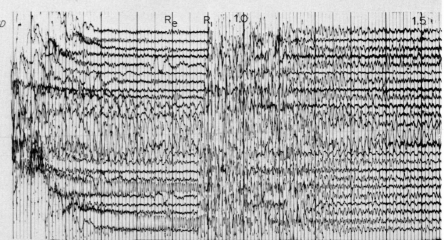

TRAVERSE: ELLSWORTH HIGHLAND
STATION: 1176
DATE: 9 FEB., 1961
RECORD NUMBER: 6446
CHARGE SIZE: 0.9 KG
CHARGE DEPTH: 4M, SPRUNG
SPREAD: L
SHOT LOCATION: CORNER
HORIZ. GEOPHONES: 2,4,6,8,10,15
SCALE FACTOR 17,19,21,23
  1-12: 1.0
  13-24: 1.0
FILTERS
  1-12: 0-215
  13-24: 0-215

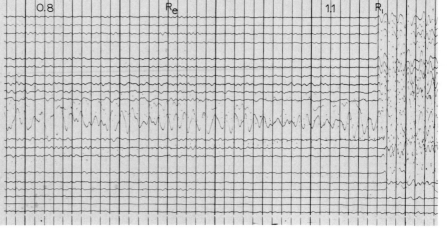

TRAVERSE: ANTARCTIC PENINSULA
STATION: 432
DATE: 28 DEC., 1961
RECORD NUMBER: 432-2
CHARGE SIZE: 0.4 KG
CHARGE DEPTH: 4M, SPRUNG
SPREAD: IN LINE
SHOT LOCATION: CENTER
HORIZ. GEOPHONES: 14,16,20,22,24
SCALE FACTOR
  1-12: ~5
  13-24: ~5
FILTERS
  1-12: 0-215
  13-24: 0-215

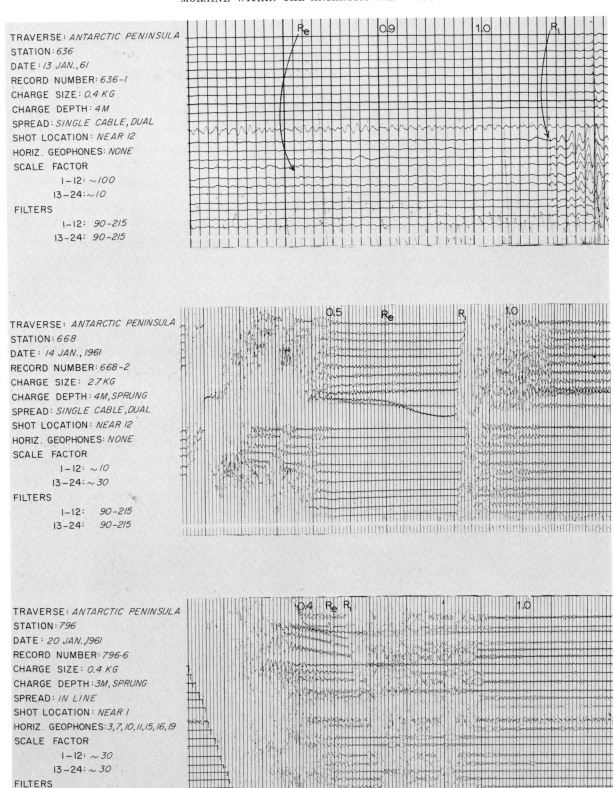

SEISMIC ANISOTROPY IN THE WEST ANTARCTIC ICE SHEET

Charles R. Bentley

University of Wisconsin, Department of Geology and Geophysics
Geophysical and Polar Research Center, Middleton 53562

Sixteen seismic profiles completed at fourteen stations between 1957 and 1961 have revealed the existence of extensive anisotropy in the West Antarctic ice sheet. Analysis based on refracted wave velocities and the angular dependence of mean velocities along reflection paths has shown that in many locations there must be significant differences from the vertical concentration of crystal c axes observed in the deep core from Byrd station. Several more speculative inferences about the nature and extent of the anisotropy are also suggested. The predominant crystal fabric appears likely to be a pronounced concentration of c axes around a single mean direction, the direction varying from place to place. The common occurrence of high seismic velocities at depths of only a few hundred meters, of converted P-to-S wave reflections at normal incidence, and of mean velocity variations that approach the maximum to be found in a single crystal of ice suggests that in much of the region anisotropic ice composes more than half, even as much as 90%, of the ice column, extending to the base of the ice. At a majority of the stations, the data suggest a mean axial direction inclined 30° or more to the vertical and are consistent with an axial azimuth lying in or near the flow plane. But at four interior stations, including Byrd station, the suggested inclination is small and apparently at a large angle with the flow plane. If the anisotropic models that are presented are even approximately correct and if the profiles are representative of the region, there is a tendency toward near-vertical axial orientation at small strain rates in the interior of West Antarctica and toward near-horizontal orientations associated with larger strain rates nearer the coast. At many stations there is evidence for a variation in mean axial orientation with depth; it is tentatively suggested that in general the fabric is relatively constant in most of the bottom half of the ice sheet, and that variations in the upper half and near the base are common.

In the course of the United States oversnow traverse program in West Antarctica, a number of seismic refraction and wide-angle-reflection profiles were completed to measure seismic wave velocities in the ice and in the rock beneath. Since the first three profiles, on the 1957–1958 Sentinel traverse, had indicated a surprisingly low mean compressional wave velocity over ray paths through the entire ice sheet, and since reflected and refracted shear waves on the same traverse and at Byrd station had shown some strong indications of anisotropic propagation [*Bentley and Ostenso*, 1961; *Bentley*, 1964], particular emphasis was placed during the 1960–1961 Ellsworth Highland traverse on the measurement of velocities in the ice. Eleven profiles were completed on the Ellsworth Highland traverse, including four that were some twenty kilometers in length, two of the latter being shot along perpendicular directions at a single station (E-348). In addition, two long, perpendicular profiles were completed at Byrd station during 1957 and 1958.

The results of these sixteen profiles confirm the hypothesis of anisotropic propagation in a major part of the ice column. The evidence for this conclusion can be summarized as follows.

1. At two stations (E-348 and E-972), very high wave velocities were observed from well-recorded refraction arrivals corresponding to depths of 800 meters or so in the ice. These velocities, $V_s = 2035$ m/sec or greater for shear (S) waves (E-348, Figures 7 and 8) and $V_p = 3940$ m/sec for compressional P waves (E-972, Figure 13), each 100 m/sec higher than the corresponding velocity in the solid ice at lesser depths, are much too high to be explained by any known phenomenon in isotropic ice (see review by *Röthlisberger* [1971]). Furthermore, velocity increases of one type of wave were not associated with increases in the other. This is natural

for anisotropic propagation, since the P-wave velocity is usually low in directions in which the S-wave velocity is high and vice versa, but is difficult to explain otherwise.

2. Several other stations exhibit increases of a few tens of meters per second in refracted wave velocities at a depth of a few hundred meters. The corresponding slope changes in the travel-time curves are typically abrupt, indicating velocity changes over depth intervals small compared with the total depth; again, a change in one wave velocity is usually not associated with a change in the other. A velocity increase could conceivably correspond to a density increase or a temperature decrease; let us consider the magnitude of the changes that might result from these causes.

At a depth of 200 meters at Byrd station, the porosity is about 0.5% [*Gow*, 1968]. The maximum possible density increase would occur if the porosity decreased to zero. Using equations for the elastic moduli derived by *Mackenzie* [1950] for a solid containing spherical holes, we find the corresponding changes in V_p and V_s to be about 15 and 5 m/sec, respectively. As a check, by extrapolating *Robin's* [1958] linear velocity-density relation for firn to solid ice, we find a change in V_p of 20 m/sec. Differences would be twice as great at a depth of about 130 meters, and half as great at about 1000 meters. If these estimates are not grossly in error, the effect of a possible density change is too small to explain any of the observations on S waves, but could account for some of the P-wave velocities at shallow depths, as will be discussed below in connection with the individual profiles.

Laboratory and field measurements [*Robin*, 1958; *Bentley*, 1964, unpublished manuscript] both indicate a temperature coefficient of V_p in cold ice equal to -2.3 m/sec °C. The corresponding coefficient for V_s has not been measured accurately, but cannot be much larger. A temperature decrease of the order of 10°C is thus needed to effect the observed velocity increases. There is, however, neither observational nor theoretical reason to expect such a temperature drop at a few hundred meters depth.

3. On all but five of the profiles, the mean velocities for reflected P waves traveling through the ice were significantly different from the velocity for refracted arrivals: usually lower, but not always. The differences are several times too large to be temperature effects, or to be attributed to a low-velocity basal layer [*Bentley*, this volume]. At least three of the longer profiles (E-972, B-58, and S-858; Figures 13, 14, and 17) show a distinct increase in mean velocity with increasing angle of incidence. These increases could not be fit by the effect of a subglacial water layer, as sample calculations showed. On the other hand, all velocity variations with angle of incidence can be fit reasonably well by anisotropic models.

4. Mean velocities of reflected S waves are also generally different from those of refracted S waves. This is true whether the incident waves are compressional or shear.

5. At a majority of stations, waves are recorded that have been converted from P to S motion upon reflection at normal incidence at the base of the ice. Such conversions are to be expected upon reflection from an anisotropic-isotropic boundary if the anisotropic principal axes are asymmetrical with respect to the boundary, but are difficult to produce consistently otherwise.

6. Most convincing of all, on four of the best-recorded profiles (E-348E, E-348N, S-360, and E-756; Figures 7, 8, 19, and 25), double sets of reflected S arrivals arising from incident P waves are seen, the two sets in each case having approximately perpendicular directions of motion. As was mentioned above, this clear evidence for birefringence has previously been reported on vertical reflection shots in the same region of West Antarctica. It is very difficult to imagine any explanation for this phenomenon other than anisotropy. It should be pointed out that clear wave splitting also occurs in reflected S waves arising from incident S waves, although these arrivals have not been analyzed in detail because of the multiplicity of wave-type combinations discussed below.

7. In addition to the seismic evidence there are, of course, direct observations on the deep core from Byrd station that indicate a high degree of preferred crystal orientation at depths below 1200 meters [*Gow et al.*, 1968]. A rapid decrease in bubble concentration at 900-meter depth was also observed [*Gow et al.*, 1968], but, as is indicated above, the corresponding velocity increases at this depth would probably be negligibly small.

There are thus many seismic phenomena that can be explained in terms of anisotropy, and in no other apparent way. Here we attempt to explain these phenomena by devising specific anisotropic models, while pointing out those cases where high P-wave velocities at shallow depths might result from density changes.

NOTATION

Since we will have repeated occasion to refer to travel times and velocities along various ray paths, it will be convenient to provide a symbolic notation at the outset. In general, we will refer to P (compressional) and S (shear) waves propagating with velocities V_p and V_s, respectively. We will designate refracted P and S waves by p and s; and P, S, and converted P to S reflections by R_p, R_s, and R_{ps}, respectively (we mean by 'refracted' waves those that propagate entirely within the ice without reflection). The corresponding travel times and velocities in the ice sheet (measured or hypothetical) will be denoted by t_p, t_s, T_p, T_s, and T_{ps}, and v_p, v_s, $\bar{V}_p$, and $\bar{V}_s$, in the same order, the bars on the last two symbols indicating mean velocities through the ice sheet. We omit a symbol for the mean velocity for the whole R_{ps} arrival, which is not significant in itself, and refer instead to the travel times and velocities measured along the downgoing and upgoing legs: T_{p-}, T_{s+}, $\bar{V}_{p-}$, and $\bar{V}_{s+}$, respectively. Where p or s arrivals correspond to more than one layer within the ice, subscripts will be used; thus, p_2 propagates with velocity $(v_p)_2$ in the second layer below the surface (ignoring the region of variable velocity near the surface). Since the mean velocity and travel time for vertically traveling waves, and time intercepts for refracted arrivals, have particular significance, we will specify them by the subscript zero, e.g. $(T_p)_0$, $(\bar{V}_s)_0$, $(t_p)_0$, etc. We will also wish to refer to the time interval at any distance x between t_p and the p arrival at the same distance that has been reflected once from the snow surface; we call this interval δt_p.

There are three distinct types of body waves that can propagate in a crystal of hexagonal symmetry or, equivalently, a transversely isotropic medium: one compressional (type 1) and two shear waves, the latter being distinguished by polarization planes containing the crystal c axis (type 2) and normal to the crystal c axis (type 3), respectively. Because of the anisotropy, the phase velocities, which we denote by v_1, v_2, and v_3, for wave types 1, 2, and 3, respectively, differ from the corresponding wave front or, for short, wave velocities V_1, V_2, and V_3, which are related to the phase velocity by the relations [*Postma*, 1955; *Bennett*, 1968]:

$$V_k(\phi + \psi) = \sec \psi \, v_k(\phi) \quad k = 1, 2, 3$$

where

$$\tan \psi = \frac{1}{v_k} \frac{dv_k}{d\phi} = \frac{1}{V_k} \frac{dV_k}{d(\phi + \psi)}$$

and φ is the angle between the c axis and the direction of propagation. (Wave fronts are determined by the criterion of stationary phase, but we prefer not to use the term 'group' velocity because phase velocity is a function of direction rather than frequency.) Energy propagates with the wave velocities, and so it is with these that we will be concerned in fitting models to the observations. When considering distributions of crystal axes around some mean direction, we will speak of v_k and V_k defined relative to that mean direction, although such distributions lead to mixing of type 2 and type 3 motion.

For reflected paths, we will refer to theoretical or model velocities by V_{11} for R_p; V_{22}, V_{23}, V_{32}, and V_{33} for R_s; and $V_{1\pm}$, $V_{2\pm}$, and $V_{3\pm}$ for single up-going or down-going legs of wave type 1, 2, and 3, respectively. A plus sign indicates an up-going wave and a minus sign, a down-going wave. Velocities in polycrystalline, isotropic ice will be indicated, where distinction is necessary, by the subscript 'iso.'

To help prevent confusion in the following sections, we note here that field values (observed or hypothetical) are designated by subscripts p and s, model values by numerical subscripts.

We choose a coordinate system fixed at the shot point in the ice sheet with the z axis upward and the x axis toward the recording spread along the line of the seismic profile. We will take θ and I to be the azimuth and inclination, respectively, of the c-axial direction, $\hat{c}$, in normal spherical coordinates, with $\theta = 0$ along the positive x axis, and positive θ counterclockwise as viewed from above (Figure 1). We denote the angle between the z axis and the ray vector, $\hat{v}$, which lies in the x, z plane, i.e. the angle of incidence, by i, and the angle between $\hat{v}$ and $\hat{c}$ by ϕ. When necessary to distinguish between them, angles of incidence for P waves will be designated i_p and those for S waves i_s. It will be useful to indicate the supplement of θ by θ':

$$\theta' \equiv \pm 180° - \theta$$

the plus or minus sign being taken accordingly as θ is positive or negative.

In addition, we will denote the ice thickness by H,

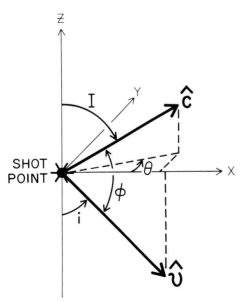

Fig. 1. Coordinate system.

the upper ice temperature by T, the ice accumulation rate by $\dot{A}$, and the snow and ice density by ρ.

FIELD METHODS AND DATA REDUCTION

A fixed recording spread was used for all profiles, all of which were unreversed. Several different types of geophones, having various sensitivities and frequency-response characteristics, were used. Amplitude information has only been used in the most general way, and so no attempt has been made to calibrate the detector combinations. A standard geophone interval of 30.8 meters was used throughout, a spread normally being about 700 meters long. On almost all profiles, some of the geophones were oriented in each of the three component directions, although the number in each orientation varied between profiles. On the longer profiles, it was common practice to connect two identical geophones in series at one position to improve sensitivity, since with a single geophone and quiet recording conditions, background noise was lower than amplifier noise. No other geophone arrays were used, nor was any form of mixing or compositing. Signal-to-noise ratios were almost universally high for all types of seismic energy arrivals through the ice.

Texas Instruments 7000B Seismograph Systems provided reliable, low-noise amplification, a wide choice of filtering, and photographic recording throughout the program. Time checks against National Bureau of Standards radio station WWV assured precision of the instrumental time measurements of 0.1 msec/sec. This precision was needed, since p arrivals could often be picked within a few parts in 10^4 (e.g., Figure 2a). Arrival times for s (Figure 2b, c, and d), and for reflections (Figure 3) could generally be determined to about 1 part in 10^3.

Distances were measured by optical triangulation on the Sentinel traverse, by taping to 10 km and using an accurate vehicle odometer at larger distances for the 1958 Byrd station profile, and by Tellurometer on the Ellsworth Highland traverse. The odometer measurements were reproducible within about 0.1%, whereas the other distance determinations were accurate to 1 part in 10^4 or 10^5 (an optical range finder used for the 1957 Byrd station profile was considerably less accurate, but only an approximate velocity from that profile has been used in this paper). Time and distance observations were thus precise enough that the corresponding errors are insignificant for our purposes; they will henceforth be ignored.

Corrections for the effect of the low wave velocities in the upper firn layers of the ice sheet have been applied by subtracting travel times and distances down to a depth of 200 meters, as calculated from the velocity-depth curves observed at Byrd station, from all reflected-wave travel times. The errors in the mean velocities that result from imperfect knowledge of the local velocity-depth functions do not exceed a few meters per second. Shot-depth corrections were applied to all arrival times by adding the vertical travel time to the surface calculated from the Byrd station velocity-depth curves.

The only significant source of error in the calculation of mean velocities from the reflection shooting lies in the effect of irregularities in the subglacial surface. To minimize this error, vertical reflection shots were made upon completion of each profile at intervals of one-half to several kilometers along the shot line.

CALCULATIONS

For each reflection shot, $\bar{V}_p$ was calculated as a function of i_p from the simple relations

$$\bar{V}_p^2 = \frac{X^2 + (V_{11})_0^2 (T_p)_0^2}{T_p^2} \qquad (1)$$

$$\cos i_p = (T_p)_0 / T_p \qquad (2)$$

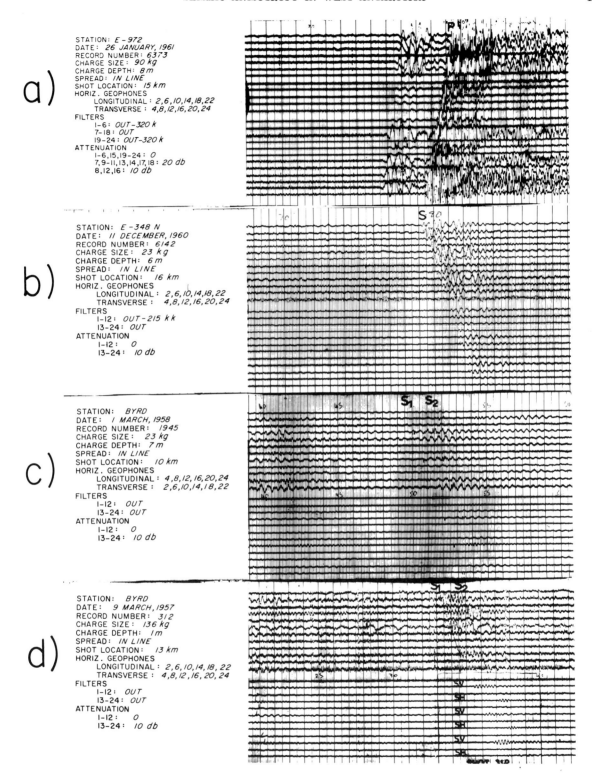

Fig. 2. Sample refraction seismograms. Timing line interval is 0.01 sec. From top to bottom, records show: (a) high-velocity p arrival, profile E-972; (b) high-velocity s arrival, profile E-348N; (c) double s arrival, profile B-58, with high frequency s_2; (d) double s arrival, profile B-57, with s_1 and s_2 of comparable amplitudes.

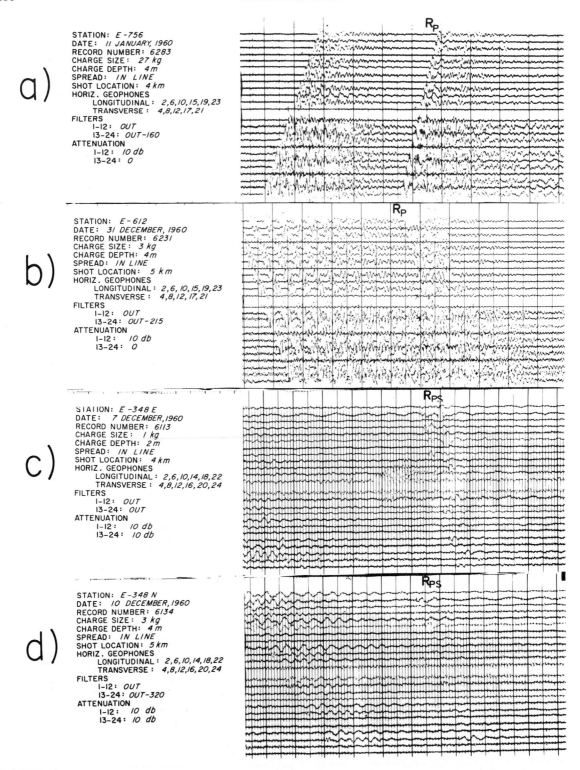

Fig. 3. Sample reflection seismograms. Timing line interval is 0.01 sec. From top to bottom, records show: (a) high-velocity R_p, profile E-756; (b) low-velocity R_p, profile E-612; (c) double R_{ps}, profile E-348E, with earlier arrival primarily transverse; (d) double R_{ps}, profile E-348N, with earlier arrival primarily longitudinal.

where X is the mean shot-detector distance for those geophones on which the reflection was observed, T_p and $(T_p)_0$ are observed reflection times, and $(V_{11})_0$ is an assumed value for the vertical velocity chosen in accordance with the anisotropic model being tested. Note that the calculated values of $\bar{V}_p$ thus vary from model to model depending upon the value of $(V_{11})_0$ associated with each (see, e.g., Figure 7b). The near-surface velocity gradient and anisotropic velocity variations have been ignored in equation 2 as being insignificant in determining i_p. R_s was treated similarly, although, as we shall see below, the interpretation of these reflections was too ambiguous to be of use in anisotropic model selection.

$\bar{V}_{s+}$ was calculated from observed values of T_{ps} by the equation

$$\bar{V}_{s+} = \frac{(T_p)_0 \sec i_s}{2(T_{ps} - T_{p-})} \bar{V}_{1-} \qquad (3)$$

For vertical reflections, which were observed on most of the profiles, equation 3 gives $(\bar{V}_{s+})_0$ as a function of the unknown real vertical velocity $(\bar{V}_{1-})_0 = (\bar{V}_{11})_0$ without any additional assumptions. For oblique angles of incidence, the procedure used was to assume a value of i_p, determine T_{p-} from the measured values of $(T_p)_0$ along the profile and an assumed $\bar{V}_{1-}$, calculate i_s assuming a $V_p : V_s$ ratio of 1.97, determine the length of the up-going S-wave path and the total shot-detector distance, and finally compute $\bar{V}_{s+}$ by comparison with the observed T_{ps} at the calculated distance. In principle, this computation could be refined by a recalculation of i_s on the basis of the computed $\bar{V}_{s+}$, but it was found by numerical example that V_{s+} was not significantly changed by a second approximation.

After several unsuccessful attempts to devise a satisfactory analytical procedure for applying topographic corrections, a graphical procedure, whereby the topography was plotted on a one-to-one scale and the reflected arrival was simply assumed to have followed the minimum time path, was resorted to for both R_p and R_{ps}. Sample calculations under extreme assumptions showed that the effect of a priori uncertainty in wave velocities on the topographic correction was negligibly small.

'Topographic errors' given in the plots of $\bar{V}_p$ versus i_p in a later section indicate the magnitude of the effect on the computed velocity, $|dV_p|$, of the estimated error in determining the relative subglacial elevations, $|dH|$, in the vicinity of the points of reflection, where

$$|d\bar{V}_p| \simeq \bar{V}_p \cos i_p (|dH|/H)$$

The elevation error, arising primarily from the imperfect coincidence between vertical and wide-angle reflection points, is difficult to evaluate, since no statistical analysis is possible. It was taken as the maximum observed local subglacial relief, but no less than 5 meters. It is important to note that vertical and wide-angle reflection points for R_p were about equally spaced, so that real topographic deviations at the wide-angle points should be essentially uncorrelated. Errors in individual estimates of $\bar{V}_p$ should therefore appear chiefly as random deviations of points from smooth curves of $\bar{V}_p$ versus i_p, rather than as displacements of the curves.

The 'topographic errors' for $\bar{V}_{s+}$ determined from R_{ps} are more difficult to evaluate. From equation 3, we can write

$$|d\bar{V}_s| \simeq (\bar{V}_{11})_0 \left(1 + \frac{\cos i_s}{\cos i_p}\right) \frac{|dH|}{H}$$

At small i_s, the coefficient of $|dH|$ is about twice that for $\bar{V}_p$, but $|dH|$ is small. As X increases, $\cos i_s / \cos i_p$ increases without bound, but the distance between successive reflection points approaches zero, so that the relative elevation error also approaches zero. Thus the random part of the 'topographic error' may remain fairly constant, but all $\bar{V}_{s+}$ points for i_s greater than, say, 25° on a single profile could very well be systematically distorted. These points have not been weighted heavily in model selection.

It is clearly important that the correct value of $\bar{V}_{p-}$ be known for the calculation of $\bar{V}_{s+}$ and that the correct value of $(\bar{V}_p)_0$ be used for determining H. In practice, a fixed value of $\bar{V}_{p-}$, independent of angle of incidence, was used for all $\bar{V}_{s+}$ calculations at a given station, corrections later being applied according to the velocities implied by the particular anisotropic model being tested. These corrections are large, as can be seen by comparing curves of $\bar{V}_{s+}$ versus i_s for different models at the same station (e.g. Figure 18c).

The next step was to select the type of anisotropic model of the ice sheet to be used in attempting to fit the observations. The only direct observations on ice crystal fabrics in deep, grounded polar ice are those on the drill cores from Byrd station, where preliminary observations indicate a single, near-vertical maximum [Gow et al., 1968]. As we will see in the discussion of the individual profiles, such a pattern is in violation of the seismic data at most of the stations. We are then left with very little evidence on which to base a guess as to the actual fabric pattern. One, two, and four maximums have all been observed in various circumstances: near the edge of a cold ice cap [Rigsby, 1960; Vallon, 1962], in temperate ice [e.g., Kamb, 1959a; Rigsby, 1960], and in Antarctic ice shelves [Schytt, 1958; Gow, 1963]. Laboratory experiments and theoretical analyses also yield various results [Steinemann, 1958; Kamb, 1959b; 1964; Brace, 1960]. There is some reason to believe that multiple maximums occur primarily by recrystallization near the melting point, so that a single-maximum pattern in the Antarctic ice sheet, where temperatures are mostly well below melting, may be reasonable. At any rate, since we have no good a priori reason to choose any other pattern, and since computations on the basis of a single maximum are the easiest, we will assume such a pattern for model fitting. We can then consider generally how the results might be altered by other assumptions.

In a number of instances, p_2 or s_2 arrivals were recorded, resulting from a velocity increase downward within the ice sheet. Although it is quite unlikely that any velocity increase in the ice is actually discontinuous (i.e., occurring over a depth interval $\ll 10$ meters), a discontinuity was assumed to simplify depth calculations. As a further simplification, the near-surface velocity gradient was approximated by 3 or 4 layers. It is worth noting that, when analyzing the refracted arrival in an anisotropic medium at the boundary with an isotropic medium, Snell's law holds if one uses the wave velocity in the anisotropic medium as determined from the refraction arrivals, although this is not true for noncritical angles of incidence [Bennett, 1968].

In a few instances, the interval of surface-reflected P waves, δt_p, has been used as an aid in interpreting refraction profiles. If v_p does not vary horizontally in the upper part of the ice sheet, then $t_p(x) + \delta t_p = 2t_p(x/2)$, and, for distances large enough that the ray path for the surface reflection reaches essentially the same depth as p, $\delta t_p = (t_p)_0$. If the former relation is found not to hold, v_p must vary horizontally, whereas failure of the data to agree with the latter could mean either that $\partial v_p/\partial x \neq 0$ or that p has penetrated a deeper layer of higher wave velocity that the surface reflection has not reached. If $\partial v_p/\partial x \neq 0$, but is still small compared to $\partial v_p/\partial z$, as it surely must be in the ice sheet, then $\delta t_p(x)$ very nearly equals the intercept, $(t_p)_0$, which would be measured by recording at distance $x/2$. This relation can provide the equivalent of a reverse point in an unreversed profile.

THEORETICAL VELOCITIES AND AMPLITUDE RATIOS

We wish to develop equations by which to predict the mean velocity of waves propagating through a layer in which the crystal axes are perfectly aligned in an arbitrary direction. Velocities as a function of the relative angle between the c axis and the propagation direction in a single crystal have been reported by several authors [Bass et al., 1957; Bennett, 1968; Brockamp and Querfurth, 1964; Jona and Scherrer, 1952; review by Röthlisberger, 1971]. The shapes and amplitudes of the several velocity curves are similar, but the mean values differ considerably. If the curves are normalized to the observed field values for randomly oriented, polycrystalline ice, it thus makes little difference which set is used. In this paper the measurements by Bennett [1968], which agree closely in the polycrystalline mean with the observations in West Antarctica, were adopted (Table 1, Figure 4).

Designating unit vectors by caps over the corresponding symbols, we take $\hat{v}$ along the direction of propagation, $\hat{c}$ along the crystal c axes, and $\hat{u}$ in the direction of particle displacement $\mathbf{u}$. We will relate variables to specific wave types either by the appropriate numerical subscripts, or by subscripts p (P wave) and s (S wave with arbitrary polarization).

We then have

$$\hat{v}_\pm = \sin i\, \hat{x} \pm \cos i\, \hat{z}$$

where the plus and minus signs indicate upward- and downward-traveling waves, respectively, and

$$\hat{c} = \sin I \cos \theta\, \hat{x} + \sin I \sin \theta\, \hat{y} + \cos I\, \hat{z}$$

Since $\phi_\pm$ is the angle between $\hat{v}_\pm$ and $\hat{c}$,

$$\cos \phi_\pm = \hat{v}_\pm \cdot \hat{c} = \sin I \cos \theta \sin i \pm \cos I \cos i$$

It follows that

$$\hat{u}_{3\pm} = \csc \phi_\pm \hat{v}_\pm \times \hat{c}, \qquad (4)$$

$$\hat{u}_{2\pm} = \hat{u}_{3\pm} \times \hat{v}_\pm = \csc \phi_\pm (\cos \phi_\pm \hat{v}_\pm - \hat{c}) \qquad (5)$$

For compressional waves, it is sufficient for the

TABLE 1. Phase Velocities (v_k) and Wave Velocities (V_k) in a Single Crystal of Ice at $-10°C$ as a Function of the Angle of Propagation Relative to the c Axis (ϕ)

ϕ	v_1	V_1	v_2	V_2	v_3	V_3
0	4077	4077	1827	1827	1827	1827
5	4071	4068	1838	1832	1828	1828
10	4055	4043	1868	1844	1831	1830
15	4028	4007	1914	1867	1836	1834
20	3993	3966	1972	1895	1842	1839
25	3953	3923	2034	1934	1850	1846
30	3910	3886	2092	1980	1859	1853
35	3870	3851	2140	2036	1868	1862
40	3837	3826	2172	2098	1878	1871
45	3813	3810	2184	2155/2184	1888	1880
50	3803	3803	2172	2098	1897	1890
55	3805	3805	2140	2036	1906	1899
60	3822	3814	2092	1980	1915	1910
65	3838	3831	2034	1934	1922	1918
70	3861	3851	1972	1895	1928	1926
75	3882	3874	1914	1867	1933	1932
80	3900	3895	1868	1844	1937	1936
85	3911	3910	1838	1832	1939	1939
90	3915	3915	1827	1827	1940	1940

For isotropic, polycrystalline ice: $v_1 = V_1 = 3871$ m/sec.
$v_2 = V_2 = v_3 = V_3 = 1951$ m/sec.

calculation of the wave velocity to know ϕ. For the shear waves, however, we obviously must also know the plane of polarization. Considering first an outgoing shear wave from the source, we resolve it into type 2 and type 3 components, each of which will, in general, give rise upon reflection to a shear wave comprising both types. To simplify the discussion, we will assume reflection from a perfectly rigid boundary at the base of the ice. This is, for our purposes, an adequate approximation to an increase of acoustic impedance at the boundary, since we are concerned only with relative phase reversals between displacement components, not with amplitudes. (Because of the very low density of ice, a higher acoustic impedance in the subglacial material is to be expected unless a layer of water, not just a wet boundary, exists beneath the ice. If there is a water layer, we would need only to interchange solutions applying at less than, and greater than, the critical angle for reflected P waves.) Because of the difficulty in calculating reflection coefficients at an anisotropic-isotropic boundary [*Ahmad*, 1967], we will further assume that anisotropic effects are unimportant in determining the direction of polarization of a reflected wave, i.e. that the basal ice is

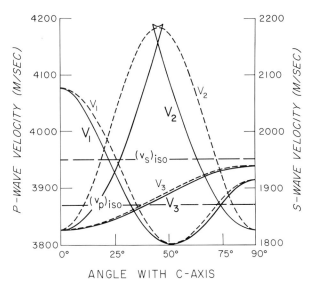

Fig. 4. Phase velocities (V_k, dashed lines) and wave velocities (V_k, solid lines) in single crystal of ice at $-10°C$ as a function of direction of propagation. Upper and lower dashed lines represent V_s and V_p, respectively, for isotropic, polycrystalline ice.

either isotropic, as at Byrd station [*Gow et al.*, 1968], or transversely isotropic with a vertical or horizontal axis of symmetry. (We will have reason to question this assumption later.)

If, then, we have incident on the bottom a shear wave of either type

$$\hat{u}_k = u_{kx}\hat{x} + u_{ky}\hat{y} + u_{kz}\hat{z} \qquad k = 2, 3$$

the reflected S-wave displacement vector $\mathbf{u}_{ks}$ will be

$$\hat{u}_{ks} = u_{kx}\hat{x} + u_{ky}\hat{y} - u_{kz}\hat{z} \qquad i \lessgtr i_c \qquad (6)$$

where i_c is the critical angle for reflected P waves, and the upper and lower signs are to be associated.

The reflected waves $\hat{u}_{ks}$ are of mixed type; we wish to analyze them into type 2 and type 3 components. If we define $\cos \beta_{23}$ as the fraction of $\hat{u}_{2s}$ that is of type 3, then

$$\cos \beta_{23} = \hat{u}_{2s} \cdot \hat{u}_{3+} \qquad (7)$$

From (4), (6), and (7), we find

$$\cos \beta_{23} = -\frac{\sin 2I \sin^2 \theta}{\sin \phi_+ \sin \phi_-} \qquad i < i_c$$

$$= \frac{\sin^2 I \sin 2\theta \cos i}{\sin \phi_+ \sin \phi_-} \qquad i > i_c \qquad (8)$$

Similarly, for the type 2 component of $\mathbf{u}_{3s}$, we find from (5) and (6) that

$$\cos \beta_{32} \equiv \hat{u}_{3s} \cdot \hat{u}_{2+} = \pm \cos \beta_{23} \qquad i \lessgtr i_c$$

This means that, upon reflection, the proportion of a down-going type 2 wave transformed to an upgoing type 3 wave, and of a down-going type 3 wave transformed to an upgoing type 2 wave, is the same:

$$\left| \frac{u_{23}}{u_{22}} \right| = \left| \frac{u_{32}}{u_{33}} \right| = |\cot \beta_{23}| \equiv \cot \beta \qquad (9)$$

the subscript not being necessary so long as we are concerned only with amplitudes.

We next consider an outgoing P wave from the source, converted to an S wave upon reflection. We analyze the reflected S wave

$$\hat{u}_{1s} = \cos i_s \hat{x} - \sin i_s \hat{z}$$

into its type 2 and type 3 components:

$$\cos \alpha \equiv \hat{u}_{1s} \cdot \hat{u}_{2+} = B \csc \phi \qquad (10)$$

$$\sin \alpha \equiv \hat{u}_{1s} \cdot \hat{u}_{3+} = -A \csc \phi \qquad (11)$$

respectively, where

$$A = \sin I \sin \theta$$

$$B = \sin I \cos \theta \cos i - \cos I \sin i$$

Our final need is to determine the transverse component u_T and the longitudinal component u_L of horizontal motion at the surface to which the geophones respond. In the absence of velocity variations near the surface, we would have for any incident wave $\mathbf{u}$:

$$u_T \equiv \hat{u} \cdot \hat{y}$$

$$u_L \equiv \hat{u} \cdot \hat{x} = |\hat{u} \times \hat{y}| \cos i$$

But the ray will be bent near the surface to an angle of incidence i', so that we replace $\cos i$ by $\cos i'$:

$$u_L = |\hat{u} \times \hat{y}| \cos i'$$

whence

$$(u_{2+})_T = -A \csc \phi_+ \qquad (12)$$

$$(u_{3+})_T = B \csc \phi_+ \qquad (13)$$

$$(u_{2+})_L = (u_{3+})_T \cos i'$$

$$(u_{3+})_L = (u_{2+})_T \cos i'$$

It is useful to express these equations in terms of the ratio of transverse to longitudinal displacement for each incoming wave. We denote this ratio by $\tan \delta$; thus

$$\tan \delta_2 \equiv \frac{(u_{2+})_T}{(u_{2+})_L} = -\frac{A}{B} \sec i' \qquad (14)$$

$$\tan \delta_3 \equiv \frac{(u_{3+})_T}{(u_{3+})_L} = -\frac{B}{A} \sec i' \qquad (15)$$

Equations 14 and 15 do not depend on the source of the incoming shear wave, but only on the definition of $\mathbf{u}_2$ and $\mathbf{u}_3$ and on the assumption that there is no rotation of polarization in the ice and snow above the anisotropic zone.

An interesting result follows from (10)–(13):

$$(u_{2+})_T \cos \alpha = (u_{3+})_T \sin \alpha \qquad (16)$$

i.e., for R_{ps} the transverse components of motion for the two shear waves should always be the same in amplitude, regardless of the crystal orientation. This is a simple result that is easily checked experimentally and that provides a good test of the validity of the assumption that the basal ice does not have anisotropic axes skew to the bottom interface.

ANALYSIS OF THE PROFILES

Expected values of the mean reflected velocities and of $\tan \delta_2$ were calculated at 15° intervals in I, θ, and i over all possible attitudes of $\hat{c}$ and $\hat{v}$ in the ice sheet (Table 2, Figures 5 and 6). Quantities calculated for a single crystal should be a reasonably good approximation to those for a random distribution of axial directions within a cone, provided that the apex angle of the cone is not too large. To get a quantitative idea of the closeness of the approximation, velocities for conical distributions with semi-apex angles of 10°, 20°, and 30° have been calculated for various values of ϕ. Velocities were determined by averaging compliances, i.e. by calculating the mean square slowness. Slownesses were closely approximated as functions of ϕ by fitting equations of the form $a + b \cos 2\phi + \cos 4\phi$ (for details, see *Bennett* [1968, pp. 100 ff.]). Simple velocity averaging did not produce significantly different results. The average velocities are presented in Table 3, along with the deviation of each velocity (δv_k) from the

TABLE 2. Velocities in an Ice Sheet at −10°C with Perfect Crystal Alignment as a Function of Axial Orientation and Angle of Incidence

I = 15

θ	i	φ_-	V_{1-}	V_{2-}	V_{3-}	φ_+	V_{1+}	V_{2+}	V_{3+}	V_{11}
0	0	15°	4007	1867	1835	15°	4007	1867	1835	4007
	15	30	3884	1980	1856	0	4077	1827	1827	3980
	30	45	3810	2155	1884	15	4007	1867	1835	3908
	45	60	3814	1980	1912	30	3884	1980	1856	3849
	60	75	3874	1867	1933	45	3810	2155	1884	3842
	75	90	3915	1827	1940	60	3814	1980	1912	3864
	90	75	3874	1867	1933	75	3874	1867	1933	3874
15	0	15	4007	1867	1835	15	4007	1867	1835	4007
	15	30	3884	1980	1856	4	4071	1830	1828	3978
	30	45	3810	2155	1884	16	3999	1874	1836	3904
	45	60	3814	1980	1912	31	3879	1991	1858	3846
	60	74	3870	1873	1932	46	3808	2149	1886	3839
	75	89	3915	1828	1940	60	3814	1980	1912	3864
	90	76	3879	1862	1934	76	3879	1862	1934	3879
30	0	15	4007	1867	1835	15	4007	1867	1835	4007
	15	29	3895	1971	1854	8	4055	1839	1829	3975
	30	44	3813	2146	1882	19	3974	1895	1839	3894
	45	58	3810	2002	1909	33	3863	2014	1861	3836
	60	73	3865	1878	1931	47	3806	2136	1888	3836
	75	88	3914	1829	1940	62	3821	1962	1916	3868
	90	77	3883	1858	1934	77	3883	1858	1934	3883
45	0	15	4007	1867	1835	15	4007	1867	1835	4007
	15	28	3898	1962	1852	11	4036	1849	1831	3967
	30	41	3822	2110	1876	22	3948	1915	1843	3885
	45	56	3806	2025	1905	36	3846	2048	1867	3826
	60	71	3856	1889	1928	50	3803	2098	1894	3830
	75	86	3912	1830	1939	65	3831	1934	1920	3872
	90	79	3891	1849	1936	79	3891	1849	1936	3891
60	0	15	4007	1867	1835	15	4007	1867	1835	4007
	15	26	3915	1943	1849	15	4007	1867	1835	3961
	30	39	3830	2086	1873	25	3923	1934	1848	3876
	45	54	3804	2048	1902	39	3830	2086	1873	3817
	60	68	3843	1911	1925	53	3803	2061	1900	3823
	75	83	3904	1836	1938	68	3843	1911	1925	3874
	90	83	3904	1836	1938	83	3904	1836	1938	3904
75	0	15	4007	1867	1835	15	4007	1867	1835	4007
	15	24	3932	1926	1846	18	3983	1888	1838	3958
	30	37	3840	2061	1869	30	3884	1980	1856	3862
	45	50	3803	2098	1894	43	3816	2136	1880	3810
	60	65	3831	1934	1920	57	3808	2014	1907	3820
	75	79	3891	1849	1936	72	3861	1884	1929	3876
	90	86	3912	1830	1939	86	3912	1830	1939	3912
90	0	15	4007	1867	1835	15	4007	1867	1835	4007
	15	21	3957	1901	1842	21	3957	1901	1842	3957
	30	33	3863	2014	1861	33	3863	2014	1861	3863
	45	47	3806	2136	1888	47	3806	2136	1888	3806
	60	61	3817	1971	1914	61	3817	1971	1914	3817
	75	76	3879	1862	1934	76	3879	1862	1934	3879
	90	90	3915	1827	1940	90	3915	1827	1940	3915

TABLE 2. (continued)

$I = 30$

θ	i	φ_-	v_{1-}	v_{2-}	v_{3-}	φ_+	v_{1+}	v_{2+}	v_{3+}	v_{11}
0	0	30°	3884	1980	1856	30°	3884	1980	1856	3884
	15	45	3810	2155	1884	15	4007	1867	1835	3908
	30	60	3814	1980	1912	0	4077	1827	1827	3946
	45	75	3874	1967	1933	15	4007	1867	1835	3940
	60	90	3915	1927	1940	30	3884	1980	1856	3900
	75	75	3874	1967	1933	45	3810	2155	1884	3842
	90	60	3814	1980	1912	60	3814	1980	1912	3814
15	0	30	3884	1980	1856	30	3884	1980	1856	3884
	15	45	3810	2155	1884	16	3999	1873	1836	3904
	30	60	3814	1980	1912	8	4055	1839	1829	3934
	45	74	3870	1973	1932	17	3991	1879	1837	3930
	60	89	3915	1928	1940	31	3879	1991	1858	3897
	75	77	3883	1958	1934	46	3808	2149	1886	3846
	90	61	3817	1971	1914	61	3817	1971	1914	3817
30	0	30	3884	1980	1856	30	3884	1980	1856	3884
	15	44	3813	2140	1882	18	3983	1884	1838	3898
	30	58	3810	2002	1909	15	4007	1867	1835	3908
	45	72	3861	1884	1929	23	3940	1918	1845	3900
	60	87	3913	1829	1940	36	3846	2048	1867	3880
	75	79	3891	1849	1936	50	3803	2098	1894	3847
	90	64	3827	1938	1920	64	3827	1938	1920	3827
45	0	30	3884	1980	1856	30	3884	1980	1856	3884
	15	42	3818	2124	1878	22	3948	1911	1843	3883
	30	55	3805	2036	1904	22	3948	1911	1843	3876
	45	69	3847	1903	1926	30	3884	1980	1856	3866
	60	83	3904	1836	1938	41	3822	2111	1876	3863
	75	83	3904	1836	1938	55	3805	2036	1904	3854
	90	69	3847	1903	1926	69	3847	1903	1926	3847
60	0	30	3884	1980	1856	30	3884	1980	1856	3884
	15	40	3826	2098	1875	26	3915	1943	1849	3870
	30	51	3803	2086	1896	29	3895	1971	1854	3849
	45	64	3827	1943	1919	38	3835	2073	1871	3831
	60	77	3883	1858	1934	49	3804	2111	1892	3844
	75	89	3915	1828	1937	62	3821	1962	1916	3868
	90	76	3879	1864	1934	76	3879	1864	1934	3879
75	0	30	3884	1980	1856	30	3884	1980	1856	3884
	15	37	3840	2061	1869	30	3884	1980	1856	3862
	30	47	3806	2136	1888	35	3851	2036	1865	3828
	45	57	3808	1991	1911	45	3810	2155	1884	3811
	60	71	3856	1889	1928	57	3808	2014	1907	3832
	75	84	3907	1834	1939	70	3851	1895	1927	3879
	90	82	3902	1838	1938	82	3902	1838	1938	3902
90	0	30	3884	1980	1856	30	3884	1980	1856	3884
	15	33	3863	2014	1861	33	3863	2014	1861	3863
	30	41	3822	2111	1876	41	3822	2111	1876	3822
	45	52	3803	2073	1898	52	3803	2073	1898	3803
	60	64	3827	1943	1919	64	3827	1943	1919	3827
	75	77	3883	1858	1934	77	3883	1858	1934	3883
	90	90	3915	1827	1940	90	3915	1827	1940	3915

TABLE 2. (continued)

I = 45

θ	i	φ_-	V_{1-}	V_{2-}	V_{3-}	φ_+	V_{1+}	V_{2+}	V_{3+}	V_{11}
0	0	45°	3810	2155	1884	45°	3810	2155	1884	3810
	15	60	3814	1980	1912	30	3884	1980	1856	3849
	30	75	3874	1867	1933	15	4007	1867	1835	3940
	45	90	3915	1827	1940	0	4077	1827	1827	3996
	60	75	3874	1867	1933	15	4007	1867	1835	3940
	75	60	3814	1980	1912	30	3884	1980	1856	3849
	90	45	3810	2155	1884	45	3810	2155	1884	3810
15	0	45	3810	2155	1884	45	3810	2155	1884	3810
	15	60	3814	1980	1912	31	3879	1991	1858	3846
	30	74	3870	1873	1932	17	3991	1878	1837	3930
	45	89	3915	1828	1940	11	4036	1849	1831	3976
	60	76	3879	1862	1934	19	3974	1889	1838	3926
	75	61	3817	1971	1914	33	3863	2014	1861	3840
	90	47	3806	2136	1888	47	3806	2136	1888	3806
30	0	45	3810	2155	1884	45	3810	2155	1884	3810
	15	58	3810	2002	1909	33	3863	2014	1861	3836
	30	72	3861	1884	1929	23	3940	1918	1845	3900
	45	86	3912	1830	1939	21	3957	1903	1842	3934
	60	80	3895	1844	1937	28	3898	1962	1852	3896
	75	66	3835	1926	1922	39	3830	2086	1873	3832
	90	52	3803	2067	1899	52	3803	2067	1899	3803
45	0	45	3810	2155	1884	45	3810	2155	1884	3810
	15	56	3806	2025	1905	36	3846	2048	1867	3826
	30	69	3847	1903	1926	30	3884	1980	1856	3866
	45	82	3902	1839	1938	31	3879	1991	1858	3890
	60	85	3910	1831	1939	38	3835	2073	1871	3872
	75	73	3865	1878	1931	49	3804	2111	1892	3834
	90	60	3814	1980	1912	60	3814	1980	1912	3814
60	0	45	3810	2155	1884	45	3810	2155	1884	3810
	15	54	3804	2048	1902	39	3830	2086	1873	3817
	30	64	3827	1943	1919	38	3835	2073	1871	3831
	45	76	3879	1862	1934	41	3822	2111	1876	3850
	60	87	3913	1829	1940	49	3804	2111	1892	3858
	75	81	3899	1841	1937	58	3810	2002	1909	3854
	90	69	3847	1903	1926	69	3847	1903	1926	3847
75	0	45	3810	2155	1884	45	3810	2155	1884	3810
	15	50	3803	2098	1894	43	3816	2136	1880	3810
	30	59	3812	1991	1911	45	3810	2155	1884	3811
	45	68	3843	1911	1925	51	3803	2086	1896	3823
	60	79	3891	1849	1936	59	3812	1991	1911	3852
	75	90	3915	1827	1940	69	3847	1903	1926	3881
	90	80	3895	1846	1936	80	3895	1846	1936	3895
90	0	45	3810	2155	1884	45	3810	2155	1884	3810
	15	47	3806	2136	1888	47	3806	2136	1888	3806
	30	52	3803	2073	1898	52	3803	2073	1898	3803
	45	60	3814	1980	1912	60	3814	1980	1912	3814
	60	69	3847	1903	1926	69	3847	1903	1926	3847
	75	79	3891	1849	1936	79	3891	1849	1936	3891
	90	90	3915	1827	1940	90	3915	1827	1940	3915

TABLE 2. (continued)

I = 60

θ	i	φ_-	V_{1-}	V_{2-}	V_{3-}	φ_+	V_{1+}	V_{2+}	V_{3+}	V_{11}
0	0	60°	3814	1980	1912	60°	3814	1980	1912	3814
	15	75	3874	1867	1933	45	3810	2155	1884	3842
	30	90	3915	1827	1940	30	3884	1980	1856	3900
	45	75	3874	1867	1933	15	4007	1867	1835	3940
	60	60	3814	1980	1912	0	4077	1827	1827	3946
	75	45	3810	2155	1884	15	4007	1867	1835	3908
	90	30	3884	1980	1856	30	3884	1980	1856	3884
15	0	60	3814	1980	1912	60	3814	1980	1912	3814
	15	74	3870	1873	1932	46	3808	2149	1886	3839
	30	89	3915	1828	1940	32	3871	2002	1859	3893
	45	76	3879	1862	1934	19	3974	1889	1839	3926
	60	62	3821	1962	1916	13	4023	1858	1833	3922
	75	43	3816	2136	1880	20	3966	1895	1841	3891
	90	33	3863	2014	1861	33	3863	2014	1861	3863
30	0	60	3814	1980	1912	60	3814	1980	1912	3814
	15	73	3865	1878	1931	47	3806	2136	1888	3836
	30	87	3913	1829	1940	36	3846	2048	1867	3880
	45	80	3895	1844	1937	28	3898	1962	1852	3896
	60	66	3835	1926	1922	26	3915	1943	1849	3879
	75	53	3803	2061	1900	32	3871	2002	1859	3837
	90	42	3818	2116	1877	42	3818	2116	1877	3818
45	0	60	3814	1980	1912	60	3814	1980	1912	3814
	15	71	3856	1889	1928	50	3803	2098	1894	3830
	30	83	3904	1836	1938	42	3818	2124	1878	3861
	45	85	3910	1831	1939	38	3835	2073	1871	3872
	60	74	3870	1873	1932	39	3830	2086	1873	3850
	75	62	3821	1962	1916	44	3813	2149	1882	3817
	90	52	3803	2067	1899	52	3803	2067	1899	3803
60	0	60	3814	1980	1912	60	3814	1980	1912	3814
	15	68	3843	1911	1925	53	3803	2061	1900	3823
	30	77	3883	1858	1934	49	3804	2111	1892	3844
	45	87	3913	1829	1940	49	3804	2111	1892	3858
	60	83	3904	1836	1938	51	3803	2086	1896	3854
	75	73	3865	1878	1931	57	3808	2014	1907	3836
	90	64	3827	1938	1920	64	3827	1938	1920	3827
75	0	60	3814	1980	1912	60	3814	1980	1912	3814
	15	65	3831	1934	1920	57	3808	2014	1907	3820
	30	71	3856	1889	1928	57	3808	2014	1907	3832
	45	79	3891	1849	1936	59	3812	1991	1911	3852
	60	87	3913	1829	1940	64	3827	1943	1919	3870
	75	85	3910	1831	1939	70	3851	1895	1927	3880
	90	77	3883	1858	1934	77	3883	1858	1934	3883
90	0	60	3814	1980	1912	60	3814	1980	1912	3814
	15	61	3817	1971	1914	61	3817	1971	1914	3817
	30	64	3827	1943	1919	64	3827	1943	1919	3827
	45	69	3847	1903	1926	69	3847	1903	1926	3847
	60	76	3879	1862	1934	76	3879	1862	1934	3879
	75	83	3904	1836	1938	83	3904	1836	1938	3904
	90	90	3915	1827	1940	90	3915	1827	1940	3915

TABLE 2. (continued)

I = 75

θ	i	φ₋	V_{1-}	V_{2-}	V_{3-}	φ₊	V_{1+}	V_{2+}	V_{3+}	V_{11}
0	0	75°	3874	1867	1933	75°	3874	1867	1933	3874
	15	90	3915	1827	1940	60	3814	1980	1912	3864
	30	75	3874	1867	1933	45	3810	2155	1884	3842
	45	60	3814	1990	1912	30	3884	1980	1856	3880
	60	45	3810	2155	1884	15	4007	1967	1835	3908
	75	30	3884	1980	1856	0	4077	1927	1827	3980
	90	15	4007	1867	1835	15	4007	1967	1835	4007
15	0	75	3874	1867	1933	75	3874	1867	1933	3874
	15	89	3915	1828	1940	61	3817	1971	1914	3866
	30	76	3879	1862	1934	46	3818	2149	1886	3844
	45	61	3817	1971	1914	33	3863	2014	1861	3840
	60	47	3806	2136	1888	20	3966	1895	1841	3886
	75	34	3857	2025	1863	15	4007	1867	1835	3932
	90	21	3957	1903	1842	21	3957	1903	1842	3957
30	0	75	3874	1867	1933	75	3874	1867	1933	3874
	15	88	3914	1829	1940	62	3821	1962	1916	3868
	30	79	3891	1849	1936	50	3803	2098	1894	3847
	45	66	3835	1926	1922	39	3830	2086	1873	3837
	60	54	3804	2054	1902	31	3879	1991	1859	3826
	75	42	3818	2124	1878	29	3895	1971	1854	3856
	90	33	3863	2014	1861	33	3863	2014	1861	3863
45	0	75	3874	1867	1933	75	3874	1867	1933	3874
	15	86	3912	1830	1939	65	3831	1934	1920	3872
	30	83	3904	1836	1938	56	3806	2025	1905	3855
	45	73	3865	1874	1931	48	3805	2124	1890	3835
	60	62	3821	1962	1916	44	3813	2149	1882	3817
	75	54	3804	2048	1902	43	3816	2136	1880	3810
	90	47	3806	2136	1906	47	3806	2136	1806	3806
60	0	75	3874	1867	1933	75	3874	1867	1933	3874
	15	83	3904	1836	1938	68	3843	1911	1925	3870
	30	89	3915	1828	1940	62	3821	1962	1916	3868
	45	81	3899	1841	1937	58	3810	2002	1909	3854
	60	73	3865	1878	1931	57	3808	2014	1907	3836
	75	66	3835	1926	1922	58	3810	2025	1909	3822
	90	61	3817	1971	1914	61	3817	2071	1914	3817
75	0	75	3874	1867	1933	75	3874	1867	1933	3874
	15	79	3891	1849	1936	72	3861	1884	1929	3876
	30	84	3907	1834	1939	70	3851	1895	1927	3879
	45	90	3915	1827	1940	69	3847	1903	1926	3881
	60	85	3910	1831	1939	70	3851	1895	1927	3880
	75	80	3895	1844	1937	72	3861	1884	1929	3878
	90	75	3874	1864	1934	75	3874	1864	1934	3876
90	0	75	3874	1867	1933	75	3874	1867	1933	3874
	15	76	3879	1862	1934	76	3879	1862	1934	3879
	30	77	3883	1858	1934	77	3883	1858	1934	3883
	45	79	3891	1849	1936	79	3891	1849	1936	3891
	60	83	3904	1836	1938	83	3904	1836	1938	3904
	75	86	3912	1830	1939	86	3912	1830	1939	3912
	90	90	3915	1827	1940	90	3915	1827	1940	3915

TABLE 2. (concluded)

I = 90

θ	i	φ₋	V_{1-}	V_{2-}	V_{3-}	φ₊	V_{1+}	V_{2+}	V_{3+}	V_{11}
0	0	90°	3915	1827	1940	90°	3915	1827	1940	3915
	15	75	3874	1867	1933	75	3874	1867	1933	3874
	30	60	3814	1980	1912	60	3814	1980	1912	3814
	45	45	3810	2155	1884	45	3810	2155	1884	3810
	60	30	3884	1980	1856	30	3884	1980	1856	3884
	75	15	4007	1867	1835	15	4007	1867	1835	4007
	90	0	4077	1827	1827	0	4077	1827	1827	4077
15	0	90	3915	1827	1940	90	3915	1827	1940	3915
	15	76	3879	1862	1934	76	3879	1862	1934	3879
	30	61	3817	1971	1914	61	3817	1971	1914	3817
	45	47	3806	2136	1888	47	3806	2136	1888	3806
	60	33	3863	2014	1861	33	3863	2014	1861	3863
	75	21	3957	1903	1842	21	3957	1903	1842	3957
	90	15	4007	1867	1835	15	4007	1867	1835	4007
30	0	90	3915	1827	1940	90	3915	1827	1940	3915
	15	77	3883	1858	1934	77	3883	1858	1934	3883
	30	64	3827	1943	1919	64	3827	1943	1919	3827
	45	52	3803	2073	1898	52	3803	2073	1898	3803
	60	41	3822	2111	1876	41	3822	2111	1876	3822
	75	33	3863	2014	1861	33	3863	2014	1861	3863
	90	30	3884	1980	1856	30	3884	1980	1856	3884
45	0	90	3915	1827	1940	90	3915	1827	1940	3915
	15	79	3891	1849	1936	79	3891	1849	1936	3891
	30	69	3847	1903	1926	69	3847	1903	1926	3847
	45	60	3814	1980	1912	60	3814	1980	1912	3814
	60	52	3803	2073	1898	52	3803	2073	1898	3803
	75	47	3806	2136	1888	47	3806	2136	1888	3806
	90	45	3810	2155	1884	45	3810	2155	1884	3810
60	0	90	3915	1827	1940	90	3915	1827	1940	3915
	15	83	3904	1836	1938	83	3904	1836	1938	3904
	30	76	3879	1862	1934	76	3879	1862	1934	3879
	45	69	3847	1903	1926	69	3847	1903	1926	3847
	60	64	3827	1943	1919	64	3827	1943	1919	3827
	75	61	3817	1971	1914	61	3817	1971	1914	3817
	90	60	3814	1980	1912	60	3814	1980	1912	3814
75	0	90	3915	1827	1940	90	3915	1827	1940	3915
	15	86	3912	1830	1939	86	3912	1830	1939	3912
	30	83	3904	1836	1938	83	3904	1836	1938	3904
	45	79	3891	1849	1936	79	3891	1849	1936	3891
	60	77	3883	1858	1934	77	3883	1858	1934	3883
	75	76	3879	1862	1934	76	3879	1862	1934	3879
	90	75	3874	1867	1933	75	3874	1867	1933	3874
90	0	90	3915	1827	1940	90	3915	1827	1940	3915
	15	90	3915	1827	1940	90	3915	1827	1940	3915
	30	90	3915	1827	1940	90	3915	1827	1940	3915
	45	90	3915	1827	1940	90	3915	1827	1940	3915
	60	90	3915	1827	1940	90	3915	1827	1940	3915
	75	90	3915	1827	1940	90	3915	1827	1940	3915
	90	90	3915	1827	1940	90	3915	1827	1940	3915

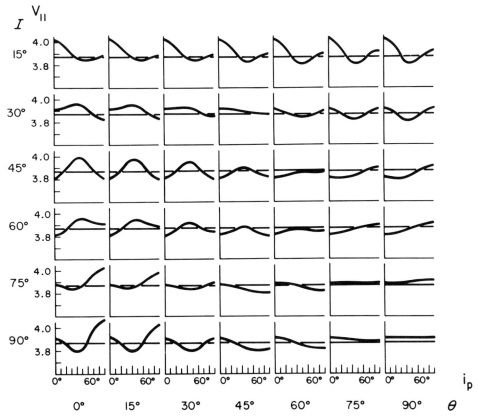

Fig. 5. Mean velocities (V_{11}) for reflected P waves in an ice sheet with perfect crystal alignment as a function of axial orientation and angle of incidence. Light, horizontal lines mark $(V_p)_{iso} = 3871$ m/sec. For greater accuracy, see Table 2.

single-crystal value. Phase velocities rather than wave velocities are shown, but the deviation for the latter would be very nearly the same.

For a 10° cone, the approximation is correct within 20 m/sec at all ϕ, and within 10 m/sec with only three exceptions: v_1 when $\phi < 15°$, v_2 when $30° < \phi < 60°$, and v_3 when $\phi < 20°$. The first two cases result from the inclusion of the respective velocity maximums within the cone, whereas the third arises from the fact that for small ϕ there is, in type 3 motion relative to the axis of the cone, a significant u_2 component of motion relative to individual c axes that lie off the plane defined by the ray and the axis of the cone.

Deviations are three to four times as great for a 20° cone, but are still less than 20 m/sec for v_1 away from its maximum and for v_3 when $\phi > 40°$. For v_2, however, the deviation is greater than 20 m/sec at most angles and reaches 66 m/sec at $\phi = 45°$. These deviations are substantial in terms of the goodness of model fit to the profile data. For a 30° cone, the differences are twice as large again, and the approximation becomes quite unsatisfactory.

In reality, the c axes are likely not to be randomly distributed within the cone, but to be more concentrated near its axis, thus tending to improve the approximation. We can conclude that if, say, 90% of the c axes are aligned within 20 or 25 degrees of their mean direction, as in the case of the most concentrated patterns found by *Rigsby* [1960] in Greenland, the single-crystal approximation is satisfactory, but that if the concentration is much less, distinct deviations from the single-crystal models should appear in the observed velocities, particularly near velocity extremes.

On all profiles, good p arrivals were recorded, and, on most, s arrivals also, giving v_p and v_s in the upper part of the ice sheet. In several instances, where s arrivals were present but with too much scatter to determine v_s from the arrivals alone, an intercept

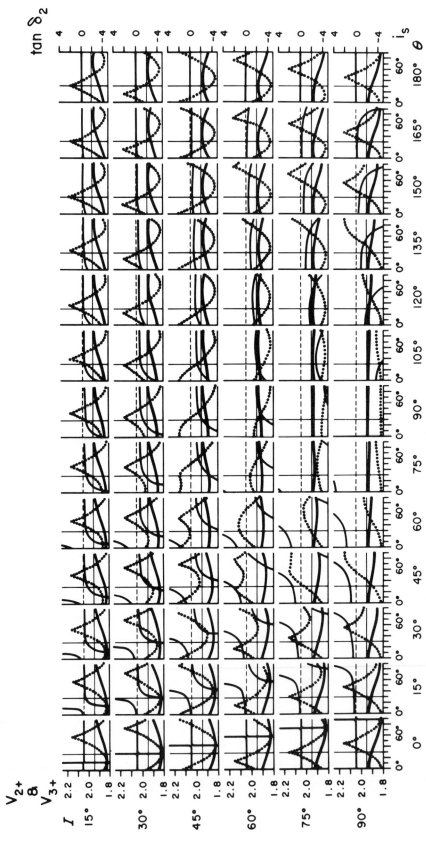

Fig. 6. Velocities and component amplitude ratios for upward-traveling S waves in an ice sheet with perfect crystal alignment, as a function of axial orientation and angle of incidence. Dotted lines, V_{2+}; heavy solid lines, V_{3+}; medium-weight solid lines, $\tan \delta_2$; light, horizontal solid lines, $(V_s)_{iso} = 1951$ m/sec; light, vertical lines, critical angle ($i_s = 30°$) assumed for R_{ps}; dashed lines, $\tan \delta_2 = 0$. For greater accuracy see Table 2.

TABLE 3. Phase Velocities in Ice at $-10°C$ Having c Axes Randomly Distributed within Cones of Various Apex Angles as a Function of the Angle of Propagation Relative to the Axis of the Cone (ϕ)

ϕ	Semi-Apex Angle	Mean Phase Velocity, m/sec			Differences from Velocities for a Single Crystal, m/sec		
		v_1	v_2	v_3	δv_1	δv_2	δv_3
0	0	4077	1827	1827			
	10	4063	1837	1837	−14	+10	+10
	20	4025	1864	1864	−52	+37	+37
	30	3974	1901	1901	−103	+74	+74
15	0	4017	1906	1834			
	10	4007	1911	1853	−10	+5	+19
	20	3982	1923	1875	−35	+17	+41
	30	3947	1939	1906	−70	+33	+72
30	0	3895	2086	1854			
	10	3895	2077	1861	0	−9	+7
	20	3892	2053	1881	−3	−33	+27
	30	3889	2020	1908	−6	−66	+54
45	0	3814	2190	1882			
	10	3818	2172	1886	+4	−18	+4
	20	3831	2124	1898	+17	−66	+16
	30	3847	2063	1915	+33	−127	+33
60	0	3820	2086	1910			
	10	3823	2077	1912	+3	−9	+2
	20	3833	2053	1916	+13	−33	+6
	30	3846	2020	1922	+26	−66	+12
75	0	3880	1906	1932			
	10	3877	1911	1931	−3	+5	−1
	20	3875	1923	1929	−5	+17	−3
	30	3870	1939	1927	−10	+33	−5
90	0	3915	1827	1940			
	10	3910	1837	1938	−5	+10	−2
	20	3899	1864	1934	−16	+37	−6
	30	3884	1901	1928	−31	+74	−12

Mean Deviations

I_f	v_1	v_2	v_3
10	2.1%	2.6%	1.8%
20	7.6	9.4	5.4
30	15.0	18.6	10.3

time, $(t_s)_0$, was assumed from $(t_p)_0$ at that station and the average $(t_p)_0/(t_s)_0$ ratio measured at neighboring stations. Only on one profile, E-756, was s not observed at all. It is worth noting that, if $(v_s)_2$ due to anisotropy is recorded, a type 2 wave is immediately implied, since $V_3(\phi) < (V_s)_{iso}$ for all ϕ.

It was assumed that $(v_p)_1$ and $(v_s)_1$ were the true values at each station of the velocities $(v_p)_{iso}$ and $(v_s)_{iso}$ in isotropic ice. (This assumption proved in subsequent analysis to be not entirely accurate, but corrections based on a more exact assumption were not significant enough to warrant replotting the finished figures.) Velocity models for the ice sheet were derived from the velocity table (Table 2) by taking the tabular differences from the theoretical isotropic velocity (ΔV_{11}, ΔV_{2+}, and ΔV_{3+}), multiplying by a factor relating to the part of the ice sheet assumed to be anisotropic and the degree of anisotropy therein, and then adding to $(v_p)_1$ or $(v_s)_1$. In addition, a temperature correction of -10

m/sec, corresponding roughly to a downward temperature increase of 20° such as might be expected in the lower half of the ice sheet [*Robin*, 1955] was applied in obtaining V_{11}. No temperature corrections were applied to calculate S-wave velocities, since the proper coefficient is not known [*Röthlisberger*, 1971; *Bentley*, unpublished manuscript]. Because of the great sensitivity of V_{2+} to changes in the direction of propagation relative to the axial direction, it is unlikely that any model would be significantly in error on account of the unknown temperature effect.

We thus have

$$V_{11} = q_p \Delta V_{11} + (v_p)_1 - 10 \quad \text{(in m/sec)}$$
$$V_{k+} = q_s \Delta V_{k+} + (v_s)_1 \quad k = 2, 3 \quad (17)$$

where q_p and q_s are factors to be chosen. They are not necessarily equal, because of the different effect of deviation from the single-crystal approximation on P- and S-wave velocities. The procedure used for their selection was an empirical one based on the velocity measurements where the best data were available. At E-348, it was possible to determine the approximate mean crystal orientation without first knowing q_p and q_s. The factors were then taken equal to the ratio of the mean observed velocity deviations from isotropic values to the expected velocity deviations for the appropriate single-crystal model. This gave $q_p = 50\%$ and $q_s = 30\%$. At station E-972, on the other hand, refraction data indicate that 85% of the ice is anisotropic, and a good fit to V_p was obtained on this basis. Hence $q_p = 85\%$ was selected as an alternate value. The ratio of q_p to q_s was kept the same, giving $q_s = 50\%$. (Here again, subsequent analysis has not borne out the computational assumption, since it turns out to be quite likely that axial concentration around a single direction actually exists in the ice, and the mean deviations for V_p and V_s would then not be very different (Table 3). However, primary emphasis in model fitting has been put on V_p, and, since V_{2+} usually changes rapidly with angle, it was not felt that the small model corrections that would result from setting $q_p = q_s$ warranted the considerable labor necessary to recalculate and replot all the models.) In the figures showing models fit at each station, '85% anisotropy' simply means $q_p = 85\%$, $q_s = 50\%$, and '50% anisotropy' means $q_p = 50\%$, $q_s = 30\%$.

In comparing models, it is useful to keep in mind that the velocity information contained in R_p diminishes to zero as i_p approaches zero from equations 1 and 2, $\partial \bar{V}_p / \partial (V_{11})_0 = \cos i_p$, so that, for small angles of incidence, $\bar{V}_p$ will simply change very nearly in accordance with the assumed value of $(V_{11})_0$. This is shown clearly in Figures 7 and 8. $\bar{V}_p$ for relatively small i_p can nevertheless be important for model selection, since the trend of values, i.e. $\partial \bar{V}_p / \partial i_p$, may be significant where the velocity values themselves are not.

Although R_s was recorded on many profiles, it has not been used in any of the model fitting. The reasons for this become clear upon examination of representative curves for the two profiles at station E-348 (Figure 9). Four different mean velocity curves exist for any single model, generally covering a wide range of velocities. Associated with the velocity curves are curves of cot β (equations 8 and 9), giving the expected relative amplitudes of transformed to untransformed reflections, and of tan δ_2 and tan δ_3 (equations 12 and 13), which give the ratios of transverse to horizontal longitudinal amplitude for an arrival of each type. The complications are great enough that it would be difficult to know which arrivals to assign to which type of reflection, even if the predicted amplitude ratios were known to be reliable.

Without a confirmed model fit to R_s, we cannot test cot β. We can, however, consider the prediction (equation 16) that follows from the same assumptions about the reflection process, that the amplitudes of the transverse components of the two types of R_{ps} arrivals are always the same. This prediction is not substantiated. Where the two types of reflection are clearly separated, the transverse amplitude associated with one is clearly larger (see, for example, record 6113 in Figure 3). In other cases, only one of the R_{ps} types may be recorded, but with approximately equal horizontal longitudinal and transverse components (e.g., E-444, Figure 23).

Because of the failure of this prediction, there is little reason to believe in the calculated values of cot β. Apparently the model of simple reflection at the base of the ice, with wave-type splitting upon passage through a perfect anisotropic layer, is incorrect. This may be due to skew anisotropic axes at the base of the ice sheet, or to failure of the single-crystal approximation, or to both.

The failure of the predictions for amplitude ratios does not mean that tan δ_2 and tan δ_3 should be

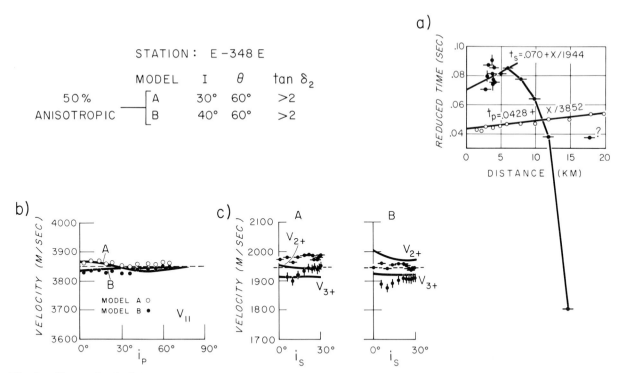

Fig. 7. Observed velocities and model fitting, profile E-348E. (a) Travel times and travel-time equations; t_p (open circles) reduced to 3860 m/sec, t_s (solid points) reduced to 1950 m/sec. (b) Observed $\bar{V}_p$ (points) and fitted models (solid lines) for R_p; dashed line marks adopted $(v_p)_{iso} = 3853$ m/sec. Maximum expected topographic error is shown at base of diagram. (c) Observed $\bar{V}_{s+}$ (points) and fitted models (solid lines) for R_{ps}; dashed line marks adopted $(v_s)_{iso} = 1946$ m/sec. Short vertical (horizontal) lines through t_s and $\bar{V}_{s+}$ points indicate a horizontal longitudinal (transverse) component of motion.

ignored. The relative transverse and horizontal longitudinal amplitudes for an individual wave approaching the surface in no way depend upon the reflection process. Thus, if the perfect anisotropic model is even approximately correct, tan δ_2 and tan δ_3 should provide a general idea as to where to expect relatively large transverse amplitudes. In the profile analyses that follow, comparison of observed and predicted values of tan δ_2 has been used as a secondary criterion in model selection (for qualitative comparison, tan δ_3 is close enough to cot δ_2 (equations 14 and 15), since, because of the low wave velocities near the surface, i' never differs greatly from 0°), and it was possible in all cases to find a satisfactory model without major disagreement.

At station E-348, because of the two perpendicular profiles and the excellent shear-wave data, the azimuth of the model axis was uniquely determined and was found to be down slope. At all other stations an ambiguity in the sign of θ necessarily remained; by analog with E-348 and in accordance with the expectations of symmetry, if one choice corresponded to a down-slope azimuth, it was preferred. Some justification for this procedure was afforded by the existence of such an alternative at most stations. (Note that ambiguity in the sign of θ is quite different from an 180° ambiguity in geographic azimuth.)

Using the procedures outlined above, the sixteen seismic profiles have been analyzed to find the best-fitting anisotropic models. We will discuss first the three stations at which the data are the most definitive. The models for these three stations are distinctly different, and are generally representative of the different geographical regions in which they are found. E-348, lying centrally in the Amundsen Sea drainage basin (Figure 29), is the site of two long perpendicular profiles. E-972, a long profile lying near the base of the Antarctic Peninsula in the Weddell Sea drainage system, is the only profile on which p arrivals propagating nearly along the crystal c axis, thus clearly defining a horizontal preferred orientation, were recorded. The model-fitting profile Byrd-58 near Byrd station, in the Ross Sea drainage basin, can be directly compared with

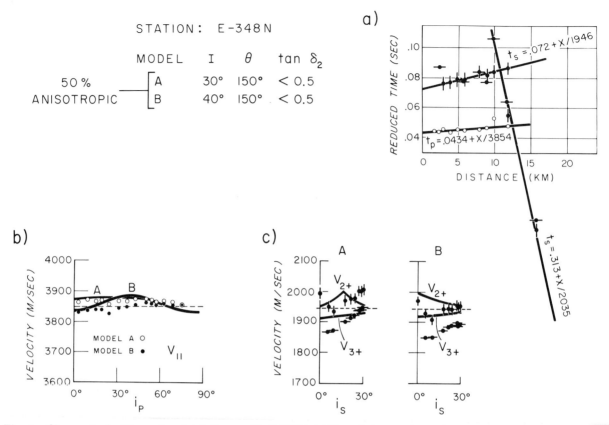

Fig. 8. Observed velocities and model fitting, profile E-348N. (a) Travel times and travel-time equations; t_p (open circles) reduced to 3860 m/sec, t_s (solid points) reduced to 1950 m/sec. (b) Observed $\bar{V}_p$ (points) and fitted models (solid lines) for R_p; dashed line marks adopted $(v_p)_{iso} = 3853$ m/sec. Maximum expected topographic error is shown at base of diagram. (c) Observed $\bar{V}_{s+}$ (points) and fitted models (solid lines) for R_{ps}; dashed line marks adopted $(v_s)_{iso} = 1946$ m/sec. Short vertical (horizontal) lines through t_s and $\bar{V}_{s+}$ points indicate a horizontal longitudinal (transverse) component of motion.

the ice fabrics observed in the cores from the deep drill hole.

Analysis of the other profiles will then be taken up, more or less in geographical order, the results already discussed being used as a guide to appropriate models.

Station E-348

At station E-348, located at the junction of the Sentinel and Ellsworth Highland traverse tracks, two complete profiles were shot, one to the east-southeast from the recording spread (348E), and one to the north-northeast (348N). Shots provided continuous coverage to a distance of 6 km, then were fired every 2 km to 12 km, and every 4 km to 24 km. Excellent p, s, R_p, and R_{ps} arrivals were recorded on almost every shot on both profiles, except that s became weak beyond 12 km on 348E. These two profiles together provided the most complete seismic evidence found at any of the stations for the anisotropy of the ice.

There are two particularly striking characteristics of the profiles: (1) unequivocal high-velocity s_2 arrivals on both (e.g., Figure 2b), whereas V_p is constant with distance, and (2) a complete, double set of R_{ps} arrivals (e.g., Figure 3c, d), with the faster having transverse motion and the slower horizontal longitudinal on 348E (Figure 7c), and the opposite being true for 348N (Figure 8c). The two t_s plots have different characteristics, that for 348N (Figure 10) being fit nicely by two straight lines with $(v_s)_1 = 1944$ m/sec, $(v_s)_2 = 2035$ m/sec, and a cross-over distance of 10½ km, whereas that for 348E (Figure 11) shows a gradual increase of apparent velocity from the same $(v_s)_1$ value beyond $x = 6$ km.

For 348N, we calculate a depth to the second layer of 850 meters. Fitting a sequence of straight lines to the 348E profile, we find the velocity starts to

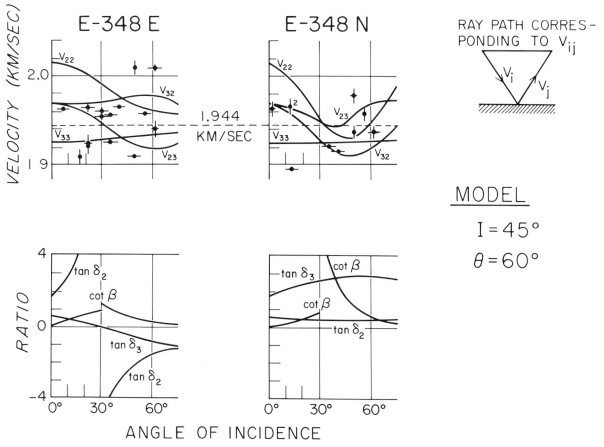

Fig. 9. Mean velocities, modeled (solid lines) and observed (points), for reflected shear waves at station E-348, together with predicted amplitude ratios, as a function of angle of incidence. Short vertical and horizontal lines on observed points represent horizontal longitudinal and transverse motion, respectively.

increase at a depth of about 250 meters and reaches 2.03 km/sec at a depth of 800 meters. The second arrivals at $x = 5$ km fall on the 2.03-km/sec line, suggesting that they may correspond to a caustic of a triplicated travel-time curve, and hence that the velocity may increase quite rapidly with depth in the region above this level. This, in turn, would imply that the depth to the 2.03-km/sec level was somewhat greater than 800 meters. In any case, this velocity is found at about the same depth on both the profiles.

That being so, and assuming for the moment perfect crystal alignment, we can see from Figure 12 that correspondingly either (a) $I = 55°$ and $|\theta|$ or $|\theta'| \simeq 45°$ ($\theta' \equiv 180° - \theta$), or (b) $I = 90°$, $|\theta|$ or $|\theta'| \simeq 35°$ or $55°$ (θ can be taken relative to either profile). For less than perfect anisotropy, it would be necessary in case (a) that $I > 55°$, and in case

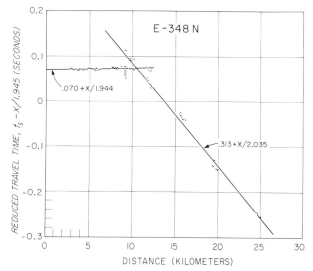

Fig. 10. Reduced s-wave travel-time plot, profile E-348N. The travel-time equation is shown for each line.

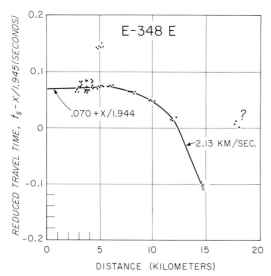

Fig. 11. Reduced s-wave travel-time plot, profile E-348E. Propagation velocities are shown for the slowest and fastest parts of the travel-time curve.

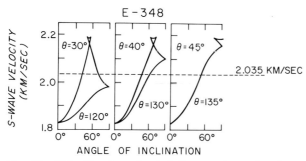

Fig. 12. Velocities for refracted S waves in an ice sheet with perfect crystal alignment, as a function of axial orientation. The two values of θ in each section differ by 90° for comparison with the two profiles E-348E and E-348N. The dashed line marks $(v_s)_2$ from profile E-348N.

(b) that $35° < |\theta|$ or $|\theta'| < 55°$. Thus in either case a relatively large inclination and an azimuth intermediate between the two profiles is indicated.

It is possible to suggest causes for the different form of the two refraction profiles. Suppose, for example, that the degree of axial alignment increases gradually with depth starting at 250 meters, and that $I = 60°$, $|\theta|$ or $|\theta'| \simeq 30°$ relative to 348E. There would be a gradual and observable increase in velocity on 348E, but an unobservable decrease on 348N. If there was then a rotation at greater depth toward $|\theta|$ or $|\theta'| = 45°$, leading to case a, in a zone in which the anisotropy was becoming pronounced, the two effects would add to cause a rapid increase in v_s on 348N, while a slower increase in the already larger value of v_s on 348E would occur. Similarly, case b could be approached with I initially about, say, 60° and increasing with depth. Other explanations involving more complicated fabric patterns are undoubtedly possible; the conclusion seems inescapable, however, that some geometric change with depth is necessary, and that a simple increase with depth in the degree of anisotropy relative to a fixed geometric pattern is insufficient to explain the refraction profiles.

Let us now consider the implications of the double set of R_{ps} arrivals. We will adopt case a for the sake of discussion; similar conclusions would follow from case b. The refraction data do not determine in which quadrant the vector **c** lies; all four are equally probable. The number of choices can immediately be reduced, however (assuming there is not a 90° rotation of axes with depth), by considering reversal in order of arrival of the transverse and longitudinal R_{ps} events on the two profiles, which would not be found if $\hat{c}$ lay between the two profiles or between their two extensions, i.e. either east or west (see Figure 29), so that θ was the same relative to both. Thus $\hat{c}$ must point either north or south. But the transverse orientation of the fast (V_{2+}) arrival on 348E means that $\tan \delta_2$ must be large on this profile; conversely, it must be small on 348N. Anticipating the results of the reflection analysis, which show that on the average $I \simeq 30°$, we find that, for large values of $\tan \delta_2$, θ must be $-45°$, not $+135°$ relative to 348E, hence that $\hat{c}$ can only lie in the north quadrant. Thus we have ascertained without ambiguity that the mean axial direction points more or less down slope, without even considering detailed models.

In fact, detailed models are difficult to fit. It seems necessary to conclude from the reflection data that further axial rotation with depth exists, since neither case a nor case b yields V_{11} values compatible with $\bar{V}_p$. We have therefore sought to fit a model to the reflection data alone. By interpolation, we see that all $\bar{V}_p$ data and $\bar{V}_{s+}$ on 348E (Figure 7) are reasonably well satisfied with $I = 35°$, $|\theta| = 60°$ relative to 348E. The fit to V_{s+} on 348N (Figure 8) is unsatisfactory, however; it could be improved if I were less than 30°. The failure to produce a single consistent model for the reflections may be another consequence of axial rotation with depth. It is worth noting that $|\theta|$ could be near 45° rather than 60°:

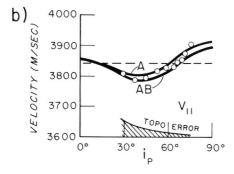
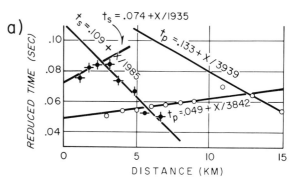
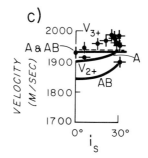

Fig. 13. Observed velocities and model fitting, profile E-972. (a) Travel times and travel-time equations; t_p (open circles) reduced to 3860 m/sec, t_s (solid points) reduced to 1949 m/sec. (b) Observed $\bar{V}_p$ (points) and fitted models (solid lines) for R_p; dashed line marks adopted $(v_p)_{iso} = 3842$ m/sec. Maximum expected topographic error is shown at base of diagram. (c) Observed $\bar{V}_{s+}$ (points) and fitted models (solid lines) for R_{ps}; dashed line marks adopted $(v_s)_{iso} = 1935$ m/sec. Short vertical (horizontal) lines through t_s and $\bar{V}_{s+}$ points indicate a horizontal longitudinal (transverse) component of motion.

it would then differ from refraction case a only by 30° in I.

Despite the imperfection of the model fitting, it does appear likely that the mean vector $\hat{c}$ is oriented approximately along a flow line, and is inclined 30° to 60° from the vertical, and that the inclination decreases with depth below 1000 meters or so.

Station E-972

Profile E-972 is remarkable in that breaks are observed on both the p and the s travel-time lines, but do not correspond nearly to the same depth of refracting horizon. The $(t_s)_1$ line (Figure 13) was determined by assuming $(t_s)_0$ to be 1½ times $(t_p)_0$, that intercept ratio being both the traverse average and the value for the closest appropriate station, E-612. This intercept leads to $(v_s)_1 = 1935$ m/sec, a reasonable value in comparison with other profiles. The calculated depth to the 1985 m/sec refracting horizon is 230 meters.

The v_p break is determined by only one arrival, but that one arrives 0.014 sec early, whereas all other arrivals at $x > 3$ km, with the exception of the late arrival at 11 km, lie well within a millisecond of the 3842-m/sec line. The arrival is an excellent one (Figure 2a) with a sharp onset. The corresponding shot distance as determined by Tellurometer was checked by Sno-Cat odometer within 10 meters, whereas 0.014 sec corresponds to more than 50 meters in distance. Because a 14-msec error would be inconsistent with the precision of v_p determinations to be expected, the v_p break is believed to be real. The late arrival at 11 km could arise from a later branch of a triplicated travel-time curve, the first arrival having been lost for reasons suggested below. The corresponding depth is 800 meters, about half the total ice thickness.

The high-velocity P wave must correspond to an axial inclination close to 90°, since in no other way can a high enough velocity be obtained. To show this, we first note that, for isotropic ice, the calculated v_p is 3871 m/sec, whereas the observed value at this station is 3842 m/sec. Thus the calculated velocities for perfect axial alignment should be about

30 m/sec more than those measured in the field. Any effect of increasing temperature with depth or of less than perfect anisotropy would increase the difference. It follows that the field value $(v_p)_2 = 3940$ m/sec corresponds to a model value $V_1 \geq 3970$ m/sec. Since velocities this high are found only within 20° of the axial direction, the mean axis in the refracting layer must lie within 20° of the horizontal.

Because of the low minimum in the $\bar{V}_p$ versus i_p curve (Figure 13), I was taken to be 90°; then, to match the minimum refracted $(v_p)_2$, $|\theta|$ was chosen as 165° (model A). Fifty per cent of the ice sheet was assumed to be anisotropic. The resulting V_{11} curve (Figure 13) is a remarkably good fit to $\bar{V}_p$, and the V_{2+} and V_{3+} curves fit the observations reasonably well; tan δ_2 for this model is about 0.5, also in fair agreement with the observed similar horizontal longitudinal and transverse amplitudes of R_{ps}. Furthermore, the anisotropic c axes, determined solely by the seismic data except for the unavoidable ambiguity in the sign of θ, point nearly down slope. In this model, $(v_p)_2$ and $\bar{V}_p$ relate to the same general part of the ice column.

It remains to consider $(v_s)_2$. Suppose that at a depth of 230 meters anisotropic ice with $I = 75\%$, $|\theta| = 145°$ is encountered. There will be an increase in the horizontal V_2 to about 2050 m/sec which, after allowance for imperfect anisotropy, could reasonably correspond to the field value of 1985 m/sec. At the same time, the expected v_p would decrease slightly. If the axial orientation shifts (gradually or suddenly) to $I = 90°$, $|\theta| \leq 160°$ at greater depth, v_s would decrease, while v_p would increase with depth. This model would explain both the velocity data and, at least qualitatively, the failure to record either s_2 beyond 7 km or p_1 beyond 9 km.

To approximate a downward-changing axial orientation, a 3-layered model was chosen. Model AB comprises 15% isotropic ice, 35% ice with $I = 75°$, $|\theta| = 145°$, and 50% ice with $I = 90°$, $\theta = 165°$. As can be seen in Figure 13, model AB is not significantly different from model A, fitting $\bar{V}_p$ very well and fitting V_{s+} without a large discrepancy. For either model we would expect a maximum refracted v_p of about 3980 m/sec; it is certainly possible that a higher velocity would have been observed in the field if the profile had been extended to a greater distance.

The success of model AB in fitting the observations, together with its uniqueness insofar as the refracted velocities are concerned, leads us to believe that there is a real rotation of the anisotropic axes downward and counterclockwise with increasing depth in the ice. It is not possible to distinguish between a gradual and an abrupt change. The azimuth of around 145° as found in the upper anisotropic zone agrees more closely with the downslope direction at E-972 than does an azimuth of 165°.

Byrd Station

Two long profiles in perpendicular directions have been completed at Byrd station, but unfortunately, owing to shot holes that were too shallow, no usable reflection data were obtained on one (Byrd-57). There is thus only one profile (Byrd-58), supplemented by some refraction results.

The important fact about this station is, of course, that crystal orientations have been measured in cores from a nearby drill hole through the ice [*Gow et al.*, 1968]. Axes were found to be generally vertical, the concentration increasing between depths of 900 and 1200 meters to a very high value, and continuing thus to a depth of 1800 meters. The bottom 340 meters are isotropic. The seismic profiles were about 10 miles west of the drill site, where the ice is 400 meters thicker. If most of this additional ice is anisotropic, as seems reasonable if the zonation is a result of flow, the assumption $q_p = 50\%$ for model fitting is appropriate.

Let us consider the reflection data first. A good fit is obtained by the near-vertical model (model B) shown in Figure 14. The fit would be poorer for $I = 0$, since V_{2+} would be decreased, whereas the high implied value of $(V_{11})_0$ would in turn imply a higher value for the observed $\bar{V}_{s+}$. Assuming c axes uniformly distributed within a vertical cone of semi-apex angle 20°, $(V_{11})_0$ would be reduced nearly to the single-crystal value for $I = 15°$ (Table 3), thus decreasing $\bar{V}_{s+}$. Since V_{2+} would also be reduced, however, $\bar{V}_{s+}$ can still be fit better by a small mean inclination with $|\theta| \simeq 105°$. This azimuth is approximately parallel to the contours. A good fit could also be obtained with $I = 90°$, $\theta = -30°$, which is down slope.

The refraction data introduce a real puzzle into the picture. What appear to be s_2 arrivals occur on both profiles one or two tenths of a second after s_1, and with amplitudes comparable to, or even greater than, s_1 (Figures 2c and d). Treating these arrivals

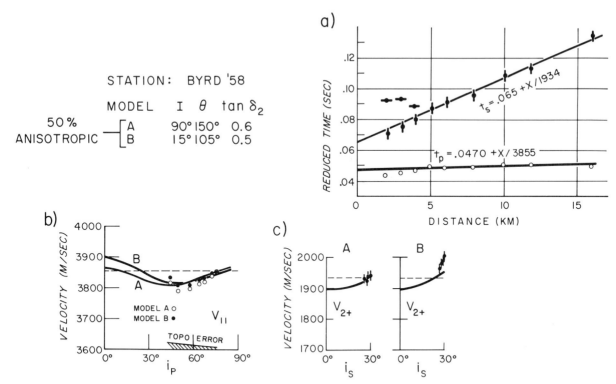

Fig. 14. Observed velocities and model fitting, profile Byrd-58. (a) Travel times and travel-time equations; t_p (open circles) reduced to 3860 m/sec, t_s (solid points) reduced to 1950 m/sec. (b) Observed $\bar{V}_p$ (points) and fitted models (solid lines) for R_p; dashed line marks adopted $(v_p)_{iso} = 3855$ m/sec. Maximum expected topographic error is indicated at base of diagram. (c) Observed $\bar{V}_{s+}$ (points) and fitted models (solid lines) for R_{ps}; dashed line marks adopted $(v_s)_{iso} = 1934$ m/sec. Short vertical (horizontal) lines through t_s and $\bar{V}_{s+}$ points indicate a horizontal longitudinal (transverse) component of motion.

as head waves from a deeper layer and fitting either of the lines shown in Figure 15 and that shown in 16, we find a depth of 1200 meters. The only anisotropic model that would cause $(v_s)_2$ to equal 1990 m/sec on both profiles is $I = 90°$, $|\theta| = 30°$, relative to one, and $|\theta| = 60°$ relative to the other. (The model for $(v_s)_2 = 1963$ on Byrd-58 would be only slightly different.) Furthermore, the initial occurrence of the arrivals is at a distance in agreement with the calculated critical distance. Interestingly enough, one of these models, $\theta = -30°$ relative to Byrd-58 and $\theta = +60°$ relative to Byrd-57 agrees with the alternative reflection solution mentioned above, which coincides with the flow line.

That solution, of course, is in direct contradiction to the observations in the deep drill hole only 10 miles away. We might suppose, instead, that it is reflections from the top of the anisotropic layer that have been recorded, the travel times being in reasonable agreement with those to be expected (Figures 15 and 16). The problem here is one of reflectivity.

It is difficult to understand how, even at angles of incidence of 75°–85°, a reflection comparable in amplitude to s_1, seemingly implying a reflection coefficient on the order of 1, could arise from a gradual decrease of velocity with depth by a maximum of about 5% through a zone which, from the drilling results, is an order of magnitude larger than a wavelength in thickness! This difficulty is accentuated by the high-frequency character of s_2 on profile Byrd-58 (Figure 2c).

There is no simple resolution to this disagreement. It would appear that either (1) there is a drastic rotation of axes throughout the ice column in the ten miles between the drill hole and the site of the profiles, (2) there is a relatively thin zone of anomalous anisotropy at Byrd-58 that either was missed in the preliminary examination of the deep drill cores or does not exist at the drill site, or (3) the arrivals in question arise from another cause that has been overlooked. Of these, the second alternative seems the most likely to the writer.

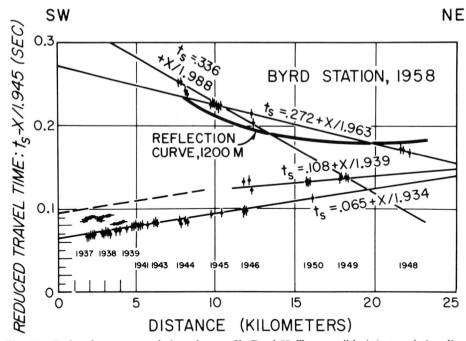

Fig. 15. Reduced s-wave travel-time plot, profile Byrd-58. Two possible $(v_s)_2$ travel-time lines and a reflection curve calculated for isotropic ice to a depth of 1200 meters are shown. Short vertical (horizontal) lines through points indicate a primarily horizontal longitudinal (transverse) component of motion. Small numbers are record numbers.

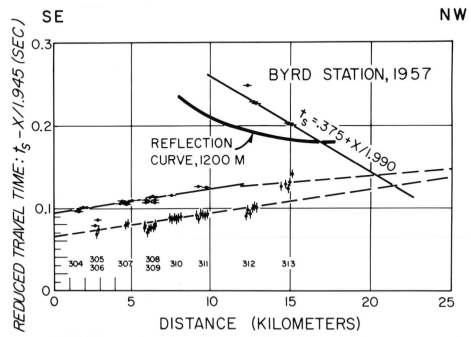

Fig. 16. Reduced s-wave travel-time plot, profile Byrd-57. One possible $(v_s)_2$ travel-time line and a reflection curve calculated for isotropic ice to a depth of 1200 meters are shown. Short vertical (horizontal) lines through points indicate a primarily horizontal longitudinal (transverse) component of motion. Small numbers are record numbers. Dashed lines are carried over from Figure 15.

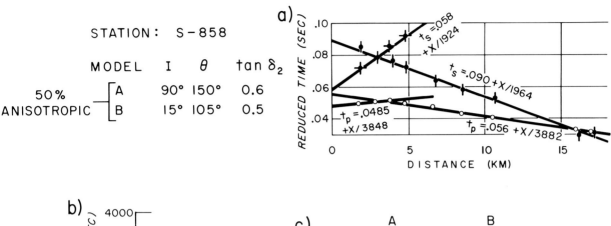

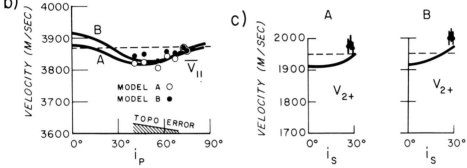

Fig. 17. Observed velocities and model fitting, profile S-858. (a) Travel times and travel-time equations; t_p (open circles) reduced to 3860 m/sec, t_s (solid points) reduced to 1950 m/sec. (b) Observed $\bar{V}_p$ (points) and fitted models (solid lines) for R_p; dashed line marks adopted $(v_p)_{iso} = 3873$ m/sec. Maximum expected topographic error is shown at base of diagram. (c) Observed $\bar{V}_{s+}$ (points) and fitted models (solid lines) for R_{ps}; dashed line marks adopted $(v_s)_{iso} = 1964$ m/sec. Short vertical (horizontal) lines through t_s and $\bar{V}_{s+}$ points indicate a horizontal longitudinal (transverse) component of motion.

Station S-858

The profile S-858 is the only one, besides Byrd-58, in the Ross Sea drainage basin, and the reflection data here are almost identical to those on Byrd-58, even to the observation of R_{ps} only at large angles of incidence. The same two models thus provide a good fit: $q_p = 50\%$ with either $I = 15°$, $|\theta| = 105°$, implying, as at Byrd station, a slight axial inclination in a direction nearly parallel to the surface contours, or $I = 90°$, $|\theta| = 150°$, meaning nearly horizontal axes oriented downslope, if $\theta = -150°$ is chosen. Because of the similarity between this profile and Byrd-58, the solution $I = 15\%$, $|\theta| = 105°$ is preferred.

No late s arrivals, such as those recorded at Byrd station, were observed here. The t_s plot (Figure 17a) exhibits an interesting feature, however. Arrivals with both components of horizontal motion occur out to $x = 5$ km, defining $(v_s)_1 = 1924$ m/sec. Starting at $x = 2$ km, a second line is defined by horizontal longitudinal arrivals only, giving $(v_s)_2 = 1964$ m/sec at a calculated depth of 150 meters. The coexistence of two sets of arrivals, and especially the occurrence at $x = 4$ km and $x = 5$ km of s_1 with transverse motion and s_2 without transverse motion, indicates a deviation from the normal, downward-decreasing velocity gradient, and suggests that the increase in velocity from $(v_s)_1$ to $(v_s)_2$ may be attributable to anisotropy.

The t_p plot also shows a possibly abrupt velocity increase at $x = 3$ km, although a gradual change cannot be ruled out. The calculated depth to the $(v_p)_2$ layer, 200 meters, may not be significantly different from that to the $(v_s)_2$ layer. Thus the possibility exists of a change in velocities due to alteration from bubbly to bubble-free ice, although both the magnitude of the change in v_s and the likelihood of a different layer depth for v_p and v_s suggest that anisotropy is more probable.

Station S-150

Profile S-150, the first one north of the Amundsen Sea-Ross Sea drainage divide, is interesting because it is the only one for which neither a model with I

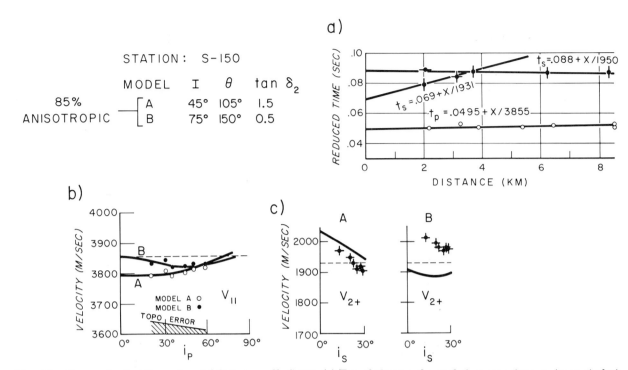

Fig. 18. Observed velocities and model fitting, profile S-150. (a) Travel times and travel-time equations; t_p (open circles) reduced to 3860 m/sec, t_s (solid points) reduced to 1950 m/sec. (b) Observed $\bar{V}_p$ (points) and fitted models (solid lines) for R_p; dashed line marks adopted $(v_p)_{iso} = 3855$ m/sec. Maximum expected topographic error is shown at base of diagram. (c) Observed $\bar{V}_{s+}$ (points) and fitted models (solid lines) for R_{ps}; dashed line marks adopted $(v_s)_{iso} = 1931$ m/sec. Short vertical (horizontal) lines through t_s and $\bar{V}_{s+}$ points indicate a horizontal longitudinal (transverse) component of motion.

small nor one with an azimuth along the flow line is satisfactory. The important features are low values of $\bar{V}_p$ for i_p between 30° and 60°, $\bar{V}_{s+}$ decreasing markedly with increasing i_s, and $\tan \delta_2 \simeq 1$. The first two characteristics eliminate all models except those for intermediate values of I and $|\theta|$ between 60° and 120°. Model A, with $I = 45°$, $|\theta| = 105°$, $\tan \delta_2 = \frac{1}{2}$ and $q_p = 85\%$, provides a good fit to $\bar{V}_p$ and a fairly good one to $\bar{V}_{s+}$; the fit to the latter could be greatly improved with only a small change in V_{11} by increasing I and perhaps decreasing $|\theta|$ by a few degrees each.

The azimuth for this model is at a large angle to the down-slope direction, which corresponds to $\theta = +150°$. It is not possible to find a model with $\theta = 150°$ that is even close. For example, model B, chosen to give the best fit to $\bar{V}_p$, could hardly be worse for $\bar{V}_{s+}$ (Figure 18). Similarly, the model (not shown) that would fit V_{s+} at this azimuth produces a high peak in V_{11} at intermediate values of i_p, exceeding the observed values by some 100 m/sec. The conclusion seems inescapable that at this station the axis of anisotropy lies far from the down-slope direction rather than along it.

There is an abrupt 20-m/sec increase in V_s at $x = 4$ km, which probably results from anisotropy. The corresponding depth is about 200 meters. If this is due to anisotropy, $|\theta|$ must be $>120°$ (or $<60°$) whatever the value of I, and I must be $>30°$. Possible solutions include $I \simeq 35°$, $\theta \simeq +150°$, giving a downslope orientation, $I \simeq 45°$, $|\theta| = 135°$, maintaining the inclination indicated by the reflection solution, and $I = 90°$, $|\theta| = 120°$, for the azimuth closest to that of the reflection solution.

Unless there is some other cause for a sudden increase in v_s at a depth of 200 meters which does not increase v_p, and it is difficult to imagine one, anisotropy commences here at a depth of only 8% of the total ice thickness, justifying the assumption of $q_p = 85\%$. (Of course, a zone of isotropic ice could lie somewhere below 200 meters, but there is no evidence for such a stratification, and reflection data require a major part of the ice sheet to be anisotropic anyway.) As at several other stations,

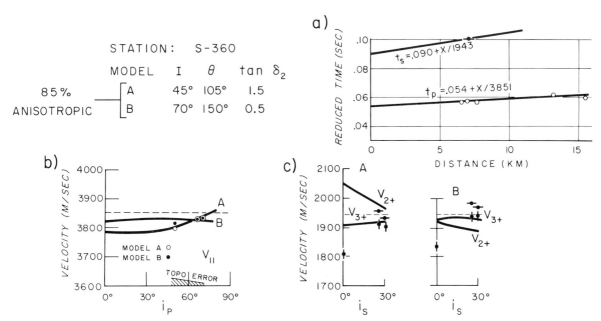

Fig. 19. Observed velocities and model fitting, profile S-360. (a) Travel times and travel-time equations; t_p (open circles) reduced to 3860 m/sec, t_s (solid points) reduced to 1950 m/sec. (b) Observed $\bar{V}_p$ (points) and fitted models (solid lines) for R_p; dashed line marks adopted $(v_p)_{iso} = 3851$ m/sec. Maximum expected topographic error is shown at base of diagram. (c) Observed $\bar{V}_{s+}$ (points) and fitted models (solid lines) for R_{ps}; dashed line marks adopted $(v_s)_{iso} = 1943$ m/sec. Short vertical (horizontal) lines through t_s and $\bar{V}_{s+}$ points indicate a horizontal longitudinal (transverse) component of motion.

there is evidence for axial rotation with depth, in this case away from the flow line.

Station S-360

Profile S-360, shot primarily to record subglacial refractions, comprises only three shots at wide angles (Figure 19). $\bar{V}_p$ was observed only for $i_p > 50°$, but the three points are enough to show $\bar{V}_p$ increasing from a low value around $i_p = 50°$. On two records, both components of R_{ps} were recorded, the transverse component being the faster. These data are enough to limit the possible models to those with intermediate values of I and θ (no model can explain the extremely low apparent value of $(\bar{V}_{s+})_0$). A good fit to all velocities is obtained with $I = 45°$, $|\theta| = 105°$, $\tan \delta_2 = 1\frac{1}{2}$ (model A), the model having been chosen to reproduce the observed velocity difference between V_{2+} and V_{3+}, rather than to match either exactly. Model B provides almost as good a fit; model A is preferred primarily because it seems to suggest better the observed trends in the velocities. $\theta = +105°$ is in reasonably good agreement with the fall line.

In this case, v_s was determined from one arrival by fixing $(t_s)_0$ on the basis of intercept ratios at neighboring stations. The value is reasonable, but cannot, of course, be considered a truly independent determination of $(v_s)_{iso}$.

Station E-300

$\bar{V}_p$ was observed at station E-300 only for $i_p < 50°$, enough to indicate a sharply decreasing velocity with increasing i_p, and consequently a fairly high value of $(\bar{V}_p)_0$. This, in turn, means truly high velocities for $\bar{V}_{s+}$. As a consequence, only one model (model B) provides a good fit: $I = 28°$, $|\theta| = 105°$, $q_p = 85\%$ (Figure 20). The value of $\tan \delta_2$ has little significance to this profile, since no transverse geophones were used.

The azimuth in model B corresponds either to 45° from the down-slope direction or to 75° from up slope. For comparison, the two best models with azimuths parallel to the flow line are presented. Model A fits $\bar{V}_p$ as well as model B, but is unacceptable because of its gross disagreement with $\bar{V}_{s+}$. Model C is better for $\bar{V}_{s+}$, and worse for $\bar{V}_p$. If, in model C, we consider a conical distribution of axes, $(V_{11})_0$ will be substantially reduced (Table 3), reducing $\bar{V}_{s+}$, and thus improving the fit to V_{2+}. At the same time, however, $\bar{V}_p$ will also be reduced,

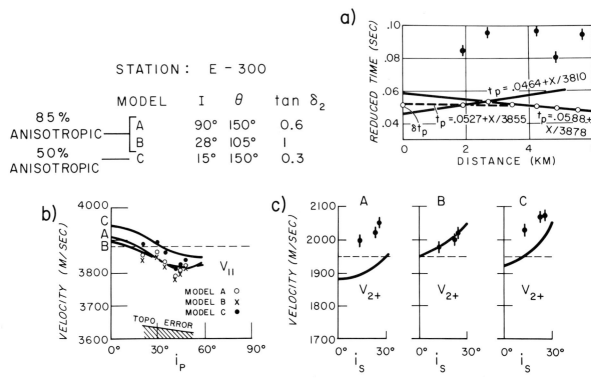

Fig. 20. Observed velocities and model fitting, profile E-300. (a) Travel times and travel-time equations; t_p (open circles) reduced to 3850 m/sec, t_s (solid points) reduced to 1950 m/sec. (b) Observed $\bar{V}_p$ (points) and fitted models (solid lines) for R_p; dashed line marks adopted $(v_p)_{iso} = 3878$ m/sec. Maximum expected topographic error is shown at base of diagram. (c) Observed $\bar{V}_{s+}$ (points) and fitted models (solid lines) for R_{ps}; dashed line marks adopted $(v_s)_{iso} = 1947$ m/sec. Short vertical lines through t_s and $\bar{V}_{s+}$ points indicate a horizontal longitudinal component of motion.

increasing the disagreement with V_{11}. A similar situation holds for the effect of moderate axial rotation, or of change in q_p; that is, a model that tends to improve the fit for one of the velocities increases the disagreement for the other. Model C is therefore also less acceptable than model B.

There are three indications that v_p, as measured at distances greater than 2.5 km, does not represent $(v_p)_{iso}$ for undisturbed firn ice. First, 3878 m/sec is a rather high value [Bentley, unpublished manuscript]; second, $(t_p)_0$ is greater than δt_p by 7 msec; and third, t_p at $x = 1.9$ km falls below the travel time line by more than 3 msec. (Unfortunately, the t_s points are too scattered to determine v_s with the accuracy needed for comparison with v_p, although t_s at $x = 1.9$ km is also relatively low.) Two possible explanations can be considered: anisotropy and abnormal densification.

Assuming anisotropy and flat layers, we find $(v_p)_1 = 3810$ m/sec from the points at $x = 1.9$ and 2.7 km, $(v_p)_2 = 3878$ m/sec, and a depth to the anisotropic ice of about 200 meters. This value for $(v_p)_2$ is just that to be expected on the basis of anisotropic model B above. However, we have ignored the rather large discrepancy between $(t_p)_0$ and δt_p. Furthermore, 3810 m/sec seems very low for $(v_p)_1$ at depths near 200 meters. We might, on the other hand, assume $(v_p)_1 = 3855$ m/sec, a representative value for $(v_p)_{iso}$ [Bentley, unpublished manuscript], but this agrees poorly with both the observed travel times and $(t_p)_0 - \delta t_p$. In fact, there is no one-dimensional model that satisfies all the data.

As was pointed out earlier, δt_p at distance x is just the intercept that would be observed if we recorded at distance $x/2$. We have recorded δt_p at 5.8 km; between this distance and 2.9 km, the travel-time line has a constant slope. We can thus formally treat half the profile as a case of a sloping, planar interface, using the standard technique for seismic refraction shooting of introducing a fictitious station at 2.9 km. The apparent velocities are then 3878 and 3842 m/sec, corresponding to a true $v_p = 3860$ m/sec.

This value is so close to $(v_p)_{\text{iso}} = 3855$ m/sec that we cannot be certain whether or not there is a velocity discontinuity. If there is not, the velocity-depth function, $V_p(z)$, must vary along the profile. Even if there is, v_p is so close to the isotropic value that it is not possible for $\partial V_p/\partial x$ to be zero down to the interface, since a physically unrealistic slope of several degrees would be implied.

What is actually implied by the refraction data is that the total travel time through the ice and firn above the refracting horizon differs along the profile. For a refracting horizon at depth z_p, $\int_0^{z_p} dz/V_p$ must differ by about 13% (the relative difference between $(t_p)_0$ and (δt_p) along the profile), and consequently, because V_p is closely proportional to ρ [*Robin*, 1958], $\partial \rho/\partial x \neq 0$. The value of ρ must average some 10% less at a given depth in the center of the profile than at the ends.

This is rather a startling result, but one for which a possible explanation is at hand. There is a step about 80 meters in height in the subglacial profile (Figure 21), facing upstream, and sharp enough to cause a discontinuity in the seismic reflection times. By simple conservation of mass, the velocity of ice flowing over a step of height h must increase by a factor close to h/H, the increase occurring in a narrow zone of large strain rate. There would correspondingly be a large decrease in the horizontal compressive stress that could decrease the densification rate. Beyond this zone, densification would

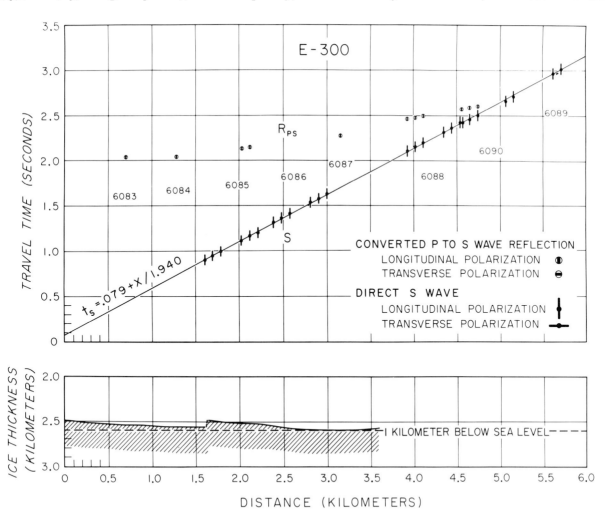

Fig. 21. Travel-time plot, s and R_{ps}, profile E-300. Lower plot shows subglacial topographic profile on a one-to-one scale. Longitudinal and transverse polarizations are in the horizontal plane.

proceed at faster than the normal rate, because the density at a given pressure level would be less than normal. The density anomaly would decrease more or less linearly, both because of this and because of accumulation of undisturbed snow.

This process would create a near discontinuity in the travel times. Arrivals from shots fired upstream from the scarp would not be affected. Those just past the scarp (allowing for the curved ray path) would be delayed the most, the delay decreasing about linearly with increasing distance of the shot beyond the anomalous zone. The apparent velocity would thus be about constant, as observed.

If such an anomalous zone exists, it can be no more than a few hundred meters in width, since the travel time at $x = 1.9$ km appears to be mostly unaffected, and it is not physically reasonable to suppose that it could be much narrower. If we estimate the flow velocity as 30 m/yr, on the basis of strain rates approximated by $\dot{A}/H$ and the distance to the ice divide, we thus find a passage time through the anomalous zone on the order of 10 years.

In 10 years, about 7 meters of snow will accumulate. The P-wave travel time through 7 meters of snow is about 7 msec, which is the delay measured. Thus, for this hypothesis to be valid, the velocity-depth function relative to the 10-year-old surface would have to be similar to that relative to the current surface in an undisturbed area, i.e. there would have to be a major reduction in the densification rate.

To see whether this could be possible, let us estimate the strain rate, using the approximate relation

$$\dot{e}_{xx} = \frac{1}{H}\left(\dot{A} - v\frac{dH}{dx}\right) \quad (18)$$

[Robin, 1967], where v is the flow velocity. From the seismic reflection results, dH/dx is no less than 1, so $\dot{e}_{xx} \simeq 0.01$ yr^{-1}. (Equation 18 is not strictly valid for such large slopes, but we seek only an order of magnitude result.) The compactive strain rate at $z = 10$ meters, $\rho = 0.5$ g/cm^3, is about 0.015 yr^{-1} [Ramseier and Pavlak, 1964]. Since the two strain rates are about equal and opposite in sign, a significant reduction in the densification is possible.

However, there are difficulties with this explanation. If $\dot{e}_{xx} = 0.01$ yr^{-1}, and $h/H = 80/2500 \simeq 0.03$, it would only take 3 years to effect the necessary velocity increase, and it is quite doubtful that there would be sufficient time to develop the density anomaly. Another problem is an unexplained 500-meter offset between the subglacial scarp and the travel-time discontinuity (compare Figures 20 and 21). We can conclude that, for this hypothesis to be acceptable, the scarp must, on a three-dimensional average, be higher and closer to the recording site than is shown by the reflection profile.

Station E-396

The reflection data at station E-396 are rather poor. R_{ps} was not observed at all, and $\bar{V}_p$ cannot be fit very well by any model, since none exists that produces a velocity minimum at $i_p < 20°$ and a maximum near 35°, as the data appear to require. It is probable that $\bar{V}_p$ is distorted by topographic error, which can be considerable at small i_p. We have thus sought a model that fits the mean of $\bar{V}_p$, not a very restrictive condition. Assuming a down-slope azimuth, $\theta = +45$, we find a fairly good fit by interpolating between models A and B (Figure 22), giving $I = 70°$, $\theta = +45°$ for the preferred model. The fit is just as good for smaller I and larger θ, e.g. $I = 45°$, $\theta = +70°$. The data do appear to exclude small and large values of I and θ.

There are some interesting anomalies in the refraction profile. For $x < 2.7$ km, $v_p = 3874$ m/sec, which is rather high to be a good isotropic value [Bentley, unpublished manuscript], whereas s disappears completely. A likely cause for this behavior is a shallow anisotropic zone in which v_p is larger, and v_s smaller, than in the ice above. To fit $(v_p)_2$ with an anisotropic model, we need $\phi = 30°$ or $80°$. In the former case, however, V_2 exceeds the isotropic value, so that only for $\phi = 80°$, i.e. for c axes nearly vertical, are the seismic results satisfactorily explained. An alternative explanation for the high value of $(v_p)_2$ is the occurrence of bubble-free ice, although in that case v_s should also increase slightly.

Another anomaly is the disagreement between δt_p on the most distant shot and $(t_p)_0$ for the high-velocity line. For a flat refracting boundary the two should agree, since half the distance to the last shot is past the break in t_p, but in fact $\delta t_p = 0.444$ sec, whereas $(t_p)_0$ is only 0.414 sec. As at station E-300, the inequality implies either a sloping boundary on the high-velocity zone or horizontal density variations of about 5% in the upper firn layers. Introducing a fictitious station at 2.9 km, we find a true value for $(v_p)_2$ of 3882 m/sec and depths, on the sloping

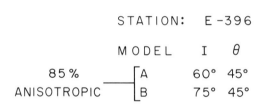
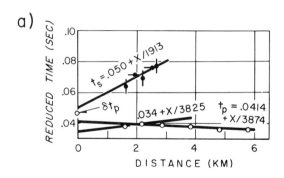
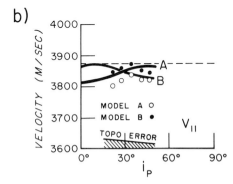

Fig. 22. Observed velocities and model fitting, profile E-396. (a) Travel times and travel-time equations; t_p (open circles) reduced to 3850 m/sec, t_s (solid points) reduced to 1950 m/sec. (b) Observed $\bar{V}_p$ (points) and fitted models (solid lines) for R_p; dashed line marks adopted $(v_p)_{iso} = 3853$ m/sec. Maximum expected topographic error is shown at base of diagram. Short vertical (horizontal) lines through t_s points indicate a horizontal longitudinal (transverse) component of motion.

boundary assumption, of 130 meters beneath the recording site and 160 meters at $x = 2.9$ km. The corresponding slope is about ⅔ of a degree.

There is nothing in the subglacial topographic profile to suggest any zone of abnormal strain rate.

Station E-444

The salient features of profile E-444 are a rapidly decreasing $\bar{V}_p$ with increasing i_p, $\bar{V}_{s+}$ increasing with i_s from a low value at vertical incidence, and comparable horizontal components of motion in R_{ps}. These features limit acceptable models to two regions: I small but not zero, with $|\theta|$ near 90°, and I near 90°, with $|\theta|$ small but not zero. Equally satisfactory velocity models can be found in each region, as is seen in Figure 23. In model A, the horizontal axes are aligned with the slope direction; in model B, the near-vertical axes are inclined in a direction approximately parallel to the contours, as with similar models at other stations. Because at other stations vertical R_{ps} arrivals are generally strong when I is large and absent when I is small, model A is preferred.

Refracted velocities are about normal, v_s having been determined by assuming $(t_s)_0 = 1.5(t_p)_0$. The disappearance of s for $x > 2$ km is consistent with the occurrence of anisotropic ice, according to either reflection model, at a depth of a few hundred meters.

Station E-708

Profile E-708 was cut short by Tellurometer failure; only three wide-angle shots were recorded (Figure 24). R_{ps} was observed only on transverse geophones, and with such scatter as to provide very little information. Relatively large $\bar{V}_p$ values for i_p between 30° and 50° do eliminate models with near vertical or near horizontal axes. The model selected is simply the best fit assuming $\theta = -150°$, approximately along the flow line. Refracted velocities are about normal.

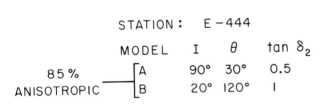
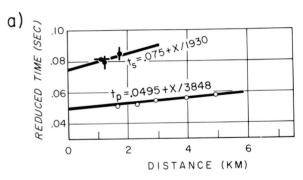
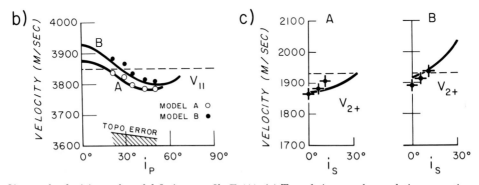

Fig. 23. Observed velocities and model fitting, profile E-444. (a) Travel times and travel-time equations; t_p (open circles) reduced to 3860 m/sec, t_s (solid points) reduced to 1950 m/sec. (b) Observed $\bar{V}_p$ (points) and fitted models (solid lines) for R_p; dashed line marks adopted $(v_p)_{iso} = 3855$ m/sec. Maximum expected topographic error is shown at base of diagram. (c) Observed $\bar{V}_{s+}$ (points) and fitted models (solid lines) for R_{ps}; dashed line marks adopted $(v_s)_{iso} = 1943$ m/sec. Short vertical (horizontal) lines through t_s and $\bar{V}_{s+}$ points indicate a horizontal longitudinal (transverse) component of motion.

Station E-756

Good R_{ps} arrivals were recorded, including a double set for $i_s > 25°$, at station E-756 (Figure 25). Good R_s arrivals were also recorded on several shots, both components of motion being represented. In view of the good R_s arrivals, which demonstrate shear wave generation at the source, the absence of any observable s waves is remarkable; the profile is unique in this regard.

Careful examination of t_p reveals a slight increase in v_p at $x \simeq 6$ km (Figure 25a). The travel times can be fit closely by two lines, with $(v_p)_1 = 3840 - 3847$ m/sec (depending on the weight given to the low point at 4.6 km), and $(v_p)_2 = 3860$ m/sec. The $(v_p)_1$ range is in keeping with the value expected for isotropic firn-ice at the temperature of this station ($-23°$C at 10-meter depth) [Bentley, unpublished manuscript], so that $(v_p)_2$ probably corresponds to either anisotropic or high-density, bubble-free ice.

The velocity increase is so small that we cannot distinguish between a gradual and an abrupt change.

The calculated depth to an abrupt change is 200 meters. If the increase were gradual, however, thus starting from a depth considerably less than 200 meters, and if v_s correspondingly decreased, the failure to record s would be explained. v_1 exceeds the isotropic value by 13 m/sec at $\phi = 30°$ and $\phi = 77°$. The corresponding differences for S waves are: for $\phi = 30°$, $\Delta V_2 = +29$ m/sec, $\Delta V_3 = -97$ m/sec, and for $\phi = 77°$, $\Delta V_2 = -93$ m/sec, $\Delta V_3 = -17$ m/sec; thus, in the latter case the required decrease in v_s does exist. If we set limits on ϕ of 75° to 80°, we find satisfactory models for v_p in the following ranges: $I = 15°$, $|\theta|$ or $|\theta'| = 0$ to $50°$; $I = 30°$, $|\theta|$ or $|\theta'| = 60°$ to $70°$; $I = 45°$, $|\theta|$ or $|\theta'| = 65°$ to $75°$; $I = 60°$, $|\theta|$ or $|\theta'| = 70°$ to $80°$; $I = 75°$ to $90°$, $|\theta|$ or $|\theta'| = 75°$ to $80°$.

Considering now the reflection data, we will first assume $q_p = 85\%$, since the depth to anisotropic ice suggested by v_p is only 10% of the ice thickness. The flatness of $\bar{V}_p$ with respect to changes in i_p, together with relatively high velocities compared to

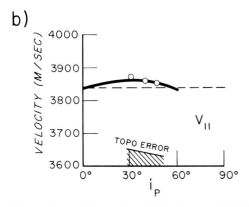
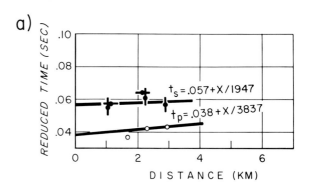
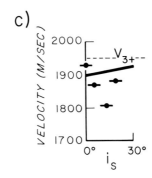

Fig. 24. Observed velocities and model fitting, profile E-708. (a) Travel times and travel-time equations; t_p (open circles) reduced to 3860 m/sec, t_s (solid points) reduced to 1950 m/sec. (b) Observed $\bar{V}_p$ (points) and fitted models (solid lines) for R_p; dashed line marks adopted $(v_p)_{iso} = 3837$ m/sec. Maximum expected topographic error is shown at base of diagram. (c) Observed $\bar{V}_{s+}$ (points) and fitted models (solid lines) for R_{ps}; dashed line marks adopted $(v_s)_{iso} = 1947$ m/sec. Short vertical (horizontal) lines through t_s and $\bar{V}_{s+}$ points indicate a horizontal longitudinal (transverse) component of motion.

$(\bar{V}_p)_0$, eliminates models with pronounced maximums or minimums or with large $(V_{11})_0$. Combined with the observed separation between the two components of $\bar{V}_{s+}$ for i_s between 25° and 30°, this restriction limits satisfactory models to $I = 30°$ – 45°, $|\theta|$ or $|\theta'|$ between 30° and 45°. Finally, the need for a large value of tan δ_2 to explain the transverse motion of the faster $\bar{V}_{s+}$ component eliminates the larger range of values for θ. A very good fit to $\bar{V}_{s+}$ and a fair fit to $\bar{V}_p$ is obtained by the model $I = 40°$, $\theta = -35°$ (Figure 25), the azimuth being in good agreement with the presumed flow line. For $q_p = 50\%$, the same model fits $\bar{V}_p$ a little better and v_{s+} a little more poorly.

Comparing the reflection and refraction solutions, we see once again that, if shallow anisotropy exists, some rotation of axes with depth must take place. From Figures 5 and 6 we can see that, if I decreases and $|\theta|$ increases upward to a model satisfying the refraction data, the fit to $\bar{V}_{s+}$ will not be changed much, while the V_{11} curve will tend to be flattened off at a level that will better fit $\bar{V}_p$. If, instead, I increases upward, a reasonable fit to most of the data could be obtained simply by considering I and $|\theta|$ to be slightly less than 40° and 35°, respectively, in the main body of the ice. However, $V_{11} < (V_p)_{iso}$ would be implied for $i_p > 60°$, whereas actually $\bar{V}_p > (V_p)_{iso}$ (Figure 25b). A model with I decreasing toward the surface is therefore preferred.

It would still be consistent with the observations if, in either model, I continued to change upward in the same sense above the depth corresponding to the p_2 arrival. If this were so, the tendency for a downward decrease in v_s would be accentuated, and the absence of s arrivals would be better explained.

Station E-828

Although profile E-828 (Figure 26) was only extended to 5 km, a clear break to a higher v_s was recorded. To delineate the break in the travel-time curve more clearly, S-wave arrivals from the individual traces on the three most distant shots have

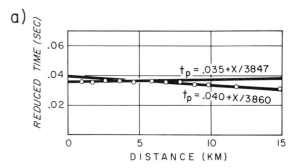
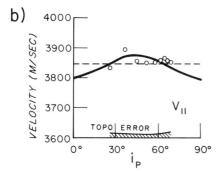
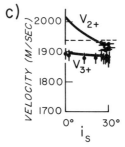

Fig. 25. Observed velocities and model fitting, profile E-756. (a) Travel times and travel-time equations; t_p (open circles) reduced to 3860 m/sec. (b) Observed $\bar{V}_p$ (points) and fitted models (solid lines) for R_p; dashed line marks adopted $(v_p)_{iso} = 3860$ m/sec. Maximum expected topographic error is shown at base of diagram. (c) Observed $\bar{V}_{s+}$ (points) and fitted models (solid lines) for R_{ps}; dashed line marks adopted $(v_s)_{iso} = 1937$ m/sec. Short vertical (horizontal) lines through $\bar{V}_{s+}$ points indicate a horizontal longitudinal (transverse) component of motion.

been plotted, rather than the usual record average. Note that the group of low arrivals between 4½ and 5 km contains points from two different shots at different central distances. The calculated depth to the layer in which $(v_s)_2 = 2015$ m/sec is 390 meters.

If we estimate that $(v_s)_2$ corresponds to a model $\bar{V}_2$ of 2050 m/sec, we find that $|\theta|$ or $|\theta'|$ must be $<55°$ and I must be $>35°$. Acceptable models include $\theta' = 0$, $I = 35°$ or $55°$; $\theta' = 30°$, $I = 40°$ or $70°$; $\theta' = 45°$, $I = 55°$; and $\theta' = 55°$, $I = 90°$. This high a value of v_s necessarily implies a decrease in v_p, so no model limitation is implied by the absence of a recorded $(v_p)_2$. Note that near-vertical axes cannot produce the observed increase in v_s.

Both $\bar{V}_p$ and $\bar{V}_{s+}$ show a very gradual change with angle of incidence, the latter at values about equal to $(v_s)_{iso}$ or greater. This combination cannot be reproduced by any single model. V_{11} provides a very good fit to $\bar{V}_p$ at an azimuth of $-120°$ (down-slope) for either of the two models shown in Figure 26b. Although an intermediate model would yield the correct mean value for V_{2+}, the slope with respect to i_s is entirely too great. A possible solution to this could be a changing inclination with depth; if we consider $I = 45°$ to be a mean between $I = 25°$ and $I = 60°$, V_{2+} would be fairly flat and slightly higher than the isotropic velocity. The refraction data, however, prove that I is already greater than $35°$ at a depth of 400 meters, and that large inclinations are needed if $|\theta|$ is near $120°$. Consequently, if the explanation just offered for $\bar{V}_{s+}$ is correct, the rotation must be from relatively large inclinations near the surface to lesser ones at depth.

The reflection data alone could be fit as well by a single bipolar model as by a rotation of axes with depth, but the high value of $(v_s)_2$ requires either a much greater concentration around one pole, or a separation between poles of no more than $20°$, neither possibility being satisfactory with respect to $\bar{V}_{s+}$. Both sets of data could be satisfied, however, if we postulated a strong, shallow concentration of c axes around $I = 60°$, $\theta = 135°$, and the downward development of a second concentration around $I = 30°$, $\theta = 105°$, at the expense of the first.

Station E-612

The large negative slope of $\bar{V}_p$ versus i_p on profile E-612 (Figure 27) can only be fitted by a model with I either near $0°$ or near $90°$. But, for $I \simeq 0°$, the im-

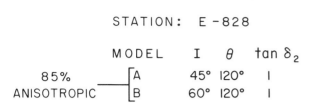
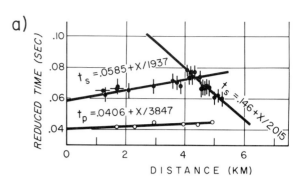
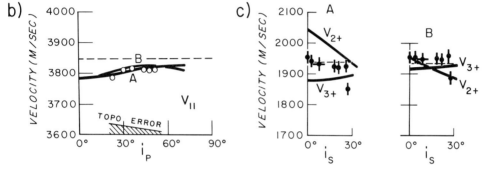

Fig. 26. Observed velocities and model fitting, profile E-828. (a) Travel times and travel-time equations; t_p (open circles) reduced to 3860 m/sec, t_s (solid points) reduced to 1950 m/sec. (b) Observed $\bar{V}_p$ (points) and fitted models (solid lines) for R_p; dashed line marks adopted $(v_p)_{iso} = 3847$ m/sec. Maximum expected topographic error is shown at base of diagram. (c) Observed $\bar{V}_{s+}$ (points) and fitted models (solid lines) for R_{ps}; dashed line marks adopted $(v_s)_{iso} = 1937$ m/sec. Short vertical (horizontal) lines through t_s and $\bar{V}_{s+}$ points indicate a horizontal longitudinal (transverse) component of motion.

plied $(V_{11})_0$ would be very high, increasing the already large values of $\bar{V}_{s+}$ at large i_s to impossibly high levels. Thus we can be quite certain that the c axes are nearly horizontal. A good fit to $\bar{V}_p$ can be found for $I = 90°$, $|\theta|$ between 135° and 150°, (models A and B), but the match to the extremely large $\bar{V}_{s+}$ versus i_s slope is poor. The only way this can be improved is by considering inclinations slightly greater than 90°, or, equivalently, $I < 90°$, $|\theta| \sim 30°$ (model C).

The models discussed, particularly $I = 90°$, $|\theta| = 135°$, would imply a large v_s. The greatest distance at which s was recorded was only 2.7 km. On the last shot, however, the transverse component arrived 0.02 sec early, suggesting a break to a very high velocity (nominally 2170 m/sec). Assuming $(v_s)_2 = 2150$ m/sec, we estimate the corresponding depth to anisotropic ice to be 340 meters, 14% of the total ice thickness of 2460 meters. However, this model does not provide an explanation for the disappearance of s for $x > 3$ km, which would seem to imply a downward velocity decrease, unless for some reason S waves were poorly generated by the more distant shots.

Station E-1020

Both v_p and v_s are substantially lower at this station than is acceptable for isotropic values. For model fitting, $(v_p)_{iso}$ and $(v_s)_{iso}$ were assumed to be 3848 and 1930 m/sec, respectively, on the basis of a study of the variation of these quantities with temperature [*Bentley*, unpublished manuscript].

The reflection velocities alone are not definitive in determining a model; either a near-vertical or a near-horizontal model axis is possible. Because of the occurrence of R_{ps} at vertical incidence, however, a model with a horizontal axis is preferred. A very good velocity fit is found for a model intermediate between the two shown in Figure 28, although the indicated $\tan \delta_2$ is too small. As at E-612, the model implies an inclination slightly greater than 90° in the downslope direction.

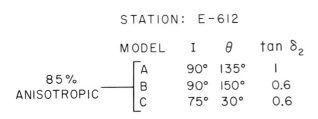
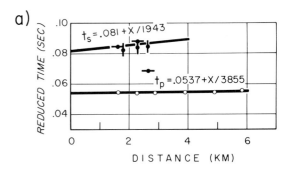
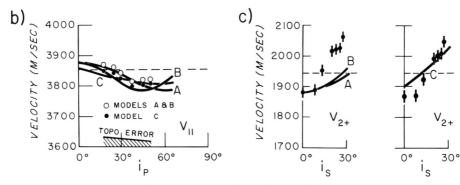

Fig. 27. Observed velocities and model fitting, profile E-612. (a) Travel times and travel-time equations; t_p (open circles) reduced to 3855 m/sec, t_s (solid points) reduced to 1950 m/sec. (b) Observed $\bar{V}_p$ (points) and fitted models (solid lines) for R_p; dashed line marks adopted $(v_p)_{iso} = 3855$ m/sec. Maximum expected topographic error is shown at base of diagram. (c) Observed $\bar{V}_{s+}$ (points) and fitted models (solid lines) for R_{ps}; dashed line marks adopted $(v_s)_{iso} = 1943$ m/sec. Short vertical (horizontal) lines through t_s and $\bar{V}_{s+}$ points indicate a horizontal longitudinal (transverse) component of motion.

Because there is no indication of velocity increase with distance beyond 1 km, the most likely explanation for the low refraction velocities is the occurrence of anisotropy, resulting in a decrease in horizontal velocities, at a depth less than that of complete densification of the firn. The near-surface model would then have to be such that ϕ is around 65 or 70° for horizontal ray paths, compared with 15 to 20° for the reflection solution. A 50° axial rotation with depth is larger than that suggested at any other station. If it is real, the goodness of fit to the reflection velocities suggests that the rotation takes place at relatively shallow depths. Such a rotation (at whatever depth) could help to explain the discrepancy between tan δ_s and the observed amplitude ratios.

DISCUSSION

The analysis we have carried out has permitted us to find a 'single-crystal' model that is in reasonably good agreement with the seismic data at most stations (summarized in Table 4). How are we to interpret these models? Strongly developed anisotropy is necessary for satisfactory model fitting at a majority of stations, since at a majority, including all with better than average data, a high value of $(v_p)_2$ or $(v_s)_2$ or a pronounced low value of $\bar{V}_p$ or $\bar{V}_{s+}$ was observed. These velocities were too different from those in isotropic ice to be consistent with a distribution of axes either within the volume, or on the surface, of a cone of large apex angle. It follows that a pattern of strong axial concentration (e.g., 90% of the axial directions within 30° of a pole) around a very few poles is required at many places and is consistent with all results. For further discussion we will therefore assume such a pattern to exist throughout the part of West Antarctica investigated. The question is then whether more than one pole generally exists.

The best evidence for the number of poles obviously comes from the Byrd drill hole. In a preliminary analysis, *Gow et al.* [1968] report no concentration of axes in any direction other than vertical. It is reasonably safe to assume by analogy that a

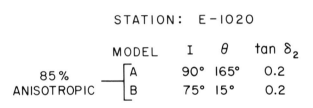
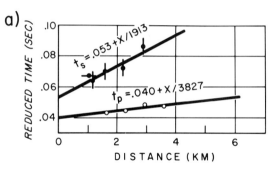
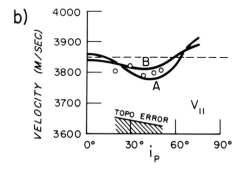
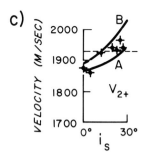

Fig. 28. Observed velocities and model fitting, profile E-1020. (a) Travel times and travel-time equations; t_p (open circles) reduced to 3860 m/sec, t_s (solid points) reduced to 1950 m/sec. (b) Observed $\bar{V}_p$ (points) and fitted models (solid lines) for R_p; dashed line marks adopted $(v_p)_{iso} = 3827$ m/sec. Maximum expected topographic error is shown at base of diagram. (c) Observed $\bar{V}_{s+}$ (points) and fitted models (solid lines) for R_{ps}; dashed line marks adopted $(v_s)_{iso} = 1913$ m/sec. Short vertical (horizontal) lines through t_s and $\bar{V}_{s+}$ points indicate a horizontal longitudinal (transverse) component of motion.

single pole describes the fabric at S-858, where the seismic data are very similar to those at Byrd-58, and perhaps also wherever I is small, but extrapolation to other stations, where the fabrics are clearly quite different in orientation, is not automatically justified. We must consider what limits are implied by the seismic data themselves.

To begin with, at all stations except E-348 an ambiguity in the sign of θ necessarily exists. We could not distinguish, therefore, between any unipolar model at azimuth θ, and a bipolar model with azimuths $\pm\theta$. Our preferred solutions, however, have been found with a wide range in θ at different stations; it is certainly not reasonable to suppose that many, if any, of the profiles were accidentally laid out midway in direction between two poles of axial concentration. We will henceforth assume, therefore, that we are concerned with either $+\theta$ or $-\theta$, but not both.

It is also generally difficult to distinguish between models with I small, θ near 90°, and those with θ small, I near 90°, since the primary differences in V_{11} (see Figure 5) occur at small i_p, where we can obtain little experimental information, and at large i_p, corresponding to distances attained only on the longest profiles. Examples of this were seen in the model fitting for Byrd-58 and S-858. Consequently, it would be even more difficult to rule out a bi-polar distribution combining these two models. The best chance of doing this would arise if it were possible to observe a very rapid velocity increase for $i_p > 60°$, such as occurs when $I \simeq 90°$, $|\theta| < 15°$ (Figure 5).

This is just the case at E-972, where the large $(v_p)_2$ points to a major concentration of axes nearly horizontal and along the line of the profile. An additional concentration near the vertical is possible without violating the refraction data, since we can maintain $(v_p)_2$ constant in the model by moving the horizontal pole closer to $\phi = 0$ as the near-vertical concentration increases, up to a limit of 50% of the axes at $I = 90°$, $\theta = 0°$, and 50% at $I = 0°$. The good agreement of V_{11} with $\bar{V}_p$ at large angles of incidence (Figure 13b), however, suggests that, if a vertical concentration exists, it is substantially smaller than the horizontal. (As was pointed out

TABLE 4. Summary of Results
Values of H were determined using listed values of $(V_{11})_0$

Station	H m	T °C	$\dot{A}$ gm/cm²yr	$\dot{A}/H$ 10^{-5}yr^{-1}	Layer No.	v_p (m/sec)	v_s (m/sec)	Depth to top of layer (m)	$(V_{11})_0$ (m/sec)	$(t_p)_0$ (millisec)	\multicolumn{5}{c}{Preferred Reflection Model}	\multicolumn{2}{c}{Refraction Model}
											q_p	I	θ	$\tan\delta_2$	Angle with Fall Line	I	θ
B-58	2590	-28.4	14.4(3)	6	1	3855	1934		3900	47.0	50%	15°	±105°	0.5	60° or 90°		
S-150	2530	-28.0	24(2)	9	1	3855	1931		3795	49.5	85	45	±105	1.5	60 or 90	>30	>120
					2		1950	200									
S-360	2970	-26.6	27(2)	9	1	3851	1943		3790	54.0	85	45	+105	1.5	0		
S-858	2820	-31.2	20(2)	7	1	3848	1924	150	3910	48.5	50	15	±105	0.5	75		
					2		1964	200									
					2	3882				56.0							
E-300	2600	-28.1	30(2)	12	1	3855*			3900	52.7	85	28	±105	1	45 or 75	(probably isotropic)	
					2(?)	3860		200(?)		58.8							
E-348 E&N	3440	-27.7	36(1)	10	1	3853	1946		3850	43.1	50	35	-60 (rel. to E)	>2	0	55	45
E					2		1950-2130	250-1240									
N					2		2035	850									
E-396	2690	-25.8	38(1)	14	1	3825	1913	130	3815	34.2	85	70	+45	-	0*	<10°	-
					2	3882				41.4							
E-444	2200	-25.5	56(1)	25	1	3848	1930		3875	49.5	85	90	-30	0.5	<30°		
E-612	2460	-27.7	34(1)	14	1	3855	1943		3860	53.7	85	105	-150	0.6	0	90	-135
					2(?)		~2170	340(?)									
E-708	1500	-23.7	29(1)	19	1	3837	1947		3840	38.0	85	30	-150	0.3	0*		
E-756	1970	-23.0	36(1)	18	1	3840-47			3795	35.0	85	40	-35	>3	0		
					2	3860		200		40.0							
E-828	2040	-24.2	37(1)	18	1	3847	1937	390	3785	40.6	85	45	-120	1	0	30	-65
							2015										
E-972	1260	-25.9	49(1)	39	1	3842	1935	230	3855	49.0	35	75	+145	0.5	0	55	-135
					2		1985	800			50	90	+165	0.2	20	75	+145
					3	3940				133.						90	+165
E-1020	1360	-25.0	51(1)	38	1	3827	1913		3850	40.0	85	95	-165	0.2	0	65°≤ ∅ ≤70°	

*Assumed

References: for T: C. Bull in Bentley, e al., (1964)
for A: (1) Shimizu (1964)
(2) Giovinetto (personal communication)
(3) Gow (1961)

earlier, $(v_p)_2$ at this station probably is representative of that part of the ice sheet which also primarily determines $\bar{V}_p$, although this is not generally the case.) Thus extrapolation to other stations at which I in the preferred model is near 90°, if justified, suggests that a unipolar distribution is typical of horizontal as well as vertical concentrations.

Consider now the high velocity $(v_s)_2$ at stations where an intermediate value of I was preferred. At E-828 and on both profiles at E-348, $(v_s)_2$ was large enough to indicate propagation within 10° or so of the single-crystal V_2 maximum at $\phi = 45°$. A second pole at E-828 would thus have to lie in a spherical annulus having an area about 25% of that of the unit sphere. At E-348, the second pole would have to lie within the intersection of two such annuli, the corresponding area being only 5%, and the a priori probability of a separated pole in the right position would only be about 2½%.

From the refraction data above, two poles symmetrically placed with respect to the horizontal plane would be indistinguishable from one. However, such an arrangement throughout the anisotropic ice appears to be inconsistent with R_{ps} at E-348 unless one pole is substantially stronger than the other.

Although bimodal distributions cannot be completely ruled out, there is no consistency between hypothetical patterns, the type of bi-polar model which would be satisfactory at a particular station being quite ad hoc, depending upon the type of velocity information being considered. Furthermore, theoretical analyses [*Kamb*, 1959a; *Macdonald*, 1960; *Brace*, 1960], even though they may not be very satisfactory in detail, do suggest that, if two poles exist, they tend either to lie 90° apart in the plane normal to the intermediate stress direction or to lie symmetrically with respect to that plane, which in the ice sheet is normally the flow plane [*Nye*, 1957]. The poles in most of our unipolar models have indeed been close to that plane (partly, of course, by preferential selection) (Figure 29), yet among the cases discussed, our analyses have indicated no satisfactory bipolar solution of either of these types. A bimodal (and, a fortiori, a multimodal) distribution with well separated poles thus appears less likely than a unipolar distribution at stations where the seismic data are best, and, therefore, if these stations are representative, in the region in general.

Considering now the resolution of two neighboring

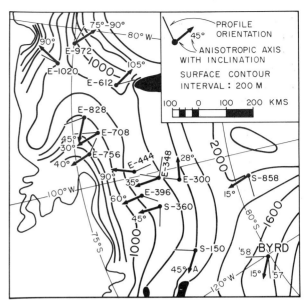

Fig. 29. Map of northern West Antarctica, showing station locations, profile directions, orientations of preferred model axes, and elevation contours on the ice surface. Solid black areas denote regions of exposed rock.

poles, examination of Figure 4 or Table 1 shows that an acceptable bimodal distribution could comprise poles no more than 20° apart at E-348 and E-828, or 30° apart at E-972. But there is no basis in theory, experiment, or observation to expect two poles so close together. Once again, extrapolating to the rest of the ice sheet, we conclude finally that the most likely crystal fabric throughout West Antarctica, as well as at Byrd station, is a pronounced concentration of c axes around a single pole, any secondary poles being relatively minor.

If a unipolar distribution does generally exist, the attitude of the pole in space must be a function of depth in most places, since, almost without exception, models that fit refraction data differ considerably from those satisfying the reflection data from the same profiles. It is to such variation that we would then presumably attribute most of the disagreement between models and observations; model improvement on this basis was specifically suggested for profiles E-972 and E-828. There is no consistent pattern to the indicated variations. The inclination can either increase or decrease with depth, and the azimuth can change either toward or away from the flow line. This suggests a dependence on the stress history as well as the local stress distribution in the ice, not a surprising result in view of the very slow recrystallization rates that might be expected in ice at −25°C.

Alternatively, we might attribute the imperfection of the models to incorrectness of the unipolar assumption, despite the difficulty in fitting specific bipolar models at several stations. This possibility could only be tested by extensive calculations.

From the discussion above, it follows that it may be fruitful to treat the perfectly anisotropic models that we have produced as reasonably close to the predominant crystal orientation. Let us therefore examine the results further to see whether we can discern any patterns of possible significance. We have already noted geographical variations in the models, but these could hardly be meaningful in themselves. If we guess that the preferred orientations result from recrystallization in an inhomogeneous stress field, then we would like to know the state of stress in the ice at, and upstream from, each station. In actuality, we have little such information, so we will simply use an indicator of changing stress conditions between stations, which is easy to calculate. Assuming plane strain and $\partial \dot{e}_{xx}/\partial x = 0$, ($x$ is now taken in the direction of flow), we can estimate $\dot{e}_{xx}$ by $\dot{A}/H$, where $\dot{e}_{xx}$ is the longitudinal strain rate. We then assume by Glen's law [*Glen*, 1955] that the longitudinal stress deviator $\tau_{xx'}$ is proportional to $(\dot{e}_{xx})^{1/4}$, so that by examining I as a function of $(\dot{A}/H)^{1/4}$ we obtain an idea of the variation of the axial inclinations as a function of stress.

There does appear to be a general increase in I with increase in $(\dot{A}/H)^{1/4}$, despite the considerable uncertainty in the values of I (Table 4). To depict this variation, we have plotted cos I against $(\dot{A}/H)^{1/4}$, the cosine having been chosen, rather arbitrarily, because it provided an approximately linear relationship over part of the range of values (Figure 30). If the models are anywhere nearly correct, then there is a clear tendency for near-horizontal inclinations to be associated with large strain rates (E-444, E-972, E-1020), and near-vertical orientations (Byrd-58, S-858) with small strain rates. At intermediate strain rates, there is a suggestion of a linear correlation between cos I and the estimated longitudinal deviatoric stress, and it is interesting that the three points in worst agreement, those for E-708, E-756, and E-828, all correspond to stations that lie in an area where there is a strong regional convergence of flow lines (Figure 29), indicating important transverse compressive stresses, with consequent deviation from plane strain.

Of course, the reality of these relationships is far from proven, and the pattern discussed could easily be altered by, for example, choosing $+\theta$ or $-\theta$ at each station in order to produce an azimuth farthest from the flow line instead of along it, or by different choices of preferred models where more than one was acceptable. On the other hand, the fact that a pattern exists which was in no way preselected provides a modicum of support for the model choices.

At any rate, the evidence is convincing for some real differences in preferred orientation between stations, since we can at the very least say that high velocities for refracted shear waves are incompatible with a vertical preferred orientation, and the evidence is at least suggestive of a regular increase in inclination with increasing strain rate from the interior of West Antarctica toward the coast. A speculative but reasonable extrapolation would suggest that for strain rates smaller than that at Byrd station the inclination (if preferred orientations develop) is

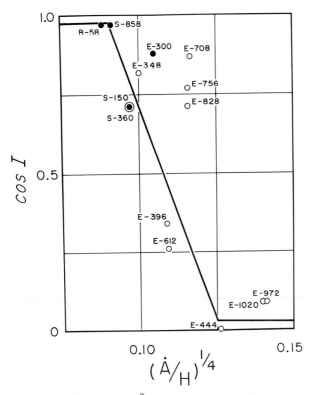

Fig. 30. Cos I versus $(\dot{A}/H)^{1/4}$. Solid points denote preferred-model azimuths at a large angle to the flow direction. Solid line is a schematic fit to the data. Station numbers are shown next to points.

small, whereas for strain rates higher than the maximum estimated here the inclination remains near horizontal.

An interesting consequence of this discussion is a possible implication for East Antarctica. If extrapolation is valid, which is particularly questionable in view of the much colder temperatures there, one would infer that near-vertical preferred orientations occur throughout most of that region. It is in any case reasonable to suppose for the central regions of the ice sheet that, if strong anisotropy exists, the crystal axes may be oriented near the vertical. Experimental support for this conclusion is found in a comparison of seismic and electromagnetic echo times at stations in the interior of Queen Maud Land, which suggest a very high value of $(V_p)_0$ [*Clough and Bentley*, 1970].

A significant feature of most of the profiles is the occurrence of $(R_{ps})_0$, i.e. R_{ps} at vertical incidence. This would not be expected for the theoretical model given above, a feature of which was an isotropic zone at the base of the ice sheet. At the boundary between isotropic mediums, no reflected S wave is produced by a P wave at normal incidence, nor should one be produced in a transversely isotropic medium with axis of symmetry either normal or parallel to the boundary. For an inclined, transversely isotropic medium, however, the symmetry is lost, and a reflected S wave (type 2) is to be expected.

Consequently, where $(R_{ps})_0$ occurs, the implication is that anisotropic ice exists at the base of the ice, and that the axial orientation is skew. At B-58, where it is known that the basal ice is isotropic and that I is near $0°$ [*Gow et al.*, 1968], R_{ps} occurs only at large angles of incidence (Figure 14). The same is true at S-858 (Figure 17), for which our preferred model is similar, and perhaps at E-396, where R_{ps} was not recorded at all, but the maximum value of i_s attained was smaller. At S-150 (Figure 18) and E-300 (Figure 20), the other two stations with preferred model azimuths at large angles to the flow line, but with larger inferred inclinations, R_{ps} occurs at moderate and large angles of incidence, but not near the vertical. $(R_{ps})_0$ was recorded at normal incidence at all other stations, including those at which the estimated strain rate is relatively high. Since these stations include several at which the preferred model inclination is nearly horizontal, the implication is not only that anisotropic ice with a skew axis of symmetry occurs at the base of the ice sheet, but also that the orientation may differ significantly from that in the ice above.

Preferred model inclinations for the five profiles on which $(R_{ps})_0$ was not observed range up to $60°$, suggesting that the reason for the absence of this phase is more likely to lie in the presence of a basal isotropic zone than in near-vertical or near-horizontal axial orientations. Since the preferred orientation on four of these profiles forms a large angle with the flow plane (and perhaps also on the fifth profile, as the azimuth for E-396 was assumed, rather than shown, to be down slope), we may speculate that the conditions conducive to the development of preferred orientations that are nearly vertical or oblique to the flow plane may also tend to favor the development of an isotropic basal layer.

SUMMARY AND CONCLUSIONS

The analysis of 16 refraction and reflection profiles in West Antarctica has confirmed the widespread existence of highly anisotropic ice throughout much of the thickness of the West Antarctic ice sheet, and has shown that the preferred c-axial direction must differ in some places from the vertical orientation observed in the deep cores from Byrd station. Since clear evidence was found on all the profiles where the data were best, it is reasonable to conclude that anisotropy is a predominant, if not universal, characteristic of the deep ice in West Antarctica. Several additional, more speculative inferences can be drawn about the nature and extent of the anisotropy.

1. The predominant crystal fabric at stations where the data are most nearly definitive, and thus perhaps throughout West Antarctica, probably comprises a pronounced concentration of c axes around a single mean direction, any secondary poles being of significantly less importance.

2. It appears likely that a down-slope azimuth of the preferred axial orientation, i.e. one in the presumed flow plane, occurs commonly where the indicated axial inclination is relatively large. This was suggested without ambiguity at only one station, where perpendicular profiles were completed. However, a down-slope azimuth is consistent with the seismic data at ten out of eleven stations where the inferred inclination is greater than or equal to $30°$. (At two of these stations, a down-slope azimuth was assumed, resulting in a preferred model inclination greater than $30°$, but the other models were chosen

without regard to geography.) On the other hand, no satisfactory model with a down-slope azimuth could be found at any of the three stations where an inclination less than 30° was indicated.

3. Relatively high-velocity refracted arrivals from indicated depths of 130 to 390 meters were observed on about half the profiles. In all instances the occurrence of anisotropic ice provides a satisfactory explanation for the high velocities, although in a few cases the high velocity could instead arise from a layer of bubble-free ice. The travel-time break is usually abrupt, indicating a rather rapid transition (i.e., in a few tens of meters) from isotropic to anisotropic ice. Discontinuities (i.e. significant changes in less than ten meters) are possible, but there is no positive evidence that they exist. On one profile (E-348E), however, a gradual velocity gradient covering a depth interval of some 1000 meters appeared, presumably resulting at least in part from a gradual increase in the directional concentration of crystal axes.

4. There is a strong presumption from the occurrence of R_{ps} at and near vertical incidence that anisotropic ice extends to the glacial bed at most stations. This conclusion is consistent with the disagreement between amplitude ratios observed for the two types of R_{ps} arrivals and those calculated on the assumption of an isotropic layer at the base of the ice. Where vertical R_{ps} reflections are absent, an isotropic basal layer probably is present. If so, there appears to be an association between such a layer and the occurrence of axial inclinations at a large angle with the flow line which may be more than coincidental, although unexplained.

5. Most of the profiles have been fit reasonably well by models in which 85% of the ice is assumed to be anisotropic, and on several of these it would not be possible to find a satisfactory fit with a much thinner anisotropic section. This is consistent with the evidence for anisotropy near the surface and at the base of the ice sheet. Consequently, it is likely that in most of northern West Antarctica at least half, and in some place perhaps more than 90%, of the ice column is anisotropic. Byrd station appears to represent about a minimum in this regard.

6. There is widespread evidence of a changing anisotropic pattern with depth. This evidence is seen both in the refracted arrivals themselves, as at E-348 and E-972, and in the common lack of agreement between models satisfying refraction and reflection data. On the other hand, the variation cannot be too great or extended over too large a depth range, or the anisotropic effects on the reflections would become blurred. A reasonable guess is that the fabric is relatively constant over most of the bottom half of the ice sheet, and that variations in mean crystal orientation of a few tens of degrees are common in the upper half and near the base.

7. Some correlation between the axial inclination I implied by the preferred model and the longitudinal strain rate or stress appears likely. For the preferred models of this study, $\cos I$ varies approximately linearly with the longitudinal deviatoric stress, as estimated from the ratio of accumulation rate to ice thickness, for strain rates between 0.5 and 2.5 × 10^{-4} yr^{-1}. Above and below this range, I appears to be relatively constant near 90°, and 0°, respectively. The data are far too few and the models too uncertain, however, to suggest any quantitative significance or general applicability of this result. Even if the apparent correlation is qualitatively real, it is highly probable that I will depend on factors other than the local longitudinal stress: dependence on temperature and stress history, for example, might well be expected. Extrapolation beyond the range and region of observation must therefore be made with great caution. Nevertheless, it seems quite possible that a vertical preferred orientation is common in East Antarctica, where the strain rates are very small. This conclusion is supported by independent information from electromagnetic sounding.

Note added in proof. Since this manuscript was submitted, a new and exciting idea of great potential importance to glaciology has been put forward by *Hughes* [1971]. Analyzing the temperature measurements in the deep drill hole at Byrd station, he has found it to be likely that the critical Rayleigh number is exceeded in the lower part of the ice sheet, and hence that diapiric flow in the ice is a distinct possibility. Whatever the details of such motion, its occurrence as a common feature in the West Antarctic ice sheet would go far toward explaining the remarkable anisotropic features described above. The large strains associated with convection would naturally result in highly developed ice fabrics. Anisotropic conditions would extend much higher above the base of the ice than seems possible by any other mechanism. Since the diapiric cells would have dimensions on the order of the ice thickness, the anisotropic patterns would be locally regular, but could be expected to vary considerably with depth, as well as in horizontal distances of a few kilometers. The difficulty in fitting consistent anisotropic models at most stations would be naturally explained, as would the discrepancy between observations in the deep drill hole at Byrd station and the seismic measurements ten miles away. In short, the seismic evidence presented in this paper strongly supports Hughes' general concept of convection in the ice sheet.

Acknowledgments. The author wishes to express his thanks for valuable field assistance to Perry Parks, Ned Ostenso, Leonard LeSchack, George Widich, Jack Long, and George Toney, and his appreciation for essential logistic support provided by representatives of the National Science Foundation Office of Antarctic Programs and U.S. Navy Air Development Squadron Six. Financial support under N.S.F. grants G-13210, G-20970, and GA-1136 is also gratefully acknowledged.

Contribution 228 from the Geophysical and Polar Research Center, University of Wisconsin.

REFERENCES

Ahmad, G., Energy distribution among the reflected and refracted elastic waves at the boundary between transversely isotropic media, 2, *Ann. Geofisi., 20,* 303–329, 1967.

Bass, R., D. Rossberg, and G. Ziegler, Die elastischen Konstanten des Eises, *Z. Phys., 149,* 199–203, 1957.

Bennett, H. F., An investigation into velocity anisotropy through measurements of ultrasonic wave velocities in snow and ice cores from Greenland and Antarctica, Ph.D. thesis, University of Wisconsin, Madison, 1968.

Bentley, C. R., The structure of Antarctica and its ice cover. *Res. Geophys., 2,* 335–389, 1964.

Bentley, C. R., Seismic evidence for moraine within the basal antarctic ice sheet, this volume.

Bentley, C. R., Temperature coefficient of seismic wave velocity in the West Antarctic ice sheet, unpublished manuscript.

Bentley, C. R., and N. A. Ostenso, Glacial and subglacial topography of West Antarctica, *J. Glaciol., 3,* 882–911, 1961.

Brace, W. F., Orientation of anisotropic minerals in a stress field, Discussion, *Geol. Soc. Amer. Mem. 79,* 9–20, 1960.

Brockamp, B., and H. Querfurth, Untersuchungen über die Elastizitätskonstanten von See- und Kunsteis, *Polarforschung, 5,* 253–262, 1964.

Clough, J. W., and C. R. Bentley, Measurements of electromagnetic wave velocity in the East Antarctic ice sheet, in *International Symposium on Antarctic Glaciological Exploration (ISAGE),* Publ. 86, International Association of Scientific Hydrology and Scientific Committee on Antarctic Research, Cambridge, England, 1970.

Glen, J. W., The creep of polycrystalline ice, *Proc. Roy. Soc. London, A, 228,* 519–538, 1955.

Gow, A. J., The inner structure of the Ross ice shelf at Little America V, Antarctica, as revealed by deep core drilling, *I.A.S.H. Comm. Snow Ice, Publ. 61,* 272–284, 1963.

Gow, A. J., Bubbles and bubble pressures in antarctic glacier ice, *J. Glaciol., 7,* 167–182, 1968.

Gow, A. J., H. T. Ueda, and D. E. Garfield, Antarctic ice sheet: Preliminary results of first core hole to bedrock, *Science, 161,* 1011–1013, 1968.

Hughes, T., Convection in polar ice sheets as a model for convection in the earth's mantle, *J. Geophys. Res., 76,* 2628–2638, 1971.

Jona, V. F., and P. Scherrer, Die elastischen Konstanten von Eis-Einkristallen, *Helv. Phys. Acta, 25,* 35–54, 1952.

Kamb, W. B., Ice petrofabric observations from Blue Glacier, Washington, in relation to theory and experiment, *J. Geophys. Res., 64,* 1891–1909, 1959a.

Kamb, W. B., Theory of preferred crystal orientation developed by crystallization under stress, *J. Geol., 67,* 153–170, 1959b.

Kamb, W. B., Glacier geophysics, *Science, 146,* 353–365, 1964.

MacDonald, G. J. F., Orientation of anisotropic minerals in a stress field, *Geol. Soc. Amer. Mem. 79,* 1–8, 1960.

Mackenzie, J. K., The elastic constants of a solid containing spherical holes, *Proc. Phys. Soc. London, 63B(1),* 2–11, 1950.

Nye, J. F., The distribution of stress and velocity in glaciers and ice sheets, *Proc. Roy. Soc. London, A, 239,* 113–133, 1957.

Postma, G. W., Wave propagation in a stratified medium, *Geophysics, 20,* 780–806, 1955.

Ramseier, R. O., and T. L. Pavlak, Unconfined creep of polar snow, *J. Glaciol., 5,* 325–332, 1964.

Rigsby, G. P., Crystal orientation in glacier and in experimentally deformed ice, *J. Glaciol., 3,* 589–606, 1960.

Robin, G. de Q., Ice movement and temperature distribution in glaciers and ice sheets, *J. Glaciol., 2,* 523–532, 1955.

Robin, G. de Q., Seismic shooting and related investigations, in *Norwegian-British-Swedish Antarctic Exped., 1949–52, Sci. Results 5, Glaciology 3,* Norsk Polarinstitutt, Oslo, 1958.

Robin, G. de Q., Surface topography of ice sheets, *Nature, 215,* 1029–1032, 1967.

Röthlisberger, H., Seismic exploration in cold regions, *Cold Regions Sci. Eng. Monograph II-A2a,* Cold Regions Res. Eng. Lab., Hanover, N.H., in press, 1971.

Schytt, V., The inner structure of the ice shelf at Maudheim as shown by core drilling, in *Norwegian-British-Swedish Antarctic Expedition, 1949–52, Sci. Results 4, Glaciology 2C,* pp. 113–152, Norsk Polarinstitutt, Oslo, 1958.

Steinemann, S., Experimentelle Untersuchen zur Plastizität von Eis, *Beitr. Geol. Schweiz, Hydrologie, 10,* esp. pp. 46–50, 1958.

Vallon, M., Contribution a l'étude structurographique de la glace froide de haute latitude, *Acad. Sci. (Paris), Compt. Rend., 257,* 3988–3991, 1963.

GRAVIMETER OBSERVATIONS ON ANVERS ISLAND AND VICINITY

Gilbert Dewart

Institute of Polar Studies, Ohio State University, Columbus   43210

Gravimeter and magnetometer observations were made in the region of Anvers Island and Bismarck Strait. The gravity data were used to estimate the regional gravity anomaly and the thickness of the Anvers Island ice cap. The gravity bases were tied to existing intercontinental networks through South America and the Antarctic Peninsula. The regional Bouguer anomaly was found to increase from +30 mgal along the southwest coast of Anvers Island to more than +50 mgal on the west coast. The ice cap in the southwestern part of Anvers Island was estimated to be largely between 300 and 600 meters thick. It is underlain by a narrow coastal plain and a bedrock plateau approximately 200 meters high. A magnetic anomaly of more than 1000 γ was found northeast of Palmer station.

During the period January through March 1967, the Institute of Polar Studies of Ohio State University conducted a gravimeter survey in the Antarctic Peninsula region. A gravity base was established at Palmer station, on the southwest coast of Anvers Island, and it was connected to existing networks of gravity stations. Gravity observations were made on Anvers Island and its vicinity, and these data were used to estimate the regional gravity anomaly and the thickness of a part of the Anvers Island ice cap.

CALIBRATION

The gravity meter used in this investigation was a La Coste and Romberg Model G geodetic gravimeter, No. 41, with a dial constant given by the manufacturer of 1.045 mgal/scale division. The journey to Antarctica and back, ranging from midlatitude to low latitude to fairly high latitude, presented an opportunity to calibrate the gravity meter in the field under favorable reading conditions. A map of the route is given in Figure 1.

Gravimeter bases of high accuracy, established by a University of Wisconsin gravimeter program at airports and harbors along the route [*Woollard and Rose*, 1963], were occupied on the trip to Antarctica, and some were re-occupied on the way back. The calibration constants for our gravimeter, obtained over the various legs of the trip from the absolute gravity values of Woollard and Rose, are presented in Table 1. On the basis of these data, the calibration constant given by the manufacturer was judged to be correct within the margin of reading error.

The last column in Table 1 shows the deviation between the gravity difference as obtained by meter observations at a dial constant of 1.045 mgal/scale division and the gravity difference obtained from the base values of Woollard and Rose. These deviations are similar in magnitude to the value 0.3 mgal given by Woollard and Rose for the average agreement between mean gravimeter and pendulum values of gravity on a worldwide basis.

In Punta Arenas, a tie was made between the base at Carlos Ibáñez del Campo airport (formerly Chabunco airport) and the Puerto Magellanes harbor base, WH1019. A difference of 24.1 mgal was found, compared to 23.3 mgal between the values for the two bases given by Woollard and Rose. This is a rather large discrepancy, considering the good agreement obtained on the long flight legs.

On the return flight, a new gravimeter base was established at the new Santiago airport, with gravity value $g = 979.4591$ gal (the reading was taken on the terrace just outside the street entrance to the airport, on the last white tile square to the left, facing the airport; see Figure 2).

CONNECTIONS BETWEEN PALMER AND THE OTHER GRAVITY NETWORKS

Gravity stations have been established on the Antarctic Peninsula and nearby islands by the University of Wisconsin [*Cohen*, 1963] and the British Antarctic Survey [*Griffiths et al.*, 1964]. The British values were revised by *Kennett* [1965]. Cohen's gravity reference base was the Punta Arenas harbor station (WH1019) mentioned above. The original British reference point was Ezeiza international airport at Buenos Aires, and the revision included the Punta Arenas harbor base (WH1019) and Montevideo. Carlos Ibáñez del Campo (Chabunco) airport (WA6136) at Punta Arenas was selected as the ref-

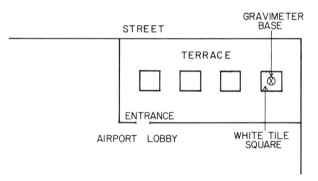

Fig. 2. Site of gravimeter base at the new Santiago airport, Chile.

erence base for the present survey because of the consistent results obtained there in comparison with the rest of the calibration range.

At Palmer station, a new temporary gravity base was established on the concrete tie-down pier near the entrance at the western corner of the British Antarctic Survey hut. Later a permanent base was established at the Astronomical Observation Point (Figure 3). The gravity differences for the links from Carlos Ibáñez del Campo to Palmer and from Palmer to Carlos Ibáñez del Campo were 995.6 mgal and 995.2 mgal, respectively. The mean was used in obtaining the absolute gravity value for Palmer station. The values for the temporary and permanent gravity bases are given in Table 2.

From Palmer station, three links were made with bases on the other gravity networks (Figure 4). Port Lockroy, on the British net, and Waterboat Point (site of the Chilean station Gabriel González Videla), which was occupied by the University of

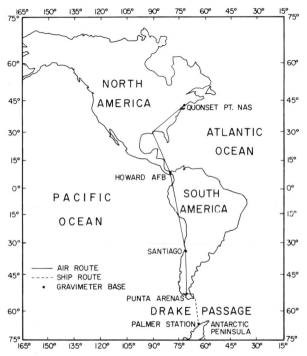

Fig. 1. Map of route to Antarctica and gravimeter bases.

TABLE 1. Calibration Constants

Base	Index No.*	Absolute Gravity Value,* gal.	Computed Calibration Constant, mgal/scale div.		Deviation, mgal
			To Antarctica	From Antarctica	
Quonset Pt. N.A.S.	WA 261	980.3122			+0.1
			1.0449		
Howard A.F.B.	WA 4005	978.2304			−0.5
			1.0454 ⎫ 1.0452	1.0451	·
Los Cerrillos airport	WA 6110	979.4500	1.0449 ⎭		+0.1
Carlos Ibáñez del Campo (Chabunco) airport	WA 6136	981.3130			

*Woollard and Rose [1963].

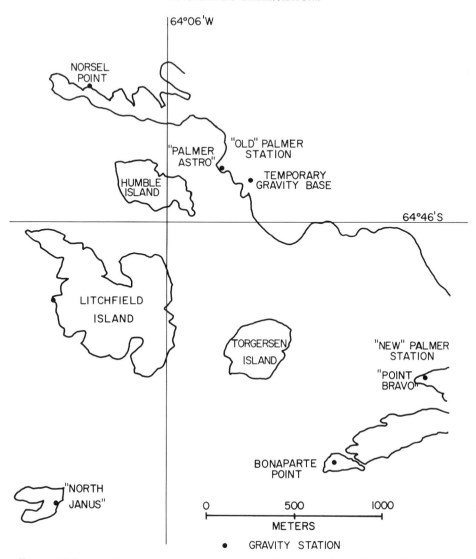

Fig. 3. Palmer station gravity bases and other gravity stations in the Arthur Harbor area.

Wisconsin and the British Antarctic Survey, were tied in during a helicopter reconnaissance. Galindez Island (site of the British station Argentine Islands) was visited on a two-day round trip by Greenland cruiser. At Port Lockroy and Galindez Island, it was possible to occupy the precise sites established by the British Antarctic Survey as gravity bases [*Griffiths et al.*, 1964]. At Waterboat Point, the observation was made at the entrance to the building containing the Chilean pendulum station [*Cohen*, 1963].

Small drift rates (<0.01 mgal/hr) were encountered during both trips, and were distributed time-proportionally for the legs of the flight to Port Lockroy and Waterboat Point. The absolute gravity values obtained for these inter-network ties and the values obtained by the preceding surveys are given in Table 2 (British values are revised according to *Kennett* [1965]).

The results for the present survey are consistently higher than the British Survey values by about 1 mgal. This may be the result of operating from different reference bases. The discrepancy from Cohen's value for Waterboat Point is more pronounced and is much greater than the discrepancy for the gravity difference between the reference base at Carlos Ibáñez del Campo airport and Cohen's reference base at Punta Arenas harbor.

TABLE 2. Comparison of Absolute Gravity Values

Base	Absolute Gravity, gal.		Brit. Antarct. Surv.	Diff: OSU−BAS Values, mgal
	Ohio State	Univ. Wisc.		
Carlos Ibáñez del Campo airport		981.3130		
Palmer Station				
Temporary	982.3084			
Astro Point	982.3092			
Port Lockroy	982.3046		982.3036	+1.0
Waterboat Point	982.3052	982.3080	982.3044	+0.8
Galindez Island	982.3402		982.3392	+1.0

REGIONAL GRAVITY ANOMALY

A series of gravimeter observations was made on the west and southwest coasts of Anvers Island and on several islands in the Bismarck Strait in order to obtain regional Bouguer gravity anomaly values. An attempt was made to find sites at sea level in low-relief environments in order to minimize elevation and terrain effects. It is estimated that the topographic error is not more than a few tenths of a milligal except at Bonnier Point station (see Table 3), where it may be several milligals. The 'zero elevation' stations were all within a few centimeters of the ocean surface. The maximum tidal range at Port Lockroy is given as 0.85 meter (U.S. Naval Oceanographic Office Map H.O. 6650, 1963) and observations at Arthur Harbor indicated a similar range. Hence the gravity values at the zero-elevation stations are probably not more than 0.2 mgal from the value at mean sea level.

Elevations not at sea level were determined with aneroid altimeters and may be subject to considerable error. However, on the basis of barometric observations at Palmer station, and considering the rapidity of the surveys and the frequency of sea-level checks, it seems unlikely that the elevation corrections for the higher stations are in error by more than a few tenths of a milligal.

Most of the station positions were located on maps, with the aid of aerial photographs (U.S. Naval Oceanographic Office Charts H.O. 6690, 1964; H.O. 6691, 1965; Gt. Brit. Directorate of Overseas Surveys, British Antarctic Territory, Sheet 46, 1960). It is unlikely that any of them are in error by more than 0.1 minute of latitude, with reference to the map coordinates. This corresponds to about 0.1 mgal in theoretical gravity value.

The gravity values and Bouguer anomalies for the gravimeter stations are presented in Table 3. Figure 5 is a contour map of the regional Bouguer anomaly. The contours correspond rather well with those obtained for adjacent regions by the British Antarctic Survey [*Griffiths et al.*, 1964]. The Anvers Island region appears to produce a second-order anomaly superimposed on a primary Bouguer anomaly gradient extending from the continental crust of the Antarctic Peninsula toward the ocean. Structurally, the Bismarck Strait and the Wauwermans Islands seem to be part of the Anvers Island unit. The gravity low near the center of the west coast may reflect the proximity of the central massif of Anvers Island, which swings westward in this region.

In the vicinity of Arthur Harbor, the 30-mgal contour coincides with a presumed contact between Andean intrusives and an adjacent assemblage of altered rocks [*Hooper*, 1962]. It may be that here the contours follow the outline of a major Andean batholith.

Over the southwestern part of the island, it is possible to project a small gravity gradient (1 mgal/km) extending from the center of the island (where the summit of the central massif is located) toward the west and southwest coasts.

INVESTIGATION OF THE ANVERS ISLAND ICE CAP

Anvers Island is approximately 70 km long by 40 km wide (Figure 6). The eastern and central parts

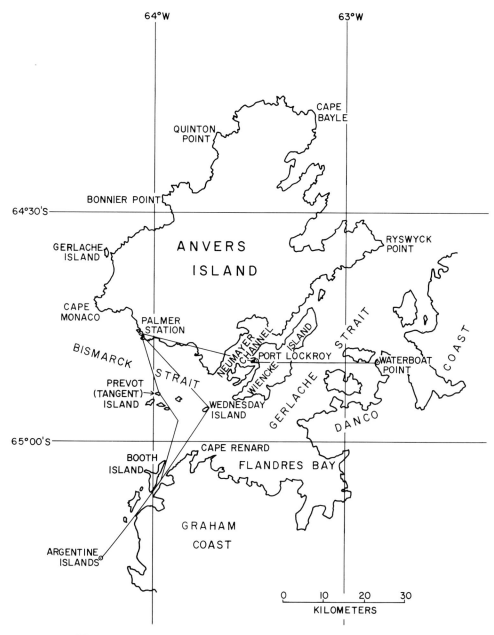

Fig. 4. Gravity links between Palmer station and other survey bases.

of the island are occupied by mountains that culminate in Mount Français (2760 meters). The western half of the island is covered by an ice cap that increases in width from about 10 km in the north to more than 20 km in the south. An ice divide, which extends from the vicinity of Mount Français west-southwest to Cape Monaco, separates the narrow northern part of the ice cap, called Marr Ice Piedmont, from the southern part, which flows generally south and southwest. This roughly triangular southern part, which covers approximately 200 km², was the subject of this investigation. It extends from sea level at the Bismarck Strait to an elevation of approximately 900 meters at the foot of the mountains near the ice divide, and it is unbroken by nunataks. Gravity measurements were made in order to esti-

TABLE 3. Gravity Values on Anvers Island and Vicinity

No.	Station	Latitude, S	Longitude, W	Height, meters	Observed Gravity, gal	Theoretical Gravity, gal	Bouguer Anomaly, mgal
1	Booth Island	65° 02.5′	63° 56.5′	0	982.3265	982.2969	+29.6
2	Wednesday Island	64° 55.5′	63° 42.7′	0	.3269	.2887	+38.2
3	Prevot Island	64° 53.2′	63° 57.0′	29	.3145	.2860	+34.0
4	Biscoe Point	64° 48.90′	63° 48.35′	0	.3124	.2810	+31.4
5	'Point Bravo'	64° 46.45′	64° 04.17′	0	.3072	.2780	+29.2
6	'Palmer Astro'	64° 45.83′	64° 05.62′	7	.3092	.2774	+33.2
7	Dream Island	64° 43.62′	64° 14.02′	0	.3110	.2748	+36.2
8	Cape Monaco	64° 42.62′	64° 18.22′	0	.3158	.2736	+42.2
9	'Monaco North'	64° 41.55′	64° 16.63′	0	.3170	.2724	+44.6
10	Gerlache Island	64° 35.8′	64° 12.4′	7	.3164	.2657	+52.1
11	'Gerlache North'	64° 33.6′	64° 06.4′	0	.3084	.2631	+45.3
12	Bonnier Point	64° 28.7′	63° 55.4′	171	.2631	.2574	+41.2
13	Giard Point	64° 26.9′	63° 51.4′	0	.2912	.2553	+35.9
14	Perrier Bay	64° 21.7′	63° 42.1′	18	.2920	.2492	+46.1
15	Quinton Point	64° 19.6′	63° 39.4′	0	.2957	.2468	+48.9
Other Arthur Harbor stations							
	'North Janus'	64° 46′ 51″	64° 06′ 38″	0	982.3110	982.2786	+32.4
	Bonaparte Point	64° 46′ 43″	64° 04′ 48″	4	.3059	.2784	+28.4
	Litchfield Island	64° 46′ 14″	64° 06′ 49″	0	.3094	.2778	+31.6
	'Palmer Temporary'	64° 45′ 53″	64° 05′ 24″	10	.3084	.2775	+33.1
	Norsel Point	64° 45′ 36″	64° 06′ 33″	0	.3106	.2771	+33.5

mate the thickness of the ice cap and the morphology of the underlying land surface.

In view of the generally flat surface of the ice cap and its small depth in relation to its lateral extent, the assumption was made that at each observation point it consisted of an infinite slab of ice resting on an infinite horizontal rock surface. The difference in the Bouguer gravity anomaly between a point on the ice cap and a base station on bedrock (Palmer station), after certain corrections, was assumed to be due to the presence of the ice and rock layers under the ice cap point. With values assumed for the ice and rock, the height of the rock surface with respect to the elevation of the base station and the thickness of the ice sheet could be found: $h = \Delta g / 0.04185 \, (\rho_r - \rho_i)$, where h is the ice thickness, Δg is the difference in Bouguer anomalies, and ρ_r and ρ_i are the densities of rock and ice, respectively.

The gravity observations on the ice cap were made, for the most part, at existing survey markers established for use in the glaciology program. These markers, mostly aluminum poles or 4 in. by 4 in. (10.16 cm by 10.16 cm) wooden beams, were located at 1- to 3-km spacings along a network of intersecting lines, and their relative positions are probably accurate to within a few meters (Figure 6). Most of the points were accessible by motor toboggan, but a few had to be approached on foot or by helicopter.

Station elevations were measured by altimeter. Pressure variation control was provided by the bar-

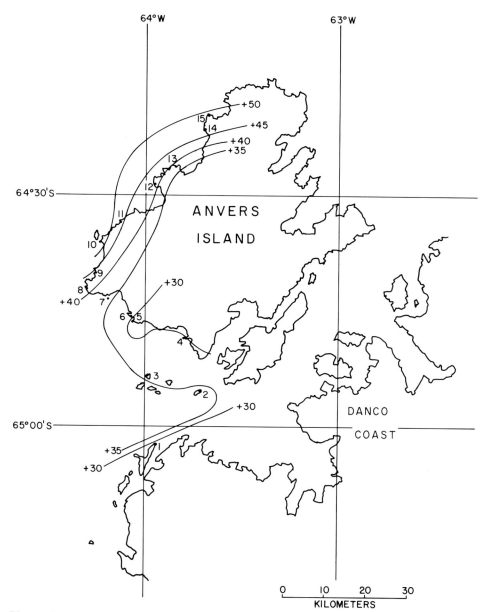

Fig. 5. Bouguer anomalies in the region of Anvers Island and Bismarck Strait. The contour interval is 5 mgal. The numbered points refer to gravimeter stations listed in Table 3.

ograph records at Palmer station. Pressure gradient effects were probably small, inasmuch as the most distant gravity station was only 25 km from Palmer station. The elevations are probably accurate to ±10 meters.

The regional gravity anomaly was estimated from the data of Figure 5. Unfortunately, the only bedrock stations were those along the coast. In the interior of the island, especially near the massif, the regional gradient may change significantly.

The topographic effect of the ice surface was generally negligible, but an estimated correction using the Hammer zone method was made on the relatively steep slopes near the coast. The sub-ice terrain effect was also estimated where there appeared to be significant bedrock features. The topographic

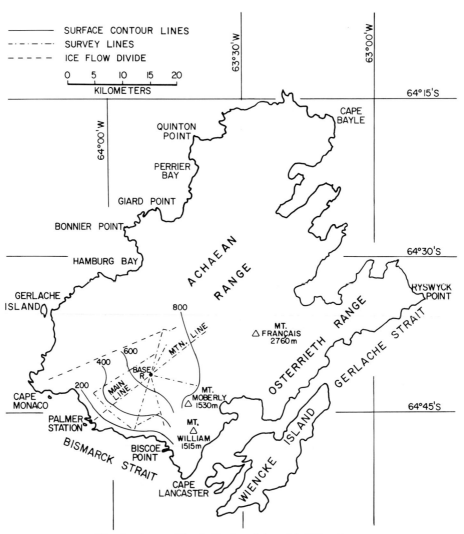

Fig. 6. Anvers Island. Contour interval is 200 meters.

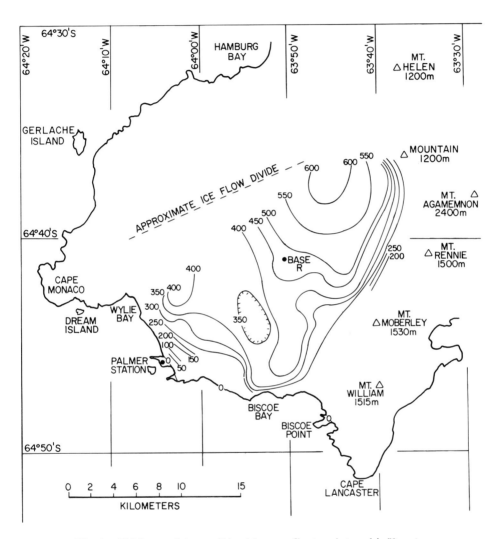

Fig. 7. Thickness of Anvers Island ice cap. Contour interval is 50 meters.

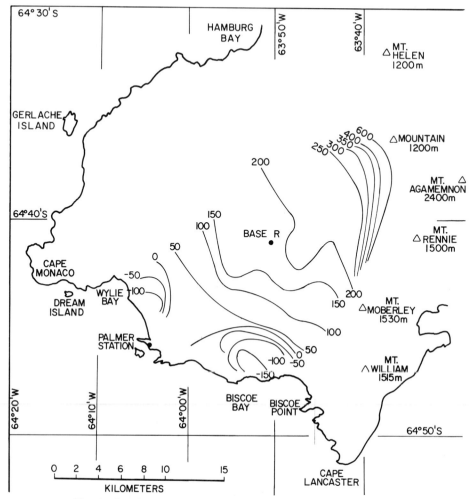

Fig. 8. Southwest Anvers Island. Bedrock elevation in meters.

effect of the mountains could also be only roughly estimated without detailed topographic maps.

The density of the ice was assumed to be 0.9 g/cm³. The mean density of forty specimens of rock collected along the southwest coast was 2.67 ± 0.12 g/cm³. Samples from the altered assemblage and from the Andean intrusive suite had densities near this mean, and these rocks apparently underlie much of the section of the ice cap under investigation.

Gravimeter drift became significant during the work on the ice cap, and it was necessary to frequently re-occupy stations to make corrections. For the main period of ice cap operations, January 30 to February 22, the mean drift rate was 0.02 mgal/hr, and double this rate was experienced for periods of one day or less.

Over-all, it is estimated that the ice thickness values may be in error by as much as ±20%, although the accuracy in the flat region between the coastal slope and the proximity of the mountains is probably better than this.

The estimated ice thickness in the southwestern part of Anvers Island is shown in Figure 7. Figure 8 is a contour map of the underlying topography.

On the basis of these data, the subglacial surface of the southwestern part of Anvers Island appears to consist largely of a narrow coastal plain and a low plateau. The over-all slope is gentle (2–3%), but the slope increases on the approach to the plateau between 4 and 10 km from the coast and again at the foot of the mountains. The coastal plain is indented by two hollows which represent the heads

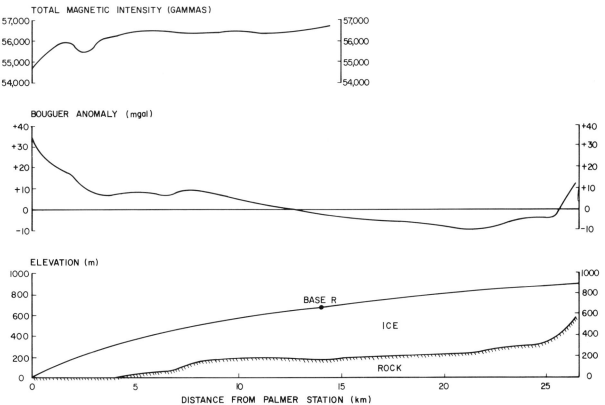

Fig. 9. Vertical section along main line and mountain line (Figure 6). The vertical scale is exaggerated five times. The Bouguer anomaly and total magnetic intensity along the profile are also shown.

of Wylie Bay and Biscoe Bay. In these bays the bedrock drops below sea level about 3 to 4 km in from the ice front, which marks the apparent coastline. However, the ice front appears to be grounded, and along the rest of the southwest coast bedrock is observed at the foot of the ice front. The bays are separated by a low saddle that meets the sea in the rocky headlands near Palmer station. The relative smoothness of the 50-meter contour (Figure 8) suggests a former long-standing position of sea level.

The bedrock terrain is covered by an ice cap that is largely between 300 and 600 meters thick. It attains its greatest thickness just below the foot of the mountains, then thins abruptly as the elevation rises. Toward the coast the ice thickness decreases over the plateau, then remains nearly constant or even increases slightly on the slope down to the coastal plain. On the plain the ice thickness decreases rapidly toward the coast.

The features described above are illustrated in the cross section of Figure 9 from Palmer station to the foot of the mountains. The Bouguer anomalies and the total magnetic intensity along this profile are also shown.

Total field measurements were made with an Elsec proton precession magnetometer. The profile from Palmer station to base R on the ice cap was recorded continuously. The data were corrected by reference to magnetograms from the Argentine Islands observatory, which is 55 km south of Palmer station. The large magnetic anomaly that appears near the coastal edge of the ice cap may be due to the contact noted above between the Andean intrusives and the altered assemblage. *Hooper* [1962] found a much higher iron mineral content in specimens from the Andean intrusive suite than in specimens from the altered assemblage.

Acknowledgments. This research was supported by NSF grant GA-529 awarded to the Ohio State University Research Foundation. The assistance of the Palmer station wintering party, 1967–1968, the personnel of the USCGC *Westwind,* and the U.S. Navy helicopter crews attached to it is gratefully acknowledged. J. C. Farman of the British Antarctic Survey and A. S. Rundle of the Institute of Polar Studies provided valuable information.

Contribution 150 from the Institute of Polar Studies, Ohio State University.

REFERENCES

Cohen, Theodore J., Gravity survey of Chilean Antarctic bases, *J. Geophys. Res., 68,* 263–266, 1963.

Griffiths, D. H., R. P. Riddihough, H. A. D. Cameron, and P. Kennett, Geophysical investigation of the Scotia arc, *Brit. Antarctic Surv. Sci. Rep. 46,* 1964.

Hooper, P. R., The petrology of Anvers Island and adjacent islands, *Falkland Islands Dependencies Surv. Sci. Rep. 34,* 1962.

Kennett, P., Revision of gravity links between South America and the Antarctic, *Brit. Antarctic Surv. Bull., 7,* 25–28, 1965.

Woollard, G. P., and J. C. Rose, *International Gravity Measurements,* George Banta, Menasha, Wis., 1963.

SECULAR INCREASE OF GRAVITY AT SOUTH POLE STATION

Charles R. Bentley

*University of Wisconsin, Department of Geology and Geophysics
Geophysical and Polar Research Center, Middleton 53562*

A regular increase in observed gravity at South Pole station between 1957 and 1967 has been well established. The conclusion that a substantial part of this change is due either to a large horizontal movement of the ice sheet or to a secular decrease in ice surface elevation, however, does not stand closer examination. Observations and calculations indicate that at least 90% of the gravity change can be attributed to the sinking of South Pole station, partly in balance with the normal snow accumulation rate and partly in response to the superimposed load of the station-induced snow drift. A small contribution may arise from motion relative to the local surface slope. No disequilibrium of the ice sheet is implied. A model for the sinking rate of the station that is in satisfactory agreement with the observations is derived.

A secular increase in observed gravity at South Pole station on the Antarctic ice sheet between December 1957 and January 1966 was reported by *Behrendt* [1967]. An additional tie between South Pole station and the base station on solid ground at McMurdo station, made with Lacoste-Romberg gravimeter 5 in November 1967, is in agreement with Behrendt's results (Figure 1). Including the new point, and excluding one old observation because of unsteady gravimeter behavior recorded at the time of observation (point in parentheses in Figure 1), the indicated rate of gravity increase at South Pole station is 0.10 ± 0.01 mgal/yr, i.e. 1.0 ± 0.1 mgal in the ten-year period of observation.

The question we wish to consider is the cause of this increase. Since the gravity station is on the 1956–1957 summer surface underlying the South Pole station buildings rather than on the surface of the ice sheet, part of the increase results simply from the motion of the upper snow layers. In an ice sheet of constant surface elevation, there will be a downward motion of any horizon at a speed just balancing the rate of snow accumulation. Thus, if the polar ice sheet is in dynamic equilibrium, the 1956–1957 summer surface should, in November of 1967, have been at the same elevation as the 11-year-old horizon at any other time. This horizon lay at a depth below the surface of 2.0 ± 0.1 meters in 1957–1958 [*Giovinetto*, 1960] and at a depth of 2.1 meters in 1962–1963 [*Picciotto et al.*, 1964]. The gravity increase corresponding to equilibrium motion would therefore be about 0.55 ± 0.05 mgal in 10 years, including the Bouguer effect of the overlying snow accumulation. There remains a change of about 0.45 mgal unexplained.

Since South Pole station was built, a snow drift several meters high and several hundred meters across has grown up over it. *Behrendt* [1967] estimated that the gravity effects of the snow drift and of the elevation decrease resulting from snow compaction beneath the station in response to the drift load are about equal and opposite, and suggested that the remaining gravity increase results from horizontal motion of the ice sheet. Estimating the regional surface slope and horizontal gravity gradient from the best data then available, comprising gravimetric and altimetric observations at 1 to 4 km intervals along several oversnow traverses radiating from the South Pole, he calculated a velocity on the order of 50 m/yr to be required. With the slightly different values for the rates of gravity increase and snow accumulation used here, the velocity would need to be about 40 m/yr.

As a possible alternative, he suggested a wasting ice sheet; the necessary secular decrease in surface elevation would be about 15 cm/yr. This alternative he did not favor because of the lack of any supporting evidence.

Horizontal ice movement anywhere near 40 m/yr, however, also implies a wasting ice sheet, as can be

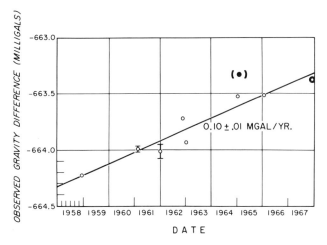

Fig. 1. Variation of observed gravity difference between South Pole station and McMurdo station as a function of time. Heavy parentheses indicate questionable value. Heavy circle denotes author's observation. Other data from *Behrendt* [1967].

shown by a simple order-of-magnitude calculation. The horizontal strain rate at any point, assumed not to be a function of depth, can be approximated by the accumulation rate divided by the ice thickness. At the South Pole the relevant values are 7.5 cm/yr [*Picciotto et al.*, 1964] and 2800 meters [*Kapitza and Sorokhtin*, 1963], respectively, yielding a strain rate of about 3×10^{-5}. The ice divide, some 200 km to the southeast (grid direction), is roughly linear [*Bentley*, 1964, Fig. 15], so that flow lines are approximately parallel. Between the South Pole and the divide, neither the mean ice thickness [*Bentley*, 1964, Fig. 18] nor the accumulation rate [*Giovinetto et al.*, 1966] appears to change substantially, so we can assume a constant strain rate along the flow lines. The equilibrium rate of movement is thus about $200 \times 3 \times 10^{-5}$ km/yr = 6 m/yr, whereas a surface lowering of some 50 cm/yr, corresponding to a gravity increase of 1.5 mgal in 10 years, would be required to balance a movement of 40 m/yr. It follows that (*a*) any model for the gravity increase depending primarily upon horizontal motion is internally inconsistent, and (*b*) either of the mechanisms suggested by Behrendt, or any combination of them, would imply a secular lowering of the ice surface of ten centimeters or more per year. The glaciological implications of this are important enough to warrant further study of the problem.

There are, in fact, two objections that can be raised to Behrendt's analysis. (1) The elevation decrease due to snow compaction and ice flow beneath the station in response to the load of the overlying snow drift was not carefully analyzed. Behrendt gives an 8-year total of 45 cm for the compaction sinking without demonstration and ignores flow altogether. Since this is not a simple problem, it seems quite possible that the sinking rate could be significantly higher. A model that does in fact lead to a faster rate is developed below. (2) The contribution of decreasing elevation from horizontal movement across the local surface slope has been ignored. It is well known that polar plateau surfaces typically exhibit undulations of a few kilometers in wavelength and of the order of 10 meters in height. Thus local slopes exist that are as large as 10^{-2} radian. Furthermore, it may well be that the surface undulations are fixed relative to the ice sheet bed [*Robin*, 1967] and that South Pole station is moving relative to them. The corresponding gravity change could be as much as one or two tenths of a milligal in ten years, even with a horizontal movement of only 5 m/yr.

OBSERVATIONS

During a brief stay at South Pole station between November 19 and 21, 1967, the author had an opportunity to carry out a short gravimetric and topographic survey of the local area to test the ideas given above. Gravity readings were made along the floor of the Pole station tunnels from the back door of the communications building (west end of the line) to within 30 meters or so of the seismograph building (east end of the line), and on the outer snow surface from the top of the entrance ramp southward about 270 meters. The meter was read three times consecutively at each station, and repeatability was within a few microgals. All readings were made within an hour of a reading at the base station. Closure was obtained on only one of three runs; failure on the other runs occurred once because of loss of power between the last field station and the return to base, and once because of an erratic base station reading. The observed change at the base station over the two-day period was +0.009 mgal, corresponding to a negligible drift rate of 0.2 μgal/hour. Tidal variations, as indicated on the UCLA earth tide recorder at the station, were also negligible.

Elevations at the gravity stations were obtained by means of a surveying transit and stadia rod. Distances were measured with a steel tape wherever the vertical angle was not zero, and simply by pacing at other stations. The accuracy of vertical angle measurements was about ±1′, resulting in a maximum relative error in free-air correction of ±0.006 mgal. At most stations, where elevations were obtained by foresight-backsight leveling, the corresponding error could be no more than 1 or 2 μgals.

Bouguer corrections were applied both for underlying snow layers corresponding to different elevations of the reading surface, and for the snow layer overlying the tunnel stations. The thickness of the latter was calculated from the surface topographic profile, the elevation of the tunnel floor, and the height of the tunnel. Since all the tunnel stations were near one side wall, the correction for the snow layer surrounding the tunnel was taken as one-half the Bouguer correction. This approximation could not cause an error of more than 0.01 mgal, and would be significant at all only in the comparison of inside and outside stations. The snow densities used were 0.40 g/cm^3 down to the level of the tunnel ceiling and 0.42 g/cm^3 below. The latter value was obtained as an average of four density determinations made by W. De Breuck and R. Smith on snow samples collected from the wall of the station entrance ramp. Both values are about 0.02 g/cm^3 greater than the corresponding densities of undisturbed snow in the area [*Giovinetto*, 1960]. An error in the choice of density of as much as 0.05 g/cm^3 would result in a noticeable discontinuity in the local gradient between inside and outside stations.

The repeatability of readings at each station, the negligible drift, and the small elevation errors all combine to suggest an accuracy of gravity anomalies relative to the base station of better than ±0.01 mgal. This estimate is supported by the fact that both anomaly profiles, including the connection between inner and outer stations, can be fit with straight lines from which only one point deviates by as much as 0.01 mgal. (Station 3 at the west end of the line appears to be in error by 0.03 mgal, probably owing to an error in leveling. The results of the survey are not affected by omitting this point.)

With an ice thickness of 2800 meters, it is clear that the gravity gradient over a distance of 300 meters must be quite smooth, once surface irregularities have been accounted for, so we are justified in taking the best linear fits to the two lines representing the components of the local gravity gradient (Figure 2). The slopes of these lines are 0.33 ± 0.04 mgal/km

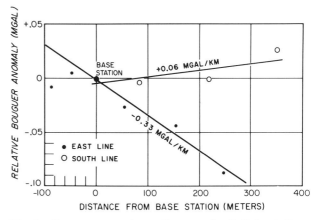

Fig. 2. Bouguer anomaly profiles near South Pole station. Anomalies are relative to value at base station next to science building. 'East line' and 'south line' refer to grid directions.

from east to west, and 0.06 ± 0.03 mgal/km from north to south. Combining components yields a Bouguer anomaly gradient of 0.34 ± 0.04 mgal/km, along grid azimuth 265° ± 5°, i.e. along meridian 95° ± 5°W.

If the ice moves northwestward at the calculated speed of 6 m/yr, the gravity change at South Pole station would be about 0.015 mgal in 10 years, a factor of 30 less than that required. It is thus quite apparent that horizontal movement cannot be responsible for a significant part of the observed gravity change.

The topographic survey was carried out by running five lines with transit and stadia rod, each extending to a distance of about 800 meters from the station. Four of the lines were approximately along the four cardinal grid directions; since the line to the south ran through an area badly disturbed by station activities (including the station dump), a fifth line was surveyed to the south-southwest. Most of the distances were determined by stadia intercepts, although the south line (2 points) and the west line (1 point) were extended by pacing.

The transit error of ±1′ corresponds to an error in surface elevation which increases away from the station to ±20 cm at a distance of 800 meters. On the north and east lines, which were leveled, this is the only error involved. If we take stadia intercepts as providing distances good within ±1%, the additional error in height determination is less than 6 cm, which is negligible. With paced distances, good to ±3%, the corresponding error in height is ±15

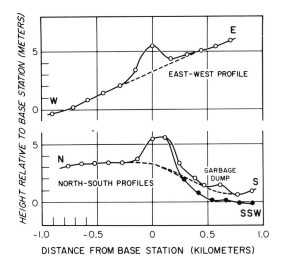

Fig. 3. Profiles of ice surface topography near South Pole station. Dashed lines indicate interpolated 'natural' topographic profiles. 'East-west' and 'north-south' refer to grid directions.

cm. We can thus take the height error as being no more than ±4 cm per hundred meters from the station (±5 cm/100 meters for extrapolated points).

The topographic profiles (Figure 3) clearly show the effect of snow drift in the station area, extending a few hundred meters in each direction. Although the exact horizontal extent of the drift is not certain, we can nevertheless estimate fairly well what the surface profiles would look like in its absence by connecting smooth curves through the more distant points. By requiring these curves to match where they cross, and noting that the curves at that point cannot satisfactorily be drawn below 3.2 meters on the east-west profile, nor above 3.4 meters on the north-south profile, we can conclude that the 'natural' surface elevation at this point is 3.3 ± 0.1 meters relative to the South Pole gravity station. Since in the absence of drift the relative surface elevation should be only 2.1 ± 0.1 meters, the implication is that the station has sunk 1.2 ± 0.2 meters relative to its 'natural' level. The corresponding increase in observed gravity would be 0.37 ± 0.06 mgal, over and above the increase due to the equilibrium downward motion of the ice.

From the topographic profiles we can produce a map of the local surface elevation in the absence of the station snow drift (Figure 4). The ice flow is generally northwest, although the exact direction is not known. If we take the flow velocity to be between 3 and 10 m/yr, and assume that the surface topography is stationary relative to the subglacial rock, the elevation decrease in 10 years (allowing for uncertainty in the contouring) can be placed with reasonable confidence between 0 and 0.5 meters, corresponding to a gravity increase of 0.08 ± 0.08 mgal.

The effects on the observed gravity are summarized in Table 1. It is clear that the total is not significantly different from the observed change. The horizontal motion down the local surface slope helps to explain the observed change, although it is not required. There is a suggested confirmation, therefore, that the local topographic variations are stationary with respect to the solid earth.[1]

CALCULATIONS OF SINKING RATE

Let us now see whether a model of the response of the ice sheet to the snow drift load can be devised. To estimate the sinking rate of the station, we start with the basic equations relating stress (τ_{ij}) and strain rate ($\dot{e}_{ij}$) in snow and ice:

$$\tau_{ij}' = 2\mu \dot{e}_{ij}' \quad (1)$$

$$\tau_{kk} = 3\eta \dot{e}_{kk} \quad (2)$$

where

$$\tau_{ij}' \equiv \tau_{ij} - \tfrac{1}{3}\tau_{kk}\delta_{ij}$$

$$\dot{e}_{ij}' \equiv \dot{e}_{ij} - \tfrac{1}{3}\dot{e}_{kk}\delta_{ij}$$

δ_{ij} is the Kronecker delta, and the summation convention is understood. In these equations, μ is the shear viscosity and η is the bulk viscosity. Laboratory experiments show that for small stresses the coefficients of viscosity are independent of the stress;

[1] Since the completion of this paper, a report has been published [*Chapman and Jones*, 1970] giving for the ice movement at South Pole station, determined from repeated astronomical observations, a value of 19 m/yr, with a standard deviation of 4 m/yr, along grid direction N37°W (standard deviation 12°). But Chapman (personal communication) cautions that standard deviations found in the reduction of astronomic observations generally do not fairly represent the standard error of estimate, which should be some 3 times as large. He thus estimates that at the 68% confidence level the ice movement rate could be anywhere between 7 and 31 m/yr, and thus does not necessarily violate equilibrium considerations.

Tripling the movement rate would triple the size of items *d* and *e* in Table 1, increasing the total by 0.2 mgal. This would have no effect on the conclusions of this paper.

μ and η can be treated as constants that depend only upon density and temperature [*Landauer*, 1957; *Mellor and Smith*, 1967]. Combining equations 1 and 2 to eliminate τ_{kk} yields

$$\tau_{ij} = 2\mu\dot{e}_{ij} + \lambda\dot{e}_{kk}\delta_{ij}$$

in which $\lambda = \eta - \tfrac{2}{3}\mu$. The stress-strain rate equation thus has the same form as the stress-strain equation for an elastic medium.

Our approximate approach will consist of adapting the known solution for the elastic deformation of a semi-infinite homogeneous elastic medium by a surface load to the ice sheet, which is highly inhomogeneous in its upper layers. Our procedure is as follows. For a homogeneous, semi-infinite ice body subjected to a vertical load distribution $P(r, \theta)$ on its upper surface $z = 0$, the vertical velocity component $\dot{u}_z$ is given, for points at depth z on the vertical axis $r = 0$, by

$$\dot{u}_z = \frac{z^2}{4\pi\mu} \iint_{\text{load}} \frac{P(r,\theta)}{(z^2+r^2)^{3/2}} r\,dr\,d\theta + \frac{\lambda+2\mu}{4\pi\mu(\lambda+\mu)} \iint \frac{P(r,\theta)}{(z^2+r^2)^{1/2}} r\,dr\,d\theta \quad (3)$$

[*Sokolnikoff*, 1956, p. 341].

We assume a parabolic cross section with rotational symmetry for the station snow drift

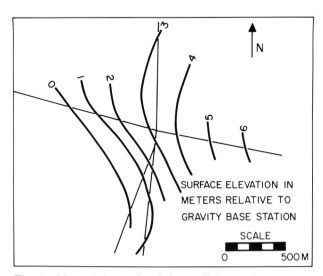

Fig. 4. Map of interpolated 'natural' ice surface topography near South Pole station. Light lines indicate profile lines of Figure 3. Elevations are relative to South Pole gravity base station. Arrow points to grid north (i.e. along Greenwich meridian).

TABLE 1. Contributions of Various Factors to the 10-Year Gravity Change at South Pole Station

Source	Est. Effect on Obs. Grav., mgal
a. Equilibrium sinking of station	+0.55 ± 0.05
b. Bouguer effect of snow drift	−0.03 ± 0.02
c. Sinking due to snow drift	+0.37 ± 0.06
d. Motion down local surface slope	+0.08 ± 0.08
e. Motion down gravity gradient	+0.02 ± 0.01
Total	+0.99 ± 0.11
Total without d	+0.91 ± 0.08
Observed change	+1.0 ± 0.1

$$P(r, \theta) = P_0(1 - r^2/r_0^2) \quad (4)$$

where r_0 is the radius of the drift, and P_0 is the load at the center. Substituting (4) into (3) and carrying out the integration, we find

$$\dot{u}_z = \frac{P_0 z}{2\mu}\left[1 + \frac{2z^2}{r_0^2} - 2\frac{z(z^2+r_0^2)^{1/2}}{r_0^2} - \frac{\lambda+2\mu}{\lambda+\mu}\left(1 + \frac{2}{3}\frac{z^2}{r_0^2} - \frac{2}{3}\frac{(z^2+r_0^2)^{3/2}}{r_0^2 z}\right)\right]$$

Differentiating:

$$\frac{d\dot{u}_z}{dz} = -\frac{P_0}{2\mu}\left[-1 - \frac{6z^2}{r_0^2} + 4\frac{z(z^2+r_0^2)^{1/2}}{r_0^2} + \frac{2z^3}{r_0^2(z^2+r_0^2)^{1/2}} + \frac{\lambda+2\mu}{\lambda+\mu}\right. \quad (5)$$
$$\left.\cdot\left(1 + 2\frac{z^2}{r_0^2} - 2\frac{z(z^2+r_0^2)^{1/2}}{r_0^2}\right)\right]$$

which is an exact expression for the homogeneous medium. To approximate $\dot{u}_z$ for an inhomogeneous medium, we now integrate equation 5, taking λ and μ to be functions of z. Although this procedure is strictly improper, we can show as follows that no serious error is introduced. For the homogeneous case, the stress solution for a concentrated force F, using cylindrical coordinates and setting $r^2 + z^2 = R^2$, is [*Sokolnikoff*, 1956]

$$\tau_{zz} = -\frac{3Fz^3}{2\pi R^5}$$

$$\tau_{rr} = -\frac{Fz}{2\pi R^3}\left[3\frac{r^2}{R^2} - \frac{\mu}{\lambda+\mu}\frac{R^2}{z(R+z)}\right]$$

$$\tau_{\theta\theta} = -\frac{F\mu[R^2 - z(R+z)]}{2\pi(\lambda+\mu)R^3(R+z)}$$

$$\tau_{rz} = -\frac{3Frz^2}{2\pi R^5}$$

$$\tau_{r\theta} = \tau_{z\theta} = 0$$

Since the shear stresses are independent of λ and μ, they are the same whether the medium is homogeneous or inhomogeneous. It follows that (5) is exact insofar as the shear strain contribution is concerned. For the compressive stresses, we find

$$\tau_{kk} = \frac{9Fz}{2\pi R^3(3+\mu/\eta)} \quad (6)$$

Implicit in our method is the assumption that the stress at each depth z_1 is what it would be for a homogeneous body with density $\rho(z_1)$. The true stress will be less, since the ratio μ/η increases upward from the depth z_1. (It is not necessary to integrate equation 6 for a distributed load, since the viscosity dependence would still be given by the factor $3 + \mu/\eta$.)

The maximum value of μ/η is about 0.6, whereas the value effectively assumed in the integration is, approximately,

$$\frac{1}{z_0}\int_0^{z_0}\frac{\mu}{\eta}dz = 0.16$$

Since, according to our calculations below, about two-thirds of the strain is due to τ_{kk}, we can say that the relative error in our method is less than

$$\frac{2}{3}\frac{[\tau_{kk}]_{\mu/\eta=0.16} - [\tau_{kk}]_{\mu/\eta=0.6}}{[\tau_{kk}]_{\mu/\eta=0.16}}$$

which, from equation 6, amounts to -9%. Our approximate calculation will therefore exceed the exact solution by only a few per cent, which is negligible in comparison with other errors.

In order to integrate equation 5, we will split the ice sheet into 3 zones, in each of which certain simplifications can be made: an upper zone, in which the density ρ varies with depth and the temperature T is constant, a middle zone in which both ρ and T are constant, and a lower zone of constant ρ and variable T.

Upper Zone

When $z \ll r_0$, (5) reduces to

$$\frac{d\dot{u}_z}{dz} = -\frac{P_0}{2\mu}\left[\frac{\mu}{\lambda+\mu} + \frac{2\lambda}{\lambda+\mu}\frac{z}{r_0}\right] \quad (7)$$

where μ and λ are functions of ρ, and hence functions of depth. To integrate (7) we must adopt a relationship between μ, η, and ρ.

The unconfined creep rate of snow samples was measured at South Pole station by *Ramseier and Pavlak* [1964]. Deformation measurements were made in an undersnow room, so the temperature of the samples never exceeded $-48°C$. Their observations yielded a stepwise linear relationship between density and the logarithm of the 'compressive viscosity,' η_c, which the author has fit by the following expression:

$$\eta_c = 1.1 \times 10^{19} e^{-35(0.9-\rho)} \quad 0.40 < \rho < 0.48 \text{ g/cm}^3$$
$$= 1.4 \times 10^{13} e^{-3(0.9-\rho)} \quad 0.48 < \rho < 0.62$$
$$= 3.8 \times 10^{17} e^{-39(0.9-\rho)} \quad 0.62 < \rho < 0.9$$

η_c being measured in g wt cm^{-2} sec. This involves an extrapolation without change in slope for $\rho > 0.65$ g/cm^3, the upper limit of the experimental range. A stepwise function of this sort has been measured by others [*Mellor and Smith*, 1967] and is probably real.

Knowledge of η_c is not enough to give us both η and μ; it is also necessary to 'interpolate judiciously between deducible limits' [*Mellor*, 1968] to obtain appropriate values of the ratio μ/η, whence μ can be found from the relation

$$\mu = \frac{\eta_c}{9}(3+\mu/\eta)$$

In accordance with Mellor's estimate [*Mellor*, 1968, Fig. 15, p. 46], we have taken $\mu/\eta = 0.6$ for $\rho = 0.45$ g/cm^3 decreasing about linearly to $\mu/\eta = 0$ for $\rho = 0.9$ g/cm^3. (The approximations used, which simplify the integration, are linear fits to $\eta_c/(\lambda+\mu)$ and $\lambda\eta_c/[\mu(\lambda+\mu)]$. Variations of μ/η within Mellor's diagrammed limits change the computed sinking rates by $\pm20\%$ or less.)

We must also choose a model for the variation of density with depth. For ease of integration, we have fit the measurements of *Giovinetto* [1960] by the function

$$\rho = 0.44 + 0.18 \log_e(z/18 + 1) \quad z < z_0$$
$$= 0.9 \quad z > z_0$$

where z is in meters, and z_0 is the depth at which ρ attains the value 0.9 g/cm^3. This provides a good fit throughout the observed density range, and a reasonable extrapolation below.

The integration is simplified by noting that, as ρ approaches 0.9 g/cm^3, η_c increases by 2 orders of magnitude above its value in the 'plateau region,' where $\rho < 0.62$ g/cm^3. Thus the contribution for z near z_0 is very small. It further transpires that $z_0 \simeq r_0$, and so we can assume that equation 7 holds throughout the region $0 < z < r_0$ without important error.

Lower Zone

In the lower part of the ice sheet, the temperature surely increases significantly as the bedrock is approached. As an approximation to the temperature structure, we take a linear gradient from $-50°C$ at $z = H/2$ to $-20°C$ at $z = H$, where H is the total ice thickness. This is close to the theoretical distribution given by *Robin* [1955]. We further assume that

$$\mu(z) = \mu(H/2) \exp\left[\frac{Q}{R}\left(\frac{1}{T} - \frac{1}{223}\right)\right]$$

where R is the gas constant, and Q is taken as 12 kcal/mole.

The model of a semi-infinite solid naturally is not valid near the base of the ice. For an order-of-magnitude result, however, we will simply assume no discontinuity at $z = h$. As we shall see, the contribution of strain in the deep ice to $\dot{u}_z$ is negligible anyway, and so more precise models would be superfluous.

When $z > H/2$, we can say that $z \gg r_0$ and we can expand equation 5, keeping only the lowest-order terms. Furthermore, for pure ice, we let $\eta \to \infty$ so that (5) becomes

$$\frac{d\dot{u}_z}{dz} \simeq \frac{P_0 r_0^2}{8\mu z^2} \qquad z > H/2 \qquad (8)$$

Middle Zone

In the range $r_0 < z < H/2$, μ is a constant, and $\eta \to \infty$, whence, from (5),

$$\int_{r_0}^{H/2} \frac{d\dot{u}_z}{dz} dz = -\frac{P_0}{2\mu}$$
$$\cdot \int_{r_0}^{H/2} \left[1 - \frac{2z^2}{r_0^2} + \frac{2z^3}{r_0^2(z^2 + r_0^2)^{1/2}}\right] dz \qquad (9)$$

which can be integrated easily without further approximations.

It remains to evaluate the sinking rate numerically. From Figure 3, we can estimate r_0 to be about 200 meters. The extra load height above the station is 3.4 meters; with a mean snow density of 0.42 g/cm^3, the corresponding axial load is 140 g wt/cm^2. To allow for the gradual buildup of the drift, we will take P_0 to be half the present value: $P_0 = 70$ g wt/cm^2. Substituting the numerical value for r_0 into (7) and (9) and the value for P_0 into (8), we find the following sinking rates:

Upper zone	15 cm/yr
Middle zone	0.4
Lower zone	0.05

Clearly the contributions from the middle and lower zones are negligible, whereas the 10-year total downward movement, 1.5 meters, is in good agreement with the observed sinking, considering the approximate nature of the calculation. We can note that two-thirds of the calculated sinking is due to compaction, one-third to shear strains.

As a check upon the numerical values assumed in this calculation, it is useful to calculate the undisturbed total compaction rate, which should equal the accumulation rate, using the same parameters. In this case, we assume that $d\dot{u}_z/dz$ at depth z is given simply by $(1/\eta') \int_0^z \rho(\xi) d\xi$, where η' is the viscosity in confined creep measured in g wt cm^{-2} sec, and is related to η_c by the expression

$$\eta' = \eta_c \frac{\eta}{3\mu}\left(1 + \frac{5\mu}{3\eta} + \frac{4\mu^2}{9\eta^2}\right)$$

The total compaction rate is $\int_0^{z_0} (dz/\eta') \int_0^z \rho(\xi) d\xi$. Carrying out the integration with a convenient approximation to η'/η_c, we find a total downward motion for a horizon initially at the surface of 15 cm/yr, twice the observed accumulation rate. This may indicate that the viscosities assumed are too small by a factor of 2, but even if the sinking rate calculated above is cut in half, the agreement with the observed sinking rate is still satisfactory.

Finally, let us consider the possible effect of high temperatures beneath South Pole station. Since no measurements are available, only hypothetical estimates are possible. For example, suppose that the mean snow surface temperature increased by a 5° step in 1956, when the station was established. After 10 years, the temperature rise would be about 4° at a depth of 5 meters, and 1° at a depth of 30 meters, the range between 5 meters and 30 meters corresponding to the 'plateau region' of the viscosity-density curve [*Ramseier and Pavlak*, 1964]. Taking activation energies of 10 kcal/mole above 5 meters and 7 kcal/mole in the 'plateau region' [*Ramseier and Pavlak*, 1964], we estimate that the sinking rate would be increased by 20%. Approximate factors by which the sinking rates would be multiplied for other temperature steps are as follows: for 10°, a factor of 1.5; for 20°, a factor of 2.5; for 30°, a factor of 4.4. The presence of a temperature anomaly beneath the station could modify the agreement between calculated and observed sinking rates; it certainly would not obviate the conclusion that such sinking could reasonably be expected in response to the load of the station snow drift.

CONCLUSIONS

1. The secular increase in observed gravity at South Pole station, 1.0 ± 0.1 mgal/yr, can be explained satisfactorily in a manner consistent with equilibrium of the ice sheet. Neither an abnormally large horizontal movement of the ice nor a secular decrease in the ice thickness is implied. At least 90% of the gravity change can be attributed to sinking of South Pole station, partly in equilibrium with the normal snow accumulation rate and partly in response to the superimposed load of the station-induced snow drift.

2. The agreement between calculated and observed sinking rates is evidence supporting the viscosity values measured for South Polar ice by *Ramseier and Pavlak* [1964], unless the unknown temperatures beneath the station are much higher than $-50°C$. Sinking rates of about 1 m/yr are obtained if the viscosities measured by *Mellor and Hendrickson* [1965] are used; on the other hand, the calculated sinking rate is only about 3 cm/yr if the variation of viscosity with density deduced by *Bader* [1962] from Sorge's law is assumed. Bader's relationship would be brought into agreement with the observed sinking rate only by a temperature elevation of some 30°C, much higher than seems reasonable.

3. There is a suggestion that the station has moved down the local surface slope. If so, the surface topographic irregularities with linear dimensions of a few kilometers must be a flow phenomenon. If they were caused by migrating 'dunes' on the surface, there would be no effect on the station elevation. This survey thus provides indirect evidence supporting *Robin*'s [1967] explanation for the surface topography.

Acknowledgments. The author wishes to thank John Clough and John Freitag for their assistance in making the gravity and elevation measurements, and to express his appreciation to Richard O'Connell and the U.S. Coast and Geodetic Survey for the loan of a transit, and to the U.S. Weather Bureau for the loan of a stadia rod. Discussions with William Budd were very helpful in devising a model for the station sinking rate, and much useful information was provided by Rene Ramseier and Malcolm Mellor, U.S. Army Terrestrial Sciences Center. The author is also grateful to Johannes Weertman for his critical review of the manuscript and suggestions for its improvement. Support from the National Science Foundation through grants GA-1126 and GA-1136 is gratefully acknowledged.

Contribution 222 from the Geophysical and Polar Research Center, University of Wisconsin.

REFERENCES

Bader, H., Theory of densification of dry snow on high polar glaciers, 2, *Res. Rep. 108*, Cold Regions Res. Eng. Lab., Hanover, N.H., 1962.

Behrendt, J. C., Gravity increase at the South Pole, *Science, 155*, 1015–1017, 1967.

Bentley, C. R., The structure of Antarctica and its ice cover, *Res. Geophys., 2*, 335–389, 1964.

Chapman, William H., and William J. Jones, Analysis of Ice Movement at Pole Station, Antarctica, *U.S. Geol. Surv. Prof. Pap. 700C*, C242–C246, 1970.

Giovinetto, M. B., Glaciology report for 1958, South Pole Station, *Ohio State Univ. Res. Found. Rep. 825-2*, part 4, 104 pp., 1960.

Giovinetto, M. B., The drainage systems of Antarctica: Accumulation, *Antarctic Snow and Ice Studies, Antarctic Res. Ser.*, vol. 2, edited by M. Mellor, pp. 127–155, AGU, Washington, D.C., 1964.

Kapitza, A. P., and O. G. Sorokhtin, On errors in interpretation of reflection seismic shooting in the Antarctic, General Assembly of Berkeley, *Intern. Assoc. Sci. Hydrol. Comm. Snow Ice, Publ. 61*, 162–164, 1963.

Landauer, J. K., The creep of snow under combined stress, *Res. Rep. 41*, Snow, Ice, Permafrost Res. Estab., Wilmette, Ill., 1957.

Mellor, M., *Cold Regions Sci. Eng. III-A3d; Avalanches*, Cold Regions Res. Eng. Lab., Hanover, N.H., 1968.

Mellor, M., and G. Hendrickson, Confined creep tests on polar snow, *Res. Rep. 138*, Cold Regions Res. Eng. Lab., Hanover, N.H., 1965.

Mellor, M., and J. H. Smith, Creep of snow and ice, in *Physics of Snow and Ice*, vol. 2, edited by H. Oura, pp. 843–855, Institute of Low Temperature Science, Hokkaido University, Sapporo, Japan, 1967.

Picciotto, E., G. Crozaz, and W. De Breuck, Rate of accumulation of snow at the South Pole as determined by radioactive measurements, *Nature, 203*, 393–394, 1964.

Ramseier, R. O., and T. L. Pavlak, Unconfined creep of polar snow, *J. Glaciol., 5*, 325–332, 1964.

Robin, G. de Q., Ice movement and temperature distribution in glaciers and ice sheets, *J. Glaciol., 2*, 523–532, 1955.

Robin, G. de Q., Surface topography of ice sheets, *Nature, 215*, 1029–1032, 1967.

Sokolnikoff, I. S., *Mathematical Theory of Elasticity*, McGraw-Hill, New York, 476 pp., 1956.

VELOCITY OF ELECTROMAGNETIC WAVES IN ANTARCTIC ICE

G. R. Jiracek[1] and Charles R. Bentley

Geophysical and Polar Research Center, University of Wisconsin, Middleton, Wisconsin 53562

The average velocity of 30-MHz electromagnetic waves in the antarctic ice sheet was determined from primary and multiple echoes using standard reflection techniques. Measurements were made on the 'McMurdo ice shelf,' the Skelton glacier, and Roosevelt Island together with the adjacent Ross ice shelf. The marked differences in the values of average velocity obtained in these areas are related to differences in average density of the ice column. Polar ice and snow can be considered as low-loss (loss tangent $\ll 1$) at 30 MHz; therefore, any measurement of the velocity of electromagnetic-wave propagation serves directly as a measurement of the relative dielectric constant. This facilitates the use of dielectric mixture theories to relate the density and geometrical structure of glacial ice to the wave velocity. The envelope of values predicted using O. Wiener's theory includes nearly all the present results and recent ice core measurements when the relative dielectric constant of solid ice ($\rho = 0.917$ g/cm^3) is taken as 3.21. The general agreement of this value with the bulk of previously reported laboratory measurements at higher frequencies supports the conclusion that the dielectric constant at 30 MHz differs little from the limiting high-frequency value. The corresponding 30-MHz wave velocity in solid ice is approximately 168 m/μsec, which is close to that (168.5 $\pm$ 1.0 m/μsec) measured through 782 meters of floating ice on the Skelton glacier, where the average density is estimated to be 0.907 g/cm^3. The lack of agreement between Wiener's envelope and the velocity value obtained on Roosevelt Island may be due to the penetration of subglacial material. At South Pole station, where no velocity measurement was attempted, the recorded echo time (33.0 μsec), together with Weiner's envelope and the estimated average density ($\bar{\rho} = 0.903$ g/cm^3), yields an ice thickness of 2800 $\pm$ 20 meters. This agrees favorably with a value of 2800 meters obtained by seismic soundings about 1 km from the station. The relative dielectric constant of glacial ice may vary significantly with temperature, pressure, and crystal orientation. The need is apparent for additional laboratory and in situ measurements of the dielectric properties of ice and subglacial materials.

Various phenomena that have been observed throughout the past thirty years have led recently to the use of electromagnetic waves to measure glacial ice thickness. In fact, since the first antarctic ice soundings were reported by A. H. Waite in 1958 [*Waite and Schmidt*, 1961], the technique has proved more successful than even the most optimistic predictions. As in any sounding method, the ultimate accuracy of the electromagnetic soundings depends upon one's knowledge of the velocity of wave propagation through the ice sheet. Some laboratory and in situ measurements of this parameter are available, but marked disagreement is too often found between the reported values.

After field trials in Greenland during the summer of 1964, the U.S. Army Electronics Laboratory electromagnetic sounding equipment was loaned to the Geophysical and Polar Research Center, University of Wisconsin, for more extensive research in Antarctica. Of prime concern were field measurements of the velocity of electromagnetic waves through glacial ice. The antarctic experiments were conducted in a variety of glaciological environments (e.g., floating and grounded ice, varying ice thickness, different temperature ranges) where previous geophysical studies had been made.

Figure 1 shows three areas designated A, B, and C where velocity measurements were undertaken during the 1964–1965 season. Studies were begun in area A, the 'McMurdo ice shelf' adjacent to New Zealand Scott Base (77°51.0'S, 166°45.0'E). Three stations were established in this area at semiper-

---

[1] Now at Department of Geology, University of New Mexico, Albuquerque, New Mexico 87106.

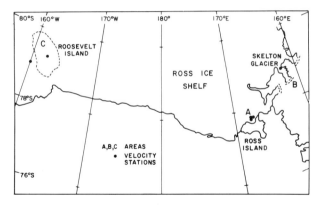

Fig. 1. Locations of 30-MHz velocity measurements in Antarctica, 1964–1965.

manent ice-movement and snow-accumulation markers recently emplaced by New Zealand scientists. The sounding equipment was next airlifted to area B, the Skelton glacier, where velocity measurements were made near station 62 (78°51.4′S, 161°54′E) of the 1958–1959 Victoria Land traverse. After these experiments, velocity measurements on grounded ice were made at Camp Wisconsin (79° 15.8′S, 162°17.5′W) on Roosevelt Island, area C, where the University of Wisconsin has undertaken extensive geophysical and glaciological investigations. The final velocity determination was made on the Ross ice shelf east of Roosevelt Island close by station K30 (78°57′S, 160°05′W) of the Wisconsin seismic network. The electromagnetic sounding project was completed at South Pole station (90°S), where bottom echoes were recorded through approximately 2800 meters of ice. In addition to the velocity measurements in the various areas, continuous profiles of ice thickness totaling more than 250 km in length were sounded.

EQUIPMENT

The equipment is a standard radar ranging system with a transmitter and receiver, together with an oscilloscope for measuring echo time delay. The primary equipment included a 30-MHz transmitter, which produced 300-watt (peak) pulses of 0.5-μsec duration at a repetition frequency of 20 kHz. Power could be reduced, if desired, by two steps of about 6-db each. The receiver incorporated two low-noise preamplifiers and a high-gain postamplifier, which together resulted in a nominal gain of 140 db. Modifications were included in the amplifiers to limit overdrive and ringing when the receiver was operated in close proximity to the high-power pulsed transmitter. The unit was built by Airborne Instruments Laboratory.

A Tektronix RM45A oscilloscope and a 35-mm camera were used to record the received signals. A transmitted pulse rise time of approximately 0.1 μsec, together with random reading errors, limited echo time accuracy to an estimated ±0.03 μsec for most of the measurements. When the echo amplitude approached the recorded noise level, this accuracy was not maintained; however, the signals received during the velocity measurements were usually well above the noise. The accuracy of the time measurements was insured by using a temperature-insensitive 0.5-MHz oscillator to check the oscilloscope sweep speeds.

Antennas used for transmitting and receiving were balanced-feed folded dipoles operated parallel to and approximately 1.5 meters above the snow surface.

In addition to the 30-MHz system, a standard SCR 718 military radio altimeter operating at a frequency of 440-MHz was used at one location in area A. This unit transmits only 7 watts peak power and is of the type Waite used when making his first soundings.

METHOD

Standard seismic reflection techniques were used to determine average velocities. This method requires that the transmitter and receiver be separated at known intervals along the surface to utilize the reflection time variations. Simple geometry yields the relationship

$$t^2 = (x^2/\bar{v}^2) + (4z^2/\bar{v}^2) \qquad (1)$$

between the surface distance x, reflection time t, ice thickness z, and the average velocity $\bar{v}$. Accordingly, a plot of t^2 versus x^2 theoretically yields a straight line, the slope of which is equal to $1/\bar{v}^2$. This simple analysis assumes the top and bottom interfaces are parallel and that the medium is homogeneous and isotropic.

Corrections can be made for known values of the bottom slope. These corrections can be important; for example, a slope of only ±0°30′ at the bottom of 500 meters of ice results in an uncorrected velocity value that is in error by ±1.5 m/μsec. The effect of bottom slope can be reduced by recording reflections in opposite directions and averaging t^2 values. Another procedure, requiring the separation of both transmitter and receiver from a central position, reduces the slope error to a second-order effect only, but can be cumbersome in practice.

In the present survey, once a location was selected, the measurements were made with the transmitter fixed while the receiver was separated at known intervals along the surface. Bottom slopes along the base lines were determined by electromagnetic profiling or occasionally by barometric altimetry on floating ice. A single mean slope correction was then applied. Since the barometric method of slope determination is less accurate, over-all average velocity values at each station were obtained using only electromagnetic measurements of slope.

The densification of glacial ice with depth results in a departure from the ideal homogeneous situation to which equation 1 applies. The increase of density with depth results in a decrease in the electromagnetic wave velocity. Figure 2 represents the ray path of an electromagnetic wave through a typical glacial profile. The low-density snow near the surface is referred to as the firn layer, and the bottom reflector is simply designated as water or rock. Subscripts refer to air, firn, ice, and so forth. In accordance with the gradual decrease in velocity with depth, an increasing refractive index $n(z)$ is assumed. This results in a ray trajectory that is concave down, eventually becoming straight when the limiting ice density ($\rho = 0.917$ g/cm³) is reached at depth.

Applying Snell's law to the path yields

$$\frac{dx}{dz} = \pm \frac{1}{[(n(z)/\sin \theta_a)^2 - 1]^{\frac{1}{2}}} \quad (2)$$

where θ_a is the angle of incidence of the ray in air on the firn surface. The total distance traveled in the glacier is

$$S_g = 2 \int_0^{z_i} \frac{n(z)}{[n(z)^2 - \sin^2 \theta_a]^{\frac{1}{2}}} \, dz \quad (3)$$

The air part of the path is

$$S_a = 2z_a/\cos \theta_a \quad (4)$$

It is apparent that, as the thickness of ice decreases or as the separation between transmitter and receiver increases, the higher-velocity parts of the path become increasingly significant. As a result, t^2 versus x^2 data in thin ice yield a curved line indicating increased average velocity with horizontal distance. The best average velocity for the vertical column is thus obtained using only those measurements

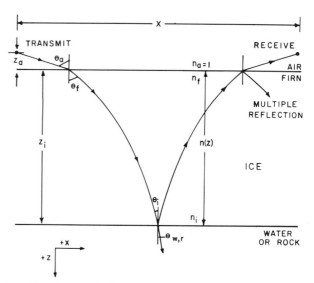

Fig. 2. Ray path of an electromagnetic wave through a typical glacial profile.

taken at short distances. Subtraction of the air travel times from the echo times would eliminate errors from this source, but the evaluation of this correction requires n_f, the refractive index near the surface. Unfortunately, large differences are found in the reported values of n_f [*Waite*, 1959; *Yoshino*, 1961; *Nottarp*, 1964]. An alternate method has been used, therefore, to correct approximately for the uncertain length of the propagation path in air.

Upon extending the ray path for the multiple reflection indicated in Figure 2, it is clear that this first multiple, M_1, would be recorded at a horizontal distance slightly less than $2x$. The air path for this multiple is equal to the air path in the primary echo E measured at x. Consequently, if the travel time for E at x is subtracted from the travel time for M_1 at $2x$, the remaining time is nearly that corresponding to just the glacial path of the primary echo. Thus velocity values determined from time differences between these reflections, rather than solely from the primary echoes, should closely equal the average velocity through the ice column. As many as four multiple reflections have been measured in thin ice. These, when evaluated alone and in conjunction with the primary echo, result in a reliable velocity determination.

Figure 3 shows a sequence of echoes recorded at station 203, area A. The transmitter and receiver separations vary from 0 to 1000 meters in steps of 50 meters. The oscilloscope trace corresponding to each

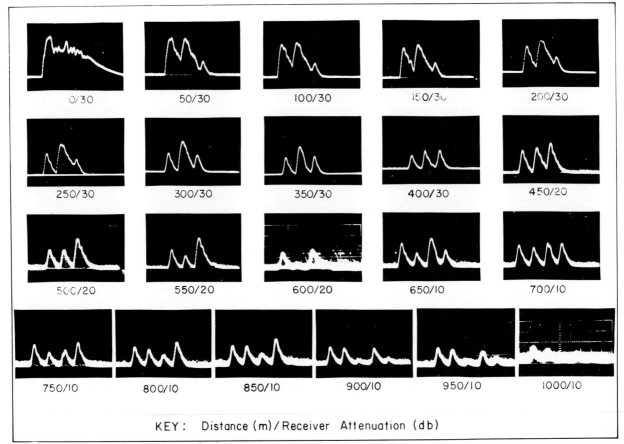

Fig. 3. Complete sequence of electromagnetic echoes recorded at station 203, area A, with transmitter-receiver separations varying from 0 to 1000 meters in steps of 50 meters.

distance shows the initial air pulse followed by the primary echo and successive multiples. It can be seen that, at zero separation, receiver ringing prevents the positive identification of any echoes. At 50 meters, however, E and M_1 are clearly distinguishable. Additional multiples are recorded (e.g., M_2 at 650 meters, M_3 at 900 meters) as transmitter-receiver separation is increased. (Multiple reflections were not consistently observed in thick ice (≥ 500 meters), but in this case the high-velocity paths in air and firn are a much smaller amount of the total path.)

Because of the relative shortness of most of the profiles, there was no significant increase in apparent velocity with distance in the t^2 versus x^2 plots (an example is shown in Figure 4) resulting from the velocity decrease downward within the ice column. We have therefore treated the results as equivalent to those for a single layer of constant velocity equivalent to the indicated mean velocity.

30-MEGAHERTZ VELOCITY RESULTS

Table 1 lists the results of the 30-MHz velocity measurements as obtained by the t^2 versus x^2 reflection method. Station elevations were determined using barometric altimetry, which was tied to previously surveyed locations whenever possible. A positive bottom slope indicates a dip downward from transmitter to receiver. A or R in the bottom slope column refers to the slope determination method, i.e. altimetry or electromagnetic sounding. Under echo type, E is the primary echo and M_1, M_2, $\cdots$, are succeeding multiple reflections. M_1-E is intended to minimize the air travel error as previously discussed. Average velocities ($\bar{v}$) and standard errors of the mean ($\sigma_{\bar{v}}$) were obtained by the method of least squares in accordance with equation 1. The ice thickness (z) derived from the intercept time of each echo plot is listed with the standard deviation of the value.

The over-all average velocity given for station 203, area A, is the mean $\bar{v}$ for all echo types except the primary echo. The anomalously high value (192.1 m/μsec) for the primary echo is probably caused by the high velocity in the air, as is indicated by close agreement between the M_1-E velocity and those from higher-order multiples. This effect is nearly absent, however, in the results from station 204, 10 km from station 203, even though the ice thickness is only 50 meters greater. On still thicker ice, there should be a negligible error from this cause even on the primary echo. This is supported by the absence of curvature in the t^2 versus x^2 plot for the Ross ice shelf area C, where the ice thickness is more than 500 meters (Figure 4).

Velocity values obtained from the primary echoes at station 68, area A, are exceptionally high from both lines, and large disagreement is evident between the multiple M_1 results. Electromagnetic profiling showed these base lines to have highly irregular bottoms, which probably resulted in systematic errors in the velocity measurements. Hence the results at this station are not reliable.

The determinations of average refractive index $\bar{n}$ and average relative dielectric constant $\bar{k}_e'$ are calculated directly from the average velocities by applying the well-known Maxwell relation

$$n = c/v = (k_e')^{1/2} \quad (5)$$

where c is the velocity of electromagnetic waves in free space (300 m/μsec). This equation only applies to electrically low-loss (loss tangent $\ll 1$) materials. That ice qualifies as such is adequately proven by the most recent loss tangent measurements performed on glacial ice cores at the Massachusetts Institute of Technology in 1961. In a sample of density $\rho = 0.902$ g/cm³, the loss tangent values at the lowest frequency (150 MHz) measured 2×10^{-3} to 5×10^{-5} in the temperature range $-1°$ to $-60°C$ (W. Westphal, personal communication, 1965). Extrapolation to 30 MHz assuming dielectric dispersion described by the Debye equations results in values larger by slightly less than 1 order of magnitude. Several authors have reported agreement of observed dispersion in ice with the Debye formulas [see *Evans*, 1965].

The ice thickness at each station was calculated both from the average time intercept of those t^2 versus x^2 regression lines used to determine the over-all average velocity and from the vertical echo time.

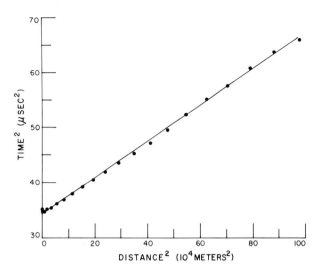

Fig. 4. Linear least squares fit of t^2 versus x^2 data obtained through 520 meters of ice on the Ross ice shelf.

The disagreement between the two determinations is always less than 15 meters (Table 1).

The average density $\bar{\rho}$ at each station was estimated from pit and core density measurements and seismic velocity-depth profiles at or near the station. Average densities in area A were determined using the reported results from station 87 of the Victoria Land traverse [*Crary et al.*, 1962]. Densities deduced from seismic velocities down to 105 meters at station K30 of the 1962–1963 University of Wisconsin survey (M. Hochstein, personal communication, 1965) were used to estimate the average density at the Ross ice shelf station in area C. The deep core results from Little America V (A. Gow, personal communication, 1965) were used for extrapolations beyond the reported values on the ice shelf in areas A and C. On the Skelton glacier, area B, density values have been measured to only 2.5 meters [*DenHartog*, 1959]. The station, however, is only 10 km 'down glacier' from an area where hard 'blue' ice ($\rho \simeq 0.85$ g/cm³) occurs at the surface [*Crary and Wilson*, 1961], and so an assumption that ice of this density is found at 10-meter depth at the station is probably not in serious error. Independent average density calculations were also made at all stations on floating ice, assuming hydrostatic equilibrium. These results agree with the integrated estimates within ± 0.001 g/cm³ in areas B and C and differ at most by 0.02 g/cm³ at station 203, area A. Density measurements to only 17-meter depth are available at Camp Wisconsin on Roosevelt

TABLE 1. Results of 30-Megahertz Velocity and Ice Thickness Measurements: Antarctica 1964-1965.

Station Location & Line Direction	Elev (m)	Bottom Slope	Echo Type	$\bar{v}$ (m/μsec)	$\sigma_{\bar{v}}$ (m/μsec)	Ice Thickness from Intercept		$\bar{v}$ (m/μsec)	$\sigma_{\bar{v}}$ (m/μsec)	Overall Station Averages						
						z (m)	σ_z (m)			$\bar{n}$	$\bar{k}_e'$	Ice Thickness (m) from				$\bar{\rho}$ (gm/cm³)
												Intercept	σ_z	Vertical Time	σ_z	
A. "McMurdo Ice Shelf"																
1. Station 203, Southwest	28	0°00' R	E	192.1	2.3	163.0	2.2	178.3	2.3	1.68	2.81	155.2	2.7	148.0	2.6	0.814
			M₁	177.1	1.3	150.2	1.4									
			M₂	177.2	1.4	153.1	1.6									
			M₃	182.8	1.7	163.4	1.9									
			M₄	178.4	3.2	157.3	3.6									
			M₁-E	176.0	3.2	152.2	3.8									
2. Station 204 North	38	0°00' R	E	178.3	1.2	179.1	1.7	175.5	2.5	1.71	2.91	187.4	3.4	194.8	3.3	0.838
			M₁	177.0	1.1	189.9	1.3									
			M₂	171.8	1.8	182.6	2.8									
			M₁-E	175.0	4.4	198.0	5.8									
West		+1°00' A	E	179.5	1.6	189.2	2.0									
			M₁	175.7	4.0	188.1	4.4									
3. Station 68 North	58	+1°05' R	E	183.4	1.2	329.7	2.8									
			M₁	171.7	1.7	312.9	3.4									
West		-1°05' R	E	185.2	2.8	326.9	5.2									
			M₁	204.3	3.2	369.9	11.8									
B. Skelton Glacier North	92	+0°25' R	E	168.5	1.0	772.3	5.0	168.5	1.0	1.78	3.17	772.3	5.0	782.2	5.2	0.907
East		-0°40' A	E	167.3	3.1	758.7	14.2									
C. Roosevelt Island																
1. Camp Wisconsin East	563	-0°10' R	E	174.8	1.0	825.4	5.1	174.8	1.0	1.72	2.95	825.4	5.1	839.0	5.4	0.894
2. Ross Ice Shelf West	73	-0°33' R	E	174.9	0.5	512.1	1.8	174.9	0.5	1.72	2.95	512.1	1.8	519.6	2.4	0.884

Island (C. Benson, personal communication, 1962). Extrapolation necessary to determine the average column density may have introduced some error in the estimate at this location.

DISCUSSION

The tabulated velocity results adequately show that a single average velocity will not hold for all locations. It may therefore be necessary to obtain local electromagnetic velocity data before dependable values of ice thickness can be determined in a given area. Ideally, an area already examined could be related to a new area by comparison with certain glaciological parameters, such as the snow-ice density distribution with depth. Unfortunately, the effect of density, plus other factors, such as high-pressure air bubbles and ice bonds, on the electrical properties of glacial ice is not yet sufficiently understood for such predictions.

Owing to the low-loss electrical properties of ice, the measured $\bar{k}_e'$ values lend themselves easily to comparison with the values predicted by dielectric mixture theories. Most of these theories result in the treatment of glacial ice as an idealized volume concentration of highly polarizable ice dispersed in relatively non-polarizable air. The theories are only satisfactory when the geometrical shapes of the dispersoid (ice) particles lend themselves to simple mathematical treatment (e.g., cylinders or spheres). Unfortunately, different theories predict different results when applied to the same geometrical distribution. *Cumming* [1952] found that the mixture formula for spherical ice particles due to *Polder and van Santen* [1946] fit his experimental values of k_e' quite well. However, the mixture theory of *Wiener* [1910] has received more attention as applied to ice and snow. Wiener found that a single parameter, the Formzahl (form number) u, can be used to describe the geometrical structure of the mixture. Accordingly, the relative dielectric constant, k_e', of a lossless air-ice mixture satisfies the equation

$$\frac{k_e' - 1}{k_e' + u} = p \left(\frac{k_i' - 1}{k_i' + u} \right) \quad (6)$$

where k_i' is the relative dielectric constant of solid ice, and p is the ratio of the density ρ of the ice-air mixture to that of solid ice (0.917 g/cm³). The form number u is allowed to take on any value from 0 to ∞; curves of these two values set the limits of Wiener's theory. *Kuroiwa* [1956] has shown that an increase in u implies an increase of lengthwise arrangement of dispersoid (ice) in the direction of applied electric field.

Of fundamental importance in defining the k_e' limits is the value of k_i' at the frequency under consideration. The contrasting values reported for k_i' at radio frequencies have been discussed by *Evans* [1965], who concludes that the relative dielectric constant of solid ice is 3.17 ± 0.07 for frequencies greater than 1 MHz. Westphal's dielectric measurements (referred to above) on four glacial ice cores yielded no significant dispersion of k_e' in the frequency range 150 to 2700 MHz. The highest value of k_e' (3.21) was measured in the sample of highest density (ρ = 0.902 g/cm³) at a temperature of −1°C. In this same sample, a value of k_e' equal to 3.13 was recorded at −60°C.

Westphal's complete results at −10° and −20°C are plotted against density in Figure 5, along with the present measurements. The average ice column temperatures at the velocity stations are estimated to be in this temperature range. The wave velocities corresponding to k_e' are also indicated in Figure 5.

Dielectric mixture theories assume that as a mixture approaches a single component the geometrical considerations vanish, resulting in a single value of relative dielectric constant. Consequently, Wiener's theory predicts an envelope of k_e' versus ρ values converging upon a single value for solid ice. The upper bound ($u = \infty$) of this envelope would just include the highest plotted value of k_e' (Westphal's sample of density 0.898 g/cm³ with $k_e' = 3.19$) if k_i' were 3.24. The value for area C of the Ross ice shelf would lie on the $u = 0$ boundary if k_i' was 3.18, but the envelope would not include the result from Camp Wisconsin unless k_i' was 3.11 or less. The former value agrees more closely with the bulk of the previously reported laboratory and in situ measurements [*Evans*, 1965]. Also, the lack of agreement between seismic and electromagnetic soundings at Camp Wisconsin (the latter indicating an ice thickness 60 meters greater than the former) suggests that the velocity measured by electromagnetic sounding at Camp Wisconsin may not be representative of the ice column alone, having perhaps been influenced by penetration of subglacial material. Some electromagnetic penetration is expected, the degree depending upon the composition and temperature of the bottom; however, it cannot at present be verified that this effect is responsible for the high-velocity meas-

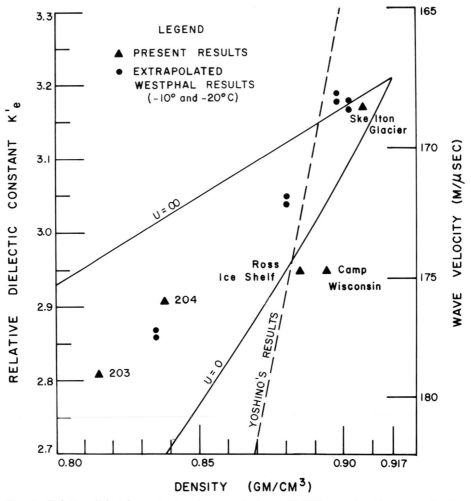

Fig. 5. Relative dielectric constant and wave velocity at 30 MHz as a function of ice density, along with Wiener's theoretical envelope.

urement at Camp Wisconsin. Using the mean of 3.24 and 3.18 (3.21) as the relative dielectric constant of solid ice, the envelope of values predicted by Wiener's theory has been calculated and is included in Figure 5. This envelope includes nearly all the plotted results, the Camp Wisconsin measurements being the only notable exception.

Recent curves published by *Yoshino* [1961] indicate k_e' values at 30 MHz, which would cross Wiener's envelope as shown by the dashed line. This line intersects $\rho = 0.917$ g/cm³ at approximately $k_i' = 3.56$. Although this high value for the relative dielectric constant of solid ice is in agreement with that reported by *Waite* [1959], it is not in accord with the present results. Also, in addition to a larger slope in Yoshino's results at high densities, his curves indicate significant dispersion above 30 MHz. For example, at the estimated density of station 203, area A ($\bar{\rho} = 0.814$ g/cm³), his curves yield 30- and 440-MHz wave velocities of approximately 195 and 201 m/μsec. These values are much higher than the comparable results obtained in the present survey ($\bar{v} = 178.3$ and 180.4 m/μsec). The latter values agree within respective errors, indicating no measurable dispersion between these two frequencies at station 203.

Some caution must be used when interpreting average in situ values on the basis of laboratory measurements, since the $\bar{v}$ and $\bar{k}_e'$ results of the present survey are harmonic means, whereas the $\bar{\rho}$ estimates are arithmetic means. However, we are dealing with velocities and densities near their limiting values, so that the harmonic and arithmetic means are nearly equal. For this reason, the average measurements obtained in the present survey may legitimately be compared with Westphal's laboratory determinations. On the basis of the present results and Westphal's data, we then conclude that k_i' equals approximately 3.21, with a corresponding wave velocity of approximately 168 m/μsec, at 30 MHz and between $-10°$ and $-20°C$. This k_i' value is only slightly higher than most reported values of the relative dielectric constant at the high-frequency limit, indicating little dispersion above 30 MHz.

The relative dielectric constant may vary significantly with temperature and other glaciological parameters, such as pressure and crystal orientation. Westphal's results demonstrate a marked temperature dependence of k_e' values which has not been reported by other investigators. For example, Lamb and Turney [1949] measured no appreciable change in k_e' in the temperature range $-10°$ to $-195°C$. Hydrostatic and nonhydrostatic pressures may also alter the dielectric properties of glacial ice. In this connection, a pressure of about 2 kg/cm² caused small but real changes in the electrical measurements on ice at 1 kHz [Brill and Camp, 1961]. In the present survey, velocity measurements were performed approximately perpendicular and parallel to known strain directions at station 204, area A, and on the Skelton glacier. The small differences in the velocity results at these stations are less than the standard deviations of the measurements, indicating no measurable stress-induced anisotropy. The investigations of Humble et al. [1953] at 1 kHz showed as much as 15% k_e' anisotropy in individual ice crystals. Anisotropy may also be the cause of the changes in wave polarization observed in the 30-MHz signals recorded through the Skelton glacier and at the South Pole. This previously unobserved phenomenon may be the consequence of preferred crystal orientation in the basal part of the ice sheet, although a depolarized reflected component would also be introduced by non-specular scattering from moraine material or by bottom roughness. In any case, an entirely new technique is suggested for studying the critical lower part of the glacier.

In conclusion, it appears that, at present, the velocities of electromagnetic waves in glacial snow and ice cannot be accurately predicted, but that the envelope predicted by Wiener's formula, using the relative dielectric constant of solid ice as 3.21, probably includes most velocity values. A value thus predicted was used at South Pole station where no velocity determination was attempted. Using the recorded echo time of 33.0 μsec and the estimated average density ($\bar{\rho} = 0.903$ g/cm³) results in an ice thickness of 2800 ± 20 meters. This agrees favorably with the seismic measurement of 2800 meters obtained approximately 1 km from the station. However, comparisons between seismic and electromagnetic soundings from a traverse in 1965–1966 into Queen Maud Land yield systematic differences in ice thicknesses determined by the two methods (J. Beitzel and J. Clough, personal communication, 1966). This serves to emphasize the need for additional laboratory and in situ measurements of the dielectric properties of ice and subglacial materials.

Acknowledgments. We wish to thank A. H. Waite (now retired) and his associates of the U. S. Army Electronics Laboratory, Fort Monmouth, New Jersey, not only for generously loaning the electromagnetic sounding equipment, but also for many hours of help in learning to use the equipment, and many hours of advice in the glaciological application of electromagnetic sounding. J. E. Nicholls of the University of Wisconsin ably assisted in the Antarctic field program. Appreciation is also extended to the personnel of the U.S. Navy Air Development Squadron Six and the administrative staff of the National Science Foundation for their assistance in the field operations. Many helpful comments were received during the data analysis from S. Evans of Scott Polar Research Institute, Cambridge, England.

The electromagnetic sounding project was performed with the support of the National Science Foundation, grant GA-148.

Contribution 186 from the Geophysical and Polar Research Center, University of Wisconsin.

REFERENCES

Brill, R., and P. R. Camp, Properties of ice, *Res. Rep. 68*, pp. 67–69, Cold Regions Res. Eng. Lab., Hanover, N. H., 1961.

Crary, A. P., E. S. Robinson, H. F. Bennett, and W. W. Boyd, Jr., Glaciological studies of the Ross ice shelf, Antarctica, 1957–1960, *IGY Glaciol. Rep. 6*, 105–119, 1962.

Crary, A. P., and C. R. Wilson, Formation of 'blue' glacier ice by horizontal compressive forces, *J. Glaciol., 3*, 1045–1050, 1961.

Cumming, W. A., The dielectric properties of ice and snow at 3.2 centimeters, *J. Appl. Phys., 23*, 768–773, 1952.

DenHartog, S. L., Snow pit work on Little America–Victoria Land traverse, 1958–1959, *Ohio State Univ. Res. Found. Rep. 825–2*, pt. 2, p. 48, 1959.

Evans, S., Dielectric properties of ice and snow: A review, *J. Glaciol., 5*, 773–792, 1965.

Humbel, F., F. Jona, and P. Scherrer, Anisotropie de Dielektrizität-skonstante des Eise, *Helv. Phys. Acta, 26*, 17–32, 1953.

Kuroiwa, D., The dielectric property of snow. *Union Géodésique et Géophysique Internationale, Association Internationale d'Hydrologie Scientifique Assemblée générale de Rome 1954, 4,* 52–63, 1956.

Lamb, J., and A. Turney, The dielectric properties of ice at 1.25 cm wavelength, *Proc. Phys. Soc. London, B, 62,* 272–273, 1949.

Nottarp, K., Dielectric measurements on Antarctic snow at 3,000 Mc/sec, *J. Glaciol., 5,* 134, 1964.

Polder, D., and J. H. van Santen, The effective permeability of mixtures of solids, *Physica, 12,* 257–271, 1946.

Waite, A. H., Ice depth sounding with ultra-high frequency radio waves in the Arctic and Antarctic and some observed over-ice altimeter errors, *U.S. Army Signal Res. Devel. Lab. Tech. Rep. 2092,* 1959.

Waite, A. H., and S. J. Schmidt, Gross errors in height indication from pulsed radar altimeters operating over thick ice or snow, *IRE Intern. Conv. Record,* Pt. 5, 38–54, 1961.

Wiener, O., Zur Theorie de Refraktionskonstanten, *Ber. Ver. Kgl. Sechs. Ges. Wiss. Leipsig, Math.-Phys. Kl., 62,* 256–268, 1910.

Yoshino, T., Radio wave propagation on the ice cap, *Antarctic Record, Tokyo, 11,* 228–233, 1961.

GLACIOLOGICAL STUDIES ON THE SOUTH POLE TRAVERSE, 1962–1963

Lawrence D. Taylor

Institute of Polar Studies, Ohio State University, Columbus 43212

Glaciological observations were made at twenty-five stations along a 1448-km over-snow traverse between the South Pole and the Queen Maud Mountains during the austral summer 1962–1963. Annual accumulation was determined from pit studies by the distribution of impermeable crusts, revealed by ink-staining pit walls, and the fluctuation in density of firn to a depth of 2 meters. Ice crusts and permeable, low-density layers of firn are formed by vapor transfer in summer (December and January) and late fall (April) during periods of large temperature gradients on the Antarctic plateau. Ice crusts may also form during the summer by radiation. Stratification is complicated by disconformities, i.e., breaks in the normal sequence of layering due to absence of one or more years' accumulation. Accumulation varies from 6.8 g/cm^2 at the Pole to 10.8 g/cm^2 near the Queen Maud Mountains. Annual accumulation, mean density between 0 and 2 meters, integrated ram hardness, and 10-meter temperatures in firn all decrease with increasing latitude between 87°08'S and 90°S as follows: accumulation, 1.2 g/cm^2/deg latitude; density, 0.037 g/cm^3/deg latitude; integrated ram hardness, 1160 joules/deg latitude; and 10-meter temperature, 1.5°C/deg latitude. Ten-meter temperatures in firn near the Pole indicate a lapse rate of −0.8°C/100 meters. At depths between 15 and 40 meters, negative temperature gradients of 0.56 to 0.64°C per 100 meters of depth were recorded. Annual accumulation, 10-meter temperature, and mean density of firn (0–2 meters) plotted against each other have linear correlations of 0.94. Within the region of the traverse, the regression lines of these plots can be used to determine any two of the values if the third is known.

Glaciological and geophysical observations were made at twenty-five stations along a 1448-km over-snow traverse between the South Pole and the Queen Maud and Horlick Mountains from December 2, 1962, to January 22, 1963 (Figure 1). The party consisted of two geophysicists and two mechanics from the Geophysical and Polar Research Center, University of Wisconsin; one geophysicist from the U.S. Coast and Geodetic Survey; and two glaciologists from the Institute of Polar Studies, Ohio State University. The party leader and chief geophysicist was Dr. Edwin Robinson of the University of Wisconsin.

The glaciological program included pit studies, measurement of firn temperatures, and the collection of firn cores for microparticle studies. In 2-meter pits, annual snow accumulation was determined from firn stratigraphy, density, hardness, and grain size. Temperature studies included the determination of the temperature gradient in 40-, 20-, and 10-meter bore holes. Thermistors, a Wheatstone bridge, and a null detector were used to measure temperatures to an accuracy of 0.01°C. Analysis of these data will contribute to the knowledge of the thermal diffusivity of firn, the heat budget, and annual mean surface temperature of the Antarctic plateau.

Table 1 is a summary of the important data obtained on the traverse. Elevations and ice thickness (measured by seismic methods) were determined by personnel from the Geophysical and Polar and Research Center [*Robinson*, 1966]. Mr. David Perkins of the U.S. Coast and Geodetic Survey determined station locations by solar navigation.

PIT STUDIES

The annual accumulation of snow in grams per square centimeter has been determined from an analysis of the physical properties of firn to a depth of 2 meters. Pits were dug (2.1 meters deep, 2.6 meters long, 1.7 meters wide) at approximately 64-km intervals along the traverse. One wall of the pit was used for the analysis, and care was taken not to disturb the surface above this wall. In each pit the

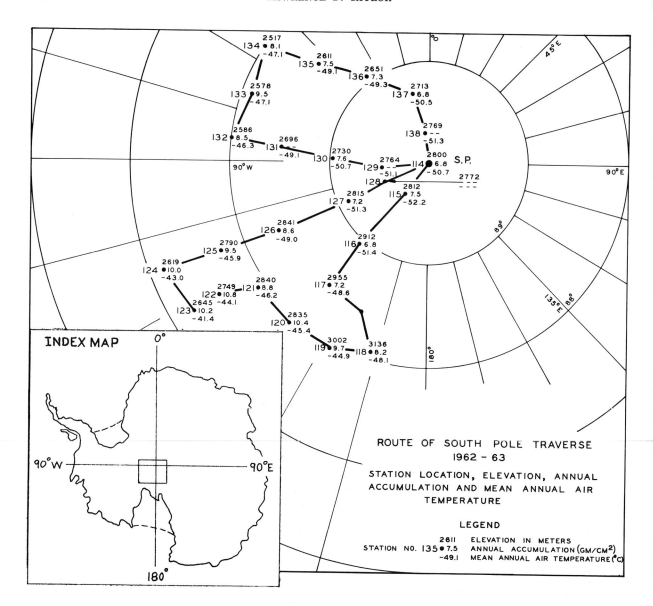

Fig. 1. Route of South Pole traverse, 1962–1963.

density and grain size were measured, and the distribution of crusts, or impermeable layers, was determined by an ink-staining method. After the wall was studied, snow hardness was measured with a Rammsonde to a depth of 3 meters parallel to the wall and ½ meter from the edge of the pit.

Interpretation of Firn Stratigraphy

The variation with depth of density, grain size, and hardness follows an irregular cycle. This cycle is the result of seasonal variations of (1) the size of snow crystals during precipitation, (2) the amount of precipitation, (3) the temperature gradient across the air-snow interface, (4) solar radiation, and (5) wind velocity and duration. An example of the cycle can be seen in Figure 2, which shows the stratigraphy of pit 136. It is assumed that each cycle represents 1 year's accumulation [*Benson*, 1962]. Other factors that tend to mask or disrupt the seasonal cycle are: compaction due to weight of overlying snow; absence of snow accumulation due to lack of precipitation or removal by wind producing a strati-

TABLE 1. Summary of Data from the South Pole Traverse, 1962–1963

Station	S	W	Surface Elevation, meters	Ice Thickness, meters	Annual Accum. g/cm^2	Mean Density 0–2 Meters, g/cm^3	Integrated Ram Hardness, joules	10-Meter Temp., °C
114	90°00′		2800	2820	6.8	0.351	1079	−50.7
115	89°38′	141°24′	2812	2922	7.5	0.360	788	−52.2
116	88°58′	140°00′	2912	2562	6.8	0.360	1049	−51.4
117	88°27′	140°24′	2955	2150	7.2	0.387	1454	−48.6
118	88°04′	163°30′	3136	2251	8.2	0.383	1522	−48.1
119	87°58′	153°00′	3002	1382	9.7	0.417	4144	−44.9
120	87°56′	139°06′	2835	2456	10.4	0.420	3874	−45.4
121	87°55′	126°12′	2840	2280	8.8	0.406	1437	−46.2
122	87°33′	122°00′	2749	2019	10.8	0.428	3742	−44.1
123	87°15′	121°30′	2645	1415	10.2	0.446		−41.4
124	87°09′	112°06′	2619	1779	10.0	0.400	2510	−43.0
125	87°46′	113°24′	2790	1780	9.5	0.399	1554	−45.9
126	88°22′	114°00′	2841	2191	8.6	0.385	1137	−49.0
127	89°07′	114°54′	2815	3185	7.2	0.365	659	−51.3
128	89°32′	112°30′	2772	2952			649	
129	89°32′	94°00′	2764	2849			659	−51.1
130	89°02′	87°06′	2730	2710	7.6	0.363	584	−50.7
131	89°32′	84°42′	2696	2531			693	−49.1
132	88°01′	83°12′	2586	2306	8.5	0.384	1157	−46.3
133	88°07′	69°42′	2578	2458	9.5	0.398	1507	−47.1
134	87°59′	56°12′	2517	2387	8.1	0.393	1103	−47.1
135	88°32′	48°54′	2611	2791	7.5	0.373	497	−49.1
136	88°57′	38°30′	2651	3041	7.3	0.382	681	−49.3
137	89°18′	14°48′	2713	2858	6.8	0.377	601	−50.5
138	89°42′	11°00′	2769	2799			490	−51.3

graphic disconformity; extreme fluctuations in the normal seasonal conditions producing physical characteristics that are not ordinarily representative of that particular season. A disruption of the normal cycle is illustrated in Figure 2 at the 60-cm depth. First it must be established that certain seasonal variations in climatic conditions do exist in the area of the traverse. Because the elevations and latitudes of all the traverse stations are nearly the same, within 267 meters and 3 deg, respectively, and for most observations not more than 2 deg, it can be assumed that seasonal variations of climate along the traverse are of the same magnitude as at the South Pole, although gradual changes in absolute values for accumulation and mean annual temperature occur away from the Pole. For example, near the Queen Maud Mountains, Rammsonde profiles show a distinct increase in hardness and in the thickness of hard layers, suggesting that wind velocities are much higher in this area than in other parts of the traverse.

Factors Responsible for Stratification

We will use the climatic and micrometeorological data obtained at the South Pole [*U.S. Weather Bureau*, 1963; *Dalrymple*, 1963] to examine the seasonal variations and their effect on properties of firn at the Pole.

Temperature gradient. It has been established that a temperature gradient in firn produces a water-vapor transfer in the direction of heat flow [*Bader et al.*, 1954; *Yosida*, 1963; *Shimizu*, 1964], which facilitates recrystallization at depth. The higher the temperature gradient, the greater is the vapor transfer. *Bader et al.* [1954] determined that there is too little moisture transferred by diffusion alone to account for the formation of depth hoar, and that air flow produced by wind is necessary. The formation of sublimation crystals, which produce a distinct stratigraphic marker, is dependent on the absolute temperature, the temperature gradient, the porosity of the firn, the permeability of the firn, and the wind velocity.

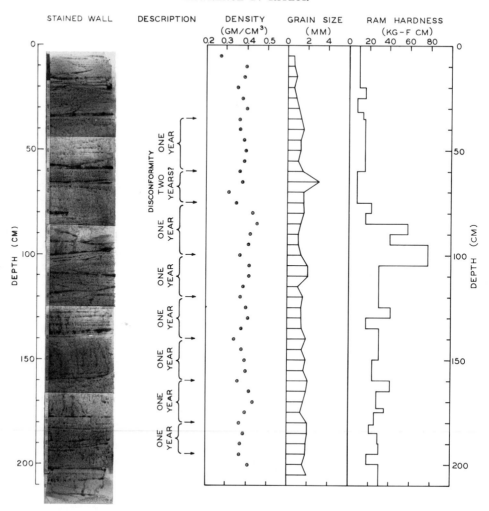

Fig. 2. Stratigraphic column of pit 136 illustrating cyclic variations in density, grain size, and ram hardness.

Sublimation crystals are formed by condensation or evaporation in the solid state. The process of condensation produces a general grain enlargement and reduction of porosity in the firn. Condensation occurs preferentially around large rather than small grains, since the former have a larger radius of curvature and thus lower vapor pressures. Evaporation generally destroys the small crystals, since they have a smaller radius of curvature and consequently higher vapor pressures. Thus condensation in the solid state produces a relatively coarse-grained impermeable layer or ice crust, and evaporation produces a fine-grained permeable layer.

It is unlikely that condensation can produce a thick ice crust, since recrystallization reduces the permeability and impedes vapor transfer. Once an impermeable layer is established, further condensation occurs on one side of the layer and evaporation occurs on the opposite side, provided that heat flow remains in the same direction. The thickness of the crust is also limited, because the variation of vapor pressure with temperature is small at the low temperatures existing on the Antarctic plateau. For example, the difference in vapor pressure between −30°C and −60°C is 0.19 mm Hg, while between 0°C and −10°C it is 2.63 mm Hg [*Dorsey*, 1940, p. 600]. Most crusts observed in firn on the South Pole traverse were less than 1 cm thick.

The direction of heat flow must be considered in determining where condensation and evaporation occur in firn. If heat transfer is downward, evaporation occurs in the permeable firn near the surface

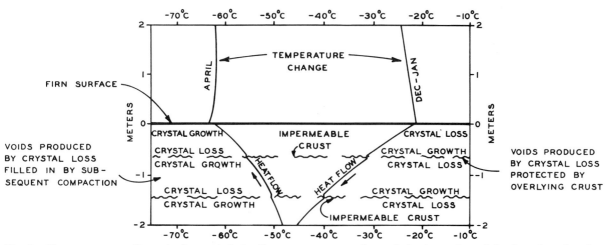

Fig. 3. Temperature gradients in firn and their effect on vapor transfer and the formation of depth and surface hoar Temperature curves are adapted from *Dalrymple* [1963, p. 53].

and vapor is transferred downward, condensing in the colder, less permeable firn below, particularly on top of a dense layer deposited the preceding winter. This condensation may produce an impermeable crust. Ice crusts already in existence below the surface may act as barriers to the downward transfer of vapor, and condensation may also occur along the top of these crusts. If heat flow is upward, then evaporation occurs in the permeable firn near the surface, as in the preceding case, but vapor is transferred upward and condenses at the firn-air interface, forming a thin crust. Condensation also occurs along the bottom of pre-existing ice crusts below the surface. Very little transfer of vapor upward or downward takes place at depths greater than two meters, for here the temperature gradient is greatly reduced and permeability of the firn is lower.

Dalrymple's analysis of data from the South Pole shows that the largest temperature gradient across the firn-air interface occurs in December and January, with heat transfer downward. These gradients are 0.125°C/cm. The largest gradient with upward heat transfer, 0.075°C/cm, occurs during April (Figure 3). According to these data, it is assumed that during December and January ice crusts form below the surface at depths probably not greater than two meters, and usually at shallow depths on top of a dense layer deposited the preceding winter. The crusts thicken just enough to prevent vapor transfer. Condensation occurs on top of the crust, and evaporation occurs on the bottom and in the firn just below the crust. The porous firn resulting from evaporation just below an ice crust may remain for some time, because the overlying crust may be strong enough to support the load of firn above so that compaction of the porous layer is extremely slow (Figure 3). During April, when heat transfer is upward, an ice crust develops at the surface. If other crusts are already present below the surface, condensation occurs along the bottom of the crust and evaporation along the top. Porous firn that forms above this crust will soon compact, unless another ice crust is present just above, supporting the load of overlying firn (Figure 3).

Solar radiation. Observations by *Kotlyakov* [1961] and *Kuznetsov* [1960] have indicated that preferential absorption of radiation produces intergranular and intragranular melting on the Antarctic plateau. The mechanism of recrystallization through radiation is not entirely understood, but recrystallization is sufficient along grain boundaries to produce a thin impermeable layer in the firn. *Kotlyakov* [1961] observed that it forms principally during periods of clear skies with little or no wind. The absence of wind reduces heat loss at the surface. *Kotlyakov* [1961] describes radiation crusts, in the vicinity of Mirnyy station, as comprising elongated pieces of ice separated by small sections of firn. Almost all crystals in the ice are oriented at an angle to the horizontal plane with c axes parallel to crystal elongation and oriented in a northerly direction. Kotlyakov indicates that the structure of these crystals is the result of radiational melting. It seems unlikely that the absorbed energy can be enough to produce melting. Solid diffusion along grain boundaries may be the mechanism whereby a crystal whose

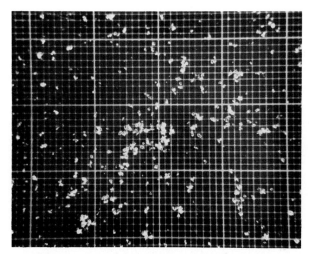 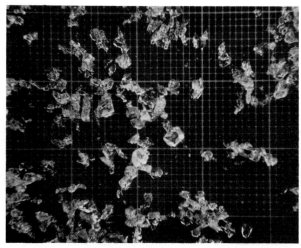

Fig. 4. Grains scraped from wall of pit (each division is 1 mm): (*left*) a winter layer; (*right*) a hoar layer.

c axis is aligned in the direction of the source of radiation grows at the expense of an adjacent crystal whose c axis is not as close to the principal direction of radiation. In any case, there seems to be agreement that the following conditions are responsible for the formation of these crusts: (1) relatively high influx of infrared radiation at high elevations during clear weather, (2) presence of foreign particles that have a considerably larger coefficient of absorption of radiant energy than does ice, and (3) concentration of radiant energy by the irregularities of the grains, which sometimes act as lenses.

In dry snow the absorption of infrared radiation decreases exponentially with depth. Less than 7.5% of the total radiation at the surface reaches a depth of five centimeters [*Dorsey*, 1940, p. 492]; therefore crystal enlargement is limited to the upper centimeter of surface. Ink-staining of the pit walls on the South Pole traverse (Figure 2) clearly shows that the layers are extremely thin, less than 5 mm. Before staining, nearly all layers contain interlocking grains as large as 3 mm in diameter. The discontinuous nature of some of the crusts can be seen in Figure 2. Without a knowledge of crystal orientation, it is difficult to establish whether these crusts are due to vapor transfer or radiation.

Firn samples containing ice crusts at Eights station have been examined for microparticles. Using the Coulter counter technique, *Taylor and Gliozzi* [1964] found the number of particles between 1 and 3 μ diameter to average about 25 per microliter of meltwater. Dirty layers of 300 or more particles per microliter are rarely found within ice crusts. The effect of foreign particles on the formation of crusts through absorption of radiant energy is therefore negligible. At Mirnyy, near ice-free areas, particle frequency is probably much higher and may account for crust formation there.

Wind velocity. According to observations at South Pole station, 1957–1962, winds are generally stronger and more persistent during October (mid-spring), June and July (early winter), and March (early fall) than during other months. Wind variations produce pressure changes and cause air movement in the permeable firn. Air movement, in addition to diffusion, is needed to transport vapor. Winds help in the formation of many ice crusts, but only if the wind occurs when a large temperature gradient exists between the air and the firn. At the Pole the strongest and most persistent winds do not occur during the December to January period of largest temperature gradients. In December and January, 1957–1962, the wind velocity averaged approximately 5.5 m/sec. It is unlikely, therefore, that ice crusts form during the mid-winter period on the Antarctic plateau, where temperature gradients in winter are small and vapor transport by air movement due to wind is small.

Although ice crusts may not form directly from the wind, high-density layers of firn may indeed result. The relatively dense but still permeable firn is formed by packing of the snow, either by the force of the wind directly or later by the load of overlying snow. Wind-borne snow is somewhat rounded and well-sorted (Figure 4, *left*), hence com-

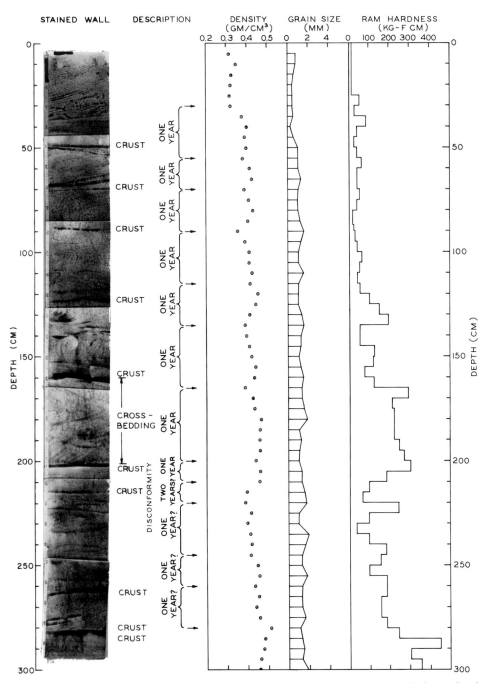

Fig. 5. Stratigraphic column from pit 124 illustrating cross bedding and unusually high ram hardness.

paction takes place easily, whereas sublimation crystals (Figure 4, *right*), which have irregularly shaped grains of different sizes, compact less readily.

Winds also produce cross-bedded structures, as is shown on the ink-stained wall of pit 124 at a depth of 160 to 200 cm (Figure 5). The layers differ in grain size, grain shape, and permeability. The character and sorting of the grains that form the layers are controlled by the wind velocity and the distance of transport by the wind.

Precipitation. Direct measurement of precipitation at South Pole station for any single year is inaccurate because of the very low accumulation and high wind velocity. If it could be proved that winter produces greater accumulation than summer, thick layers of high-density snow could be used as a winter indicator. Since there is no evidence from direct measurements that precipitation varies seasonally, this cannot be used as a criterion for determining a year's accumulation. *Shimizu* [1964] has shown that crystal size is generally smaller for snow precipitated in winter at Byrd station. As far as the author is aware, no such observations have been made at the South Pole. Although seasonal temperature differences are of the same order at the two stations, the mean temperature at the Pole is so much lower than at Byrd that temperature differences at the Pole may have a negligible effect on the size of precipitated snow. Pit studies show a variation in grain size that to some extent parallels the variation in density. The changes in grain size is, however, not a carry-over from the original size of the precipitated snow, but rather the result of compaction and recrystallization controlled by seasonal variations in temperature gradient, wind velocity, and radiation.

Density. Density measurements were made at 5-cm intervals using standard density tubes (500 cm^3) placed along two vertical rows with an overlap of approximately 0.8 cm between succeeding tubes. The tubes were excavated and weighed on a triple beam balance to 0.1 gram. The density profile is the most reliable indicator of seasonal variations in the firn (Figure 2). In the zone of compaction, density generally decreases with increasing grain size in the firn. Since grain size is controlled mainly by temperature gradient, the largest crystals, excluding those comprising ice crusts, are associated with low-density hoar formed by recrystallization mainly in December and January. Mean density plotted against latitude, however, varies according to a curvilinear regression (Figure 6).

Grain size. Grain size was determined every 5 cm by scraping the wall of the pit and catching the grains on a millimeter grid card (Figure 4) and visually estimating the mean diameter of the aggregate. This is by no means as accurate as the technique of magnification recommended by *Koerner* [1964], but time did not permit a careful examination of grain size. The size usually varied inversely with density, but measurements were not accurate enough in many pits to be helpful in delineating seasons. Grain size varied from less than 0.25 mm to 5.0 mm (Figure 4). The largest grains were found just below and sometimes above ice crusts. No relationship exists between mean grain size and change in latitude or elevation.

Hardness. A SIPRE Rammsonde was used to measure firn hardness at approximately 5-cm intervals to a depth of 2.5 meters, following the technique described by *Bull* [1956] and *Benson* [1962]. The results are generally unsatisfactory for determining seasonal layers and cannot be correlated from one pit to another. Thin ice crusts are difficult to identify (Figure 2); cyclic variations that are clearly shown in the density profile are not obvious in the hardness profile. The ram measurements were useful in revealing a region of extremely hard-packed firn along 87°45'S latitude about 40 miles south of the Queen Maud Mountains, where values of 800 kg-force cm were measured (Figure 5), compared to values of 400 kg-force cm elsewhere.

Permeability. The distribution of impermeable layers on the wall of the pits was determined by using an ink-staining method proposed by *Shimizu* [1964]. One wall of the pit is smoothed by scraping and then brushing to make the crusts stand out. The smoothed surface is prepared alongside the excavation left from density measurements. The surface is sprayed evenly with warm dilute ink from a spray gun and allowed to freeze. A blowtorch flame is passed lightly across the inked surface, and the melted ink is absorbed differentially by the firn before freezing again (Figures 2, 5, and 7). Recrystallization occurs on the surface of the wall, but the ink penetrates beyond the thin recrystallized surface and reveals gross changes in permeability of the firn. A layer of black ink accumulates just above each impermeable zone. If the heat is used sparingly and for the proper length of time, cross-bedded textures are revealed (Figure 5). The stained wall is photographed in natural light with a 4 × 5 Speed Graphic and 35-mm Contaflex camera with Panatomic-X film. The Speed Graphic is mounted on a sliding sleeve attached to a vertical tubular metal pipe pushed into the bottom of the pit and held in place equidistant from the pit wall by a board extending across the top of the pit.

ANNUAL ACCUMULATION

The annual accumulation varies from 6.8 g/cm^2 at the South Pole to 10.8 g/cm^2 near the Queen Maud

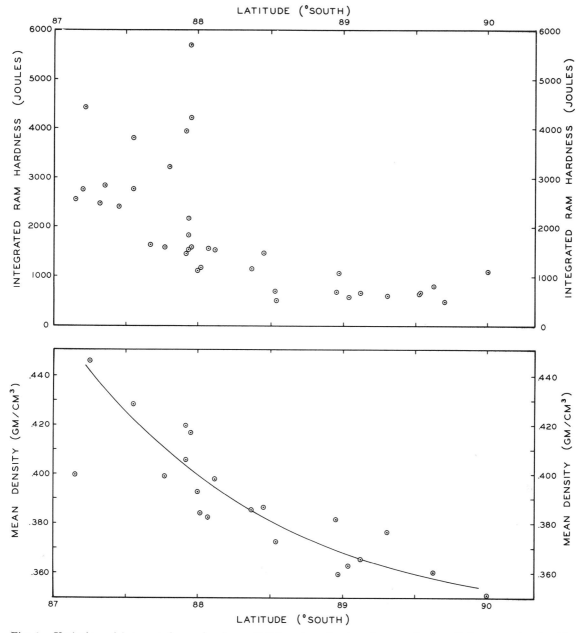

Fig. 6. Variation of integrated ram hardness (0–2.5 meters deep) and mean density (0–2 meters deep) with latitude.

Mountains (Table 2). Accumulation values near the Pole agree with those obtained by *Giovinetto* [1963] and *Picciotto* [1964]. Values are based on the average total number of cycles identified for each of the properties of firn measured in the pits, where each cycle represents a year's accumulation. An interval on the wall of the pit is chosen commencing with the first good cycle in the density profile just below the surface, and ending usually at the bottom of the 2-meter pit.

Variations in density, grain size and hardness, and the frequency of single crusts or sets of closely spaced crusts are counted over the same interval. The number of years from all four firn properties are combined and averaged (Figures 2 and 5). The property that shows the clearest cycle in each pit is

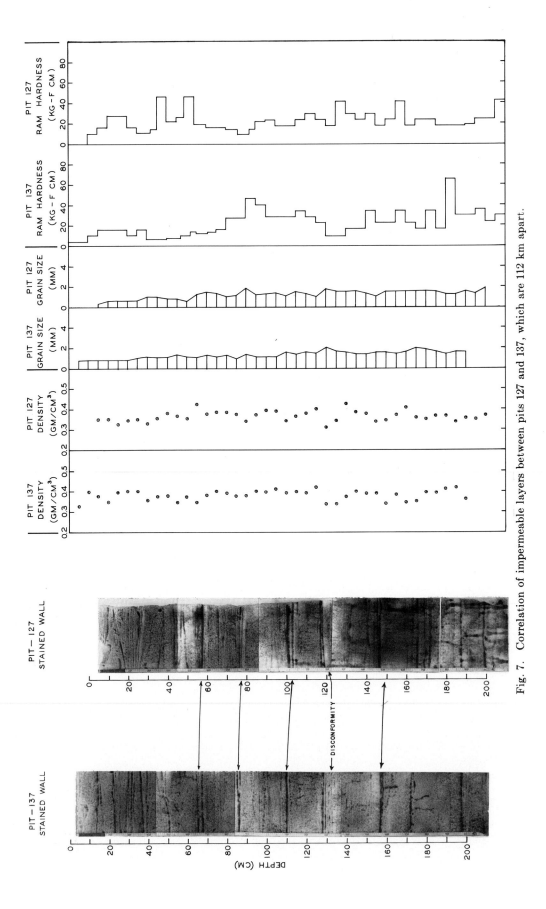

Fig. 7. Correlation of impermeable layers between pits 127 and 137, which are 112 km apart.

often given preference. It is assumed from the previous discussion that ice crusts, whether from radiation or vapor transfer, are produced during December, January, and April. A porous, low-density layer produced in association with the crust is usually present below the crust, and below this a thick layer of cross-bedded high-density firn occurs representing the preceding winter period, as illustrated in Figure 5 between 160 and 200 cm. The high-density layer of firn is sometimes missing, as at 65 to 75 cm in pit 136 (Figure 2), or a series of ice crusts is present separated by only porous layers of firn, as is shown at 132-cm depth in pit 137 (Figure 7). It is believed that this represents a missing winter layer. After the number of years is assigned to the section, the average density of the entire interval is multiplied by the thickness of the intervals to obtain the water equivalent, which is then divided by the number of years or cycles to obtain the average accumulation per year. The value for average annual accumulation is a maximum value, since there is no way of determining how many years are missing at a disconformity.

All properties of the firn are used in the interpretation of a year's increment. For example, in pit 136 (Figure 2), a series of impermeable crusts occur between 100 and 140 cm in the photograph. This section may represent three or more years with very low winter accumulation between the summer crusts. However, an examination of both density and grain-size profiles reveals only two cycles or two years, suggesting that the crusts were probably distributed over two summers.

It must be emphasized that the date of a given horizon or annual layer cannot be precisely determined, because of the extremely low accumulation and the numerous disconformities representing erosion intervals or periods of no accumulation. On the basis of the number of cycles of density, grain size, ram hardness, and the distribution of crusts, one can only estimate the date, taking into account what appear to be disconformities. Disconformities are believed to exist at horizons indicated in Figures 2, 5, and 7. But do these disconformities represent one, two, three, or more missing years? Are they truly disconformities? In this area, the stratigraphic method of determining annual accumulation has definite limitations.

An attempt was made to correlate the hard and impermeable layers using photographs of stained walls, hardness profiles, and cycles in density values. In a few cases the distribution and spacing in layers show resemblance, particularly between pit 127 and 137 (Figure 7). These pits are 112 km apart, have a difference in elevation of 102 meters, and a difference in latitude of 11.3′. Although the similarity is remarkable, considering the great variety of layers, it is indeed possible that this correlation is a coincidence. Many impermeable layers and layers representing a winter period are discontinuous and grade into disconformities. Accurate correlation on the basis of firn stratigraphy is impossible. *Koerner* [1964] has shown from trench studies at Byrd station the great problem of correlation even over extremely short distances.

TABLE 2. Annual Accumulation, South Pole Traverse, 1962–1963

Station	Accum., g/cm^2	No. Years Represented	Total Depth Interval, cm
114	6.8	9	20–195
115	7.5	7	55–195
116	6.8	9	35–203
117	7.2	10	60–205
118	8.2	8	25–190
119	9.7	8	20–200
120	10.4	7	35–205
121	8.8	7	40–190
122	10.8	7	20–190
123	10.2	5	40–160
124	10.0	7	30–200
125	9.5	7	40–200
126	8.6	8	30–200
127	7.2	9	30–200
130	7.6	9	20–200
132	8.5	7	50–200
133	9.5	7	30–190
134	8.1	8	40–200
135	7.5	9	25–200
136	7.3	9	35–200
137	6.8	10	25–200

TEMPERATURE MEASUREMENTS

Firn temperatures (±0.01°C) were determined with thermistors in 10- and 40-meter bore holes using a Leeds and Northrup 4735 guarded Wheatstone bridge and a 9834 null detector. Calibrated thermistors were supplied by the Cold Regions Research and Engineering Laboratories.

A plot of 10-meter temperature with latitude is shown in Figure 8 (*top*). The 10-meter temperature

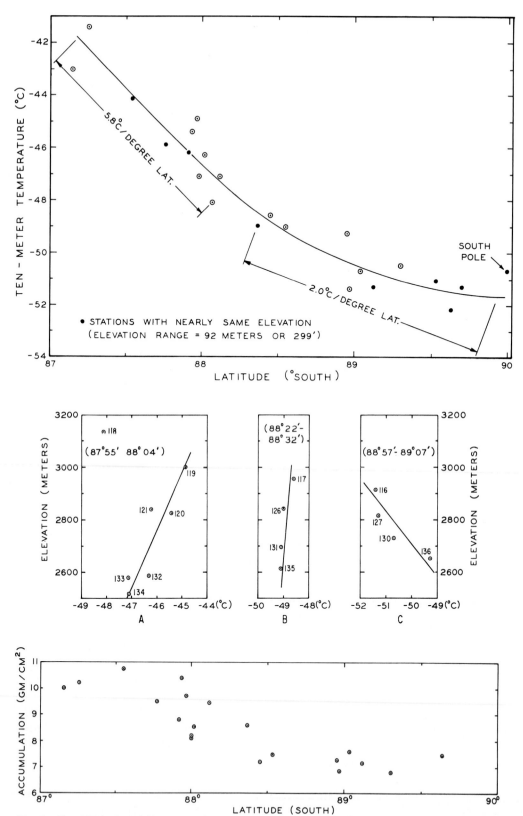

Fig. 8. *Top:* Variation of 10-meter temperature with latitude. *Middle:* 10-meter temperature versus elevation for stations with the same latitude. *Bottom:* Variation of annual accumulation with latitude.

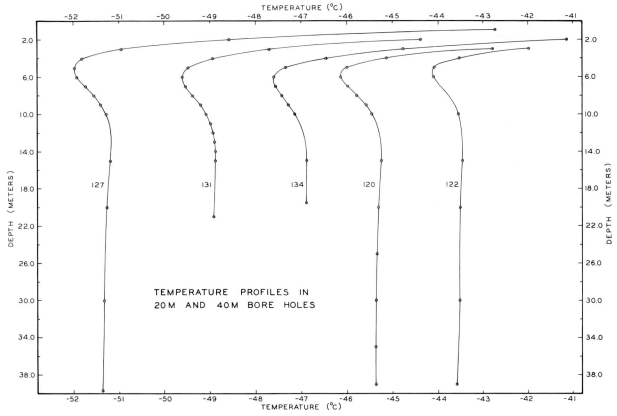

Fig. 9. Temperature profiles in 20-meter and 40-meter bore holes.

values (Table 1) are equivalent to the mean annual air temperature to within 0.1°C. A change of −2.0 to −5.8°C/1° latitude is considerably higher than values obtained by *Langway* [1961] for dry snow facies in Greenland, −1.0°C/1° latitude. The regression line for 10-meter temperatures versus latitude is not linear. The temperature may be influenced by the Queen Maud and Horlick Mountains, particularly in the vicinity of stations 119 to 125 (Figure 1).

A further plot of 10-meter temperatures at the same latitude, but for different elevations (Figure 8, *middle*) indicates that elevation does influence the temperature in the vicinity of the South Pole. The lapse rate at latitude 88°57′–89°07′S is 0.8°C/100 meters. This value is slightly lower than the mean value calculated for eastern Antarctica of 1.1°C/100 meters, and nearly twice the value for western Antarctica, 0.35°C/100 meters [*Bentley et al.*, 1964].

In Figure 8 (*middle*), the 10-meter temperatures at latitude 87°55′–88°04′S increase with increase in elevation. This anomaly is probably due to the effect of katabatic winds. Stations 119–121, which have higher temperatures than other stations at the same or even lower elevations, are near the crest of a ridge whose axis follows the 88°S parallel. Cold air flowing down the flanks of this ridge is replaced at the crest by warmer air above. Without this influence, the temperature at these stations would probably be nearly equal to that at station 118. Winds in the vicinity of stations 119–121 are stronger than elsewhere on the traverse, as is indicated by the high ram hardness and well-developed cross-bedding at these stations (Figure 5).

At depths between 15 and 40 meters, negative temperature gradients of 0.56 to 0.64°C/100 meters were recorded (Figure 9). Similar negative gradients have been reported by *Bogoslovski* [1958], *Wexler* [1959], and *Mellor* [1960]. Calculations by *Mellor* [1960] indicate that climatic warming best explains the negative gradients. Temperature gradients could be determined for only three holes to 40 meters. Not enough temperature measurements are available to relate these negative temperature gradients to ice thickness or ice flow in the area of the traverse.

EFFECT OF LATITUDE AND ELEVATION

Accumulation, mean density, integrated ram hardness, grain size, and 10-meter temperatures are plotted against latitude and elevation. All but grain size vary as a function of latitude over the traverse range of 2°52′. The regression of each property on latitude is shown in Figures 6 and 8. There is little correlation between either accumulation or firn properties with elevation except for 10-meter temperatures near the South Pole (Figure 8, *middle*), mainly because elevation changes are small compared to changes in latitude.

Density and integrated ram hardness change exponentially with distance from the Pole in the direction of the Queen Maud and Horlick Mountains. Upon approaching these mountains and lower latitude, the increase in density and hardness is caused by a higher mean annual air temperature, which facilitates recrystallization, and a stronger and more persistent wind, which transports, rounds, and sorts the snow.

The decrease in annual accumulation with increase in latitude, as shown in Figure 8 (*bottom*), is probably the result of cooling of the principal air masses transporting moisture from Marie Byrd Land to the South Pole region. The drop in mean annual air temperature (determined from 10-meter bore-hole measurements) between 87°S and 90°S, the limits of the traverse, is 9°C, which is sufficient to remove a significant amount of moisture from the air mass.

Annual accumulation, 10-meter temperatures, and mean density of firn (0–2 meters) are plotted against each other (Figure 10), with a linear correlation of 0.94. Within the general area of the traverse, the regression lines of these plots determined from a least-squares reduction can be used to determine any two values if the third is known.

SUMMARY

The annual accumulation for 21 of the 25 stations on the 1962–1963 South Pole traverse has been determined from an analysis of stratigraphy in 2-meter pits. The entire traverse is between 2517- and 3137-meter elevation on the Antarctic plateau. The following assumptions are used as a basis for determining a year's accumulation.

1. Ice crusts and permeable, low-density layers of firn are formed by vapor transfer in summer (December and January) and late fall (April) during periods of large temperature gradients on the Antarctic plateau. Ice crusts, possibly formed by radiation, may also form during the summer period.

2. High-density layers consisting of fine-grained, somewhat permeable firn, and sometimes containing cross-bedded structures but no ice crusts, are formed during midwinter. Strong winds are responsible for the formation of many high-density layers, for they break apart snow crystals, produce rounded grains, and, depending on the wind velocity and the nature of the undulating surface, deposit grains of approximately the same size. The resulting wind-blown deposits pack efficiently and, through compaction due to burial, form high-density but still permeable layers. Strong persistent winds cannot form ice crusts because of the low air vapor pressure unless accompanied by a high temperature gradient, in which case the vapor is derived from the firn. Since high temperature gradients generally do not occur in midwinter, the high winds during this time cannot produce crusts.

3. Accumulation during some years has been very low or has been removed by wind erosion, resulting in a disconformity in the normal stratigraphic sequence. This is represented by either an abrupt change in the normal cyclic variation in density or by the presence of many closely spaced ice crusts separated by coarse-grained permeable hoar layers.

Accumulation varies from 6.8 g/cm^2 at the South Pole to 10.8 g/cm^2 near the Queen Maud Mountains. Annual accumulation, mean density between 0 and 2 meters, integrated ram hardness, and 10-meter temperatures all decrease with increasing latitude between 87°08′S and 90°S as follows: accumulation, 1.2 g/cm^2/deg latitude; density, 0.037 g/cm^3/deg latitude; integrated ram hardness, 1160 joules/deg latitude; and 10-meter temperature, 1.5°C/deg latitude. At depths between 15 and 40 meters, negative temperature gradients of 0.56 to 0.64°C per 100 meters of depth were recorded. Elevation changes over the traverse were not more than 267 meters; other factors such as latitude, ice topography, and proximity to mountain ranges had a dominating influence on accumulation, firn density, and firn temperature.

Annual accumulation, 10-meter temperature, and mean density of firn (0–2 meters) plotted against each other have linear correlations of 0.94. Within the region of the traverse, the regression lines of

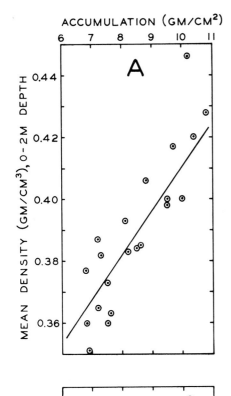
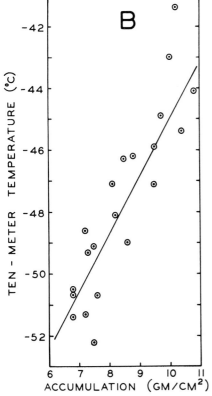
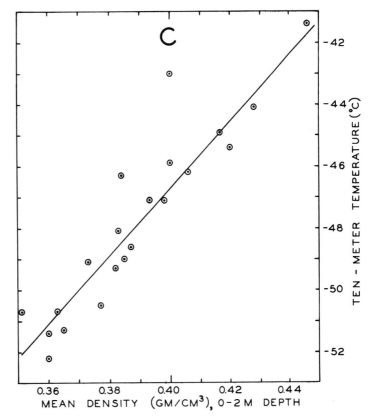

Fig. 10. (A) Mean density (0–2 meters deep) versus annual accumulation; (B) 10-meter temperature versus annual accumulation; (C) 10-meter temperature versus mean density (0–2 meters deep).

these plots can be used to determine any two of the values if the third is known.

CONCLUSIONS

Future studies should be directed toward a means of determining absolute dates of firn independent of accumulation rates or stratigraphy. Stratigraphic methods are limited to shallow depths, 10-meters or less, and depend on the observer's interpretation of disconformities, ice crusts, and hoar layer distribution. It is hoped that techniques of oxygen-isotope [*Sharp and Epstein*, 1962; *Dansgaard et al.*, 1969], deuterium-hydrogen, and lead 210 analyses [*Picciotto et al.*, 1964] will supercede the stratigraphic approach of determining age and accumulation. Laboratory experiments that reproduce the seasonal conditions of the Antarctic plateau are needed to further clarify the processes of firnification, particularly in the production of radiation crusts.

Future pit studies should include ink-staining for revealing the distribution of impermeable crusts. Accurate density profiles are most helpful in determining the annual layer.

Acknowledgments. The glaciological program was supported by the National Science Foundation under grant G 23005 to the Institute of Polar Studies through the Ohio State University Research Foundation. The author gratefully acknowledges the logistic support offered by the Geophysical and Polar Research Center in charge of traverse operations. The author wishes to thank Mr. Henry Brecher for his excellent assistance, both in the field and in the reduction of data, and Mr. Anthony Gow and Mr. Lyle Hansen of the Cold Regions Research and Engineering Laboratories for advice on firn stratigraphy and temperature measurements and for supplying calibrated thermistors. The author appreciates the many suggestions offered by Dr. Colin Bull.

REFERENCES

Bader, H., R. Haefeli, E. Bucher, J. Neher, O. Eckel, and Chr. Thams, Snow and its metamorphism (Der Schnee und Seine Metamorphose, 1939), *Transl. 14,* U.S. Army Corps of Engineers, Snow, Ice, and Permafrost Res. Estab., Wilmette, Ill., 313 pp., 1954.

Benson, C. S., Stratigraphic studies in the snow and firn of the Greenland ice sheet, *Rep. 70,* U.S. Army Corps of Engineers, Snow, Ice, and Permafrost Res. Estab., Wilmette, Ill., 93 pp., 1962.

Bentley, C. R., R. L. Cameron, C. Bull, K. Kojima, and A. J. Gow, *Antarctic Map Folio Series 2, Physical Characteristics of the Antarctic Ice Sheet,* American Geographical Society, New York, 1964.

Bogoslovski, V. N., The temperature conditions (regime) and movement of the Antarctic glacial shield, in *Chamonix Symposium, Publ. 47,* pp. 287–305, Intern. Assoc. Sci. Hydrol., Gentbrugge, 1958.

Bull, C., The use of the rammsonde as an instrument for determining the density of firn, *J. Glaciol., 2*(20), 714–719, 1956.

Dalrymple, P. C., South Pole micrometeorology program, 2, Data analysis, *Tech. Rep. ES-7,* Quartermaster Res. Eng. Center, Natick, 93 pp., 1963.

Dansgaard, W., S. J. Johnsen, J. Moller, H. C. Oersted, and C. C. Langway, Jr., One thousand centuries of climatic record from Camp Century on the Greenland Ice Sheet, *Science, 166,* 377–381, 1969.

Dorsey, N. E., *Properties of Ordinary Water-Substance,* Reinhold, New York, 673 pp., 1940.

Giovinetto, M. B., Glaciological studies on the McMurdo–South Pole traverse, 1960–61, *Rep. 7,* Inst. Polar Studies, Ohio State University, Columbus, 38 pp., 1963.

Koerner, R. M., Firn stratigraphy studies on the Byrd-Whitmore Mountain traverse, 1962–1963, in *Antarctic Snow and Ice Studies, Antarctic Res. Ser.,* vol. 2, pp. 219–236, AGU, Washington, D.C., 1964.

Kotlyakov, V. M., Results of a study of the processes of formation and structure at the upper layer of the ice sheet in East Antarctica, in *Symposium on Antarctic Glaciology, Publ. 55,* Intern. Assoc. Sci. Hydrol., Gentbrugge, 1961.

Kuznetsov, M. A., Characteristic forms of radiational melting of snow and ice and of the infiltration of water into the snow sheet, in *Soviet Antarctic Expedition,* vol. 10, *Glaciological Investigations,* Second Continental Edition 1956–1958, Leningrad, 1960.

Langway, C. C., Accumulation and temperature on the inland ice of north Greenland, 1959, *J. Glaciol., 3*(30), 1017–1044, 1961.

Mellor, M., Temperature gradients in the Antarctic ice sheet, *J. Glaciol., 3*(28), 773–782, 1960.

Picciotto, E., G. Crozaz, and W. De Breuck, Rate of accumulation of snow at the South Pole as determined by radioactive measurements, *Nature, 203,* 393–394, 1964.

Robinson, E. S., On the relationship of ice-surface topography to bed topography on the South Polar Plateau, *J. Glaciol., 6*(43), 43–54, 1966.

Seligman, G., *Snow Structure and Ski Fields,* Macmillan, London, 555 pp., 1936.

Sharp, R. P., and S. Epstein, Comments on annual rates of accumulation in West Antarctica, in *Symposium of Obergurgl,* pp. 273–285, *Pub. 58,* Intern. Assoc. Sci. Hydrol., Gentbrugge, 1962.

Shimizu, H., Glaciological studies in West Antarctica, 1960–1962, in *Antarctic Snow and Ice Studies,* vol. 2, *Antarctic Res. Ser.,* pp. 37–64, AGU, Washington, D.C., 1964.

Taylor, L. D., and J. Gliozzi, Distribution of particulate matter in a firn core from Eights Station, Antarctica, in *Antarctic Snow and Ice Studies,* vol. 2, *Antarctic Res. Ser.,* pp. 267–277, AGU, Washington, D.C., 1964.

U.S. Weather Bureau, *Climatological Data for Antarctic Stations,* U. S. Government Printing Office, Washington, D.C., 1963.

Wexler, H., Geothermal heat and glacial growth, *J. Glaciol., 3*(25), 420, 1959.

Yosida, Z., Physical properties of snow, in *Ice and Snow,* edited by W. D. Kingery, chap. 35, pp. 492–496, MIT Press, Cambridge, Mass., 1963.

A STRATIGRAPHIC METHOD OF DETERMINING THE SNOW ACCUMULATION RATE AT PLATEAU STATION, ANTARCTICA, AND APPLICATION TO SOUTH POLE–QUEEN MAUD LAND TRAVERSE 2, 1965–1966

R. M. KOERNER

Institute of Polar Studies, Ohio State University, Columbus 43210

During December and January 1966–1967, the snow accumulation at Plateau station, Antarctica (79°15′S, 40°30′E), was examined at an array of 99 stakes, 114 shallow pits, three 1-meter pits, one 2-meter pit, and one 10-meter pit. The annual layering in the snow stratigraphy can be divided into two parts. The summer section generally consists of a thin, hard, fine-grained layer, often associated with a hoar layer and a bonded grain crust. The winter section is softer, coarser-grained, and more homogeneous than the summer section. The stake measurements and shallow-pit estimates of the 1966 winter and 1966–1967 summer accumulation are used to examine the validity of the deeper-pit interpretations. It is concluded that the mean accumulation is 2.8 g cm^{-2} yr^{-1} (± 0.4). There is no evidence from the 10-meter pit of any over-all change in the accumulation rate in the past 127 years. The same stratigraphic criteria are applied to the snow stratigraphy observations from the South Pole–Queen Maud Land traverse 2 to determine the annual snow accumulation between the Pole of Relative Inaccessibility and Plateau station. The results are compared with accumulation rates based on the depth of the 1955 radioactive layer. The stratigraphic results generally overestimate the accumulation rate. However, the closeness of many of the results in the two sets indicates that a thorough investigation of the stratigraphy at every site can produce a reasonably accurate estimate of the accumulation rate.

PREVIOUS DRY-SNOW STRATIGRAPHIC WORK IN ANTARCTICA

The bulk of the studies of dry-snow stratigraphy by investigators of the U.S. Antarctic Research Program have been made in West Antarctica, where accumulation is more than 10 g cm^{-2} yr^{-1} and the mean annual temperature is above $-30°$C. (In this paper the term accumulation is synonymous with the term balance (b) now being used in mass balance studies.) At Byrd station, and in the dry-snow zone of Marie Byrd Land in general, stratigraphic analysis has relied on seasonal variations of density, hardness, and grain size in the snow-firn pack to distinguish annual layers. The depth-hoar layer, which generally develops every fall, has also been used as a datum [*Shimizu*, 1964]. At Amundsen-Scott South Pole station, *Giovinetto*'s [1960] analyses of pit and snow-mine stratigraphy were again based on the concept of seasonal characteristics of density, hardness, and grain size. His work is the first to have been subjected to a thorough statistical analysis [*Giovinetto and Schwerdtfeger*, 1966]. *Taylor*'s [this volume] snow-pit work on traverses from the South Pole relied basically on density variations between winter and summer. *Gow*'s [1965a] South Pole analyses depended on the depth-hoar layer that he observed in each annual layer. *Kotlyakov* [1966], in the low-accumulation region of central East Antarctica, observed the classical variation of snow texture from winter to summer, and used it to detect annual boundaries. *Picciotto et al.* [this volume] observed an unexpected variety of accumulation conditions on the South Pole–Queen Maud Land traverses (SPQMLT) 1 and 2 (Figure 1) and noted very indistinct stratigraphy at the Pole of Relative Inaccessibility [*Picciotto et al.*, 1968].

RESULTS

Stratigraphy in the Plateau station area had been examined only once before. Picciotto (personal communication), by measuring lead 210 to 10 meters

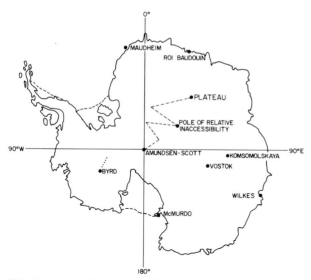

Fig. 1. Antarctica, showing location of stations mentioned in the text. The dashed line is the route of the U. S. South Pole–Queen Maud Land traverses 1 and 2, and the dotted line is the location of the Byrd–Whitmore Mountain stake network.

deep and radioactive fallout to 2 meters deep, determined an accumulation rate of 2.8 g cm^{-2} yr^{-1}. Apart from this information, there were no other investigations except those at Vostok (3.0 g cm^{-2} yr^{-1} [*Kotlyakov*, 1961]) and at the Pole of Relative Inaccessibility (3.0 g cm^{-2} yr^{-1} [*Picciotto et al.*, 1968]) on which to base an analysis. It was necessary, therefore, to develop stratigraphic criteria from on-site pit investigations. For this reason, the surface snow was thoroughly investigated before summer warmth and radiation had a chance to effect marked changes on the snow grain structure of the preceding winter snow. Sixty-four shallow pits (less than 50 cm deep) were excavated on December 10 and 11, 1966, and gave a good approximation of the variability of accumulation within a small (10 km^2) area. The preceding winter's accumulation

was easily recognized by its fine, angular grains (0.2 mm), which contrasted with the coarser, rounder grains underneath ($\geq$0.4 mm) belonging to the preceding budget year. These pits gave a mean winter accumulation of 8.2 cm of snow (2.4 g/cm^2). Measurement of 99 stakes placed the preceding February 2, 3, and 4 gave a mean winter accumulation value of 7.5 cm snow (2.2 g/cm^2). These stakes were 0.3–1.0 km apart in a T-shaped pattern, each arm measuring about 26.0 km.

To examine the effect of the summer (December and January) on accumulation and stratigraphy, stakes were remeasured and 50 shallow pits were dug on January 23, 1967. A typical January pit is shown in Figure 2, and the changes in hardness, density, and accumulation in December and January are shown in Table 1. Besides the development in December and January of a thin, fine-grained hard

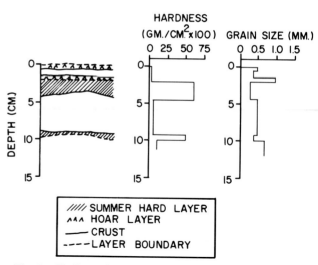

Fig. 2. Shallow pit stratigraphy, Plateau station, Antarctica, January 23, 1967. The snow between the two hoar layers is very loose. When buried, the top 2 cm will probably appear as a single hoar layer with two crusts. The density between 2 and 8 cm is 0.302 g/cm^3.

TABLE 1. Summary of Results from Stakes and Shallow Pits

Date	Density g/cm^3	Pits Mean Depth Ann. Layer, cm	Stakes, Mean Snow Depth, cm (Feb. 2, 1966–)	Mean Pit Accum., g/cm^2	Mean Stake Accum., g/cm^2	Std. Dev., g/cm^2 Pits	Std. Dev., g/cm^2 Stakes	Surface Hardness, g/cm^2 × 100
Dec. 10–11	0.288	8.2	7.5	2.36	2.15	1.6	1.8	1
Jan. 23	0.316	8.1	7.9	2.45	2.29	1.5	2.1	6

The January water equivalent values take into consideration 1.0–2.0 cm of very loose surface snow. A density of 0.1 g/cm^3 has been used for this layer. In column 3 the mean depth is of snow accumulated above the preceding summer-affected layer.

layer near the surface, the grain size of the winter snow had doubled to 0.4 mm, and considerable rounding of the grains had occurred.

The December 1966 and January 1967 shallow-pit studies and stake measurements allowed stratigraphic criteria to be established by which the deeper pits could be analyzed. The shallow pits showed an annual layer thickness generally varying between 0 and 20 cm with a mode of about 8–9 cm ($p = 0.288$ g/cm³). The summer section includes a thin (0.3–4.0 cm), hard, fine-grained layer often associated with one or more coarse-grained hoar layers and bonded grain crusts. The winter section consists of a homogeneous, coarse, soft layer free of crusts and hoar layers. This pattern was applied in the interpretation of three 1-meter pits, one 2-meter pit, and one 10-meter pit.

One-Meter and Two-Meter Pit Analyses

The technique used in all the 1-meter and 2-meter pit studies was as follows. Two pits were dug alongside each other, with a wall not more than 30 cm thick between them. The working pit was covered with a board or canvas so that the stratigraphy on the thin wall between the pits stood out clearly owing to differences in the amount of light transmitted through the various layers. Grain size was measured in situ with a Leitz 8× hand lens with a scale graduated in 0.1 mm. On each grain, one axis (that parallel to the snow surface) was recorded. From 20 to 60 grains from each layer were measured, and the mean grain size for each layer was calculated. The hardness of each layer was measured with Canadian hardness gages, and the density was measured with standard 500-cm³ tubes weighed to 0.1 gram on a triple beam balance.

Annual boundaries were determined in the field, and photographs of the thin wall (using transmitted light) were later used to corroborate the field and quantitative analyses.

The stratigraphy can be described best with reference to pit 3 (Figure 3) where the repetition of summer features proved to be most regular. The clearest features are the hard, fine-grained layers occurring at 13–15 cm, 24 cm, 29 cm, 36 cm, 46 cm, 59–66 cm, 71 cm, and 78 cm. These layers wedge out into thin, bonded-grain crusts that in places are very irregular (e.g., 58–64 cm, Figure 3). In other pits it was often necessary to examine the entire 2- to 3-meter-long pit wall to find the year's hard layer, since in some years it is discontinuous. Hard lenses similar to those in Figure 3 were found in 70% of what were determined to be annual layers in the other pits. A hard layer is usually associated with a bonded-grain crust generally found in the upper boundary of the hard layer or a few millimeters above it. A loose hoar layer is also frequently associated with the hard lenses, and in 80% of the cases it lies on the surface of the hard layer (e.g., at 28 cm depth, Figure 3). The close association of crusts, hoar layers, and hard layers, and the development of the same features during the 1966–1967 summer, identifies the entire association as a summer one. Generally, if only one of these summer features is found in what has been determined to be an annual layer, a degree of uncertainty enters the analysis. While thin crusts were observed in the winter layer of 5% of the shallow pits in December 1966, no hard layers or hoar layers were observed there. The placing of an annual boundary along a single crust, unassociated with either a hard or a hoar layer, may therefore introduce some error into the analysis. Thirteen percent of the annual boundaries in the pits were determined in this way (Table 2).

The grain size and hardness profiles (Figure 3) show the contrasts between summer and winter layers. It must be emphasized, however, that, for identification of annual layers in this area, a visual analysis of the thin wall texture is adequate. Although grain size, hardness, and density measurements illustrate quantitatively the contrast between summer and winter layers, they are not essential to the identification of annual boundaries. Table 3 gives the accumulation, variability of annual accumulation, and number of annual layers in each pit.

Ten-Meter Pit Analysis

A pit 1.5 meters by 1.5 meters was excavated to a depth of 1020 cm, 30 meters from the summer hut at Plateau station. The surface annual layer was ignored in the analysis because drift caused by the hut had increased its thickness to 28 cm. Techniques of stratigraphic analysis were similar to those described previously, except that ink staining (ink sprayed on the snow wall percolates down and into the wall under the heat of a blow torch and demarcates the different layers and crusts) was used on one wall and the opposite wall was illuminated by a 200-watt bulb lowered down a hole augered 30 cm beyond it. A stratigraphic diagram was drawn from the stained and illuminated walls and an attempt

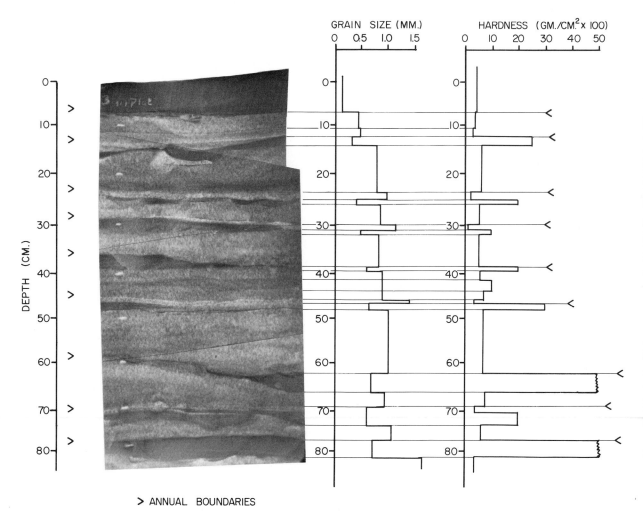

Fig. 3. Snow stratigraphy, pit 3, Plateau station, Antarctica, December 12, 1966. Hard lenses appear as dark layers.

was made, not always successful, to trace the stratigraphy from one of these walls to the other across the intervening walls.

The cycle of layers and crusts described previously applied throughout the 10-meter section. The results are shown in Figure 4, which indicates that there has been no noticeable trend of increasing or decreasing accumulation since about 1830. *Giovinetto* [1960] showed an increasing accumulation at the South Pole over the same period, but his percentage increase (8%) would represent only 0.2 g/cm², at Plateau station, a value within the standard error of the mean of annual accumulation estimates.

ACCURACY OF THE ANALYSIS

The validity of this analysis can be tested by comparison with results obtained in similar environments elsewhere, with Picciotto's results in the same area, and by statistical comparisons among the stake, shallow-pit, and deep-pit results at Plateau station.

Previous Work in the Area

Soviet investigators working in East Antarctica in areas with elevation and temperature regimes similar to those at Plateau station have determined accumulation values of 8.0 g cm^{-2} yr^{-1} at Komsomol'skaya [*Kotlyakov*, 1966] and 3.0 g cm^{-2} yr^{-1} at Vostok [*Kotlyakov*, 1961]. Of these two stations, Vostok, with a mean annual temperature of −56°C and an elevation of 3488 meters above sea level, compares more closely in environment with Plateau station (−58.4°C and 3624 meters). The Plateau

TABLE 2. Percentage of Annual Layers Determined by Crusts, Hard Layers, Hoar Layers, or Combination of Two or More of These Features

Pit	Depth of Pit, cm	Percentage Annual Layers Determined by:				Number Ann. Layers
		Crust (A)	Hard Layer (B)	Hoar Layer (C)	Comb. of 2 or 3 of A, B, C	
1	210	17	9	22	50	18
2	100	20	30	0	50	10
3	90	11	0	0	89	9
4	86	10	10	20	60	10
5	150	18	12	6	65	17
6	1020	11	15	16	58	118
Mean		13	14	14	59	

TABLE 3. Summary of Pit Analysis at Plateau Station, Antarctica, December 1966 and January 1967

Source	No. Years, Stakes or Shallow Pits	Annual Accumulation, g cm^{-2}		
		Mean $\langle A \rangle$	Max. Range	Variability $\hat{\sigma}_{\langle A \rangle}$*
1966–1967 stakes	99	2.29	0–11.1	2.06
Shallow pits	60	2.45	0–6.75	1.58
Pit 1	24	2.64	0–5.07	1.31
Pit 2	11	2.93	0–6.18	1.63
Pit 3	9	2.84	0–5.94	1.82
Pit 4	11	2.69	0–5.13	1.42
Pit 5	16	2.87	0–5.22	1.64
10-meter pit	127	2.78	0–9.43	1.48
Mean, pits 1–5	71	2.76	0–6.18	1.55

*The variability is the 'best estimate' of the standard deviation of the annual accumulation estimates and stake measurements. As each sample standard deviation applies to different numbers of observations, Bessel's correction $\hat{\sigma} = \{[n/(n-1)]S^2\}^{1/2}$, where $\hat{\sigma}$ is the best estimate of the population standard deviation and S^2 is the sample variance, has been applied to give the standard deviations a common meaning.

station analysis has indicated an annual accumulation 0.2 g cm^{-2} yr^{-1} less than Kotlyakov's value for Vostok. Results from pit investigations of the Soviet Komsomol'skaya–South Pole traverse [*Ukhov*, 1963] showed that accumulation decreases from 3.6 g/cm^2 at Vostok station to a minimum of 1.3 g/cm^2 393 km south of the station. At the Pole of Relative Inaccessibility (82°07'S, 55°02'E, elevation 3718 meters, mean annual temperature −57°C), *Picciotto et al.* [1968], from isotopic and stake measurements, concluded a mean annual accumulation of 3 g cm^{-2}. Picciotto (personal communication and *Picciotto et al.* [this volume]) found an accumulation at Plateau station, from two analyses of the distribution of radioactive fallout with depth, of 2.6–2.8 g cm^{-2} yr^{-1}. All these results are similar to the accumulation estimates derived from the stratigraphy at Plateau station.

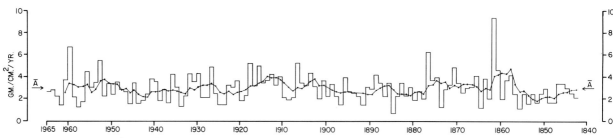

Fig. 4. Estimates of annual accumulation and 5-year running mean of estimates, 10-meter pit, Plateau station, Antarctica, January 1967.

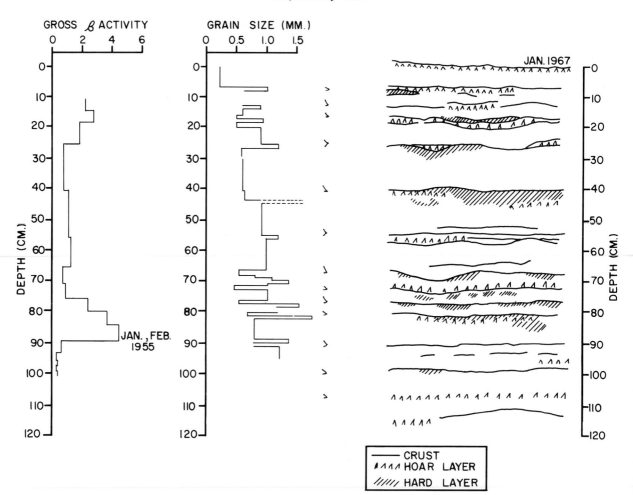

Fig. 5. Gross β activity and stratigraphy in the snow, Plateau station, Antarctica, January 3, 1967.

Comparison between Annual Accumulation Determined by Stratigraphic Interpretation and Distribution of Radioactive Fallout in a Single Pit

After the stratigraphy in one pit was interpreted, samples were taken from each major layer by the author and were sent to Dr. Picciotto at the Free University of Brussels, Belgium. His results (personal communication) of the distribution of gross β activity with depth are shown in Figure 5. The stratigraphy is also shown with the annual boundaries as determined by the basic pattern discussed previously. *Picciotto and Wilgain* [1963] and *Wilgain et al.* [1965] showed that fission products in the firn at Roi Baudouin station, Antarctica, reached a

TABLE 4. Variance Test for Plateau Station Accumulation Series

Source of Variation	Sum of Squares	Degrees of Freedom	Variance Estimate	F
	Stakes, Shallow Pits, 1-Meter, 2-Meter, and 10-Meter Pits			
Total	966	335		
Between groups	68	3	22.7	8.7*
Within groups	898	332	2.6	
	Three 1-Meter Pits, One 2-Meter Pit, and One 10-Meter Pit			
Total	273.50	179		
Between groups	2.15	4	0.54	2.85†
Within groups	271.35	175	1.55	

* Significant at <1% level.
† Significant at <5% level.

maximum in firn deposited as snow between January and February 1955. This maximum was caused by the Castle bomb tests of March–April 1954. This same maximum of fission products was found in the Plateau station pit at a depth of 85–90 cm. In the same pit, the author found 11 annual layers between the surface and 90 cm. As will be seen later, stake measurements indicate that snow accumulation is missing at about 1 out of 12 sites. Thus it is reasonable to expect that, at one site, 1 out of every 12 years of accumulation might be missed. One missing year in this pit would give a value corresponding with the isotopic dating obtained by Picciotto.

Variance Analyses

It has been assumed that, in any small area (less than 10 km²), any horizontal variation of annual accumulation in one year will quantitatively be similar to the variation of annual layer thickness over many years at any one point in this area. Thus some measure of the accuracy of any single site interpretation will be given by a comparison of the ranges, standard deviations s, and means $\langle A \rangle$ of annual accumulation estimates from that site and from surface measurements in an area which includes that site. In this context, variability of stake measurements and shallow-pit and deep-pit estimates of annual accumulation has been represented in each case by the standard deviation from the mean of the estimates.

The stake and shallow-pit measurements together showed that 8.5% of the area investigated lacked any accumulation between February 1966 and late January 1967. If 8.5% of the surface misses accumulation over year-long periods, it can reasonably be assumed that any one point in that area will miss accumulation in 8.5% of years. Therefore, to compensate for missing years in the stratigraphy of the deep pits, 12 annual layers have been taken to represent 13 years.

Variance test of the mean annual accumulation estimates. To test whether the mean accumulation values from the stake, shallow-pit, and deep-pit measurements are statistically similar, a variance analysis has been made, first, among the deep-pit measurements themselves, and, second, among the deep-pit, stake, and shallow-pit measurements. The results are shown in Table 4. The tests indicate that, though mean annual accumulation estimates from the five deep pits are not significantly different from each other, the deep-pit estimates, shallow-pit estimates, and stake measurements are significantly different. This can mean either that the deep-pit interpretation is incorrect or that the 1966–1967 accumulation was very low, as the shallow-pit interpretation and stake measurements give a lower mean annual accumulation than the deeper-pit interpretations.

To test whether the 1966–1967 accumulation at Plateau station was less than a long-period mean, stake measurements from Amundsen-Scott South Pole station, Byrd station, and the Byrd–Whitmore Mountains stake network (Figure 1) have been examined. At Amundsen-Scott station, the 1966–1967 accumulation was the lowest on record and was 22% less than the 9-year mean. At Byrd station, the 1966–1967 accumulation was 15% less than the 1963–1966 mean [*Cameron*, 1968], and, on the 360-km Byrd–Whitmore Mountains network, the 1966–1967 accumulation was 29% less than the 1963–1966 mean [*Brecher*, 1967]. These results support the conclusion that the mean annual accumulation value for 1966–1967 from the Plateau station stake and shallow-pit interpretation is significantly lower than that from the pit analysis, as accumulation was very low over a large part of Antarctica for the year 1966–1967.

Variance ratio test of the variances of the estimates of annual accumulation. A variance ratio test of the variances of annual accumulation estimates for the 10-meter and the 1- and 2-meter pits shows that the observed difference between the standard deviations is slightly less than the standard error of the difference between the standard deviations of the samples. Therefore, as far as the variability of layer thickness (annual accumulation) from pit to pit is concerned, the interpretation is consistent. However, when the variance ratio test is applied to the two sets of surface measurements, the observed difference between the standard deviations of the stake measurements and shallow-pit estimates of annual accumulation is three times the standard error of the difference. This indicates a difference between the two sets of surface observations, although the means are similar (Table 3).

Therefore, to determine where the error in the shallow-pit analysis lies, the standard deviation of the stake measurements has been recalculated allowing stake measurements of negative accumulation (i.e., deflation) to equal zero (as negative accumulation cannot be measured by the pit technique) and assuming that each of the two largest stake accumulations (33.2 cm and 35.3 cm of snow) equal two years of accumulation. The latter assumption has been made, as it is felt that, in the examination of pit-wall stratigraphy, exceptionally thick layers may be misidentified as representing two years instead of one year of accumulation. It is also possible that two anomalously thick layers like these did not occur in the pit-sampling area. The recalculation reduced the standard deviation of the stake measurements by 17% to 1.71 grams. Ninety-nine percent of this change was due to subdivision of the two large annual accumulation measurements, and only 1% to elimination of negative accumulation values. This reduces the difference between the shallow-pit and stake standard deviations to 0.18 gram. The stake measurements refer to an area about 20 times larger than the shallow-pit area, so that the 10% difference in variability between the stake measurements and shallow-pit estimates is probably real and not due to sampling errors. The recalculation suggests that the author's snow stratigraphy interpretation may be prone to error wherever unusually thick annual layers occur. However, while the error introduced by misrepresenting two annual layers as four changes the standard deviation of the stake measurements of annual accumulation by almost 17%, it alters the mean by only 1%, from 2.49 to 2.47 g cm^{-2} yr^{-1}.

Total variability of accumulation. So far, horizontal variability of accumulation has been compared between two sets of observations, and vertical variability of layer thickness has been compared among five sets of observations. Horizontal variability of accumulation in any one year in a local area is the result of relationships between topography, both on a large and small scale, and wind velocity and turbulence. Variability of annual layer thickness in a pit is a combination of the effects of horizontal and temporal variability of accumulation at the surface each year.

The standard deviation of stake measurements and shallow-pit estimates of annual accumulation reflect local horizontal variability of annual accumulation (*Giovinetto*'s [1964] local areal variability). The standard deviation of annual accumulation estimates for any one pit is a measure of total variability that includes both local horizontal variability and variability of accumulation from year to year (*Giovinetto*'s [1964] temporal variability). Therefore, the standard deviation of the annual accumulation estimates from the pit interpretations should be greater than the standard deviation for the shallow-pit estimates of the 1966–1967 accumulation.

The temporal variability of annual accumulation at Plateau station has been estimated from stake and pit measurements at several stations in Antarctica, one in Greenland, and one on a Canadian ice cap (Figure 6). Some limitations to this way of determining the temporal variability at Plateau station should be borne in mind. Other variables besides the amount of annual accumulation affect temporal variability, and the most important of these is the position of each station relative to general atmospheric circulation and in particular to cyclonic activity. However, it is beyond the scope of this paper to consider these relationships. Only Gow's [1965b] and Landon-Smith's [1965] data in Figure 6 cover large enough areas to exclude the effect of local horizontal variability of annual accumulation. The data for the Amery ice shelf [Landon-Smith, 1965] and for Maudheim [Schytt, 1958] are from pit investigations. Schytt's pit work and stake measurements [Swithinbank, 1958] led Schytt [1958] to state that local variability of accumulation each year is very low on the Quar (Maudheim) ice shelf, and his value for the standard deviation of the estimates of annual accumulation has been taken as representing temporal variability. The same assumption has been made for Landon-Smith's results. The regression equation gives a value for the temporal variability at Plateau station of 0.38 g/cm². This can be compared with a value of 0.65 g/cm² taken from a similar analysis based mainly on local areal variability of accumulation in Antarctica [Giovinetto and Schwerdtfeger, 1966, Fig. 4, p. 240].

The standard deviation of the annual accumulation estimates from the shallow-pit studies has been combined with the standard deviation due to temporal variability of accumulation from the regression analysis discussed above to give a total standard deviation of 1.63 g/cm². This value represents the theoretical total variability of annual accumulation in the Plateau station area. The standard deviation of annual accumulation estimates from the 1-meter, 2-meter, and 10-meter pits are within 0.32 gram of 1.63 grams (Table 3), and the standard deviation of all the annual accumulation estimates from the deep pits is only 0.08 gram less. The conclusion is that, as far as variability of annual accumulation estimates is concerned, the 1-meter, 2-meter, and 10-meter pit interpretations are consistent with the shallow-pit interpretations. The limitations of the deep-pit estimates of annual accumulation are therefore the same as those that apply to the shallow-pit work. Any errors are probably due to incorrect identification of some of the crusts, unassociated with hard or hoar layers, as summer datum levels. In addition, a few very thin accumulation layers may have been missed and not compensated for by the correction factor whereby 12 annual layers are taken to repre-

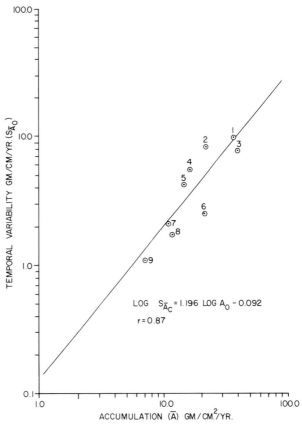

Fig. 6. Temporal variability of accumulation: (1) Maudheim, Antarctica [Schytt, 1958]; (2) Devon Island ice cap, 1787 meters [Koerner, 1966]; (3) Camp Century, Greenland [Langway, 1967]; (4) Wilkes S-2 [Black and Budd, 1964]; (5) Old Byrd station stake network; (6) Amery ice shelf [Landon-Smith, 1965]; (7) New Byrd station [Cameron, 1968]; (8) New Byrd station [Gow, 1965b]; (9) Amundsen-Scott South Pole station (U.S. Weather Bureau stake network).

sent 13 years (10% of the shallow-pit annual layers show an accumulation less than 1.0 g cm⁻² yr⁻¹, whereas only 1% of the annual layers in the deep pit fall within this range). The errors resulting from interpreting a single annual layer as representing two years and failing to recognize very thin annual accumulation layers partly compensate each other. Judging from the variance analysis, the two errors introduce an inaccuracy of about 10% into the identification of annual layers, e.g., 1 annual layer in 20 is omitted, and 1 annual layer in 20 is taken to represent 2 years. From this it is concluded that the over-all trend of accumulation during the 128-year period represented by the 10-meter pit and shown in

Figure 5 is acceptable, but individual parts of the analysis may not be.

Final Test and Summary

As a final test of the stratigraphic method and also of the techniques used to test its validity, the most straightforward of all the pit sections (Figure 3) is presented. The surface layer (0–7 cm) was omitted, as a full accumulation year, including the months of December and January, had not elapsed by the time of observation. Each annual layer is quite clear from the figure. The mean accumulation for eight layers is 3.20 g/cm². If it is assumed that one year with no accumulation is represented in the column, and the surface analysis gives some justification for this, the mean annual accumulation becomes 2.84 g/cm². The standard deviation of annual accumulation estimates in this analysis is 1.62 g/cm², which is in close agreement with the total standard deviation, comprising horizontal and temporal variability, of the shallow-pit estimates of the 1966–1967 accumulation (1.63 grams).

In summary, the validity of the estimates of annual accumulation from deep pits has been assessed by comparisons with stake measurements and shallow-pit estimates of the 1966–1967 accumulation. Results from the 1- to 2-meter pits and the 10-meter pit give similar annual accumulation results of 2.76 and 2.78 g/cm², respectively, where the standard error of the mean is 0.13 and 0.18 g/cm², respectively. Interpretation of the most straightforward stratigraphic section in the pits gives a mean annual accumulation of 2.84 g/cm², where the standard error of the mean is 0.54 g/cm². It is concluded that the mean annual accumulation for the last 10–12 years is 2.8 g/cm² (±0.4 gram).

DISCUSSION OF THE STRATIGRAPHY

The snow stratigraphy in the Plateau station area differs markedly from dry-snow stratigraphy elsewhere in Antarctica and Greenland. Generally, Antarctic investigations in the dry-snow zone have shown that winter snow is characterized by high density and hardness and by fine grains. Summer snow, on the other hand, is of low density and hardness, and is coarse-grained. This broad contrast is complicated at most inland stations by high local areal variability of accumulation so that, although missing years occur only occasionally (about 1 in 12 at Plateau station, 1 in 30 at Amundsen-Scott South Pole station, and none at Byrd station), missing seasons may occur. If a summer-affected layer is eroded, two adjacent winter layers may be mistaken for one. At Plateau station, the textural variation is from coarse, low-density winter snow to fine-grained, high-density summer snow. Although crusts and hoar layers form in the summer, this sequence is generally the reverse of that described for the rest of Antarctica.

Table 1 shows the increase in hardness and density at Plateau station over the 1966–1967 summer. Though the winter snow was undergoing considerable grain growth between early December and late January, a new layer was deposited where the grain size was about 50% smaller than that of the preceding winter. Studies in the 10-meter pit showed that subsequent grain growth in the hard layer is slower than in the loose layer. After about 25 years, the grain size of the summer layer increases to about 0.8 mm and that of the winter layer to 1.25 mm, thereby increasing the textural contrast.

Figure 7 summarizes weather conditions at Plateau station in December 1966 and January 1967. These two months can be divided into two distinct parts. The first part occupied the month of December, during which there were fairly constant light precipitation and three periods of relatively high wind speeds. Precipitation at Plateau station consists dominantly, but not exclusively, of columnar ice crystals, and on the very even surface (Figure 8), these begin drifting at wind speeds as low as 4 m/sec. Drifting snow was therefore very common during December, and it was during this period that the hard summer layer was formed. No regular measurements were made of the hardening of this layer, but it was noted that after only a few hours some of the new snow deposits were harder than any of the previous winter snow pack. No variations of hardness with depth could be detected in any of these drifts with the Canadian hardness gages. During the month in which the hard layer formed, temperature was increasing so that the snow surface was colder than the air.

Thin sections of the hard summer layers were made by the aniline technique [*Kinosita and Wakahama*, 1960]. An example is shown in Figure 9. This sample was taken two weeks after the drift formed. The columnar structure of the original precipitation is still largely preserved, but several ice bridges have formed between the grains. The angular shape

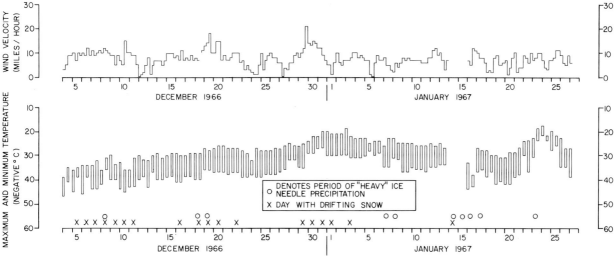

Fig. 7. Weather during December 1966 and January 1967, Plateau station, Antarctica.

of the grains suggests that the length of drift transport was short, and that little rounding had occurred. The snow-hardening process (sintering) is largely dependent on temperature [*Ramseier and Sander*, 1966] and on the vapor pressure of the air flowing over the surface [*Seligman*, 1936]. *Ramseier and Keeler* [1967] have shown that evaporation-condensation is the most important process in the sintering phenomenon. *Gow and Ramseier* [1963], in work at the South Pole, showed that sintering proceeds most rapidly at surfaces exposed to radiation and the atmosphere. In December and January, the temperature and radiation at Plateau station reach their maximum, and absolute humidity of the air is probably higher as well. Therefore, in a period of drift (as in December), it might be expected that a new layer will be of greater hardness than in winter. It does not explain, however, why similar summer layers do not form at South Pole or Vostok.

In January, wind velocities were generally low, precipitation was moderate, and drifting snow was rare (Figure 7). Ice crystals therefore collected at the surface to form a very loose layer, and by January 26, 1967, this layer had undergone intense metamorphism. A crust divided this new layer into two parts. Hoar crystals were most common at the base of the layer but were occasionally present under the crust in the middle of the layer. This entire layer, when buried, will form a single hoar layer lying over the December hard layer. In regions of low temperatures (less than −50°C) and low accumula-

Fig. 8. Surface relief, Plateau station, Antarctica, January 20, 1967. Note the absence of sastrugi. The two arrows point to very hard surface layers.

tion (less than 8.0 g cm^{-2} yr^{-1}), hoar layers are formed mostly at the surface. On subpolar and temperate glaciers, the classic depth-hoar layer forms in the fall at the base of the new balance-year accumulation, and its development is dependent on the steep temperature gradient in that accumulating snow. At Amundsen-Scott and Plateau stations, the rate of accumulation is too low (less than 2.0 cm snow/month) for a similar process of depth-hoar formation to take place. The hoar layer probably develops throughout the summer at Amundsen-Scott station [*Gow*, 1965a] and at Plateau station in summer periods of low wind velocities and relatively moderate accumulation. Moreover, it develops pri-

Fig. 9. Thin section of surface hard layer formed in December 1966. Photograph was taken January 14, 1967, Plateau station, Antarctica.

marily at the surface and should be termed a *hoar* layer rather than a *depth-hoar* layer.

Hard Surface Layer

An exceptionally hard layer with a hardness of more than 60 kg/cm^2 and a density of 0.5 g/cm^3 covered about 5% of the surface in December 1966. The grains in this layer were slightly rounded and measured 0.4 mm, which identified them as having been deposited before the winter of 1966, as the grains in the winter snow measured only 0.2 mm and were angular. In one area, the layer represented an accumulation of 25.0 g/cm^2 at its thickest part. The vertical homogeneity of this layer suggests that it was laid down during a single storm. Similar, very hard, dense layers occurred in the 10-meter pit at depths of 500, 530, 560, 690, and 745 cm. The hard layer at a depth of 560–580 cm represented an accumulation of 11.0 g cm^{-2} but showed a very thin hoar layer 6.0 cm above its base, which divides it into at least two periods of accumulation. Any thick, anomalously dense layer of this type increases the mean annual accumulation excessively if it appears in a pit section covering less than ten years.

The surface hard layers did not appear to undergo any destructive metamorphism during December and January 1967, although they remained at the surface. The grain size remained at 0.4 mm and the hardness was not measurably altered. Destructive metamorphism of the type that occurs at Amundsen-Scott South Pole station [*Gow*, 1965a] is not readily apparent at Plateau station, and it was detected only in large drifts deposited leeward of vehicles. The mode of formation of the very hard layers is not yet known, and samples have been brought back from Antarctica for cold-room fabric analysis. Very hard layers, 5.0–15.0 cm thick, were observed in pit walls on the SPQMLT 1 between the South Pole and the Pole of Relative Inaccessibility [*Cameron et al.*, this volume]. However, these layers may be similar in origin to the thin summer layers observed in the Plateau station stratigraphy.

SNOW ACCUMULATION ON SPQMLT 2: STRATIGRAPHIC INTERPRETATION

On the U.S. Antarctic Research Program South Pole–Queen Maud Land traverse 2 (SPQMLT 2, Figure 1) from the Pole of Relative Inaccessibility to Plateau station, E. E. Picciotto and R. E. Behling dug twenty-two snow pits 1–2 meters deep. The pits were used primarily to obtain specimens for geochemical and radioactivity studies, from which the rate of snow accumulation could be determined. However, the investigators also noted the depth of crusts, hoar layers, and changes of texture in each profile.

Picciotto [1968] compared the accumulation rate determined by variations of gross β activity with depth with the accumulation rate as read from the stratigraphy. He concluded that the stratigraphic method was basically unreliable in regions of very low accumulation and that it overestimated the accumulation rate.

To examine whether the criteria determined from the Plateau station stratigraphy apply to results from the snow-pit studies on SPQMLT 2, the stratigraphic diagrams and notes of Picciotto and Behling have been examined. The SPQMLT 2 pit profiles were primarily dug for chemical sampling. As there was insufficient time to complete a thorough study of the stratigraphy, only the features in one pit could be examined in each locality, and some of the less well developed but important features may have been missed.

One of the main features of Plateau station snow stratigraphy is the formation of hard, fine-grained layers in summer. However, very few thin hard layers occur in any of the SPQMLT 2 snow profiles. The features used to distinguish annual boundaries in the SPQMLT 2 pits are therefore generally limited to crusts and hoar layers. The work at Plateau station, however, indicates that these two features

TABLE 5. Accumulation Rate on South Pole–
Queen Maud Land Traverse 2*

Traverse Mile	Accumulation $\langle A \rangle$, g cm^{-2} yr^{-1}	
	Picciotto †	Koerner ‡
0§	3.2 ± 0.2	2.9
15	3.4 ± 0.2	3.6
40	2.5 ± 0.2	3.3
55	3.3 ± 0.3	3.6
80	2.3 ± 0.2	2.4
120	3.5 ± 0.4	3.2
160	3.2 ± 0.2	3.8
200	2.3 ± 0.2	4.3
220	2.2 ± 0.1	3.3
240	0.7 ± 0.2	2.9
279	3.8 ± 0.3	3.1
323	3.8 ± 0.2	2.9
363	2.4 ± 0.3	2.7
403	3.4 ± 0.2	2.8
495	4.4 ± 0.2	3.2
535	2.4 ± 0.2	3.4
575	1.8 ± 0.2	2.8
615	2.6 ± 0.2	2.8
655 ‖	2.6 ± 0.2	2.8

* At the time of writing, diagrams of pit stratigraphy at miles 175, 423, 480, and 515 were not available to the author. At miles 443 and 455, the structure was noted to be 'highly metamorphic' and 'ultrametamorphic,' respectively. Stratigraphic variations were not noticeable, and in these cases the stratigraphic methods cannot be used. At each station, Picciotto determined the accumulation rate to be less than 1.0 g/cm^3.
† Determined by the variation of gross β activity with depth [Picciotto et al., this volume].
‡ Determined by the stratigraphic method.
§ Pole of Relative Inaccessibility.
‖ Plateau station.

form only in summer, so that they do serve as definite dating criteria.

The pit profiles on SPQMLT 2 were made in covered pits so that the snow stratigraphy stood out due to variable light transmission through the layers. However, with increasing depth, the lower level of light penetration reduces the usefulness of the technique, and an increasing number of faint features may be missed. A preliminary examination of the recorded features in the pits showed that there are more features between the surface and 1-meter depth than between 1 and 2 meters. In the following discussion, only the top 1 meter is considered. The accumulation rates for each pit as determined by both the stratigraphic and geochemical methods are shown in Table 5 and are plotted against the accumulation rates determined by Picciotto et al. [this volume] in Figure 10. The stratigraphic method overestimates the accumulation rate and gives a mean accumulation rate for the whole traverse that is 0.44 g cm^{-2} yr^{-1} higher than the geochemical determination. However, the stratigraphic method is not accurate enough to determine any important variations in the accumulation rate along the traverse route (the coefficient of correlation r between the two sets of results is 0.25, which is not significant). There are two main reasons for this. First, crusts do not form over the whole surface every summer, and occasionally they may be too faint to be noticed except in a very thorough investigation. In each case, an annual layer is missed. Second, the criteria formulated at Plateau station may not apply to more than a few stations. The most notable feature of the Plateau station stratigraphy is the thin, hard, fine-grained layers formed in summer, yet they were seldom noticed in the SPQMLT 2 profiles, although occasional 'bonded layers' and 'thick crusts' referred to by Behling and Picciotto may be synonymous with the author's 'hard layer.'

Use of the Plateau station snow stratigraphy criteria in interpreting the SPQMLT 2 pit results gives the greatest errors relative to the interpretation of β

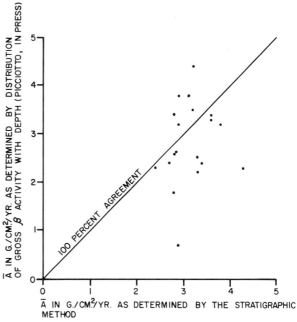

Fig. 10. A comparison of the rates of accumulation obtained by the stratigraphic method and by the variation of gross β activity with depth [Picciotto et al., this volume].

activity data in regions of very high and very low accumulation (e.g., miles 240 and 495, Table 5). In these areas, it is quite obvious that the stratigraphy must differ from that at Plateau station. However, apart from these very high or low accumulation regions, there is a sufficient degree of agreement between the geochemical and stratigraphic results on both the traverse and at Plateau station to indicate that, when a thorough analysis of the pit stratigraphy is made at each location, the stratigraphic method can produce reasonably accurate accumulation rates in East Antarctica.

Acknowledgments. I would like to acknowledge discussions with E. Picciotto of the Free University of Brussels, and O. Orheim and R. Behling of the Institute of Polar Studies, Ohio State University. E. Picciotto made the results of his investigations on SPQMLT 2 and at Plateau station available. The author is grateful to C. Bull of the Institute of Polar Studies, Ohio State University, for a particularly thorough constructive criticism of the manuscript.

The work has been supported throughout by National Science Foundation grant GA-530 (Ohio State University Research Foundation project 2262).

REFERENCES

Black, H. P., and W. Budd, Accumulation in the region of Wilkes, Wilkes Land, Antarctica, *J. Glaciol., 5*(37), 3–15, 1964.

Brecher, H., Letter to the editor, *J. Glaciol., 6*(48), 959, 1967.

Cameron, R. L., Glaciological studies at Byrd station, Antarctica, 1963–1965, this volume.

Cameron, R. L., E. E. Picciotto, H. S. Kane, and J. Gliozzi, Glaciology on the Queen Maud Land traverse, 1964–65, South Pole–Pole of Relative Inaccessibility, *Ohio State Univ., Inst. Polar Stud. Rep.,* 1968.

Giovinetto, M. B., Glaciology report for 1958, South Pole station, *Ohio State Univ. Res. Found. Rep. 825-2,* part 4, 104 pp., 1960.

Giovinetto, M. B., The drainage systems of Antarctica: Accumulation, in *Antarctic Snow and Ice Studies, Antarctic Res. Ser.,* vol. 2, pp. 127–155, AGU, Washington, D.C., 1964.

Giovinetto, M. B., and W. Schwerdtfeger, Analysis of a 200 year snow accumulation series from the South Pole, *Arch. Meteorol., Geophys., Bioklimatol., A, 15*(2), 227–250, 1966.

Gow, A. J., On the accumulation and seasonal stratification of snow at the South Pole, *J. Glaciol., 5*(40), 467–478, 1965a.

Gow, A. J., On the relationship of snow accumulation to surface topography at Byrd Station, Antarctica, *J. Glaciol., 5*(42), 843–847, 1965b.

Gow, A. J., and R. O. Ramseier, Age hardening of snow at the South Pole, *J. Glaciol., 4*(35), 521–536, 1963.

Kinosita, S., and G. Wakahama, Thin sections of deposited snow made by use of aniline, *Contrib. Inst. Low Temp. Sci., A, 15,* 35–45, 1960.

Koerner, R. M., A mass balance study: The Devon Island ice cap, Canada, Ph.D. dissertation, London Univ., 340 pp., 1966.

Kotlyakov, V. M., Outline of the nourishment intensity of the ice sheet of Antarctica, *Sov. Antarctic Exped. Inform. Bull., 25,* 19–22, 1961.

Kotlyakov, V. M., The snow cover of the Antarctic and its role in the present-day glaciation of the continent, *Section 9, IGY Glac. Program, 7,* Israel Program for Scientific Translations, Jerusalem, 256 pp., 1966.

Landon-Smith, I. H., Glaciological studies at Mawson and on the Amery ice shelf, 1962, M.Sc. thesis, Univ. of Melbourne, 1965.

Langway, C. C., Stratigraphic analysis of a deep ice core from Greenland, *U.S. Army Cold Regions Res. Eng. Lab., Res. Rep. 77,* 130 pp., 1967.

Picciotto, E., and S. Wilgain, Fission products in Antarctic snow, a reference level for measuring accumulation, *J. Geophys. Res., 68,* 5965–5972, 1963.

Picciotto, E., G. Crozaz, and W. De Breuck, Accumulation on the South Pole–Queen Maud Land traverse, 1964–1968, this volume.

Picciotto, E., R. Cameron, G. Crozaz, S. Deutsch, and S. Wilgain, Determination of the rate of snow accumulation at the Pole of Relative Inaccessibility, East Antarctica: A comparison of glaciological and isotopic methods, *J. Glaciol., 7*(50), 273–287, 1968.

Ramseier, R. O., and G. W. Sander, Temperature dependence and mechanism of sintering, *U.S. Cold Regions Res. Eng. Lab., Res. Rep. 189,* 15 pp., 1966.

Ramseier, R. O., and C. Keeler, The sintering process in snow, *J. Glaciol., 6*(45), 421–424, 1967.

Seligman, G., *Snow Structure and Ski Fields,* Macmillan, London, 555 pp., 1936.

Schytt, V., *Glaciology 2, A,* Snow studies at Maudheim; *B,* Snow studies inland; *C,* The inner structure of the ice shelf at Maudheim as shown by core drilling, Norwegian Polar Institute, Oslo, 151 pp., 1958.

Shimizu, H., Glaciological studies in West Antarctica, 1960–61, in *Antarctic Snow and Ice Studies, Antarctic Res. Ser.,* vol. 2, pp. 37–64, AGU, Washington, D.C., 1964.

Swithinbank, C. W. M., *Glaciology 1, A,* The morphology of the ice shelves of West Dronning Maud Land; *B,* The regime of the ice shelf at Maudheim as shown by stake measurements, Norwegian Polar Institute, Oslo, 74 pp., 1958.

Taylor, L. D., Glaciological studies on the South Pole traverse, this volume.

Ukhov, S. B., Engineering investigation of the snow cover between Komsomol'skaya and Amundsen-Scott stations, *Sov. Antarctic Exped. Inform. Bull., 44,* 69–73, 1963.

Wilgain, S., E. Picciotto, and W. De Breuck, Strontium 90 fallout in Antarctica, *J. Geophys. Res., 70,* 6023–6032, 1965.

SNOW ACCUMULATION AND FIRN STRATIGRAPHY ON THE EAST ANTARCTIC PLATEAU

Arthur S. Rundle

Institute of Polar Studies, Ohio State University, Columbus 43210

The problem of determining annual accumulation rates, by stratigraphic analysis, on the East Antarctic plateau is discussed with reference to 168 shallow-depth pit studies and 18 radiochemical analyses. Results from these investigations are compared. A stratigraphic interpretation, based on known values of average annual accumulation, has been made, and a system of stratigraphic criteria has been established. Accumulation rates on the Antarctic plateau appear to be below the limit that permits a straightforward stratigraphic analysis, without information from stake and radiochemical measurements.

An over-snow traverse, part of the U.S. Antarctic Research Program, 1967–1968, left Plateau station (79°14.5′S, 40°30′E, elevation 3620 meters) on December 5, 1967, proceeded to 75°56.0′S, 7°13′E, and then, after turning, to 78°42.2′S, 6°52′W, where it terminated on January 29, 1968. It constituted the third phase of the traverse program South Pole–Queen Maud Land traverse (SPQMLT 3).

Previous traverses in this program had followed zigzag routes, the first, 1964–1965, from the South Pole to the Pole of Relative Inaccessibility (82°07′S, 55°02′E, elevation 3718 meters), and the second, 1965–1966, from the Pole of Relative Inaccessibility to Plateau station (Figure 1). The overall objectives of these traverses have been the geophysical and glaciological exploration of this previously unknown region of East Antarctica.

Mr. Norman Peddie, U.S. Coast and Geodetic Survey, was leader of the 1967–1968 traverse and was responsible for navigation and geomagnetic observations. Personnel from the University of Wisconsin investigated ice thickness, using radio-echo sounding and seismic techniques [*Clough et al.*, 1968], and personnel from the Institute of Polar Studies, Ohio State University, pursued several aspects of glaciological research. Mr. Yngvar Gjessing, University of Bergen, conducted a program of surface meteorological observations.

The objectives of the Ohio State University glaciological research were (1) the recording of ice temperatures to a depth of 40 meters by means of quartz thermometers, (2) the recording of ice density to a depth of 40 meters by means of a neutron scattering device, (3) the dating of ice samples and the assessment of annual accumulation rates by means of radioisotope techniques, and (4) snow stratigraphy investigations.

This paper is concerned with the snow stratigraphy investigations, and particularly with the problems raised by the considerable differences between the assessed annual accumulation rates based on stratigraphic analysis and those based on radiochemical analysis.

PREVIOUS WORK IN EAST ANTARCTICA

Previous analyses of dry snow stratigraphy have relied on seasonal variations in density, hardness, and grain size as the means of distinguishing annual horizons. *Giovinetto* [1960] examined snow pit and snow mine stratigraphy at the South Pole and based his analysis on those criteria. *Taylor* [1965] did similar work on a traverse from the South Pole to the Queen Maud Mountains and based his analysis on density variations between summer and winter firn. *Gow*'s [1965] analysis of South Pole stratigraphy depended mainly on the depth hoar layer that generally forms each fall.

Cameron et al. [1968] found a variety of accumulation conditions on the first two phases of the South Pole–Queen Maud Land traverse and found very indistinct stratigraphy at the Pole of Relative Inaccessibility. They concluded that the limit of dating by stratigraphic techniques might be in areas where snow accumulation is about 18 cm/yr. *Kotlyakov* [1961b], however, reported a classical seasonal

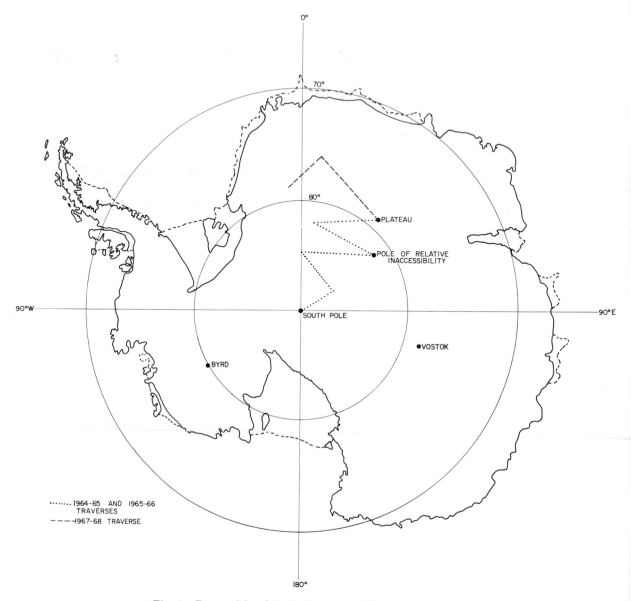

Fig. 1. Route of South Pole–Queen Maud Land traverses.

variation of snow texture in East Antarctica; he determined average annual accumulation to be 3.0 g/cm² at Vostok [*Kotlyakov*, 1961a]. *Cameron et al.* [1968] determined the average accumulation at the Pole of Relative Inaccessibility to be 3.6 g cm^{-2} yr^{-1}.

It is apparent that, as the high polar plateau is approached, accumulation rates decrease markedly and the snow stratigraphy becomes highly complex. In his Plateau station analysis, *Koerner* [this volume] emphasized this stratigraphic complexity and particularly the possibility that annual layers are missing in individual pit sections. He concluded that the classical sequence of dry snow stratigraphy, with hard winter and soft summer layers, is reversed. Over the summer period (early December 1966 to late January 1967), he found that the upper 7 cm increased in density from 0.288 g/cm³ to 0.316 g/cm³, that the grain size doubled, and that the hardness increased sixfold. From this he concluded that the 'remarkable homogeneous annual accumulation' undergoes a 'sintering' process near the surface and that the hard, relatively fine-grained layers are representative of summer conditions, whereas the rest

TABLE 1. Average Annual Accumulation Based on Shallow-Depth Snow Stratigraphy, SPQMLT 3

Mile	Accumulation, g/cm²	Mile	Accumulation, g/cm²	Mile	Accumulation, g/cm²	Mile	Accumulation, g/cm²
5	2.99	217.5	4.13	430	7.46	640	14.36
10	3.00	222.5	4.23	435	6.72	645	9.81
15	2.00	227.5	5.07	440	7.73	650	9.75
20	2.83	232.5	4.88	445	6.51	655	11.79
25	2.88	237.5	5.76	450	6.78		
30	2.62	242.5	5.07	455	5.98	660	9.88
35	2.83	247.5	5.19			665	8.48
40	1.89	252.5	5.96	460	5.90	670	9.03
45	1.78			465	6.66	675	9.11
50	2.95	257.5	5.08	470	5.87	680	9.39
		262.5	5.92	475	6.74	685	9.23
55	2.65	267.5	6.28	480	6.32	690	9.89
60	2.93	272.5	5.72	485	8.22	695	9.80
65	3.29	277.5	6.33	490	6.99	700	9.87
70	3.61	282.5	5.26	495	6.09	705	10.46
75	2.01	287.5	4.42	500	7.50		
80	1.83	292.5	4.37	505	8.85	710	8.80
85	2.88	300	5.73			715	8.81
90	2.63	305	5.11	510	6.64	720	8.31
95	3.24			515	7.89	725	7.47
100	3.07	310	4.63	520	8.85	730	8.17
		315	6.19	525	8.05	735	7.66
105	2.96	320	5.67	530	6.50	740	7.98
110	3.58	325	6.30	535	7.02	745	9.15
115	3.27	330	6.33	540	6.45	750	8.57
120	3.58	335	5.70	545	6.58	755	9.49
125	3.24	340	5.98	550	9.07		
130	3.28	345	5.41	555	9.01	760	9.07
135	3.01	350	5.99			765	8.97
140	3.29	355	5.53	560	7.29	770	9.65
145	3.34			565	6.59	775	9.09
150	3.58	360	4.70	570	7.12	780	9.35
		365	6.05	575	7.19	785	8.80
155	3.95	370	6.20	580	9.28	790	9.35
160	3.73	375	4.44	585	8.26	795	9.87
165	3.32	380	5.17	590	10.38	800	9.91
170	3.90	385	5.28	595	8.05	805	9.95
175	4.41	390	5.58	600	11.68		
180	4.05	395	6.13	605	9.38	810	10.00
187.5	4.38	400	6.35			815	9.93
192.5	4.94	405	6.28	610	9.89	820	9.13
197.5	4.48			615	10.25	825	9.12
202.5	5.55	410	6.99	620	9.92	830	9.87
		415	7.26	625	10.03	835	11.32
207.5	4.85	420	6.62	630	10.44	840	10.38
212.5	4.36	425	6.48	635	10.32		

of the annual layer is softer, coarser, and almost hoarlike in texture.

On the basis of these criteria, the annual accumulation rate was found to be 8.2 cm of snow, yielding 2.4 g/cm². By comparison, measurements at 99 stakes placed there the preceding February gave a mean winter accumulation of 7.5 cm of snow (2.2 g/cm²).

SNOW STRATIGRAPHY AND ACCUMULATION STUDIES AT PLATEAU STATION, 1967

During late November and early December 1967, stratigraphic investigations were carried out at Plateau station, the 99-stake network established there in February 1966 was remeasured, and 10 shallow pits were examined. These studies were undertaken

TABLE 2. Average Annual Accumulation, 1955–1967, Based on Radioisotope Analysis (with Corresponding Stratigraphic Results from Table 1), SPQMLT 3

Mile	Accumulation, g/cm²	
	Radioisotope 1955–1967	Stratigraphic (Rundle)
50	3.3 ± 0.2	2.95
100	3.6 ± 0.2	3.07
150	3.5 ± 0.2	3.58
187.5	3.6 ± 0.2	4.38
247.5	2.6 ± 0.2	5.19
300	2.2 ± 0.2	5.73
350	3.0 ± 0.2	5.99
400	4.6 ± 0.2	6.35
450	5.2 ± 0.2	6.78
500	4.7 ± 0.2	7.50
550	5.1 ± 0.2	9.07
585	4.6 ± 0.2	8.26
620	5.0 ± 0.2	9.92
675	4.4 ± 0.2	9.11
710	4.2 ± 0.2	8.80
750	5.9 ± 0.2	8.57
790	5.2 ± 0.2	9.35
840	4.5 ± 0.2	10.38

to determine the accumulation rates at the station site and to establish a stratigraphic reference for the projected traverse investigations.

Remeasurement of the stake network showed an average snow accumulation between January 21, 1967, and November 25, 1967, of 5.7 cm, but accumulation range from +17.8 cm to −11.1 cm (Y. Gjessing, personal communication, 1968). These stakes, when again inspected by Picciotto on January 17, 1968, showed an additional average increment of 0.3 cm of snow, but this increment ranged from +7.5 cm to −3.6 cm (E. Picciotto, personal communication, 1968).

The snow stratigraphy in 10 shallow pits (50–80 cm deep) revealed a surface layer of hard, fine-grained snow extending to an average depth of 6.5 cm but varying between pits from a trace to 11.0 cm. This layer has not been interpreted as a 'sintered' layer representing a summer horizon, as described by Koerner [this volume], but is considered to be the result of storm winds prevalent during mid-November 1967. The density of this upper layer, which accounted for more than the accumulation recorded by the stake network, was 0.364 g/cm³. On the basis of this value, the average accumulation between January 21, 1967, and November 25, 1967, has been computed to be 2.07 g/cm², but values range from 6.48 g/cm² to a deficit of 4.04 g/cm². Assuming no change in density by January 17, 1968, the average accumulation from January 21, 1967, to January 17, 1968, was 2.18 g/cm², and values ranged from 7.10 g/cm² to a deficit of 4.26 g/cm².

The stratigraphic sequence of the upper layers in late November 1967 did not conform to Koerner's concept, because the surface hard layer was in existence before summer conditions had really begun, even though air temperatures were as high as −29°C before the stratigraphy was examined. The average temperature for November was −44°C and ranged from −29°C to −64°C.

High winds from November 13 through November 15 were the much more likely cause of the surface hard layer. During this period, the average wind speed was 20.9 mph (31 km/hr), almost twice the monthly average, and a maximum speed of 55.3 mph (85 km/hr) was reached on November 13. Koerner [this volume] states that snow at Plateau station begins to move with wind speeds as low as 9 mph (14 km/hr), so that these high, persistent winds might well be expected to have completely re-sorted all the 1967 snow cover.

In all the pits examined, this surface hard layer was underlain by soft material, in which the grains were generally 0.3 mm in diameter and slightly rounded. In 8 of the 10 pits, this soft material was underlain by a second hard layer. The average depth of the top of this second hard layer was 14.7 cm, but the depth varied between pits from 12.0 cm to 17.5 cm. The average density of this upper section, from the surface to the top of the second hard layer, was 0.345 g/cm³.

If the second hard layer is taken to be the 1966–1967 summer horizon, the accumulation for the period January through November 1967 would average 5.07 g/cm², ranging from 6.04 to 4.14 g/cm², which is so large compared with the accumulation measured directly on the stakes, that this interpretation is almost certainly wrong. The interpretation that has been accepted is that the second hard layer represents the 1965–1966 summer, meaning that the mid-November 1967 storm removed and redistributed all the 1967 accumulation and part of the 1966 accumulation. This interpretation gives an average annual accumulation for the two-year period of 2.53 g/cm², the values ranging from 3.02 to 2.07 g/cm², similar to the values obtained by Pic-

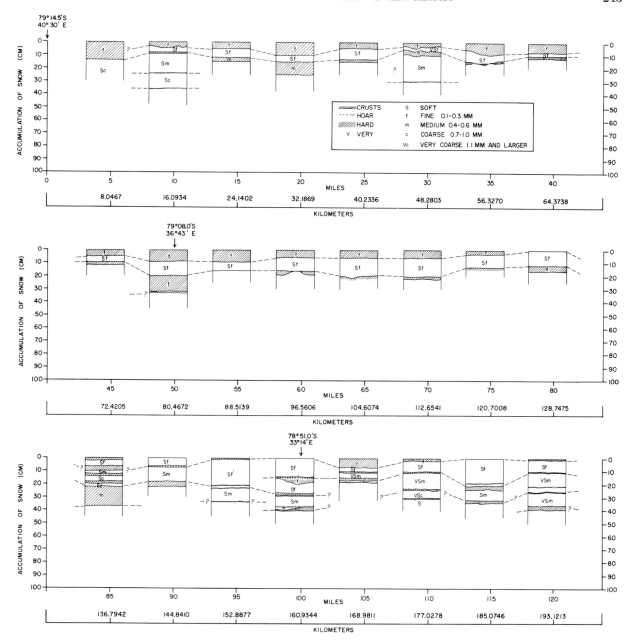

Fig. 2. Snow stratigraphy in shallow pits dug on South Pole–Queen Maud Land traverse 3 (continued on following pages).

ciotto and Koerner. This interpretation also indicates that, in this case, classical stratigraphic techniques are unreliable as a means of determining annual accumulation values. It confirms the conclusion of previous workers that, without complementary stake measurements, there is a tendency to underestimate the number of years involved in a particular section and, therefore, to overestimate the annual accumulation rates. It further confirms Koerner's [this volume] conclusion that examination of the stratigraphy in several shallow pits, as well as in deeper pits, is most desirable in areas of very low annual accumulation. This preliminary study at Plateau station also emphasizes the complexity of snow stratigraphy in such areas and particularly the problem of missing horizons.

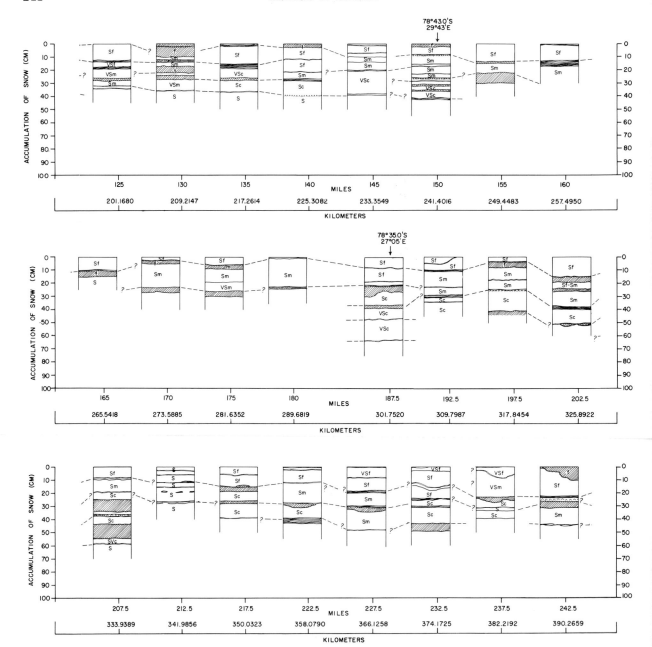

Fig. 2. (continued)

SNOW STRATIGRAPHY AND ACCUMULATION STUDIES ALONG THE TRAVERSE ROUTE

Snow accumulation along the route of SPQMLT 3 has now been determined by two methods. *Rundle* [1968a, b] assessed annual accumulation rates from shallow-depth snow stratigraphy investigations at 5-mile (8-km) intervals along the traverse route (Table 1). At approximately 50-mile (80-km) intervals, W. De Breuck collected series of snow samples to greater depths in an attempt to detect, by later radiochemical analysis, the radioactive fallout layer resulting from the 'Castle' hydrogen-bomb test of 1955. Annual accumulation rates based on these analyses (Picciotto, personal communication, 1968) are given in Table 2. A second radioactive

SNOW ACCUMULATION AND FIRN STRATIGRAPHY

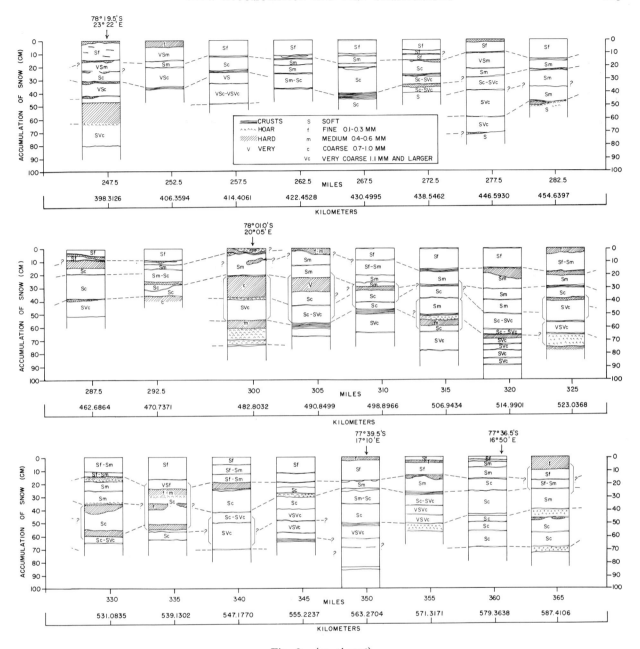

Fig. 2. (continued)

layer, dating from 1963, was also located at several stations. Comparison of the two sets of results shows that the stratigraphic assessment is almost invariably high.

The reasons for this discrepancy are clear, and, to emphasize the problems involved, the stratigraphy and its interpretation are presented diagrammatically in Figure 2. The series of sections shows the original interpretation of annual horizons. These data, together with Picciotto's data and a revised interpretation, are tabulated in Table 1 and plotted in Figure 3 [*Rundle*, 1968a, b].

Criteria for Stratigraphic Interpretation

Each pit (30–100 cm deep) was examined for stratigraphy, grain size, and hardness, and several density samples were taken from each pit. Deeper pits and additional shallow pits were dug at several lo-

246　ARTHUR S. RUNDLE

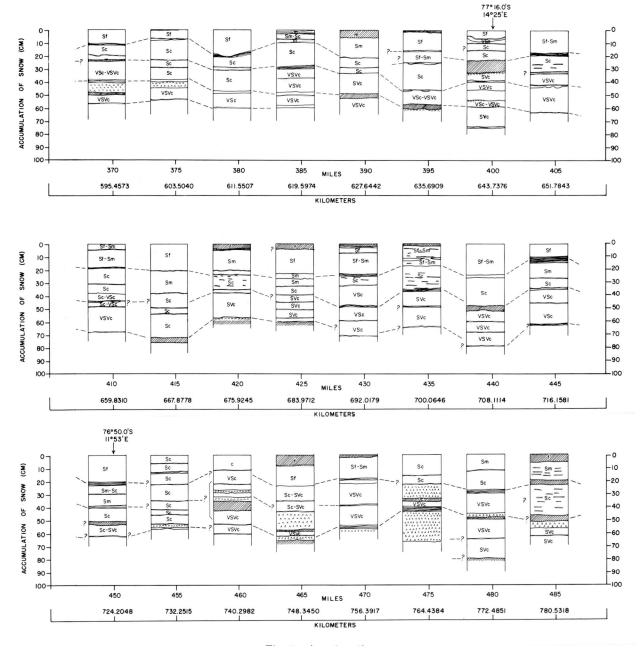

Fig. 2. (continued)

cations where extended geophysical work and vehicle repairs necessitated prolonged stays.

The general character of the stratigraphy changed markedly as the traverse progressed. Almost everywhere the stratigraphy was more complex than the more familiar dry snow stratigraphy like that at Byrd and South Pole stations. The stratigraphic sequence rarely showed a similar pattern over any appreciable distance, and the criteria used in determining annual horizons changed accordingly.

In the original interpretation, the most reliable indicator was regarded as a sudden and quite considerable change in grain size and firn texture, but other criteria were thin hard layers, crusts and thin layers of bonded crusts, depth hoar layers, and alternating layers of crusty and relatively crust-free firn. The stratigraphy and the revised interpretation

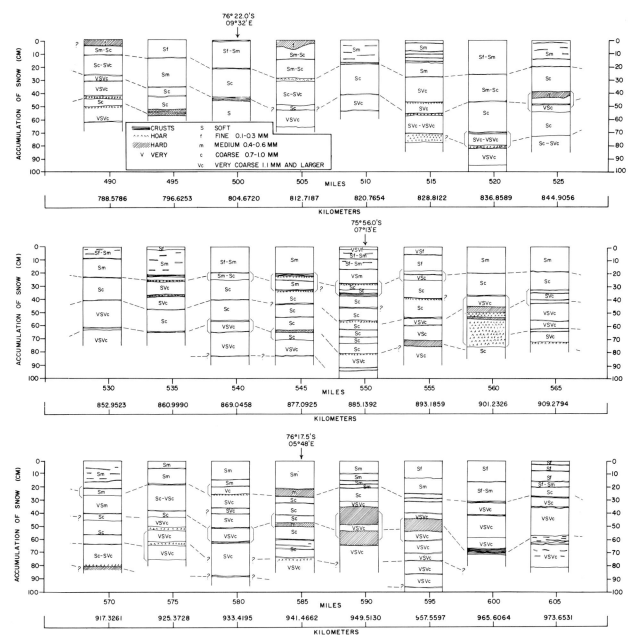

Fig. 2. (continued)

can best be described with reference to the series of sections in Figure 2.

Plateau Station to Mile 75

Over the first 75 miles (120 km) of the traverse route, the stratigraphic sequence seen at Plateau station was strongly evident. Except at mile 5, a hard surface layer was underlain by soft, coarser material, which in turn was underlain by a second hard layer. Over-all, the interpretation is that the second hard layer represents the 1965–1966 summer surface and could be a metamorphosed hard layer as described by *Koerner* [this volume]. The more recent accumulation was re-sorted by high winds in mid-November 1967. At mile 5, it appears that the accumulation at least to, and possibly beyond, the 1965–1966 horizon had been removed and redeposited as a uniform hard layer. This pattern was seen

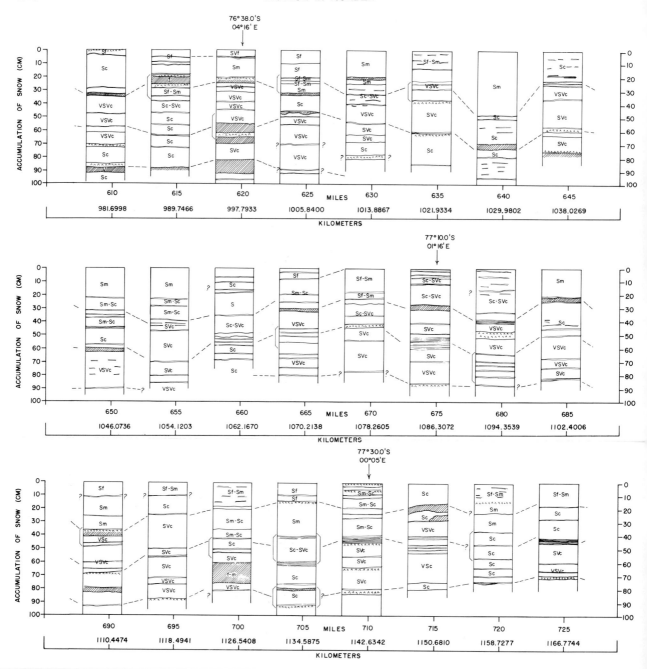

Fig. 2. (continued)

in two of the ten pits excavated at Plateau station, and the hard material is taken as containing two years' accumulation. Mile 75 appears to be close to the edge of the wind system presumably responsible for this particular stratigraphic sequence.

Mile 75 to Mile 300

At about mile 75, a change in the upper stratigraphy was evident, and the different sequence was recognized as far as about mile 300. Beyond mile 75, hard material at the surface was little more than a

SNOW ACCUMULATION AND FIRN STRATIGRAPHY

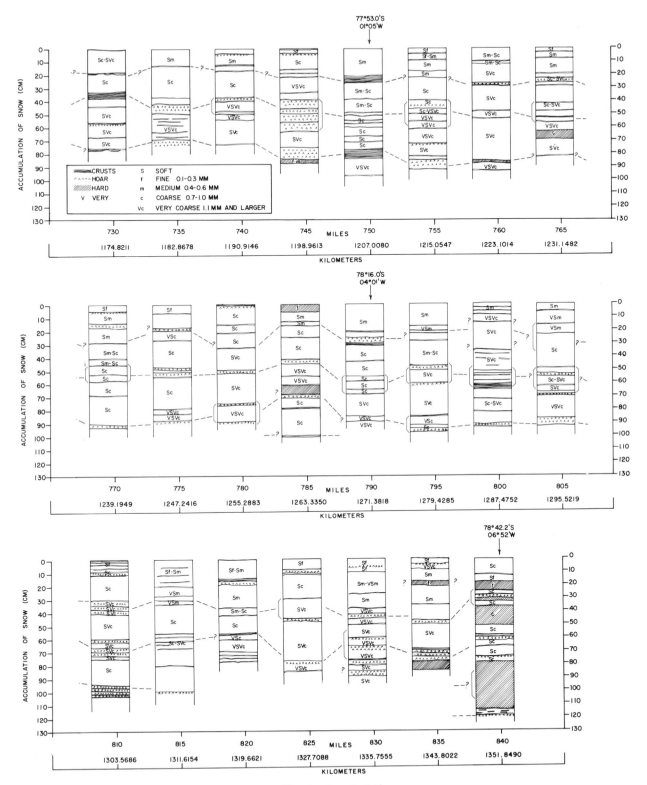

Fig. 2. (concluded)

crust, and at only six stations did it reach 5 cm in thickness. The uppermost snow layer was generally soft and fine grained and was interpreted as being undisturbed 1967 accumulation. This was separated from lower, slightly coarser material by a thin hard layer, a series of fine crusts, or, occasionally, by a single crust. In some places these layers and crusts alone were regarded as significant stratigraphic features, but frequently they were difficult to interpret. The thin hard layers and series of crusts could be interpreted as having a metamorphic origin similar to that described by Koerner; namely, during summer periods of higher solar radiation, the combination of wind action and radiation is effective in bonding the surface snow. However, they could simply be the result of wind action alone and, therefore, could have a winter origin. This is particularly the case with the single crusts.

Mile 300 to Mile 500

Beyond about mile 300, the stratigraphy became more complex, with more intensive as well as extensive crust development, and a similarity in the stratigraphic sequence was evident to the region of about mile 500. The surface layer continued generally to be fine grained but varied considerably in hardness between pits, from very hard wind-packed layers to extremely soft, undisturbed fresh snow. Evidence of mixing and re-sorting at depth was often pronounced, and some sections, for example

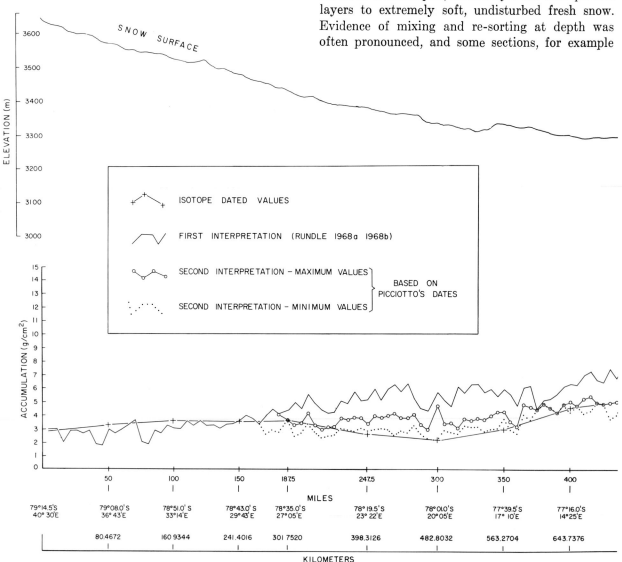

Fig. 3. Average annual accumulation along South Pole–Queen Maud Land traverse 3.

mile 300 and mile 460, were so mixed and complex as to be almost meaningless in terms of stratigraphic interpretation. The development of depth hoar also increased and, in some sections, the hoar was so well developed that it obliterated much of the lower stratigraphy. As the structure of individual crusts could not be determined, the principal criterion for establishing an annual horizon was a sudden change in grain size and firn texture.

Mile 500 to Mile 840

From about mile 500 to the end of the traverse, the upper stratigraphy again showed a change. Frequently, there was no fine material at the surface. The surface layer was often medium and occasionally coarse in texture. Strong crust development was also evident at the surface in many sections, for example between mile 660 and mile 680. This upper layer was interpreted as metamorphosed 1967 snow. These sections were examined in mid-January to late January, by which time air temperatures were relatively high. Confirmation of this interpretation seemed to be the occasional instances of fine surface material with slight crust development within it.

In the lower stratigraphy, crust development became so extensive that interpretation in the field was extremely difficult. In many cases, for example mile 720, mile 755, and mile 760, there was no noticeable difference in grain size and texture between the various crusts. Hard material at depth was less evident than in earlier parts of the traverse, but thick deposits of depth hoar continued to be a prominent feature.

DISCUSSION OF THE STRATIGRAPHY IN LIGHT OF THE RADIOCHEMICAL ANALYSIS

The values of average annual accumulation presented by *Rundle* [1968a, b] are based on the interpretation shown in the series of pit sections in Figure 2. From the radiochemical results, it is obvious that this first interpretation began to break down at about mile 175. The main causes of this failure appear to be that: (1) in many sections, one or more years' accumulation was missing; (2) in many sec-

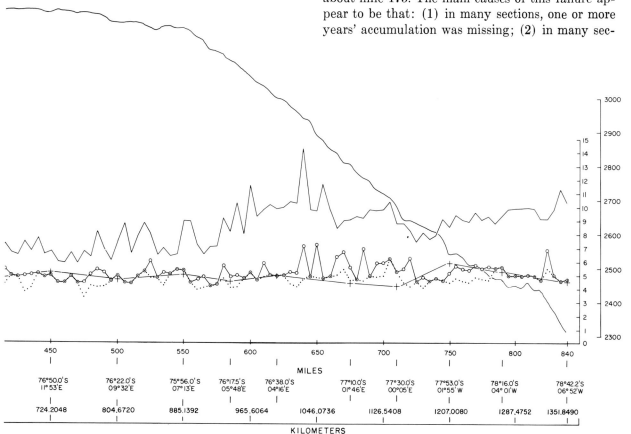

tions, erosion and redeposition, often in thick massive layers, had taken place, and probably obliterated or combined several years' accumulation, for example, the last station at mile 840; (3) thick layers of depth hoar, which have the same obliterating effect, had developed; (4) the stratigraphy itself became increasingly complex and the crusts could not be identified; and (5) too much emphasis was placed on macroscopic changes in grain size and firn texture.

The snow densities used in the radiochemical and stratigraphic assessments are comparable (Table 3). Furthermore, the radiochemical results indicate, approximately, the number of annual horizons that should exist (or be accounted for) in each pit section. In the second analysis, a search has been made for these possible horizons. As an example, at mile 500, the original analysis regarded the surface crust as the current, metamorphosed surface, i.e. 1967–1968, and assessed a little more than 20 cm of snow at an average density of 0.332 g/cm^3 down to the next (1966–1967) surface; the accumulation was, therefore, calculated as approximately 6.6 g/cm^2. Likewise, the lowest series of crusts was regarded as the 1965–1966 surface, giving 8.4 g/cm^2 from 1966 to 1967. This gives an average annual accumulation of 7.5 g cm^{-2} yr^{-1} [*Rundle*, 1968a, b]. In reality, the upper surface was one year old and the average accumulation must be obtained over a three-year period, i.e. 5.0 g cm^{-2} yr^{-1} (compared with 4.7 g cm^{-2} yr^{-1} obtained by Picciotto).

This is where the discrepancy between stratigraphic and radioisotope analysis is explained. The stratigraphy does not show the real age of the horizons, but the radiochemical analysis does. To obtain the real average annual accumulation, zero accumulation during one or more years has to be accounted for. If more accumulation occurred than could be removed in one year or if annual variations of climate were severe enough to produce strong stratigraphic features, the problem would not be so acute. If the stratigraphy could be interpreted to greater depth, the discrepancy between the two assessments is likely to be less, as it appears that perhaps one out of five years' accumulation is likely to be missing in any pit section. The results of several deeper pit investigations are not presented here, as they add little to the argument, except that they emphasize the rapid diagenesis of the firn cover into depth hoar and the total unreliability of firn stratigraphy at depths in excess of 1.5–2.0 meters (contrary to *Koerner* [this volume]). What the stratigraphy may reveal is the maximum accumulation in some years and the great annual variability of net accumulation.

All possible horizons were analyzed in the series of pit sections in Figure 2, and, from these, possible maximum and minimum values for average annual accumulation have been derived (Figure 3). For example, slight similarity can be detected in the pit sections at mile 247.5 and mile 300, which have a common sequence: hard, wind-packed material; depth hoar (at the 7th horizon); and soft, very coarse material. Extreme local meteorological effects may be responsible for the mixing and redeposition of firn in the upper stratigraphy at mile 300.

Frequently there is only one feasible value for average annual accumulation, but when maximum and minimum values have been interpreted, they often are similar and approximate closely the radiochemically dated results (Figure 3).

Some confirmation of the validity of the final interpretation can be derived from a comparison between Picciotto's values, based on the 1963 surge in gross β activity, and the stratigraphic interpretation, where the 1962–1963 summer surface has been located (Table 4).

From the radiochemical analysis, it is apparent that the average accumulation from 1963 to 1967 is about the same as that from 1955 to 1967. The very

TABLE 3. Comparison of Firn Densities Used for Accumulation Determination

Mile	Density, g/cm^3	
	Picciotto	Rundle
50	0.360	0.342
100	0.347	0.315
150	0.348	0.330
188	0.358	0.348
248	0.350	0.330
300	0.352	0.346
350	0.349	0.339
400	0.335	0.339
450	0.358	0.328
500	0.359	0.332
550	0.357	0.340
585	0.363	0.335
620	0.354	0.364
675	0.366	0.325
710	0.353	0.330
750	0.372	0.334
790	0.336	0.330
840	0.369	0.361

TABLE 4. Average Annual Accumulation, 1963–1967, Based on Stratigraphic and Radiochemical Analysis, SPQMLT 3

Mile	Depth to 1962–1963 Surface, cm		Accumulation, g/cm²	
	Stratigraphic	Radiochemical	Stratigraphic	Radiochemical
187.5	47		3.27	
202.5	51		3.33	
207.5	43		2.83	
227.5	48		3.04	
232.5	50		3.41	
242.5	44		3.07	
247.5	47	35 ± 5	3.08	2.4 ± 0.3
257.5	49		3.12	
262.5	37		2.37	
267.5	40		2.51	
277.5	39		2.51	
282.5	47		3.14	
287.5	38		2.65	
310	46		3.12	
315	38		2.57	
320	50		3.44	
325	40		2.69	
340	50		3.42	
345	49		3.42	
350	36		2.44	
355	44		2.92	
360	56		3.67	
365	58		3.80	
370	57		3.70	
375	54		3.55	
380	60		3.88	
385	59		3.74	
390	52		3.48	
440	79		5.61	
450		75 ± 5		5.1 ± 0.3
475	68		4.58	
480	65		4.11	
490	63		4.20	
515	71		4.80	
525	72		4.97	
535	64		4.21	
540	83		5.16	
545	54		3.55	
550	56	55 ± 5	3.81	3.9 ± 0.4
555	71		4.83	
560	75		4.75	
565	72		4.59	
570	80		5.42	
575	65		4.32	
585	75		5.02	
595	76		5.03	
605	81		5.36	
610	70		4.88	
615	64		4.47	
625	70		4.75	
630	69		4.66	
635	85		5.85	

TABLE 4. (continued)

Mile	Depth to 1962–1963 Surface, cm		Accumulation, g/cm²	
	Stratigraphic	Radiochemical	Stratigraphic	Radiochemical
640	74		4.87	
645	75		5.00	
650	63		4.09	
655	70		4.72	
665	80		5.09	
670	76		5.35	
675	60		3.90	
680	63		4.08	
685	66		4.51	
690	78		4.99	
695	72		4.86	
700	60		4.44	
705	80		5.52	
710	63		4.16	
715	52		3.66	
720	63		4.25	
725	70		4.48	
730	76		4.97	
735	70		4.58	
740	73		4.79	
745	75		4.84	
750	76	70 ± 5	5.08	5.1 ± 0.4
755	85		5.69	
760	85		5.64	
765	82		5.38	
770	90		5.76	
775	88		6.00	
780	90		6.08	
785	78		5.15	
790	90		6.14	
795	85		5.59	
800	65		4.28	
805	69		4.43	
810	75		4.74	
815	80		5.29	
820	75		4.89	
825	80		5.47	
830	80		5.31	
835	78		5.19	
840	63		4.55	

low value at mile 550 is most probably explained by erosion of the 1966, 1965, and 1963 firn layers (compare the stratigraphy at miles 545 and 550 with that at miles 540 and 555).

However, the interpretation at mile 540 may still be in error and may stand as an example of the failure of macroscopic textural characteristics as a stratigraphic criterion. The surface may in fact be one year old, as at mile 545, even though the uppermost layer did consist of 'soft and fine to soft and medium' material. A faulty interpretation here would explain the high value for average accumulation from 1963 to 1967, derived from the stratigraphy.

With the exception of mile 790 and the vicinity of mile 400, there is, however, a good comparison between the stratigraphically derived 1963–1967 values and Picciotto's 1955–1967 values. The final interpretation does, therefore, seem reasonably valid.

These results do not show the significant increase

in average annual accumulation with decreasing elevation that was previously reported, though gradual variations may be related to local elevation differences and meteorological effects. A significant increase in the accumulation rate does seem to occur in the vicinity of mile 375 and may be related to effects of a break of surface slope. The evenness of the values toward the end of the traverse, in spite of a continued and rapid loss of elevation, may, however, be the result of decreasing proximity to the coast.

CONCLUSIONS

The most significant fact to emerge from this analysis is that, in any pit section examined, at least one year's accumulation is likely to be missing. This results from the very low annual increment, which is not sufficient to offset erosional effects. Erosion, redeposition, and mixing of layers, together with massive depth hoar development, tend to obliterate several years' stratigraphy. Therefore, the age of a particular layer cannot be determined from the stratigraphy. Also, seasonal differences of climate may not be severe enough to produce consistent and prominent horizons that can be easily identified in the field. A reliable interpretation, without basic values from stake measurements or isotope analysis, does not seem possible, and a straightforward interpretation may well be limited to areas where annual accumulation exceeds 18 cm of snow per year [*Cameron et al.*, 1968].

Acknowledgments. The author is grateful to Mr. John W. Clough and Mr. Carl K. Poster, Geophysical and Polar Research Center, University of Wisconsin, for information on surface elevations along the route of the traverse; to Professor E. E. Picciotto and Dr. William De Breuck for allowing me to discuss their results; and to Dr. Colin Bull, Mr. John F. Splettstoesser, and Mr. Robert Behling, Institute of Polar Studies, Ohio State University, who have reviewed this and other manuscripts relating to the traverse glaciology investigations.

This program was supported by National Science Foundation grant GA-1076 awarded to the Ohio State University Research Foundation.

Contribution 151 from the Institute of Polar Studies, Ohio State University.

REFERENCES

Cameron, R. L., E. Picciotto, H. S. Kane, and J. Gliozzi, Glaciology on the Queen Maud Land traverse, 1964–1965, South Pole–Pole of Relative Inaccessibility, *Ohio State Univ. Res. Found., Inst. Polar Stud., Rep. 23*, 136 pp., 1968.

Clough, J. W., C. R. Bentley, and C. K. Poster, Ice-thickness investigations on SPQMLT 3, *Antarctic J. U.S., 3*(4), 96–97, 1968.

Giovinetto, M. B., Glaciology report for 1958, South Pole station, *Ohio State Univ. Res. Found. Rep. 825-2, Part 4*, 104 pp., 1960.

Gow, A. J., On the accumulation and seasonal stratification of snow at the South Pole, *J. Glaciol., 5*(40), 467–477, 1965.

Koerner, R. M., A stratigraphic method of determining the snow accumulation rate at Plateau station, Antarctica, and application to South Pole–Queen Maud Land traverse 2, 1965–1966, this volume.

Kotlyakov, V. M., Outline of the nourishment intensity of the ice sheet of Antarctica, *Inform. Bull. Sov. Antarctic Exped.*, no. 25, 177–181, 1961a.

Kotlyakov, V. M., The snow cover of the Antarctic and its role in the present-day glaciation of the continent, *Akad. Nauk SSSR, IX razdel progr. MGG, Glyatsiol.*, no. 7, 246 pp., 1961b.

Rundle, A. S., Snow stratigraphy and accumulation (SPQMLT 3), *Antarctic J. U.S., 3*(4), 95–96, 1968a.

Rundle, A. S., Queen Maud Land traverse, *Ice*, no. 27, p. 4, 1968b.

Taylor, L. D., Glaciological studies on the South Pole traverse, 1962–1963, *Ohio State Univ. Res. Found., Inst. Polar Stud. Rep. 17*, 13 pp., 1965.

ACCUMULATION ON THE SOUTH POLE–QUEEN MAUD LAND TRAVERSE, 1964–1968

E. PICCIOTTO AND G. CROZAZ

Service de Géologie et Géochimie Nucléaires
Université Libre de Bruxelles, Brussels 5, Belgium

W. DE BREUCK

Laboratorium voor delfstofkunde en aardkunde, Rijksuniversitair Centrum, Antwerp, Belgium

Snow accumulation rates have been determined at 75 stations along the route of the South Pole–Queen Maud Land traverse (1964–1968), covering the area between approximately 0–55°E and 74°–90°S. Elevations ranged from 2220 to 3700 meters. Estimates of the accumulation were based on laboratory measurements of the fission product radioactivity, identifying the layers deposited in early 1955. The lead 210 method was also applied at a few selected stations. The principles of these radiometric methods, as well as the experimental procedure, are fully described. The interpretation of the stratigraphy in terms of annual layers was found to be often ambiguous and sometimes impossible. It leads to overestimated values of accumulation. The 10-year average annual accumulation was found to vary from almost zero to 7 g cm^{-2} yr^{-1}. The average accumulation over the whole sector investigated is 3.7 g cm^{-2} yr^{-1}.

Knowledge of the rate of snow accumulation is one of the basic requisites for a complete understanding of the dynamics, the mass budget, and the heat budget of the Antarctic ice sheet. Up to 1965, there were practically no direct measurements of this parameter in the central region of the East Antarctic ice sheet, the region above the 3000-meter contour line and enclosing about 50% of the area of the East Antarctic ice sheet. This region is particularly interesting because it includes the highest, the most continental, the coldest, and presumably the driest area of the Antarctic ice sheet. It is therefore understandable that accumulation measurements were an important objective of the glaciological program of the South Pole–Queen Maud Land traverse.

The South Pole–Queen Maud Land traverse (SPQMLT) was a major oversnow traverse program sponsored by the Office of Antarctic Programs, National Science Foundation. The objective was to investigate, along a zig-zag route, the sector between 40°E and 10°W in the central region of the East Antarctic ice sheet. The SPQMLT covered a total distance of approximately 4200 km in three summer seasons: 1964–1965 (SPQMLT 1), 1965–1966 (SPQMLT 2), and 1967–1968 (SPQMLT 3). The trips were made in three Tucker Sno-Cats by 9 or 10 USARP personnel. General information on the schedule of the traverse is given in Table 1 and Figure 1. More detailed accounts of the traverse operations have been given by *Picciotto* [1966], *Cameron et al.* [1968], and *Peddie* [1968]. The scientific program included:

The determination of the surface topography.
Measurements of the ice thickness and determination of the subglacial rock topography.
Studies of the physical properties of the ice sheet.
Observation of the geomagnetic field.
Weather observation.
Glaciological studies, including detailed observations of the physical properties of the upper firn layers and attempts to measure accurately the accumulation rates.

Estimates of accumulation rates on oversnow traverses are generally based on pit stratigraphy studies leading to the identification of annual layers in the firn. Objective criteria for the identification of annual layers have been established for dry-snow facies in Greenland and in Antarctica, and they have yielded a large number of reliable accumulation values. However, in the central region of the East Antarctic plateau, these criteria are usually

TABLE 1. Description of the South Pole–Queen Maud Land Traverse

	SPQMLT 1	SPQMLT 2	SPQMLT 3
Departed from	South Pole, Dec. 4, 1964	Pole of Relative Inaccessibility, Dec. 15, 1965	Plateau station, Dec. 5, 1967
Arrived at	Pole of Relative Inaccessibility, Jan. 28, 1965	Plateau station, Jan. 29, 1966	78°42'S, 6°52'W, Jan. 26, 1968
Distance covered	800 n. mi.	725 n. mi.	840 n. mi.
No. of glaciological stations	29	28	18
Traverse leader	C. Bentley and R. Cameron	E. Picciotto	N. Peddie and R. Cameron

not applicable, and accumulation rates derived from pit stratigraphy alone involve personal and subjective interpretation leading generally to unreliable results. These difficulties are accounted for by the very low and often very variable annual accumulation and by the strong metamorphism of the upper firn layers, resulting in a partial or sometimes total obliteration of the annual layering. Detailed discussions and related bibliography have been given by *Kotlyakov* [1961], *Giovinetto* [1964], *Gow* [1965], *Cameron et al.* [1968], and *Picciotto et al.* [1968].

We did hope that these difficulties could be overcome by the use of the recently developed methods based on measurements of radioactive isotopes. The method based on the seasonal variation of stable oxygen isotope ratios seems to be subject to the same errors as the stratigraphic method in this region [*Picciotto et al.*, 1968], and was therefore not used.

Two radiometric methods appeared particularly promising and compatible with the severe logistic problems associated with traverse operations in Antarctica: the limitations on available time, load, and power. They are based, respectively, on: (1) the distribution of fission products with depth in the firn; and (2) the decay of lead 210.

Since the measurements cannot be made in the field, the samples must be taken to the laboratory; both methods require only small quantities, at least in the region investigated, where the annual accumulation was lower than 6 g cm^{-2} yr^{-1}. Depths of samples are limited to 10 meters, obtained with the standard digging and drilling equipment used in traverse operations (see section 'Sampling Procedures').

Although both methods involve measurements of radioactive nuclides, they are based on quite different principles. The first one relies on the fact that the fallout of artificial radioactive nuclides released by the first large thermonuclear bomb test formed a well dated reference horizon over the whole Antarctic ice sheet [*Picciotto and Wilgain*, 1963; *Vickers*, 1963; *Woodward*, 1964; *Wilgain et al.*, 1965]. This horizon corresponds to the summer 1954–1955 and thus provides a criterion for measuring the average accumulation since 1955. The lead 210 method [*Goldberg*, 1963; *Crozaz et al.*, 1964], on the other hand, is a dating method based on the decay of a natural radioactive nuclide. It is similar in its principle to the radiocarbon method, but it spans a much shorter time interval, of the order of 100 years.

We report here the results of accumulation measurements carried out at 75 stations spaced approximately every 55 km along the route of the South Pole–Queen Maud Land traverse. The geographic position, the elevation, and the mean annual temperature of the stations are reported in Table 2.

Stations will be designated in this paper by the traverse number followed by the station number and mile number.

For the first time, the two radiometric methods were systematically used in a traverse program. The fission products method was used at the 75 glacio-

SNOW ACCUMULATION

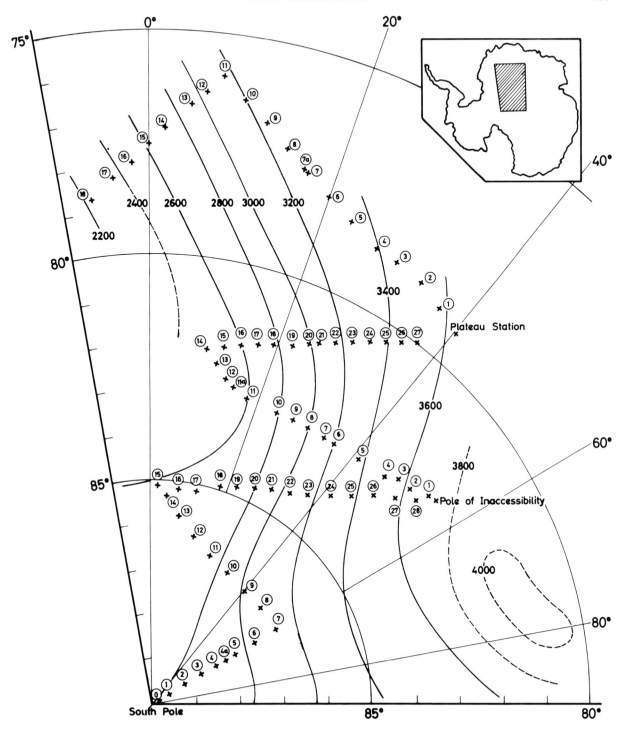

Fig. 1. The glaciological stations on the South Pole–Queen Maud Land traverse.

logical stations with excellent results. The lead 210 method requires deeper samples and elaborate radiochemical separations, and was used as a control on a smaller number of selected stations; only partial success was achieved because of contamination in the sampling.

The field work for this program included pit-stratigraphy studies and firn sample collection for the

TABLE 2. Glaciological Stations on the South Pole–Queen Maud Land Traverse

Station	Mile	Date	Pit Depth, meters	Lat. S	Long. E	Elev., meters	Annual Temp., -°C	Core, meters
			SPQMLT 1, Summer 1964–1965					
Pole		Dec. 4, 1964		90°00'				
0	8	4	2	89°52'	60.0°			
1	24	6	2.4	89°36'	60.0°	2820	50.5	
2	48	7	2	89°10'	58.5°	2850	51.3	
3	72	9	2.4	88°45'	58.0°	2890	49.6	
4	96	11	2.2	88°19'	58.9°	2970	51.2	
4a	110	12	2.2	88°05'	58.9°			9
5	125	14	2.2	87°48'	59.2°	2990	52.3	
5a	146	14	2.4	87°33'	59.2°			
6	155	15	2.0	87°17'	58.9°	3030	51.5	
7	185	17	2.4	86°45'	58.6°	3110	51.5	11.3
8	215	21	2.2	86°48'	49.2°	3100	51.9	
9	245	24	2.2	86°44'	40.0°	3020	49.3	
10	275	26	2.6	86°38'	30.6°	2860	47.9	12.2
11	305	28		86°30'	22.2°	2760	48.0	3
12	338	29	2.4	86°09'	14.8°	2730	48.4	
13	370	31	2.6	85°46'	8.7°	2690	50.0	10.5
14	395	Jan. 1, 1965	2.2	85°26'	4.7°	2670	49.0	
15	415	3	2.2	85°10'	1.6°	2630	48.6	12.2
16	445	7	2.4	85°13'	7.7°	2680	49.0	
17	470	8	2.8	85°11'	13.2°	2690	50.0	
18	496	10	2.4	84°58'	17.9°	2680	49.0	10.3
19	519	14	2.4	84°51'	21.8°	2680	48.9	
20	545	15	2.4	84°42'	26.0°	2790	47.0	
21	570	17	2.4	84°32'	30.1°	2870	46.3	
22	595	18	2.4	84°22'	33.9°	3050	47.3	
23	620	19	2.4	84°10'	37.6°	3170	48.8	10.2
24	650	22	2.2	83°53'	41°18'	3310	51.5	
25	680	23	2.4	83°32'	44°33'	3400	53.1	
26	710	24	2.4	83°10'	47°28'	3500	54.1	
27	740	25	2.4	82°50'	50°31'	3580	55.7	
28	770	26	2.2	82°27'	52°52'	3660	56.1	
29	797	28	2.4	82°07'	55°02'	3720	57.2	4.5
			SPQMLT 2, Summer 1965–1966					
0	0	Dec. 15, 1965	6	82°07'	55°06'	3720		
1	15	15	2	82°12'	53°40'	3680		
2	40	16	2	82°27'	51°05'	3620	54.8	
3	55	17	2	82°29'	48°30'	3570		
4	80	18	2	82°42'	46°25'	3520	54.4	
5	120	21	2.2	82°48'	41°05'	3390	51.7	8.4
6	160	23	2	82°52'	35°45'	3200	48.7	
7	175	24		82°53'	34°00'			2
8	200	25	2	82°53'	30°20'	2990	46.8	
9	220	27	2	82°54'	27°10'	2890		8
10	240	28	2	82°55'	24°10'	2750	45.6	
11	279	30	15	82°52'	18°10'	2610	49.2	
12	323	Jan. 6, 1966	2	82°33'	13°20'	2570	48.2	
13	343	7		82°17'	11°24'	2540		2
14	363	8	4.2	82°00'	9°35'	2510	46.7	10.2
15	383	11		81°54'	12°06'	2530		2
16	403	12	2	81°45'	14°40'	2600	45.1	

TABLE 2. (continued)

Station	Mile	Date	Pit Depth, meters	Lat. S	Long. E	Elev., meters	Annual Temp., -°C	Core, meters
17	423	Jan. 13, 1966		81°40′	17°00′	2650		2
18	443	14	2	81°34′	19°35′	2760	44.9	
19	455	16	4	81°30′	20°30′	2870		10
20	480	19		81°17′	24°00′	2990		2
21	495	20	2.6	81°10′	25°30′	3050	48.4	
22	515	21		81°00′	27°38′	3150		2
23	535	22	2.8	80°48′	29°50′	3260	50.8	8.8
24	555	24		80°36′	31°50′	3330		
25	575	25	2	80°23′	33°45′	3410	53.6	
26	595	26		80°10′	35°30′	3470		2
27	615	27	2	79°57′	37°10′	3510	55.4	
28	667	28	2.2	79°15′	40°30′	3620	58.4	
SPQMLT 3, Summer 1967–1968									
1	50	Dec. 7, 1967	2.5	79°08′	36°43′	3550	58.0	
2	100	12	2.5	78°51′	33°14′	3510	56.5	
3	150	13	2.6	78°43′	29°43′	3460	57.0	
4	188	15	2.5	78°35′	27°05′	3410		
5	248	17	3.0	78°19′	23°22′	3360	49.5	10.0
6	300	23	2.5	78°01′	20°05′	3320	51.0	
7	350	26	3.3	77°39′	17°10′	3310	51.5	
7a	360	28		77°36′	16°50′	3300		7.0
8	400	Jan. 1, 1968	3.0	77°16′	14°25′	3280	52.5	
9	450	3	3.0	76°50′	11°53′	3260	50.5	
10	500	4	3.0	76°22′	9°32′	3230	48.5	10.0
11	550	16	3.0	75°56′	7°13′	3210	48.5	
12	585	18	3.0	76°17′	5°48′	3110	47.0	
13	620	19	3.0	76°38′	4°16′	3000	45.5	
14	675	21	3.0	77°10′	1°46′	2800	43.5	5.2
15	710	22	3.0	77°30′	0°05′ Long. W	2690	42.5	5.0
16	750	23	3.0	77°53′	1°55′	2540	41.5	9.0
17	790	24	3.0	78°16′	4°01′	2470	42.0	5.0
18	840	26	3.0	78°42′	6°52′	2310	38.0	6.0

radioactive measurements. It was conducted by R. Cameron and E. Picciotto, with the assistance of J. Gliozzi on SPQMLT 1, by R. Behling and E. Picciotto on SPQMLT 2, and by W. De Breuck on SPQMLT 3. A similar program was conducted at Plateau station during the summer season 1966–1967 by R. Koerner and E. Picciotto; some of the results are included in the present report. The laboratory analyses were carried out at the University of Brussels with the assistance of C. Weyers, J. Saufnay, and J. P. Mennessier.

Before describing the experimental procedures and reporting the results, it may be useful to recall briefly the basic principles on which the two methods used here rely. Special emphasis is placed on the production and the properties of the radioactive debris from nuclear-bomb tests, since we feel that this method will be of great help in future investigations on the Antarctic plateau.

RADIOACTIVE DEBRIS FROM NUCLEAR BOMB TESTS IN THE ATMOSPHERE

Nuclear explosive devices derive their power from two processes: the fission of heavy nuclides such as uranium and plutonium isotopes on the one hand, and fusion of two light nuclides such as deuterium and tritium on the other. The energy released in a nuclear detonation is generally expressed in ton-equivalents of chemical explosive, the energy yield of one ton of explosive being 10^9 calories. The energy yield of a pure fission bomb (A bomb) is limited by critical size considerations to a few tens of kilotons, but there is no theoretical limit to the energy yield of a fusion bomb.

Most of the 'H bombs' that have been detonated since 1952 with energy yields of several megatons belong to the 'fission-fusion-fission' class. In such a device, a fission bomb is used as detonator to set off a thermonuclear fusion reaction, the fast neutrons released as by-products of the fusion reaction induce in their turn the fission of an outer shell of uranium 238. The relative contributions of the fusion reaction and of the uranium 238 fission to the over-all energy yield depend strongly on the technical characteristics of the device. On the average, in the 'standard' H bombs detonated up to the present, approximately half the power was derived from the fusion and half from the fission of uranium 238.

The detonation of an H bomb releases an enormous amount of radioactivity. The radioactive debris belongs to three main classes: the fission products, the tritium (which is the only radioactive product of the fusion reactions), and radioactive nuclides formed by neutron reactions on the weapon materials or on elements of the environment. By far the major part of the radioactivity released in a 'standard' H-bomb detonation is associated with the products of the fission of uranium 238 by fast neutrons, and this is the only class we shall consider here.

The fission of a heavy nuclide like uranium 238 leads to the formation of about 400 radioactive β emitters, which are isotopes of the 36 elements between $Z = 30$ (zinc) and $Z = 65$ (terbium), with mass numbers A between 72 and 161. The individual fission yields range from an extremely small fraction up to a few per cent. The half-lives range from a small fraction of a second up to several million years. Because of successive β decays, all these nuclides can be grouped in about 90 isobaric chains. The maximum chain yields do not exceed 7%, and appear in the neighborhood of the mass numbers 96 and 140. Fission yield curves have been calculated or measured by several authors [*Dolan*, 1959; *Katcoff*, 1960; *Hallden et al.*, 1961; *Harley et al.*, 1965]. Table 3 sets out a list of fission products restricted to nuclides with half-lives greater than 250 days, and that pertain to chains with a production yield greater than 0.1%.

The total β activity of the fission products mixture resulting from a nuclear detonation decreases in a complicated manner with time. It can be represented approximately by the empirical relation of Way-Wigner:

$$A_t = A_1 \times t^{-1.2}$$

where A_1 is the activity at unit time after the explosion, and A_t is the activity at time t after the explosion. Figure 2 shows the Way-Wigner curve and an experimental decay curve after *Hallden et al.* [1961] based on direct observations of H-bomb debris.

Relative contributions of selected nuclides to the total β activity are shown in Figure 3, also from *Hallden et al.* [1961]. It is seen that, a few years after the detonation, the residual activity is due mainly to strontium 90 (in equilibrium with its daughter yttrium 90) and to caesium 137 and successively to prometheum 147, samarium 151, and antimony 125.

The relationship shown in Figures 2 and 3 allows, in principle, for the estimation of the 'age' of a fission product mixture and the time when it was released. However, such estimates are complicated by the fact that worldwide radioactive fallout is actually a mixture of products from a number of detonations set off at various times.

In problems related to radioactive fallout, it is current practice to express the amount of fission products by the activity of strontium 90 because of the health hazard represented by this nuclide, associated with yttrium 90. However, caesium 137 is easier to measure without chemical separation

TABLE 3. Principal Long-Lived Fission Products

Nuclide	Half-Life	Production Yield, %
Cerium 144	284 days	4.7
Ruthenium 106	367 days	2.4
Antimony 125	2.0 years	0.3
Prometheum 147	2.6 years	2.7
Strontium 90	28.1 years	3.5
Caesium 137	30.0 years	5.6
Samarium 151	87.0 years	1.0

through its γ radiation; strontium 90 and yttrium 90 are pure β emitters.

Rounded figures for the average production yields of these nuclides in H-bomb explosions are 0.1 megacurie of strontium 90 and 0.18 megacurie of caesium 137 per megaton of energy yielded by fission reactions.

Types of Fallout

Within a very short time after a nuclear detonation, all the materials of the weapon and the products of the nuclear reactions are completely vaporized and form a rising and expanding fireball. A few minutes later, this matter condenses into the form of an aerosol cloud, which is then slowly dispersed in the atmosphere and eventually deposited on the ground after a lapse of time. The behavior in the atmosphere of the radioactive fallout from a nuclear detonation can be classified into three categories [*Libby*, 1956].

Local fallout. Large particles that settle in less than a day, in the vicinity of the explosion site, are associated only with surface bursts.

Tropospheric fallout. Particles ranging in size from 1 to 10 μm are confined below the tropopause. Being too small to settle under the action of gravity, they remain more or less associated with the air mass in which they were condensed until they are brought to the ground, mainly by the scavenging effect of atmospheric precipitation. The residence time of tropospheric aerosols is of the order of one month. As a result of the fast zonal tropospheric circulation, tropospheric fallout is carried around the world several times before being deposited on the earth's surface, but usually remains confined in a narrow latitude belt centered around that of the explosion site.

Stratospheric fallout. Matter condensed above the tropopause consists of particles in the range of a few hundredths of a micron. Because of their small size and of the vertical stability of the stratosphere, they spend a long time (from a few months to several years) in the stratosphere before being transferred into the troposphere, where they are rapidly scavenged out by precipitation. The stratospheric aerosols are dispersed over great distances and form the worldwide component of fallout from nuclear-bomb tests.

The distribution of the radioactive debris among

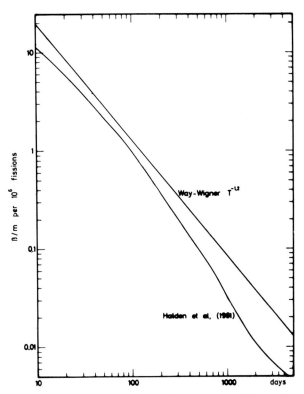

Fig. 2. Evolution in time of the gross β activity of a mixture of fission products.

the three types of fallout depends essentially on the energy yield of the detonation. A-bomb surface bursts whose energy yields are of the order of a few tens of kilotons only produce local and tropospheric fallout. Thermonuclear devices with yields in the multimegaton range always project an important fraction of their product into the stratosphere. Of course, when the detonation is set off in the stratosphere, 100% of the fallout belongs to the stratospheric type.

Stratospheric Fallout

Because of its worldwide dispersion, stratospheric fallout has attracted much attention, and it is the most important component from the point of view of glaciological applications. There are uncertainties about the exchange in the stratosphere and between stratosphere and troposphere. However, it is possible at present to predict with good approximation the fate in space and in time of nuclear bomb debris injected into the stratosphere, knowing the time of the year and the space coordinates of the injection. For this purpose, *Machta* [1962] has proposed a

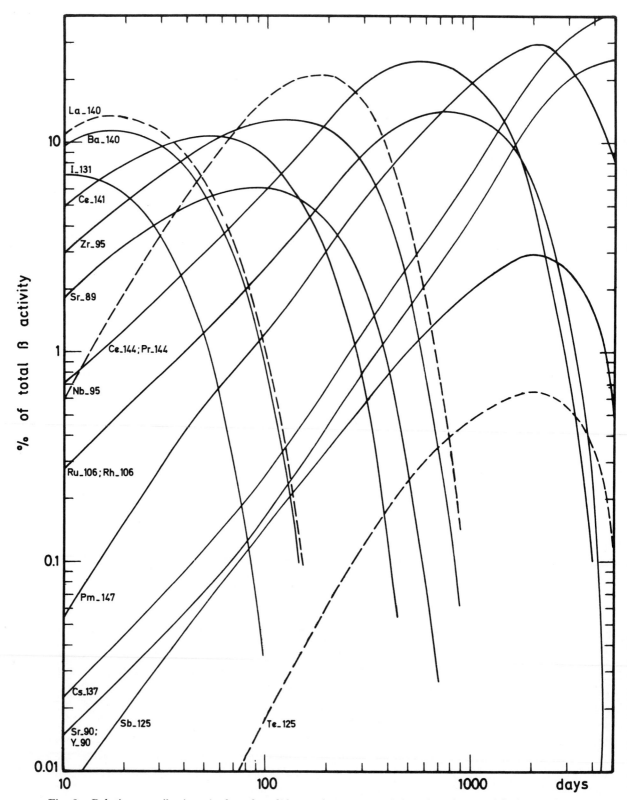

Fig. 3. Relative contribution of selected nuclides to the gross β activity of a mixture of fission products.

schematical, but very convenient, subdivision of the stratosphere into five compartments (see Figure 4 and Table 4).

Whatever its compartment of origin, stratospheric fallout is finally deposited on the ground according to a well-defined seasonal pattern: in both hemispheres, the deposition reaches a maximum in spring (actually in summer in the southern hemisphere), and is minimum in winter. Thus radioactive debris injected in compartments I, II, III, and IV will appear in significant amount in the troposphere and in the precipitations in the course of the summer season after the injection.

Strontium 90 inventory. The global strontium 90 inventory in January 1967 is given in Table 5. Because of radioactive decay, the total amount produced since 1945 is higher by approximately 5 megacuries than the amount found in 1967. The strong asymmetry in the deposition between the two hemispheres reflects the fact that almost all the high-yield nuclear tests up to 1967 have been conducted in the northern hemisphere. Because of several meteorological factors, the stratospheric fallout is deposited preferentially in the 40–50° latitude belt in both hemispheres.

Sites and Chronology of Nuclear Bomb Tests in the Atmosphere

Up to January 1967, five nations had carried out nuclear weapon tests: the United States, the United Kingdom, the Soviet Union, France, and China. However, thermonuclear tests with yields in the multimegaton range have been carried out only by the United States, the United Kingdom, and the Soviet Union. More recently, thermonuclear devices were detonated by the Chinese in June 1967, and by the French in August–September 1968.

The high-yield tests up to January 1967 were carried out in two regions of the northern hemisphere: (a) the Pacific Proving Ground, including Bikini Island (11°N, 165°E), Eniwetok Island (11°N, 162°E), and Christmas Island (2°N, 157°E) by the United States and the United Kingdom, and (b) the Arctic test site at Novaya Zemlya (75°N, 55°E) by the Soviet Union.

Low-yield A-bomb tests were carried out at the following sites: Nevada (37°N, 116°W), Siberia (52°N, 78°E), Algerian Sahara, Australia, and Lop Nor (40°N, 90°E).

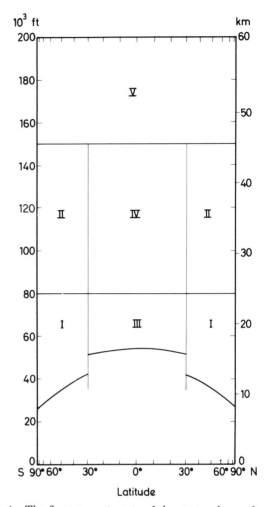

Fig. 4. The five compartments of the stratosphere, after *Machta* [1962].

The chronology of the announced tests can be summarized as follows [*Glasstone*, 1964]:

1945–1951. Only fission A bombs were detonated, all of them at northern hemisphere sites. The total amount of fission products released during this time corresponds to approximately 0.1 megacurie of strontium 90, which is quite negligible as compared to the 20 megacuries produced by the later H bombs.

1952. In October 1952, the first nuclear test in the southern hemisphere was carried out in Australia (20°S, 115°E). Only one A bomb with a fission yield of less than 20 kT was detonated at ground level. In November 1952, the first thermonuclear devices, with an energy yield in the megaton range, were detonated by the United States at Eni-

TABLE 4. Stratospheric Injections and Resulting Fallout [after Machta, 1962]

Stratosphere Compartment	Approx. Res. Time, months	Worldwide Deposition	Main Sources from Known Test Series in Multimegaton Range
I Lower polar	4–12	On the hemisphere of injection	USSR high yield tests in the Arctic
II Upper polar	12	Mostly in the hemisphere of injection, but partly in the other	USSR very high yield tests in the Arctic, 1961
III Lower tropical	10	Over both hemispheres, preference in the injection hemisphere	U.S. and U.K. tests in the Pacific Proving ground
IV Upper tropical	20	Over both hemispheres	U.S. very high yield tests in the Pacific, 1954 (?)
V Very high Stratosphere	60	Over both hemispheres	Few experimental devices

TABLE 5. Strontium 90 Distribution, January 1967*

Area	Amount, Mc
Stratosphere	0.3
Troposphere	0.1
Tropospheric fallout	2.1
Stratospheric fallout	13.0†
Total accountable	15.5

*After *Hardy et al.* [1968].
†10.7 Mc in northern hemisphere, 2.3 Mc in southern hemisphere.

wetok (Ivy test series). They injected, for the first time, a certain amount of fission products into the stratosphere; this might have resulted in moderate worldwide contamination, including the Greenland and the Antarctic ice sheets.

1953. The first thermonuclear device was detonated by the Soviet Union in 1953.

1954. In March 1954, a high-yield thermonuclear device was detonated in the Castle test series at Bikini. The total yield announced was 15 MT, but the contribution from the fission process was exceptionally high (maybe 85%), resulting in a most important worldwide contamination by radioactive debris, mainly fission products. Up to the present, this event represents the most important single injection of fission products into the lower and upper equatorial stratosphere, and consequently the first intense source of contamination of the southern hemisphere, including the Antarctic continent.

1954–1958. Up to October 1958, H-bomb and A-bomb tests were continued by the United States, the Soviet Union, and the United Kingdom.

1958–1961. From November 1958 to October 1961, all high-yield nuclear tests in the atmosphere were interrupted (nuclear moratorium).

1961. In October 1961, tests in the atmosphere were resumed by the Soviet Union at Novaya Zemlya. They included devices with very high energy yield, up to 60 MT.

1962. High-yield devices were detonated both by the Soviet Union in the Arctic and by the United States at Christmas Island. The treaty of Moscow resulted in the suspension of all nuclear tests in the atmosphere by the United States, the United Kingdom, and the Soviet Union.

1962–1967. Low-yield A bombs were detonated by France and by China.

1967. In June 1967, a thermonuclear device was detonated by China at the Lop Nor site; the reported energy yield was several megatons. Fission yield of the order of 2 MT can be estimated from the results reported by *Volchok and Kleinman* [1969].

1968. Thermonuclear devices with reported energy yields of 0.5 and 2 MT were detonated by France at the Mururoa site (22°S, 139°W) in August and September 1968, respectively. No information is available on the fission yield. These thermonuclear tests were the first to be carried out in the southern hemisphere.

Stratospheric Injections

In Figure 5, we have attempted to summarize the available data on the injection of fission products into the stratosphere up to 1967. Injections into the polar stratosphere (Soviet Union tests) and into the tropical stratosphere (United States and United Kingdom tests) are shown separately. This figure is based on public information about the announced nuclear detonations [*Glasstone*, 1964] and on published estimates of the fission yields and of the stratospheric injections [*Libby*, 1959; *Martell*, 1959; *Fowler*, 1960; *Friend et al.*, 1961; *Federal Radiation Council*, 1963; *Tompkins*, 1963; *HASL*, 1964]. We stress the approximate character of these estimates and the uncertainties involved in our interpretation of some of the reported data. Nevertheless, the general picture given by Figure 5 will be helpful in the discussion of the next section.

RADIOACTIVE FALLOUT AND REFERENCE HORIZONS IN THE ANTARCTIC ICE SHEET

Radioactive debris from nuclear bomb tests can be used as markers, allowing the identification of well-dated horizons in glaciers and thus affording a reliable method for measuring mean accumulation rates [Drevinsky et al., quoted by *Sharp and Epstein*, 1962; *Picciotto et al.*, 1962; *Picciotto and Wilgain*, 1963; *Vickers*, 1963].

Because of the high solubility of this debris, a stable horizon can be expected to be formed only in an area of dry-snow facies or in an area with mod-

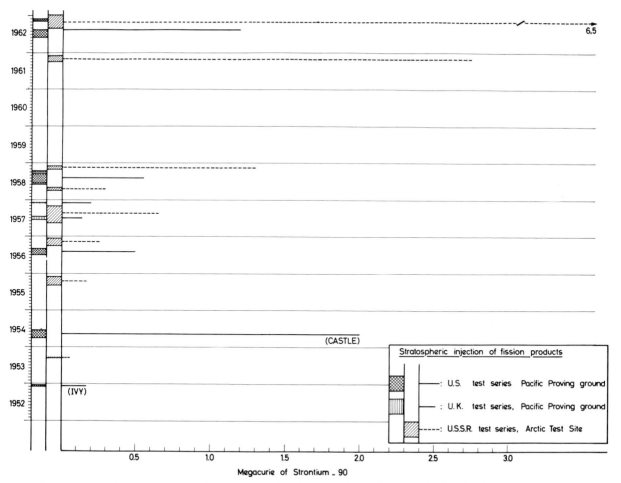

Fig. 5. Chronology of the main thermonuclear bomb test series and stratospheric injections of strontium 90.

erate percolation of melt water. Among the various types of radioactive debris, the fission products are best suited for this purpose, owing to their high production yield and to their low volatility, which ensure the absence of vertical migration in the firn layers, as long as they remain dry.

Tritium is also produced at high yield in thermonuclear explosions, but its detection and its measurements are much more difficult. Moreover, interpretation of the results is complicated by the background of naturally produced tritium. The original distribution of tritium is likely to be disturbed by exchanges of water vapor across the firn, even in dry conditions.

Radioactive Fallout over Antarctica

The glaciological applications of such marker horizons require an accurate knowledge of the evolution of artificial radioactivity in precipitation. From 1959 onward, the network of stations monitoring the radioactivity in the air and in precipitation has been dense enough to enable us to retrace this evolution with reasonable precision at any point on the earth. For the earlier years, data are very scarce, particularly for the southern hemisphere.

In Antarctica, the first systematic survey of atmospheric radioactivity was started in April 1956 at Little America V. Similar programs were carried out at Base Roi Baudouin, South Pole station, Base Dumont d'Urville, and other stations. The longest continuous record belongs to the South Pole station, which has been in operation since February 1959 and has provided very valuable information on this matter. A synthesis of the results up to 1963 has been published by *Lockhart et al.* [1965]. Later results are published in the reports of the Health and Safety Laboratory, U.S. Atomic Energy Commission. These results are shown in Figure 6.

The history of radioactive fallout over Antarctica before 1956 can be reconstructed from our general

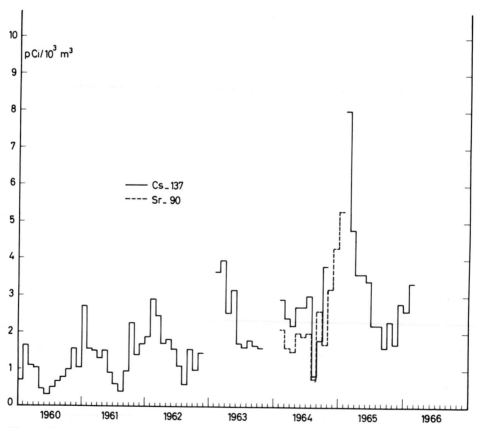

Fig. 6. Fission products in the air at the South Pole, after *Lockhart et al.* [1965] and HASL reports.

knowledge of the behavior of radioactive debris in the atmosphere, but a more reliable picture can be obtained by studying well-dated firn samples.

From the preceding discussion, it is clear that only stratospheric debris could have reached the Antarctic in detectable amounts.

The first appearance of fission products in Antarctic precipitation could be expected several months after the Ivy test series of November 1952. However, the subsequent Castle test series in March 1954 injected into the stratosphere a much larger amount of fission products, probably larger by a factor of 10 or more. Most of the stratospheric debris from the Castle test must have fallen over Antarctica during the next summer season, between November 1954 and March 1955, reaching its peak intensity in January or February 1955.

The continuation of high-yield bomb tests in the atmosphere until October 1958 would have maintained the radioactivity of Antarctic precipitation at an appreciable level up until 1959.

The nuclear moratorium, followed by the resumption of high-yield tests in 1961 and 1962, would have resulted in a minimum fallout in 1961, followed by an increase reaching a maximum in 1964 or in 1965.

Studies on well-dated firn samples have yielded results in perfect agreement with these expectations. Detailed measurements were carried out on samples from the glaciological pit at Base Roi Baudouin, carefully dated both by stratigraphic and by stable isotope ratio studies. They have demonstrated the presence of a striking surge in radioactivity in a layer attributed to February 1955 with a maximum uncertainty of ±2 months [*Picciotto and Wilgain*, 1963]. Similar conclusions were reached for other locations in Antarctica [*Vickers*, 1963; *Woodward*, 1964; *Wilgain et al.*, 1965].

Figure 7 summarizes the results obtained in the South Pole area. It shows schematically the evolution of the yearly average strontium 90 concentration in the snow from 1952 to 1965. The activities were tentatively corrected for radioactive decay since the year of deposition of the snow.

The data up to 1963 are based on the strontium 90 measurements published by *Wilgain et al.* [1965] and on unpublished results obtained on samples collected at the South Pole station in November 1964, before the departure of the SPQMLT 1 party.

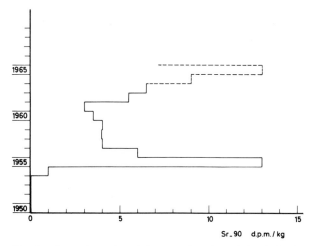

Fig. 7. Strontium 90 activity as a function of depth in the firn at the South Pole.

The concentrations for 1964 and 1965 have been estimated from the results reported in Figure 6 by assuming an activity ratio $^{137}Cs/^{90}Sr$ equal to 1.8 and a proportional relationship between the concentrations in the air and in the snow.

The main features of the profile in Figure 7 are typical of the earlier results, and they are visible on most of the profiles reported here. They can be summarized as follows:

Before 1952, there was no strontium 90 detectable (<0.1 dpm/kg).

Small amounts were present in 1953 and 1954, probably originating from the Ivy test.

The layer 1955 is characterized by a striking surge in strontium 90 concentration (and, of course, of all the other radioactive debris), reaching values of 12 to 15 dpm/kg in this area.

The strontium 90 activity varied little around 3 dpm/kg in the layers deposited after 1955, but increased from 1963 onward to reach a maximum value of around 15 dpm/kg in 1965.

The 1955 radioactivity surge represents a very convenient index horizon for measuring accumulation, and it has been used as such in the present work. We consider that the agreement with accumulation values derived from surface measurements at Base Roi Baudouin [De Breuck, quoted by *Crozaz et al.*, 1964] at the South Pole station (present work) leaves no doubt about the general validity of the method in Antarctica. Its application to the

Greenland ice sheet has been discussed by *Crozaz et al.* [1966].

The 1965 surge, in spite of being more gradual and occurring over a background of residual strontium 90 from the earlier bomb tests, can be of interest as an index horizon [*Crozaz*, 1969].

It should be pointed out that the striking activity surge in the 1955 Antarctic snow is due to a fortuitous combination of several factors: (1) the very high power of the first detonation of the Castle test series, coupled with an exceptionally high ratio of fission to fusion yields; (2) the location of the test near the equator; and (3) the time of the year at which it was set off, allowing enough time for a significant fraction of the debris to be present in the polar stratosphere at the next summer season, when it could be rapidly transferred into the troposphere and onto the ground.

THE LEAD 210 DATING METHOD

The lead 210 method for dating glacier ice, proposed by *Goldberg* [1963], was subsequently developed and applied to firn and ice samples from Antarctica, from Greenland, and from temperate glaciers by *Crozaz et al.* [1964], *Nezami et al.* [1964], *Crozaz and Langway* [1966], *Crozaz* [1967a, b], *Picciotto et al.* [1967, 1968], and *Windom* [1969].

Lead 210 (radium D in the old nomenclature) is a natural β emitter with a half-life of 22 years and belonging to the uranium 238 family. It is present in a minute amount in the atmosphere as a result of the radioactive decay of radon 222 (Figure 8). Shortly after its formation, it becomes attached to

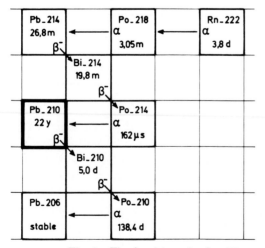

Fig. 8. The daughters of radon 222.

TABLE 6. Lead 210 Activity in Surface Snow from Antarctica*

Station	Ann. Accum. g cm^{-2}	^{210}Pb Activ., dph/kg
South Pole	6.7	97 ± 7
1-10-275	5.5	93 ± 5
1-13-370	4.2	100 ± 6
1-15-415	6.1	110 ± 6
2-0-0	3.1	108 ± 12
2-11-279	3.6	112 ± 16
Dumont d'Urville	14	98 ± 10
McMurdo	16	156 ± 6
New Byrd	18	110 ± 18
Roi Baudouin	40	74 ± 3

*After *Crozaz* [1967a].

the aerosol particles. Its residence time in the troposphere ranges from 5 to 20 days. It is brought to the ground principally by the scavenging action of precipitation.

In the northern hemisphere, the average concentration of lead 210 in tropospheric air is of the order of 3×10^5 atoms (equivalent to an activity of 0.02 dpm) per kg, its average concentration in rainfall is approximately 8×10^7 atoms/liter or 5 dpm/liter. In the southern hemisphere and particularly in Antarctica, these concentrations are lower by an order of magnitude (Table 6). This difference is accounted for by the scarcity of continental masses, which are the sources of radon, and by the thick ice cover overlying the Antarctic continent, which prevents the release of the radon into the atmosphere.

The dating of firn or ice layers by lead 210 is based on the fact that the initial concentration or activity of lead 210 in the snow (A_0) will decrease exponentially with time after its deposition on the ground, provided that the sample to be dated had remained a closed system for lead since its deposition. This implies that the changes in lead 210 content result exclusively from radioactive decay and not from exchanges by diffusion or by percolation of melt water.

The initial activity A_0 is not known directly. It is generally assumed that A_0 has remained constant at a given place, a reasonable assumption if average values over 3 to 10 years are considered in order to smooth out seasonal or other short-range variations.

In such a case, the specific lead 210 activity of a sample lying at a depth z will be given by

$$A_z = A_0 \times e^{-\lambda t} \tag{1}$$

and its age t, defined as the time elapsed since the snow was deposited on the surface, by

$$t = (1/\lambda) \ln (A_0/A_z) \tag{2}$$

where λ is the radioactive constant for lead 210 = (0.032 ± 0.001) yr^{-1}.

For a sample of finite thickness, z can be taken as the middle of the depth interval. As long as the corresponding time interval is shorter than 10 years, and provided that the yearly accumulation is not very irregular, the resulting relative error in t will be less than 3%.

The time lapse spanned by the method is limited to 5 or 6 times the half-life of lead 210; that is to say, to approximately 120 years from the present. Indeed, after 6 half-lives, the lead 210 activity of ice has decayed to 1.5% of its initial value, and it becomes very difficult to estimate with reasonable accuracy.

Equation 2 is valid under the assumptions of a closed system and of a constant initial A_0 activity; however, it does not require a constant rate of snow accumulation or of lead 210 deposition.

If the multiannual mean accumulation rate has remained constant during the time interval considered (120 years at the maximum), we can write

$$t = \rho z/a = h/a \tag{3}$$

where ρ is the density of firn or ice, and h is the depth of the sample of age t, expressed in water equivalent, and equation 2 becomes

$$\ln A_z = \ln A_0 - (\lambda/a)h$$

A semilogarithmic plot of A_z as a function of h will yield a straight line with a slope $-\lambda/a$ and an ordinate intercept at A_0.

Thus the accumulation rate a can be easily derived from such a plot. This is the method used in the present work. We recall that it relies upon three basic assumptions:

1. The evolution of the ice as a closed system for lead.
2. A constant mean lead 210 content in the freshly fallen snow.
3. A constant mean rate of accumulation a.

We consider that an internal check of the simultaneous validity of these three assumptions is afforded when a straight line is obtained by plotting $\ln A_z$ against h. It is indeed very unlikely that such a straight line could result from a fortuitous compensation of variations of A_z or a loss or gain of lead 210.

Table 6 shows that the mean lead 210 activity in Antarctic precipitation (as measured by the activity of firn layers near the surface) varies little from place to place. It is everywhere close to 100 dph/kg, although a slight inverse correlation between a and A_0 is indicated. These observations strongly support the assumption of a constant A_0 with time. Figure 9 shows an array of straight lines corresponding to various accumulation rates and to an assumed initial activity A_0 of 100 dph/kg.

SAMPLING PROCEDURES

Sampling for Fallout Measurements

The requirements are the following:

To collect a continuous section of firn from the surface down to a level corresponding approximately to the 1950 layer.

The section should be cut in individual samples covering approximately 1-year intervals or less (10- to 20-cm thick in the region investigated here).

A minimum amount of 100 grams is needed for each measurement of gross β activity; optimum sample weight would be of the order of 500 grams.

In the area of the South Pole–Queen Maud Land traverse, the 1955 level was generally expected to lie between 1 and 2 meters from the surface. Coring of the near-surface snow turned out to be impossible because of the softness and the porosity of the firn. Coring with a good yield was feasible only below 3 or 4 meters from the surface. Therefore two procedures were developed, both involving the excavation of a pit to a depth of 2 or 3 meters.

Procedure a. A rigid tube open at both ends, made of aluminized cardboard, was pushed down vertically from the surface, not far from the pit wall. Gentle hammering was sometimes necessary. The filled tube was recovered from the pit wall by sawing out the firn around it. The tube was 100 cm long and had an inner diameter of 8 cm, and thus provided a continuous firn section from 0 to −1 meter. A farther section from −1 meter to −2

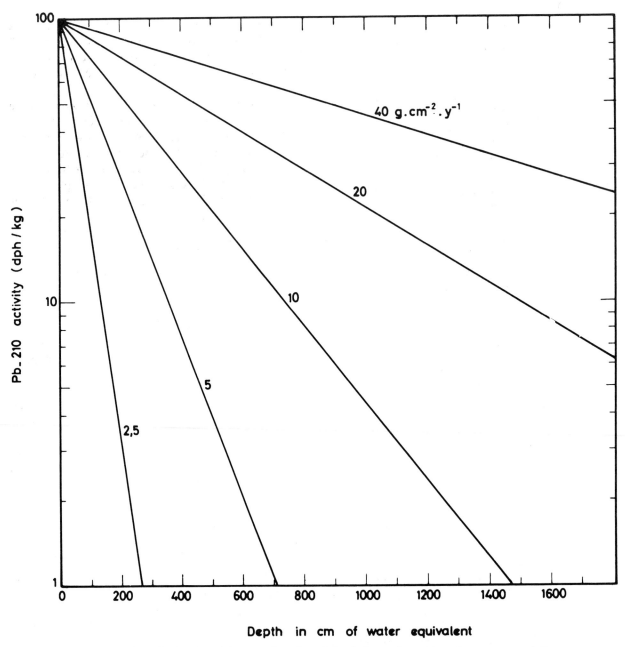

Fig. 9. Idealized lead 210 activity as a function of depth for various rates of snow accumulation.

meters was obtained in the same way by pushing down a tube from a horizontal shelf set up in the pit, not far from the end of the first tube. The tube was then slid into an outer cardboard cylinder, which was sealed at both ends. Generally two parallel two-meter sections were sampled. The tubes were brought back frozen and were cut at 10-cm intervals in the laboratory.

This method is rapid (once the pit is excavated) and allows the collection of two 2-meter sections in less than an hour. However, it presents three defects:

It generally results in an appreciable compaction of the firn, the depth interval sampled being longer than the length of the tube, sometimes by as much as 25%. In drawing the profiles, the compaction was supposed to have uniformly affected the whole section, but this assumption is certainly not correct and introduces an uncertainty as to the true depth of the samples.

Occasionally layers of very hard compacted firn are encountered in the profile, and it is impossible to penetrate them without seriously damaging the tube.

This method necessitates the transportation of the filled tubes in a frozen state, which means the total loss of the samples in case of accidental melting.

Procedure b. These three defects are overcome by the second method, which is more tedious and time consuming: blocks are sawed out from a wall of the pit, care being taken that each block represents a constant sectional area, even if the vertical thickness is variable. The samples were cut out into blocks 10 cm thick when they were supposed to be near the 1955 horizon, and in thicker blocks when they were farther from this horizon. The blocks were then broken up, and the fragments were packed in wide-necked 0.5-gallon plastic bottles.

Procedures *a* and *b* were both used on SPQMLT 1 and 2.

Procedure *b* only was used on SPQMLT 3.

Samples for Lead 210 Measurements

The requirements are:

To sample a time interval going back to about 100 years, preferably on a continuous section.

Each sample should cover a time interval of several years in order to average out the annual or seasonal variations in lead 210 fallout.

Each sample should weigh 1–2 kg (thanks to technical improvements in measuring lead 210 through the activity of polonium 210 [*Nezami et al.*, 1964; *Crozaz and Fabri*, 1966], it is now feasible to obtain good results on 1- to 2-kg samples, whereas our first measurements at the South Pole [*Crozaz et al.*, 1964] required samples of 20 to 30 kg).

Extreme precautions must be taken against contamination by lead 210 or polonium 210 because of the very low activity to be measured (1 to 0.05 dpm/kg).

Since the activity of lead 210 decays exponentially with depth, the size of the samples should be increased with depth if a constant precision in the results is wanted. Unfortunately, the difficulty of retrieving ice samples increases exponentially with depth, so we had to content ourselves with samples of constant section with depth.

Generally, samples for lead 210 in the depth intervals from 0 to 3 meters were collected by the same method as for the fallout samples. Deeper samples were obtained by coring with a Sipre hand-auger from the bottom of the pit. Most of the time, two parallel 8-cm-diameter cores were obtained by these procedures, from the surface down to a depth of 10 meters. The cores were grouped and analyzed by 1-meter increments.

Contamination was not well controlled and gave poor results for most of the lead 210 cores. This will be discussed in more detail.

At station 2-11-279, a 15-meter-deep trench had to be dug in order to retrieve drilling tools lost in a bore hole; samples for lead 210 were recovered by sawing 1-meter-thick blocks out of the wall.

All the samples were transported and stored frozen. They were melted only in the laboratory, under well-controlled conditions and after the addition of suitable stable carriers or radioactive tracers. These precautions seem necessary to prevent losses by absorption or contamination by the walls of the containers.

Should the contamination problems be well under control, the samples could probably be transported and stored in the liquid state, provided that stable carriers and radioactive tracers were added in the field before the ice was melted.

MEASUREMENT OF THE FISSION PRODUCTS

As was mentioned above, the best method for expressing the amount of fission products in a sample would be to report the specific activity of selected nuclides such as strontium 90 and caesium 137. However, this method implies lengthy radiochemical separations. In favorable cases, caesium 137 can be measured directly by γ spectrometry, but in Antarctic precipitation, samples of the order of 10 kg or more would be required. Since we were faced with the problem of analyzing more than 2000 samples weighing a few hundreds of grams, we had to develop a rapid but sensitive method. This problem is made relatively easy in Antarctica by the very low level of natural radioactivity in the Antarctic atmosphere (see bibliography in *Lockhart et al.* [1965]).

In firn older than a few months, the natural radioactivity is due principally to the lead 210 in equilibrium with its daughters bismuth 210 and polonium 210, and, to a very minor extent, to potassium 40, to members of the uranium and thorium families, and to cosmic-ray-induced nuclides such as silicon 32 and beryllium 10. Table 7 shows an estimate of these various contributions in the upper firn layers at the South Pole.

It can be seen that, in firn layers deposited later than 1955, more than 90% of the β radiation originates from the fission products. The only appreciable natural contribution is lead 210 (actually bismuth 210, since the β radiation of lead 210 has too low an energy to be detected by usual β detectors).

Therefore, a convenient and easy-to-measure parameter is afforded in our case by the gross β activity of the firn. The drawbacks are obvious from discussions above: the composition and gross β activity of a mixture of fission products decays in complicated and often unpredictable ways. Thus, the data are valid only for the date of counting; moreover, it is difficult to express the absolute activity because of calibration problems.

However, the objective aimed at here was confined to the identification of the 1955 layer, and the method of measuring the variation of the gross β activity with depth is quite suitable for this purpose.

Before describing the experimental procedure, we should point out that the radioactivities to be measured are at a very low level and require low-level counting equipment. Moreover, great care must be exercised to prevent radioactive contamination of the sample.

All the operations were carried out in a specially equipped dust-free laboratory, using wares of quartz, Teflon, or polythene.

The liquid chemicals (water and nitric and hydrochloric acids) were reagent grade redistilled in a quartz still.

The salts of caesium, strontium, and yttrium used as stable carriers were spectrographically pure substances.

The β activity of each new batch of reagents was controlled before use.

Chemical Procedure

In earlier investigations [*Picciotto and Wilgain*, 1963; *Wilgain et al.*, 1965], the gross β activity was measured by evaporating the melted sample to dryness. Here we have used a quicker procedure developed in Brussels by C. Weyers. It is based on the coprecipitation of the fission products with suitable carriers.

It will be recalled that, after a few years, the residual activity of a mixture of fission products is mostly due to strontium 90 (in equilibrium with yttrium 90) and to caesium 137, with a minor contribution from samarium 151, prometheum 147, and antimony 125.

Salts of strontium, caesium, and yttrium were added as carriers just before melting the sample. Caesium was precipitated as caesium tetraphenylboryl; strontium, yttrium, prometheum, and samarium as phosphates and fluorides. No attempt was made to quantitatively recover the antimony.

The recovery yields of strontium, caesium, and yttrium were checked by atomic-absorption flame photometry; they always amounted to more than 90%.

The detailed procedure is as follows:

Prepare and store in a plastic bottle a carrier solution for about 500 samples by dissolving in 500 ml of water approximately 13 grams of strontium carbonate, 11 grams of caesium carbonate, and 11 grams of yttrium nitrate ($+6H_2O$).

To a firn sample weighing between 150 and 350 grams,

TABLE 7. Natural and Artificial Radioactive Nuclides in the Firn at the South Pole

Nuclide	Specific Activ., dpm/kg	References and Remarks
K 40	0.01	From a measured K abundance of 6×10^{-9} g/g [*Hanappe et al.*, 1968]
U, Th	5×10^{-4}	Assuming U = 6×10^{-13} g/g
Pb 210	1.5	*Crozaz et al.* [1964]
Si 32, Be 10	0.01	
Sr 90 (1955)	13	*Wilgain et al.* [1965] (corrected for decay)
Sr 90 (1961)	3	*Wilgain et al.* [1965] (corrected for decay)
Gross β (1955)	55	*Wilgain et al.* [1965] (measured in 1964)
Gross β (1961)	12	*Wilgain et al.* [1965] (measured in 1964)

add 2.3 ml hydrochloric acid, 6 M, and 1 ml of the carrier solution.

Let melt and mix thoroughly.

Add slowly 1 ml of a solution containing, per 100 ml, 5 grams of sodium orthophosphate mono-H (Na_2HPO_4 + $12H_2O$), 5 ml of hydrofluoric acid at 40%, and approximately 5 grams of solid sodium hydroxide.

Mix until total dissolution; pH should be well above 7.

Add slowly, with constant stirring, 5 ml of a freshly prepared solution containing 1.2 g of sodium tetraphenylboryl in 100 ml.

Stir thoroughly for 10 min and let stand 12 hours.

Discard the supernatant liquid and filter on a millipore filter of 0.45 μ porosity and 25 mm diameter.

Discard the filtrate. Wash the precipitate with 100 ml of water.

It is important to avoid pulling too much air through the filter, which might deposit aerosol-bound daughters of the radon isotopes, including lead 210 and bismuth 210.

Counting Procedure

The millipore filter is mounted on a lucite holder and is covered with a thin ($\sim$1 mg/cm^2) Mylar foil. The counting is delayed 3 or 4 days so that any lead 212 present will completely decay.

The β activity of the precipitate is measured with a low-level proportional counter protected by an anticoincidence guard and by a 10-cm-thick lead shield. The detector itself is a flat flow-counter with a window of 0.8 mg/cm^2 thickness and with a sensitive area of 8 cm^2. Pulse-amplitude analysis permits the simultaneous counting of α and β particles in two distinct channels. Although not necessary, this procedure affords a convenient way of controlling the contamination by natural radioactive nuclides, which are always accompanied by α emitters.

The background over a blank millipore filter was 12 counts per hour in the β channel and 0.2 counts per hour in the α channel. The gross β activity from a 150-gram sample was found to vary from 0.5 to 5 counts per minute. The over-all detection efficiency was measured to be 40% for the β radiation from a strontium 90 + yttrium 90 equilibrium mixture. The efficiency of the counter was controlled twice a day with a standard β source. Self-absorption in the precipitate (about 10 mg/cm^2 thick) was almost negligible.

Blank tests were run on each batch of reagents, since significant variations of the β activity were found from batch to batch in material from the same origin. Every month a total blank test was run by applying the complete procedure to firn samples definitely deeper than the 1955 layer.

These blanks yielded counting rates varying from 0.7 $\pm$ 0.1 to 0.3 $\pm$ 0.1 counts per minute for a 300-gram sample, including the background of the detector. For all the samples, counting time was set in order to obtain a maximum error of $\pm$10% on the counting statistics; the accuracy was generally much better on the post-1955 samples.

On a routine basis, with two detectors in operation, it was possible to run 30 samples a week. A total of 1886 firn samples, including about 400 duplicates, were processed, resulting in 2400 radioactive measurements if one adds blank tests, standards, and background runs. A general rule was to run successively all the samples from a given station, starting from the deepest and less radioactive levels.

It must be stressed that the collection and the measurements programs extended over three years and that both the sampling method and the laboratory procedure improved at each new campaign. Therefore, the results from SPQMLT 3 are more accurate and easier to interpret in terms of accumulation.

ACCUMULATION FROM THE FISSION PRODUCTS

Profiles of the fission product activity versus depth were established at 75 stations of the South Pole–Queen Maud Land traverse. Pit stratigraphy studies were carried out at 61 of them from the surface to the 2-meter depth. They included the usual observations and measurements of the physical properties of the firn, such as density, hardness, texture, grain size, and shape. The main objective was to identify annual layers and to estimate accumulation.

This part of the program has been discussed by *Koerner and Kane* [1967], *Cameron et al.* [1968], *Picciotto et al.* [1968], *Rundle* [1968], and *De Breuck* [1969]. Only a schematic description of the stratigraphy will be given for each pit, as well as its interpretation in terms of annual layers.

The figures in the Appendix show the results of the gross β activity measurements. For each station, the following information is given:

1. A schematic stratigraphic profile shows the position of the possible indicators of annual boundaries layers (mainly ice crusts and depth hoar layers) and information on hardness and grain size. For SPQMLT 3, the hardness is expressed by the

Rammsonde data; for SPQMLT 1 and 2, by semiquantitative four-grade scale (soft, medium, hard, and very hard).

2. The interpretation of the stratigraphy is shown in terms of annual layers as it was made in the field, by R. Cameron and E. Picciotto on SPQMLT 1, E. Picciotto and R. Behling on SPQMLT 2, and by W. De Breuck on SPQMLT 3. Sometimes no interpretation at all was possible (for instance, at stations 1-17-470, 2-4-80, 2-18-443, 2-19-455). Usually several interpretations were possible, and we have generally reported the two extremes leading, respectively, to the maximum and the minimum number of years.

3. The distribution of gross β activity is shown as a function of depth. For each depth interval sampled, the β activity is expressed in counts per minute normalized to a 150-gram sample (to 300 grams for the SPQMLT 3); no background correction was made. The reported error interval represents the

TABLE 8. Average Annual Accumulation on SPQMLT 1

Station	Depth of β Surge, cm	d,* kg/m^3	Accumulation,† g cm^{-2} yr^{-1}	
			β‡	Stratigraphy§
Pole	180 ± 5	373	6.7 ± 0.2	7.4
0-8				7.5
1-24	175 ± 5		6.5 ± 0.2	7.5
2-48	170 ± 5		6.4 ± 0.2	8.2
3-72	110 ± 10		4.2 ± 0.4	5.1
4-96	120 ± 5		4.8 ± 0.2	4.6–6.6
4a-110	140 ± 10		5.6 ± 0.4	7.0–8.0
5-125	80 ± 10	366	2.9 ± 0.4	2.8–6.7
5a-140	120 ± 5	378	4.5 ± 0.2	5.0
6-155	115 ± 8	412	4.7 ± 0.3	7.2–8.4
7-185	113 ± 5	381	4.3 ± 0.2	5.5–7.3
8-215	120 ± 10		4.8 ± 0.4	3.9–6.6
9-245	100 ± 5	384	3.8 ± 0.2	6.0
10-275	135 ± 5	404	5.5 ± 0.2	7.4
12-338	130 ± 5		5.0 ± 0.2	5.7–6.4
13-370	108 ± 5	386	4.2 ± 0.2	6.8–8.1
14-395	125 ± 5	403	5.0 ± 0.2	4.8–5.4
15-415	155 ± 5	395	6.1 ± 0.2	6.7–10
16-445	105 ± 5	368	3.9 ± 0.2	4.0–5.0
17-470	35 ± 5	305	1.1 ± 0.2	?
18-496	118 ± 5	363	4.3 ± 0.2	4.7–7.0
19-519	121 ± 5	380	4.6 ± 0.2	4.8–6.8
20-545	<42	373	<1.6	?
21-570	55 ± 5	351	1.9 ± 0.2	>3.5
22-595	65 ± 5	370	2.4 ± 0.2	>4.8
23-620	<42	409	<1.7	?
24-650	79 ± 5	383	3.0 ± 0.2	?
25-680	112 ± 5	384	4.3 ± 0.2	5.0–5.8
26-710	70 ± 5	366	2.6 ± 0.2	5.4
27-740	100 ± 5	366	3.7 ± 0.2	6.1–9.2
28-770	70 ± 5	361	2.5 ± 0.2	7.2
29-797	75 ± 5	355	2.7 ± 0.2	3.6–5.1

*Measured average density to depth of β surge (when the density was not directly measured, the results of the two nearest measured stations were used).

†For the time interval January 1955 to January 1965.

‡The error is understood as the maximum uncertainty on the depth of the β surge. This uncertainty is usually due to the finite thickness of the sample.

§Accumulation derived from the interpretation of the stratigraphy. The supposed annual layers were counted between the surface and the depth of the β surge.

standard deviation on the number of counts. An arrow shows the position of the β activity surge ascribed to January 1955. The positioning of this level is generally straightforward. At a few stations, the β activity profile is more ambiguous, and the position selected for the arrow involves some personal interpretation.

From the depth of the β surge and the measured densities, the average annual accumulations since 1955 were computed. Tables 8, 9, and 10 report the results for each of the three traverses. In Table 10, we give two independent estimates of the accumulation made, respectively, by W. De Breuck and by A. Rundle [from *Rundle*, 1968]. They do not cover the same time interval. De Breuck's estimate is based on the number of 'annual layers' between the surface and the depth of the β surge; Rundle's estimate represents the average over 2 to 5 supposed annual layers.

MEASUREMENT OF THE LEAD 210

Experimental Procedure

Owing to the very low energy of its β radiation (maximum energy, 18 kev), lead 210 is very difficult to measure directly. It is generally measured through the activity of its daughters: bismuth 210 (a β emitter with a maximum energy of 1.16 Mev; half-life, 5 days) or polonium 210 (an α emitter; energy, 5.305 Mev; half-life, 138 days). Radioactive equilibrium with lead 210 is obtained at 99% after 35 days for bismuth 210 and after 2.6 years for polonium 210. Thus, in polar firn, except in the very

TABLE 9. Average Annual Accumulation on SPQMLT 2

Station	Depth of β Surge, cm	d,* kg/m³	Accumulation,† g cm⁻² yr⁻¹	
			β‡	Stratigraphy§
0-0	95 ± 2.5	355	3.1 ± 0.1	>2.8
1-15	105 ± 5	354	3.4 ± 0.2	4.4
2-40	105 ± 5	366	3.0 ± 0.5	5.0
3-55	95 ± 5	363	3.1 ± 0.2	4.9
4-80	65 ± 5	382	2.3 ± 0.2	3.0
5-120	90 ± 5	384	3.1 ± 0.2	2.7–3.5
6-160	95 ± 5		3.3 ± 0.2	3.9
7-175	150 ± 5		5.3 ± 0.2	
8-200	66 ± 6	418	2.5 ± 0.2	3.2–3.6
9-220	100 ± 5		3.2 ± 0.2	3.3
10-240	25 ± 5	332	0.8 ± 0.2	1.7–2.6
11-279	115 ± 5	349	3.6 ± 0.2	3.2–3.5
12-323	115 ± 5	361	3.8 ± 0.2	2.9–4.5
14-363	(80 ± 5)	330	(2.4 ± 0.3)	?
16-403	105 ± 5	363	3.5 ± 0.2	3.2–5.7
17-423	110 ± 10		3.6 ± 0.4	
18-443	20 ± 5	352	0.6 ± 0.2	?
19-455	20 ± 5	346	0.6 ± 0.2	?
20-480	45 ± 5		1.6 ± 0.2	
21-495	125 ± 5	387	4.4 ± 0.2	5.0
22-515	140 ± 10		5.0 ± 0.3	
23-535	65 ± 5	407	2.4 ± 0.2	4.1
25-575	60 ± 5	364	2.0 ± 0.2	2.3
27-615	75 ± 5	363	2.5 ± 0.2	2.7–3.3
28-667	85 ± 5	354	2.7 ± 0.2	3.2

*Measured average density to depth of β surge (when the density was not directly measured, the results of the two nearest measured stations were used).
†For the time interval January 1955 to January 1966.
‡The error is understood as the maximum uncertainty on the depth of the β surge. This uncertainty is usually due to the finite thickness of the sample.
§Accumulation derived from the interpretation of the stratigraphy. The supposed annual layers were counted between the surface and the depth of the β surge.

TABLE 10. Average Annual Accumulation on SPQMLT 3

Station	Depth of β Surge cm	d,* kg/m³	Accumulation,† g cm⁻² yr⁻¹		
			β‡	Stratigraphy §	
				Rundle	De Breuck
1-50	125 ± 5	361	3.5 ± 0.2	2.9	4.3
2-100	140 ± 5	348	3.7 ± 0.2	3.1	4.4
3-150	130 ± 5	348	3.5 ± 0.2	3.6	4.6
4-188	130 ± 5	358	3.6 ± 0.2	4.4	5.3
5-248	95 ± 5	343	2.5 ± 0.2	5.2	5.9
6-300	70 ± 5	353	1.9 ± 0.2	5.7	4.6
7-350	110 ± 5	348	2.9 ± 0.2	6.0	5.5
8-400	175 ± 5	336	4.5 ± 0.2	6.3	5.5
9-450	190 ± 5	355	5.2 ± 0.2	6.8	5.2
10-500	165 ± 5	356	4.5 ± 0.2	7.5	5.2
11-550	185 ± 5	357	5.1 ± 0.2	9.1	4.4
12-585	170 ± 5	363	4.7 ± 0.2	8.3	5.0
13-620	185 ± 5	356	5.1 ± 0.2	9.9	5.0
14-675	155 ± 5	362	4.3 ± 0.2	9.1	5.5
15-710	155 ± 5	354	4.2 ± 0.2	8.8	4.9
16-750	205 ± 5	354	5.6 ± 0.2	8.6	6.3
17-790	200 ± 5	338	5.2 ± 0.2	9.3	5.1
18-840	160 ± 5	366	4.5 ± 0.2	10.4	5.6

*Measured average density to depth of β surge (when the density was not directly measured, the results of the two nearest measured stations were used).

†For the time interval January 1955 to January 1968.

‡The error is understood as the maximum uncertainty on the depth of the β surge. This uncertainty is usually due to the finite thickness of the sample.

§Accumulation derived from the interpretation of the stratigraphy. The supposed annual layers were counted between the surface and the depth of the β surge.

upper layers, lead 210 can be considered in radioactive equilibrium with polonium 210, assuming, of course, that a firn sample behaves as a closed system not only for lead but for polonium also.

A description of procedures based on the measurement of bismuth 210 regrowing from lead 210 has been given by *Goldberg* [1963], *Crozaz et al.* [1964], and *Crozaz* [1967a].

In the present work, we have also measured the lead 210 through the α activity of polonium 210 contained in the ice. This procedure relies on the assumption that both nuclides are in radioactive equilibrium in the firn samples analyzed, which implies mainly that the samples behave as closed systems not only for lead, but also for polonium. The validity of this assumption, at least in the area we are dealing with, is supported by the agreement found between the distribution of lead 210 and of polonium 210 with depth at the Pole of Relative Inaccessibility [*Crozaz and Fabri*, 1966; *Picciotto et al.*, 1968].

The polonium 210 procedure requires much more elaborate counting equipment than the bismuth 210, but its sensitivity is about 50 times higher, an appreciable advantage when keeping the samples at a minimum size is a stringent requirement.

Polonium 210 Procedure

Polonium 210 measurements in glacier ice were successfully carried out by *Nezami et al.* [1964]. Their procedure was improved by *Crozaz and Fabri* [1966], who used polonium 208 as a tracer in order to control the recovery yield of polonium 210. The method described here is very similar, but includes minor improvements.

Reagents

Polonium 208 tracer stored in HCl, 2 M, with a specific activity of the order of 50 dpm/ml. Check carefully by α spectrometry for the absence of polonium 210 in detectable amount.

Lead carrier, as $PbCl_2$ devoid of any detectable Pb 210

activity. If possible, make it from lead at least 200 years old.

Hydrochloric acid 0.1 M made from distilled HCl.

Thioacetamide 0.7 M.

Ammonia.

Aqua regia.

Hydrazine chlorohydrate 0.05 M.

To a 1- to 2-kg ice sample, add just before melting a precise aliquot of the carrier solution containing a polonium 208 activity of approximately 5 dpm. As long as identical aliquots of the tracer solution are added to each sample, the accumulation can be derived from the activity ratio ^{210}Po/^{208}Po (duly corrected for polonium 208 decay) without knowing the absolute activities of either isotopes. Add at the same time 30 mg of stable lead carrier and 20 ml of HCl.

After complete melting, stir carefully, add ammonia until the solution is alkaline, add 2 ml of the thioacetamide solution in order to slowly precipitate PbS. Let stand overnight and filtrate. Discard the solution. Dissolve the PbS precipitate with aqua regia, evaporate to a few milliliters in a centrifuge tube, neutralize carefully with ammonia, heat slowly and let precipitate Pb(OH)$_2$. Centrifuge, discard the supernatant, wash with water.

Dissolve the Pb(OH)$_2$ precipitate in 1 ml of HCl, 8 M; add 2 ml of hydrazine chlorohydrate in order to reduce any trace of ferric ion that would prevent the deposition of polonium. Bring the volume to about 20 ml in order to have a medium HCl, 0.5 M. The polonium is extracted by spontaneous deposition on silver. In the solution, maintained at 60°C, rotate a silver disk 16 mm in diameter for three hours. In order to ensure the deposition of polonium on one side only, the disk is mounted in a lucite holder.

After 3 hours, almost all the polonium present is spontaneously deposited; recovery yield is approximately 80%.

The silver disk is gently dried with filter paper. Its α activity is recorded with a solid-state detector (silicon, surface-barrier type of 300 mm^2 area) followed by a charge amplifier and a multichannel analyzer. In order to measure accurately the ^{210}Po/^{208}Po activity ratio, excellent peak stability for several days must be achieved since the energies of polonium 208 and of polonium 210 are very close (200 kev from peak to peak). A peak stabilizer device or a constant-temperature conditioning of the room are helpful in this respect.

The over-all counting yield under our conditions was of the order of 30%. As was mentioned above, an absolute calibration is not necessary, since the accumulation rate is derived from the ^{210}Po/^{208}Po activity ratio.

Blank tests. In the course of the program, the full procedure was carried out 6 times on a 2-kg sample of water tridistilled in a quartz still, used as a blank. There was little variation from run to run. The average final polonium 210 activity, including the detector background, the laboratory contamination, and the polonium 210 possibly present in the water, was found to be 0.012 ± 0.002 counts per minute, or 0.05 ± 0.01 disintegrations of polonium 210 per minute, which is our limit of significant detection as far as laboratory contamination is concerned.

This limit of 3 dph of polonium 210 corresponds in East Antarctica to the activity of polonium 210 contained in a 1-kg sample of approximately 100-year-old ice. In principle, older ages could be measured by increasing the size of the sample, but the limit is mainly set by the contamination level both in the laboratory and in the field operations.

Our limit was really much higher, since it is set by the contamination level, not in the laboratory procedure, but in the procedure for sampling and storing the samples.

Contamination

The first results obtained were clearly inconsistent, but they could be explained by assuming a general level of contamination much higher than the one indicated by the laboratory blank tests.

In order to evaluate the real level of contamination, including the sampling and the storage stages, the whole procedure was carried out on 13 of the deepest samples, most of them being older than 100 years. The results are shown in Table 11. They are expressed again in activity ratio ^{210}Po/^{208}Po normalized to a sample weight of 2 kg. A constant aliquot of the tracer solution was added to each sample. Since the samples did not have exactly the same weight, we had to make corrections in order to normalize the results to a uniform weight of 2 kg. We assumed, therefore, that the total polonium 210 content of the sample was proportional to the weight of ice. This must be true for the polonium 210 naturally present, but not necessarily for the polonium 210 introduced by contamination.

The polonium 208 content of course must be constant from sample to sample. We have carefully checked to see that the specific activity of the polonium 208 solution did not vary with time for other reasons than the radioactive decay.

Under the conditions we established, an activity ratio ^{210}Po/^{208}Po = 1 corresponds to an absolute polonium 210 activity of approximately 100 disintegrations per hour and per kg, which is the initial lead 210 activity in the firn from this region. (Consequently, in a 100-year-old sample, we should ex-

TABLE 11. Polonium 210 in Deep Firn Samples as 'Total Blank' Tests

Sample No.	Traverse and Station No.	Depth Interval, cm	Month of Collection	Month of Measurement	Storage Time, months	Estimated Age Interval,* yr	^{210}Po/^{208}Po†
1	1-7	1035–1135	Dec. 1964	July 1964	43	100–110	0.37 ± 0.03
2	1-23	920–1020	Jan. 1965	May 1968	40	220	0.38 ± 0.03
3	2-5	640–740	Dec. 1965	Apr. 1968	29	90–105	0.35 ± 0.02
4	2-14	920–1020	Dec. 1965	July 1968	31	160–180(?)	0.33 ± 0.02
4′‡	2-14	920–1020	Dec. 1965	July 1968	31	160–180(?)	0.30 ± 0.02
5	2-19	900–1000	Jan. 1966	Aug. 1968	31	570–640	0.47 ± 0.02
6	PS§	1000–1100	Dec. 1966	Nov. 1967	11	150–164	0.10 ± 0.01
6′‖	PS§	1000–1100	Dec. 1966	May 1968	17	150–164	0.10 ± 0.01
7**	PS§	1000–1200	Jan. 1967	June 1968	17	150–180	0.024 ± 0.002
8**	PS§	1200–1400	Jan. 1967	July 1968	18	180–215	0.027 ± 0.004
9	3-5	900–1000	Dec. 1967	July 1968	7	135–153	0.04 ± 0.01
10	3-10	900–1000	Jan. 1968	Sept. 1968	8	77–87	0.31 ± 0.01
10′‖	3-10	900–1000	Jan. 1968	Sept. 1968	8	77–87	0.29 ± 0.03
L.B.††							0.012 ± 0.002

*Age of the layer at the middle of the depth interval, estimated on the basis of measured densities and of the accumulation rates of Table 8.
†Activity ratio ^{210}Po/^{208}Po normalized to a 2-kg sample.
‡Analysis 4 and 4′ were made on samples from the same couple of cores, but the outer part (sample 4′) was taken apart by scraping and cutting the surface with lucite tools. Sample 4 represents the inner part.
§Samples 6, 7, and 8 were collected at Plateau station during the summer season 1966–1967.
‖ Duplicate analysis made on neighboring cores from the same depth interval.
**Cores collected by R. Koerner and stored in plastic bags at Plateau station.
††L.B., laboratory blank. Average of 6 blank tests performed on 2-liter samples of purified water.

pect to find a theoretical activity ratio of the order of 0.03 and a measured ratio of 0.04, if we take into account the results of the blank tests in the laboratory.)

Table 11 shows that the activity ratios found are generally ten times higher, with one notable exception: samples 7 and 8 show the expected activity. These two samples were collected at Plateau station by R. Koerner in January 1967 with the same tools and the same methods as the ones used for all the other samples. However, there is one significant difference in their history: they were stored, not in the usual cardboard tubes, but in plastic bags. Moreover, they remained at Plateau station in a shallow pit from January 1967 to January 1968; they were then shipped to the United States and to Belgium, where they arrived in May 1968 and where they were analyzed the following month.

These circumstances indicate strongly that the high content of polonium 210 in our deep samples is not due to failure of the basic assumptions of the method, but to contamination originating from the storage conditions.

The source of this contamination is not on the inner surface of the cardboard containers. This is proved by preliminary tests made in the laboratory and by results of samples 4 and 4′, showing that the excess of polonium 210 is not localized on the surface, but seems to be homogeneously distributed in the volume of the cores.

In view of the general consistency of the results and of the agreement between duplicate samples, we can exclude a source of accidental or erratic contamination.

The only explanation we can suggest is that the contaminating agent is the radon 222 permeating the firn and decaying in situ. Whether the radon was produced by the radium present in the containers, or was introduced by exchange with the atmosphere of the cold storage rooms, is not clear. If we assume that 50% of the radon produced during 20 months by the container decayed in the stored core, the average radium concentration required in the container material would be of the order of 10^{-14} g/g, a quite plausible value.

Pending further information, it is not possible to give a definite conclusion, but we stress the following points:

- These results cast no doubt on the polonium 210 procedure for measuring lead 210 on small samples. Its validity was already supported by the results of *Nezami et al.* [1964] and *Crozaz and Fabri* [1966]. However, it must be repeated that we are dealing here with exceedingly low activities, and that lead 210 and polonium 210 are widespread contaminants.

- Our sampling and our laboratory procedures seem quite correct and correspond indeed to the limits of detection, but our samples were stored under the wrong conditions, either for too long a time in loose containers, or in containers too rich in radium. This stage should be carefully checked in any future work. The containers should be air tight or kept under pressure of a radon-free gas and should be made of a material with a radium content as low as possible (possibly less than 10^{-16} g/g). In any case, it would be highly advisable to keep the storage time to a minimum.

- Since the contamination seems more or less uniform, we could still use our experimental results by subtracting a constant contamination estimated not from the laboratory blank tests but from the activity found in the deepest samples of each firn section. We shall see that the results are generally consistent, but such a correction amounts to about 30% of the initial maximum activity and introduces of course a large uncertainty, much larger than any other experimental source of error.

ACCUMULATION VALUES FROM LEAD 210

Results

The accumulation was measured by the lead 210 method at the seven stations listed in Table 12. The samples were either cores cut at 1-meter depth intervals or blocks cut out of the walls of deep pits. Except at station 2-11-279, the polonium 210 procedure was used for all the measurements.

For station 2-0-0 (Pole of Relative Inaccessibility), the measurements were carried out by both procedures on parallel firn sections cut from the same pit. The results for this station have already been published and discussed in detail in *Crozaz and Fabri* [1966] and *Picciotto et al.* [1968].

All the results obtained are reported in Table 13. They are expressed in activity ratio $^{210}Po/^{208}Po$ normalized to a 2-kg sample. For each sample, we report both the measured ratio and the ratio corrected by subtracting the ratio value measured in the deepest sample of the profile. For stations 2-0-0 and 2-11-279, the results are expressed as absolute lead 210 specific activity.

The results for each station are reported on the usual semilog graph in Figure 10. We present the positions of both uncorrected (open circles) and corrected (filled circles) results, with the corresponding uncertainty due to the counting statistics. The best straight line fitting the corrected values was drawn wherever it was possible, and the corresponding accumulation was deduced from the slope.

TABLE 12. Stations Where the Lead 210 Method was Applied

Station No.	Type of Sample	Sampled Depth Interval, meters	Month of Sampling	Radiochemical Procedure
1-23-620	Cores	10.2	Jan. 1965	^{210}Po
2-0-0 (Pole of Relative Inaccessibility)	Blocks	4	Dec. 1965	^{210}Bi
	Blocks	6	Dec. 1965	^{210}Po
2-5-120	Cores	8.4	Dec. 1965	^{210}Po
2-11-279	Blocks	15	Dec. 1965	^{210}Bi
2-28-667	Cores	11	Dec. 1965	^{210}Po
3-5-248	Cores	10	Dec. 1967	^{210}Po
3-10-500	Cores	10	Jan. 1968	^{210}Po

TABLE 13. Results of Lead 210 Measurements

Station*	Depth Interval, cm	Depth Interval, cm water equiv.	Sample Weight, g	$^{210}Po/^{208}Po$ (2 kg)	
				Uncorrected	Corrected
1-23-620	0–105	0–41	1.50	0.76 ± 0.03	0.38 ± 0.04
	105–218	41–87	1.92	0.53 ± 0.03	0.15 ± 0.04
	220–320	87–128	1.15	0.49 ± 0.02	0.11 ± 0.04
	920–1020	380–428	1.92	0.38 ± 0.03	0
2-5-120	0–100	0–38	1.37	0.68 ± 0.03	0.33 ± 0.04
	100–200	38–78	1.41	0.58 ± 0.02	0.23 ± 0.03
	240–340	95–137	2.11	0.53 ± 0.03	0.18 ± 0.04
	340–440	137–183	2.73	0.39 ± 0.02	0
	440–540	183–231	2.74	0.35 ± 0.02	0
	540–640	231–279	2.98	0.33 ± 0.02	0
	640–740	279–327	2.95	0.35 ± 0.02	0
2-28-667	8–120	3–43	1.70	0.45 ± 0.02	0.35 ± 0.02
(Plateau	8–120	3–43	1.99	0.43 ± 0.01	0.33 ± 0.02
station)	125–235	45–87	2.05	0.34 ± 0.01	0.24 ± 0.02
	125–235	45–87	2.07	0.32 ± 0.01	0.22 ± 0.02
	205–315	76–116	2.04	0.31 ± 0.01	0.21 ± 0.02
	330–430	122–163	2.55	0.22 ± 0.01	0.12 ± 0.01
	1000–1100	410–454	2.71	0.10 ± 0.01	0
3-5-248	0–100	0–36	2.35	0.51 ± 0.02	0.47 ± 0.02
	100–200	36–73	1.80	0.43 ± 0.02	0.39 ± 0.02
	200–300	73–109	3.55	0.36 ± 0.02	0.32 ± 0.02
	300–400	109–143	1.86	0.18 ± 0.02	0.14 ± 0.01
	400–500	143–178	2.13	0.14 ± 0.02	0.10 ± 0.01
	500–600	178–216	2.57	0.11 ± 0.01	0.07 ± 0.01
	900–1000	338–383	1.85	0.04 ± 0.01	0
3-10-500	0–100	0–34	1.62	0.85 ± 0.03	0.56 ± 0.03
	100–200	34–72	1.79	0.69 ± 0.02	0.38 ± 0.02
	200–300	72–110	2.72	0.51 ± 0.02	0.20 ± 0.02
	300–400	110–144	2.45	0.44 ± 0.01	0.13 ± 0.02
	400–500	144–181	2.88	0.42 ± 0.01	0.11 ± 0.02
	600–700	221–263	1.83	0.38 ± 0.01	0.07 ± 0.02
	900–1000	349–394	1.98	0.29 ± 0.03	0
	900–1000	349–394	1.88	0.31 ± 0.01	0

Station*	Depth Interval, cm	Depth Interval, cm water equiv.	^{210}Bi Procedure		^{210}Po Procedure	
			Sample Weight, kg	^{210}Pb, dph/kg	Sample Weight, kg	^{210}Pb, dph/kg
2-0-0 (Pole of	0–100	0–37	3.21	83 ± 9	1.61	109 ± 15
Relative	100–200	37–77	3.50	66 ± 7	1.86	78 ± 14
Inaccessibility)	200–300	77–119	3.45	49 ± 5	1.81	50 ± 7
	300–400	119–161	6.93	27 ± 3	1.86	27 ± 8
	500–600	206–251	-	-	3.67	13 ± 2
2-11-279	0–100	0–35	1.66	128 ± 13		
	100–200	35–70	1.69	68 ± 18		
	200–300	70–107	7.47	52 ± 3		
	400–500	145–183	7.23	38 ± 4		
	600–700	222–263	6.64	22 ± 4		

*Traverse, station number, mile.

The accumulation values are reported in Table 14. The error represents the standard deviation of the corrected points.

Because of the large background correction, the values of accumulation represent averages over a time interval much shorter than the theoretical 120-year interval spanned by the method. We have evaluated and reported in Table 14 these time intervals for each station as the time elapsed since the deposition of the layer in which the lead 210 activity is not significantly higher than the contamination level.

Discussion

For two stations out of seven, it is not possible to fit a straight line through the corrected values. They are stations 3-5-248 and 3-10-500.

At 3-10-500, the nonlinear array could easily result from a slightly variable contamination from sample to sample. The values for station 3-5-248 are more disturbing, since this is one of the profiles for which the sample at 10-meter depth shows no evidence of contamination above the level of the laboratory blank and for which the background correction was consequently very low. Nevertheless, it is not possible to draw a straight line through the 6 experimental points. However, the observed distribution of lead 210 with depth could be accounted for by an average accumulation of 6.0 g cm^{-2} yr^{-1} between, say, 1953 and 1967 with an abrupt change to 3.5 g cm^{-2} yr^{-1} between approximately 1885 and 1953. Such an important and sudden change in the accumulation rate is difficult to accept. Moreover this interpretation would be in complete disagreement with the distribution of fission products with depth, which indicates clearly an average accumulation of (2.5 ± 0.2) g cm^{-2} yr^{-1} for the time interval 1955–1967. This is the only case where the results are clearly inconsistent with the model described.

At stations 2-0-0 and 2-11-279, for which the samples were not cores stored in the standard cardboard tubes, the lead 210 method yields results in good agreement with the fission products and the surface measurements. For the other stations, the agreement is good despite the large uncertainty due to the background correction. Of special interest is the very low accumulation indicated at station 1-23-620, in good agreement with the fission products distribution and the pit stratigraphy.

COMPARISON WITH SURFACE MEASUREMENTS AND WITH THE STRATIGRAPHY

Direct surface measurements of the accumulation are available at three stations of the South Pole–Queen Maud Land traverse: the Amundsen-Scott South Pole station, the Soviet Union station at the Pole of Relative Inaccessibility, and Plateau station.

Amundsen-Scott South Pole Station

In earlier work [*Picciotto et al.*, 1964], we had shown that accumulation values derived from fission products and lead 210 measurements were in good agreement with the surface measurements made up to December 1962.

Thanks to the cooperation of L. Aldaz, Station Scientific Leader in 1964, one of us (E.P.) was able to compile all the data recorded at the station up to December 1964. They were checked and processed by Mrs. S. Wilgain. A similar compilation up to December 1963 has been reported by *Giovinetto and Schwerdtfeger* [1966]. We report here the results of measurements at two stake networks that are far enough from the station to exclude any influence of the station buildings on the observed accumulation:

- The 'old field' stake network, erected in January 1958, consists of 42 poles placed 400 meters apart in a pentagonal pattern at approximately 5 km grid north of the station.
- The 'new field' stake network, erected in February 1962, is composed of 140 poles 400 meters apart, forming a cross centered on the station. Each branch extends 13 km from the station.

The two networks are measured each summer. The observed average annual accumulations are reported in Table 15. An average density for the upper 20 cm of firn of 0.38 was used [*Cameron et al.*, 1968]. The standard deviation on the individual annual accumulation is ±10 cm, or 3.5 g/cm^2, for both networks.

Pole of Relative Inaccessibility

At the end of SPQMLT 1, a stake network was set up on January 30, 1965. It consisted of 21 poles placed in a cross pattern 250 meters northwest of the station. It was measured on December 14, 1965. The mean accumulation between January 30 and December 14, 1965, was 8.9 cm (standard deviation

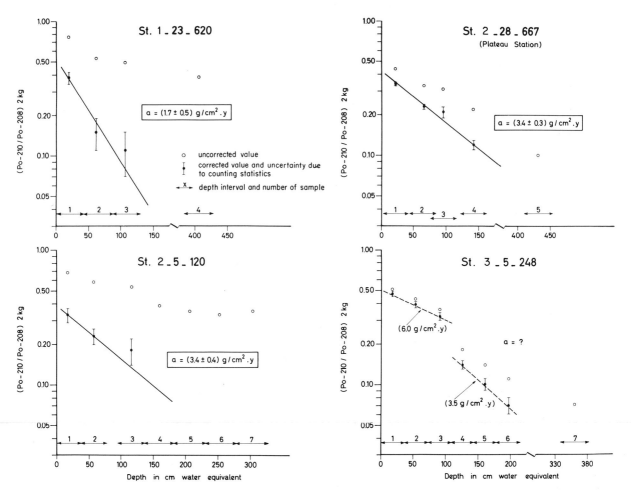

Fig. 10. Profiles of lead 210 activity as a function of depth.

of ±1.4 cm). Assuming that the monthly increment is almost constant throughout the year, and using a measured value of 0.33 for the density of firn at the surface, an annual accumulation of 3.1 g cm^{-2} was derived. The detailed results are reported by *Picciotto et al.* [1968].

Plateau Station

At Plateau station, measurements were made at two stake networks:

The 'Black Field' network, erected in October 1966, consisted of 49 dowels covering a square area of 5000 m² at 400 meters west-northwest of the station. The dowels were measured each month. The average accumulation between December 31, 1966, and December 31, 1967, was 9.11 cm, corresponding to 3.0 g/cm² (R. Dingle, private communication). This value might be slightly too high because of the closeness of the station.

The 'Behling stake network' erected on February 2–4, 1967, consisted of 99 poles disposed in an L pattern along two 10-km long lines. They were measured in December 1966 and January 1967 (by O. Orheim), in December 1967 (by I. Gjessing), and in January 1968 (by E. Picciotto). The following accumulations were observed:

Feb. 2, 1966, to Jan. 21, 1967	8.2 cm, or 2.5 g cm^{-2}
Jan. 21, 1967, to Jan. 17, 1968	6.0 cm, or 1.9 g cm^{-2}

The total average accumulation for the two years 1966 and 1967 was 13.9 cm, corresponding to an annual average of 2.16 g cm^{-2} yr^{-1} (we used an

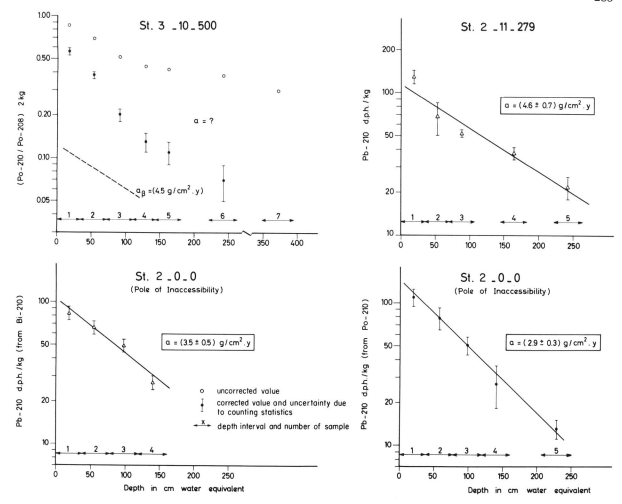

average density of 0.31 measured on the upper 10 cm of the firn cover). The data for the three stations are summarized in Table 15, and are compared with the accumulation values derived from the fission products measurements. The agreement is excellent if allowance is made for the differences in time and location between the two sets of data.

Comparison with Stratigraphy

In each glaciological pit from which samples were collected for fission products measurements, an attempt was made to identify annual layers on the basis of the criteria generally used (mainly ice crusts and depth-hoar layers). It must be stressed that the interpretations of the stratigraphy were made in the field, before the results of the laboratory measurements were known. The results are reported in the Appendix. The corresponding accumulation values computed by counting the number of 'annual layers' between the surface and the depth of the β radioactivity surge are reported in Tables 8, 9, and 10, where they can be compared with the results of the fission product measurements.

The following conclusions can be drawn:

- The identification of annual layers was usually very uncertain. Sometimes it was completely impossible because of either the obliteration of the layering by an intense metamorphism or the presence of multiple ice crusts or depth-hoar layers representing probably several consecutive years [*Gow*, 1965].
- The accumulation values derived from the stratigraphy are systematically higher, indicating a trend for missing years in counting the annual boundaries. The disagreement remains even if one considers only the minimum estimates of the accumulation corresponding to the maxi-

TABLE 14. Accumulation from Lead 210

Station No.	Accumulation ^{210}Pb	Time Interval	Accumulation β*	Time Interval
1-23-620	1.7 ± 0.5	1934–1964	1.7	1955–1964
2-0-0†	2.9 ± 0.3	1885–1965	3.1 ± 0.1	1955–1965
2-5-120	3.4 ± 0.4	1940–1965	3.1 ± 0.2	1955–1965
2-11-279	4.6 ± 0.7	1915–1965	3.6 ± 0.2	1955–1965
2-28-667‡	3.4 ± 0.3	1925–1965	2.7 ± 0.2	1955–1965
3-5-248	Not interpretable		2.5 ± 0.2	1955–1967
3-10-500	Not interpretable		4.5 ± 0.2	1955–1967

*From Tables 8, 9, and 10.
†Pole of Relative Inaccessibility.
‡Plateau station.

TABLE 15. Average Accumulation, in g cm^{-2} yr^{-1}, Derived from Surface Measurements and from the Distribution of Fission Products with Depth

Station	Surface Measurements		Fission Product Measurements	
	Avg. Accum.	Period	Avg. Accum.	Period
Amundsen-Scott South Pole	6.8*	Jan. 1958 to Nov. 1964	6.7 ± 0.2	Jan. 1955 to Dec. 1964
	7.7†	Feb. 1962 to Nov. 1964		
Pole of Relative Inaccessibility‡	3.1	Jan.–Dec. 1965	2.7 ± 0.2	Jan. 1955 to Jan. 1965
			3.1 ± 0.1	Jan. 1955 to Dec. 1965
Plateau	3.0§	Jan.–Dec. 1967	2.7 ± 0.2	Jan. 1955 to Jan. 1968
	2.2‖	Feb. 1966 to Jan. 1968		

*'Old field' stake network.
†'New field' stake network.
‡From Picciotto et al.
§'Black field' (R. Dingle, personal communication).
‖'Behling network.'

TABLE 16. Average Accumulation from Pit Stratigraphy Studies ($\bar{a}_s$) and from Fission Products Measurements ($\bar{a}_\beta$), in g cm^{-2} yr^{-1}

Δ = average deviation = $\Sigma_o^n (a_s - a_\beta)/n$

Traverse	$\bar{a}_s$	$\bar{a}_\beta$	Δ
SPQMLT 1	5.6	4.0	+1.22
SPQMLT 2	3.4	2.9	+0.53
SPQMLT 3*	6.9	4.1	+2.8
SPQMLT 3†	5.1		+1.0

*After *Rundle* [1968].
†After De Breuck's field observations.

mum number of recognizable annual boundaries. Table 16 shows that, even in this case, pit stratigraphy studies lead to accumulation values that are, on average, 25% too high.

An illustration of the subjective character of pit stratigraphy interpretations in this region is given by the disagreement between the two independent sets of results for SPQMLT 3. It is true that the values given by Rundle represent 2- to 5-year averages, whereas De Breuck's values span a longer time interval (as do the values obtained in the present work), but it seems impossible to assign the observed disagreement only to a change in the rate of accumulation with time.

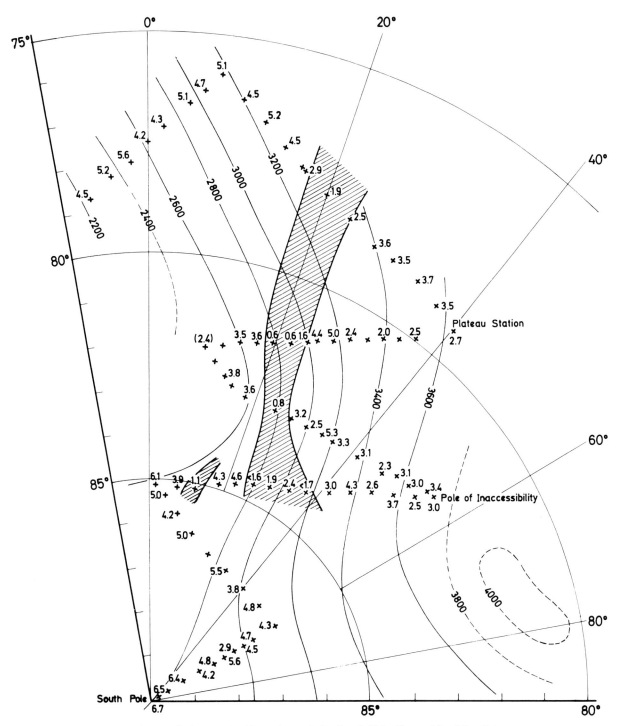

Fig. 11. Accumulation at the 77 stations of the South Pole–Queen Maud Land traverse.

Geographic Distribution

Figure 11 summarizes the results of this work and shows the geographical distribution of the average annual accumulation along the route of the South Pole–Queen Maud Land traverse.

In Figure 12, we have plotted the accumulation

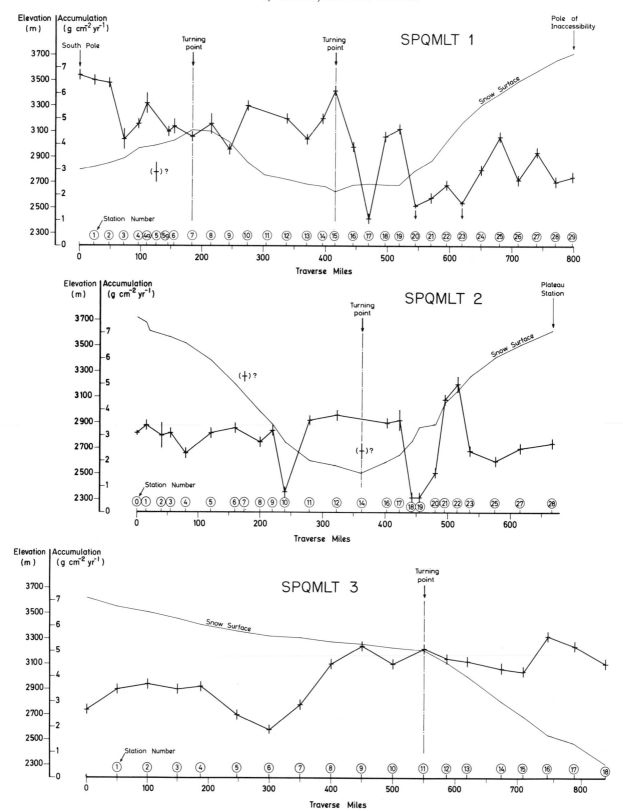

Fig. 12. Surface topography and accumulation.

and the elevation of the station (preliminary figures communicated by the Department of Geology and Geophysics, University of Wisconsin) along the traverse route.

It can be seen that the accumulation is very low over the whole area investigated. The values range from almost zero to 7 g cm^{-2} yr^{-1}. Rather unexpectedly, the highest accumulations are found at the South Pole, which is at a higher elevation and a greater distance from the coast than the last stations of the SPQMLT 3. The South Pole area thus does not seem really representative of the average conditions of the East Antarctic plateau, as far as accumulation is concerned.

Figure 13, in which accumulation is plotted against elevation, shows no general correlation between accumulation and elevation, nor between accumulation and continentality. However, if the points corresponding to an accumulation smaller than 2.5 are discarded, a general trend for an inverse correlation between accumulation and elevation can be recognized.

The discarded points of low accumulation are grouped in a definite geographical pattern. They define an elongated zone cutting across the contour lines and with no visible correlation with the topography (hatched area in Figure 11). There is a faint indication that the controlling factor could be the slope, the station with minimum accumulation being found in the area of maximum slope. Nevertheless, the existence of such a zone of such lower accumulation in an area of uniform and gentle topography remains difficult to explain. Further detailed investigations in this region are needed in order to solve this problem.

Conclusion

In conclusion, it could be said that the objective of this study, to obtain reliable accumulation data on the area investigated by the South Pole–Queen Maud Land traverse, has been achieved successfully, thanks to the method based on the 1955 fallout horizon. This method seems to be the best suited for obtaining a large number of data on traverse operations.

The lead 210 method was not as successful as had been hoped, but the difficulties encountered here should be easily overcome in future studies. In spite

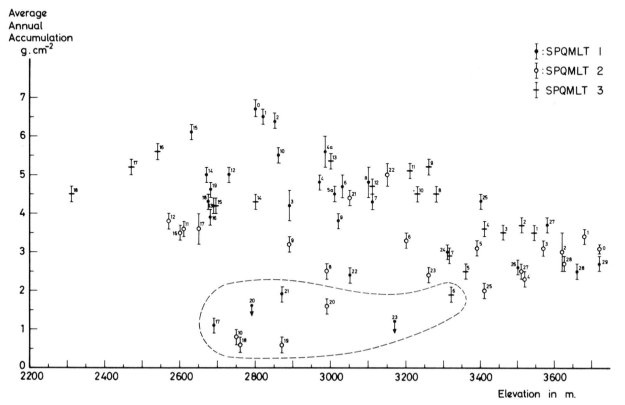

Fig. 13. Accumulation versus elevation.

of its experimental difficulties, the lead 210 method should remain an important tool for measuring average accumulation; it provides an excellent confirmation of data obtained by other methods. It spans a longer time interval than the fallout method, and the simultaneous use of both methods should make it possible to obtain information on the recent variations of the snow accumulation.

Acknowledgments. We wish to thank all the participants of the South Pole–Queen Maud Land traverse for their helpful cooperation in the field. The logistic support afforded by the U.S. Antarctic Research Programs of the National Science Foundation and by the U.S. Navy, Task Force 43, is gratefully acknowledged. We wish also to thank C. Weyers, J. Saufnay, and J. P. Mennessier for assistance in the laboratory, and J. M. Gilot and Mrs. A. Van Keer for their help in preparing the manuscript.

Financial support for this work was provided by the National Science Foundation (through grant GA.0184 to the Université Libre de Bruxelles, and grants GA.530 and GA.1076 to the Ohio State University) and by the Belgian Fonds de la Recherche Fondamentale Collective (program 243).

REFERENCES

Cameron, R. L., E. Picciotto, H. S. Kane, and J. Gliozzi, Glaciology on the Queen Maud Land Traverse 1964–1965, South Pole–Pole of Relative Inaccessibility, *Inst. Polar Studies, Ohio State Univ., Rep. 23, RF 1838*, 136 pp., 1968.

Crozaz, G., Mise au point d'une méthode de datation des glaciers basée sur la radioactivité du plomb-210, Thèse, Université Libre de Bruxelles, 1967a.

Crozaz, G., Datation des glaciers par le plomb-210, in *Radioactive Dating and Methods of Low-Level Counting*, pp. 512–522, International Atomic Energy Agency, Vienna, 1967b.

Crozaz, G., Fission products in Antarctic snow, an additional reference level in January 1965, *Earth Planet. Sci. Lett., 6*, 6–8, 1969.

Crozaz, G., and P. Fabri, Mesure du polonium à l'échelle de 10^{-13} curie, traçage par le Po-208 et application à la chronologie des glaces, *Earth Planet. Sci. Lett., 1*(6), 446–448, 1966.

Crozaz, G., and C. C. Langway, Jr., Dating Greenland firn ice cores with Pb 210, *Earth Planet. Sci. Lett., 1*(4), 194–196, 1966.

Crozaz, G., E. Picciotto, and W. De Breuck, Antarctic snow chronology with Pb 210, *J. Geophys. Res., 69*(12), 2597–2604, 1964.

Crozaz, G., C. C. Langway, Jr., and E. Picciotto, Artificial radioactivity reference horizons in Greenland firn, *Earth Planet. Sci. Lett., 1*(1), 42–48, 1966.

De Breuck, W., Queen Maud Land Traverse III 1967–1968, Glaciology, Snow sampling program of the Center of Polar Studies, Field Rep., Ohio State University, Columbus, 28 pp., 1969.

Dolan, P. J., *Defense Atomic Support Agency Rep. DASA-525*, Washington 25, D.C., 1959.

Federal Radiation Council, Estimates and evaluation of fallout in the United States from nuclear weapons testing conducted through 1962, *Rep. 4*, 1963.

Fowler, J. M., The bombs and their products, in *Fallout*, edited by Fowler, pp. 11–25, Basic Books, New York, 1960.

Friend, J. P., H. W. Feely, P. W. Krey, J. Spar, and A. Walton, The high altitude sampling program, *Defense Atomic Support Agency Rep. DASA-1300*, Washington, D.C., 1961.

Giovinetto, M. B., The drainage systems of Antarctica: Accumulation, in *Antarctic Snow and Ice Studies, Antarctic Res. Ser. 2*, pp. 127–155, AGU, Washington, D.C., 1964.

Giovinetto, M. B., and W. Schwerdtfeger, Analysis of a 200 year snow accumulation series from the South Pole, *Archiv Meteorol. Geophys. Bioklimatol., A, Meteorol. Geophys., 15*(2), 227–250, 1966.

Glasstone, S., *The Effects of Nuclear Weapons*, pp. 671–681b, Atomic Energy Commission, Washington, D.C., 1964.

Goldberg, E. D., Geochronology with lead-210, *Symposium on Radioactive Dating, I.A.E.A.*, Athens, Nov. 1962, pp. 121–130, Vienna, 1963.

Gow, A. J., On the accumulation and seasonal stratification of snow at the South Pole, *J. Glaciol., 5*, 467–478, 1965.

Hallden, N. A., I. M. Fisenne, L. D. Y. Ong, and J. H. Harley, Radioactive decay of weapons debris, *U.S. Atomic Energy Comm., HASL Rep. 117*, pp. 194–199, 1961.

Hanappe, F., M. Vosters, E. Picciotto, and S. Deutsch, Chimie des neiges antarctiques et taux de déposition de matière extraterrestre, 2e article, *Earth Planet. Sci. Lett., 4*(6), 487–496, 1968.

Hardy, E. P., M. W. Meyer, J. S. Allen, and L. T. Alexander, Strontium-90 on the earth's surface, *Nature, 219*(5154), 584–587, 1968.

Harley, N., I. Fisenne, L. D. Y. Ong, and J. Harley, Fission yield and fission product decay, *U.S. Atomic Energy Comm., HASL Rept. 164*, pp. 251–260, 1965.

HASL, Announced nuclear detonations, 1945–1962, United States, United Kingdom, Republic of France, Union of Soviet Socialist Republics, *U.S. Atomic Energy Comm., HASL Rep. 142*, pp. 218–239, 1964.

Katcoff, S., Fission-product yields from neutron-induced fission, *Nucleonics, 18*(11), 201–208, 1960.

Koerner, R. M., and H. S. Kane, Glaciological studies at Plateau station, *Antarctic J. U.S., 2*(4), 122–123, 1967.

Kotlyakov, V. M., Snow cover in the Antarctic and its role in modern glaciation of the continent, *Acad. Sci. USSR, Glaciol. Res.*, Section 9 of the IGY Program, 246 pp., 1961.

Libby, W. F., Current research findings on radioactive fallout, *Proc. Nat. Acad. Sci., 42*, 945–962, 1956.

Libby, W. F., Radioactive fallout particularly from the Russian October series, *Proc. Nat. Acad. Sci., 45*, 959–976, 1959.

Lockhart, L. B., Jr., R. L. Patterson, Jr., and A. W. Saunders, Jr., Atmospheric radioactivity in Antarctica 1956–1963, *U.S. Naval Res. Lab. Rep. NRL 6341*, 1965.

Machta, L., *Statement to the Joint Committee on Atomic Energy, U.S. Congress*, part I, pp. 54–81, U.S. Government Printing Office, Washington, D.C., 1962.

Martell, E. A., Atmospheric aspects of Sr-90 fallout, *Science*, *129*(3357), 1197–1206, 1959.

Nezami, M., G. Lambert, C. Lorius, and J. Labeyrie, Mesure du taux d'accumulation de la neige au bord du Continent Antarctique par la méthode du plomb 210, *Compt. Rend., 259*, 3319–3322, 1964.

Peddie, N. W., South Pole–Queen Maud Land Traverse III, *Antarctic J. U.S., 3*(4), 93–95, 1968.

Picciotto, E., The South Pole–Queen Maud Land Traverse II, 1965–1966, *Antarctic J. U.S., 1*(4), 129–131, 1966.

Picciotto, E., Geochemical investigations of snow and firn samples from East Antarctica, *Antarctic J. U.S., 2*(6), 236–240, 1967.

Picciotto, E., and S. Wilgain, Fission products in Antarctic snow, a reference level for measuring accumulation, *J. Geophys. Res., 68*(21), 5965–5972, 1963.

Picciotto, E., G. Crozaz, and W. De Breuck, Rate of accumulation of snow at the South Pole as determined by radioactive measurements, *Nature, 203*(4943), 393–394, 1964.

Picciotto, E., S. Wilgain, P. Kipfer, and R. Boulenger, Radioactivité de l'air dans l'Antarctique en 1958 et profil radioactif entre 60°N et 70°S, in *Radioisotopes in the Physical Sciences and Industry*, pp. 45–56, International Atomic Energy Agency, Vienna, 1962.

Picciotto, E., G. Crozaz, W. Ambach, and H. Eisner, Lead-210 and strontium-90 in an alpine glacier, *Earth Planet. Sci. Lett., 3*(3), 238–242, 1967.

Picciotto, E., R. L. Cameron, G. Crozaz, S. Deutsch, and S. Wilgain, Determination of the rate of snow accumulation at the Pole of Relative Inaccessibility, Eastern Antarctica: A comparison of glaciologic and isotopic methods, *J. Glaciol., 7*(50), 273–287, 1968.

Rundle, A., Snow stratigraphy and accumulation (South Pole–Queen Maud Land Traverse III), *Antarctic J. U.S., 3*(4), 95–96, 1968.

Sharp, R. P., and S. Epstein, Comments on annual rates of accumulation in West Antarctica, paper presented at the IUGG Symposium, Intern. Assoc. Sci. Hydrol., Committee of Snow and Ice, Obergurgl, September 1962.

Tomkins, C., *Statement to the Joint Committee on Atomic Energy, U.S. Congress*, part 1, pp. 4–27, U.S. Government Printing Office, Washington, D.C., 1963.

Vickers, W. W., Geochemical dating techniques applied to Antarctic snow samples, General Assembly, Berkeley, August 1963, *Intern. Assoc. Sci. Hydrol. Publ. 61*, 199–215, 1963.

Volchok, H. L., and M. T. Kleinman, Strontium 90 yield of the 1967 Chinese thermonuclear explosion, *J. Geophys. Res., 74*(6), 1694–1696, 1969.

Wilgain, S., E. Picciotto, and W. De Breuck, Strontium 90 fallout in Antarctica, *J. Geophys. Res., 70*(24), 6023–6032, 1965.

Windom, H. L., Atmospheric dust records in premanent snowfields: Implications to marine sedimentation, *Geol. Soc. Amer. Bull., 80*, 761–782, 1969.

Woodward, R. N., Strontium 90 and caesium 137 in Antarctic snows, *Nature, 204*(4965), 1291, 1964.

APPENDIX

The following graphs show the stratigraphy and gross β activity at the 77 stations of the SPQMLT as a function of depth.

∧∧∧∧∧∧ depth-hoar layer

――― icy crust

········· bonded-grain layer

⦵⦵⦵⦵ strongly metamorphosed firn; loose aggregate of large (1 to 5 mm) angular crystals

▨▨▨▨ very hard fine-grained layer

s soft

m medium hard

h hard

vh very hard

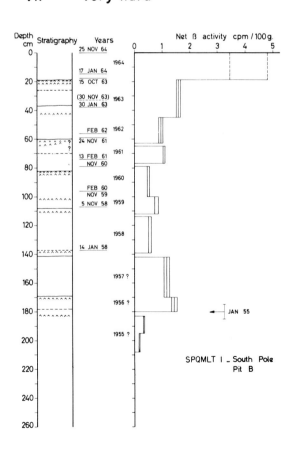

SPQMLT I – South Pole Pit B

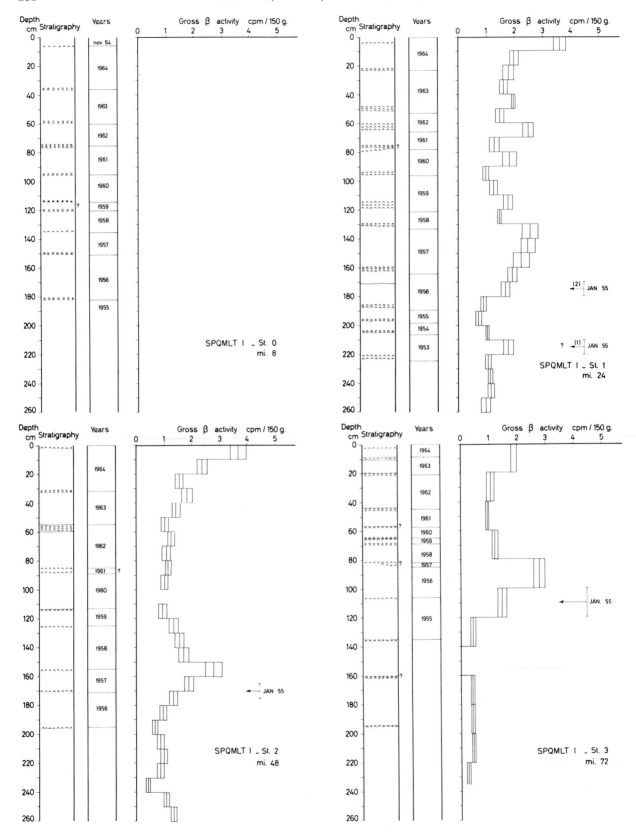

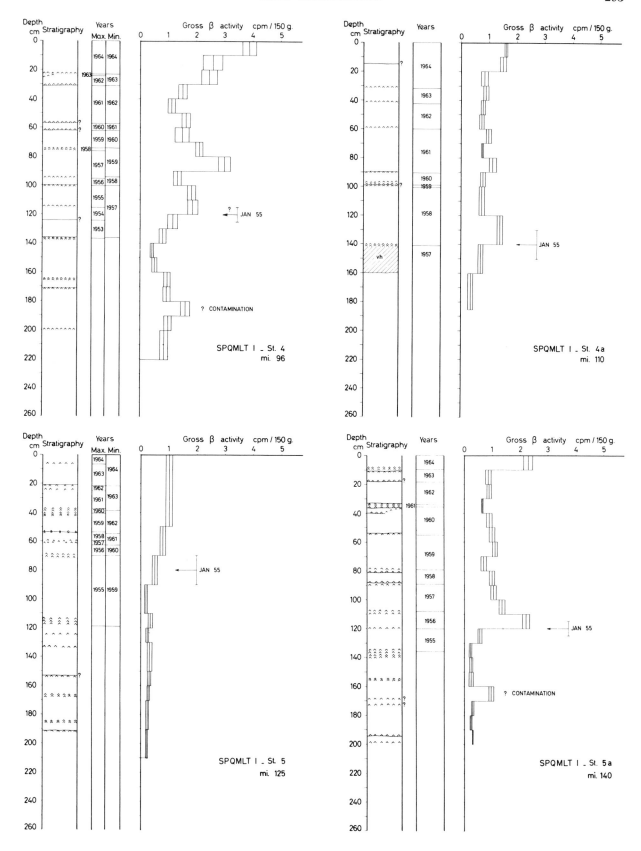

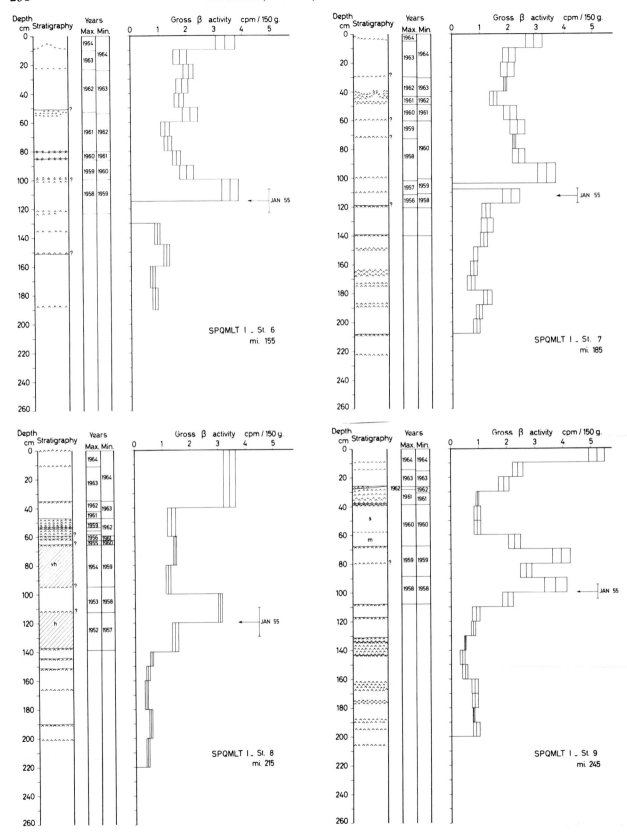

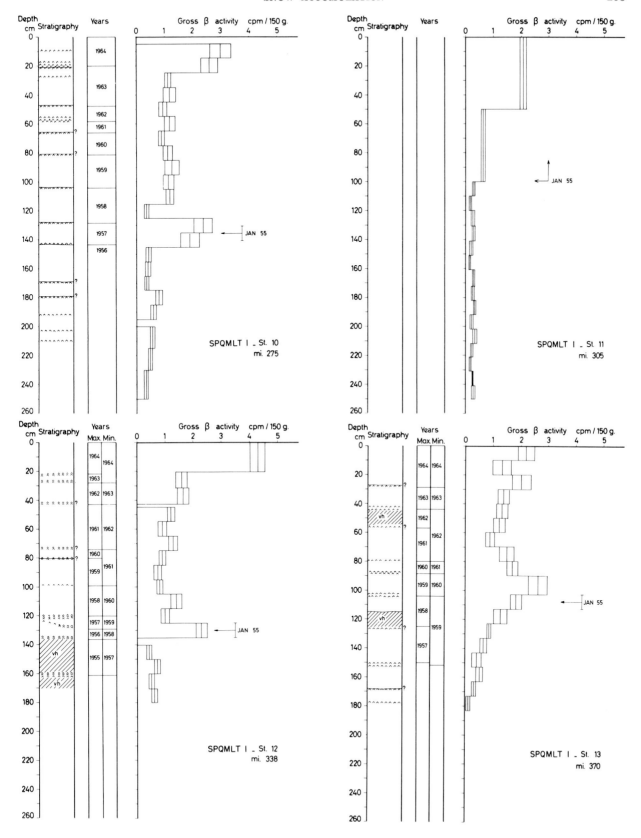

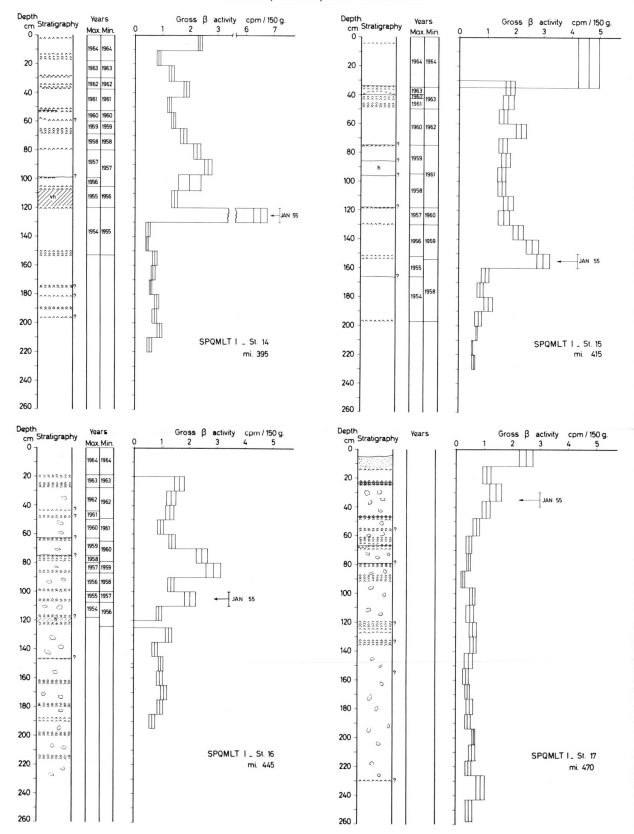

SNOW ACCUMULATION

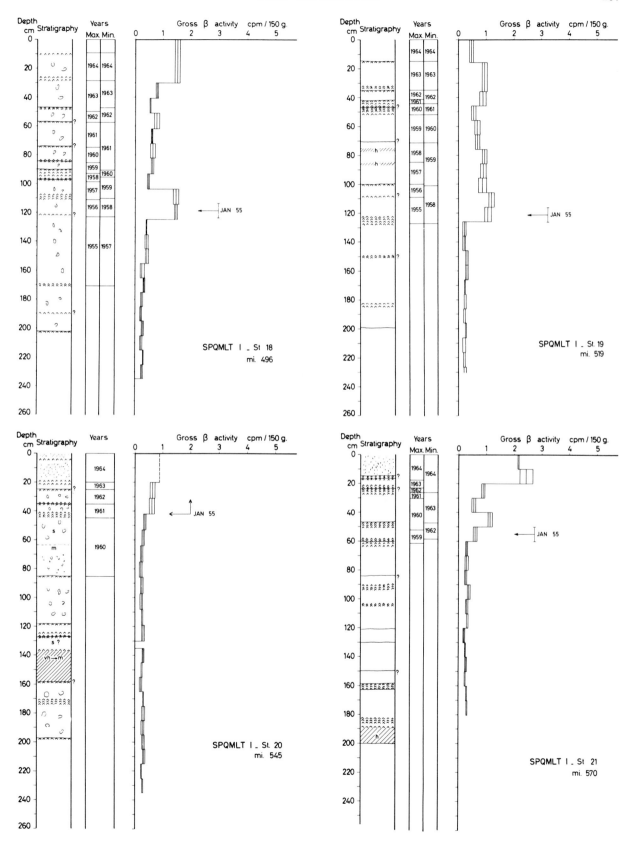

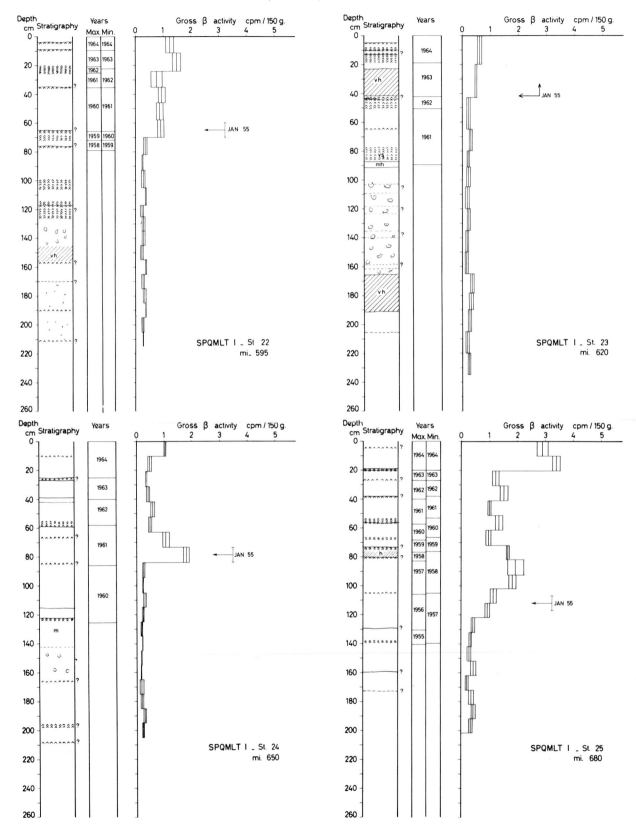

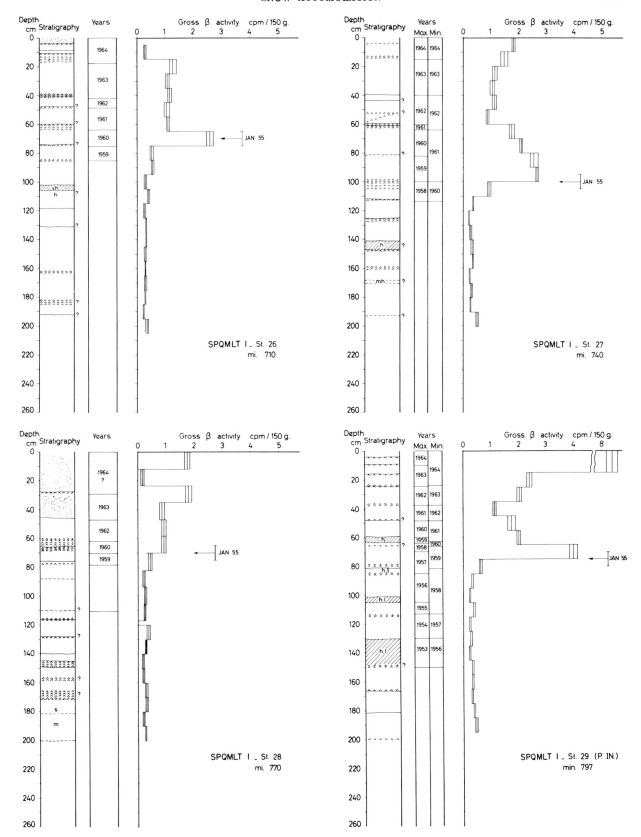

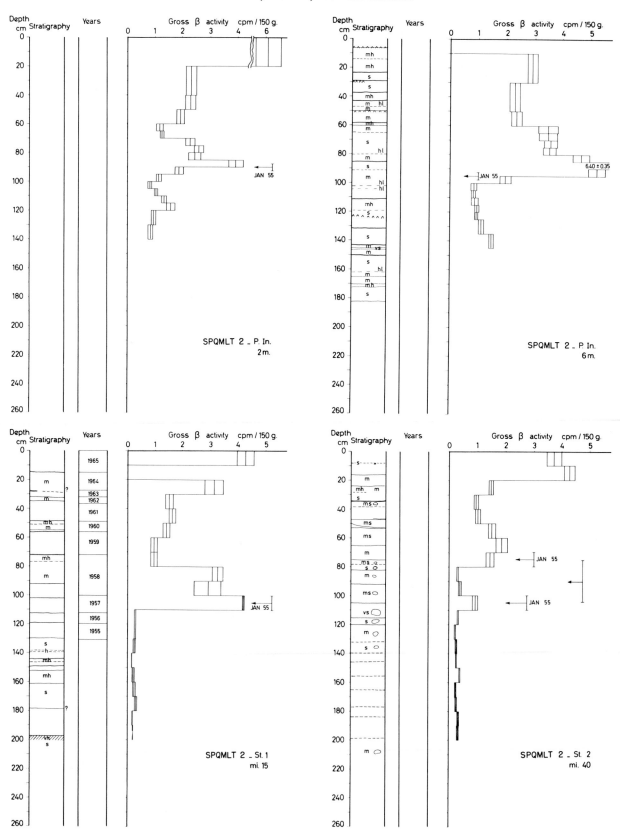

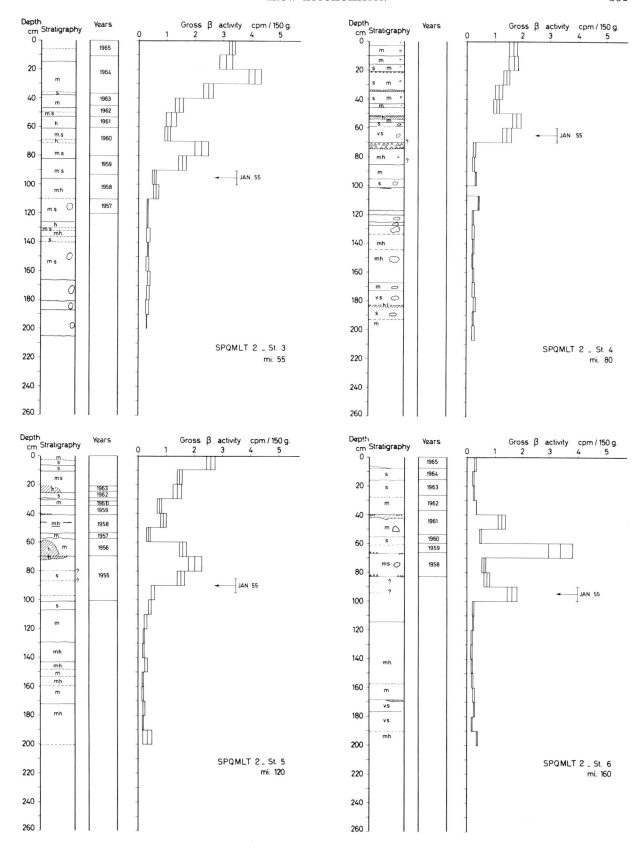

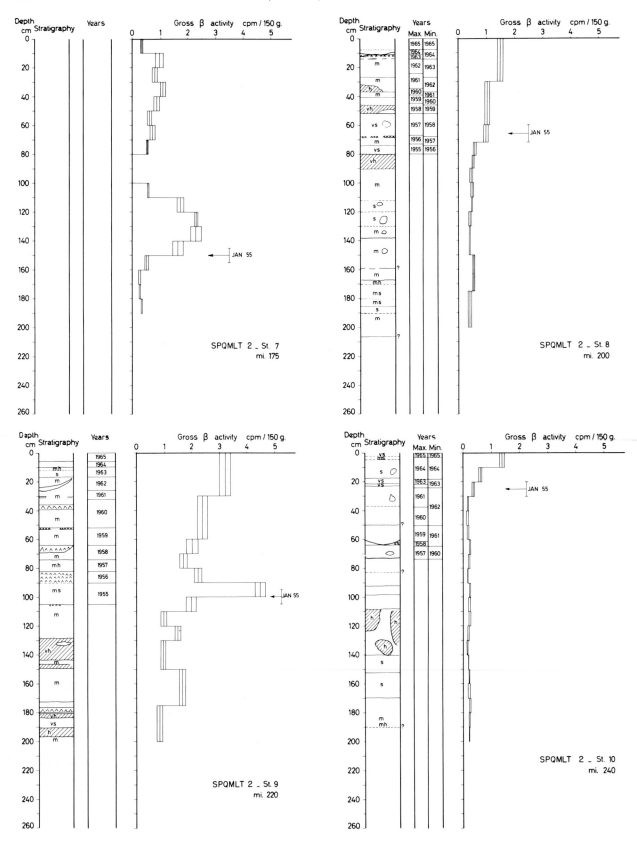

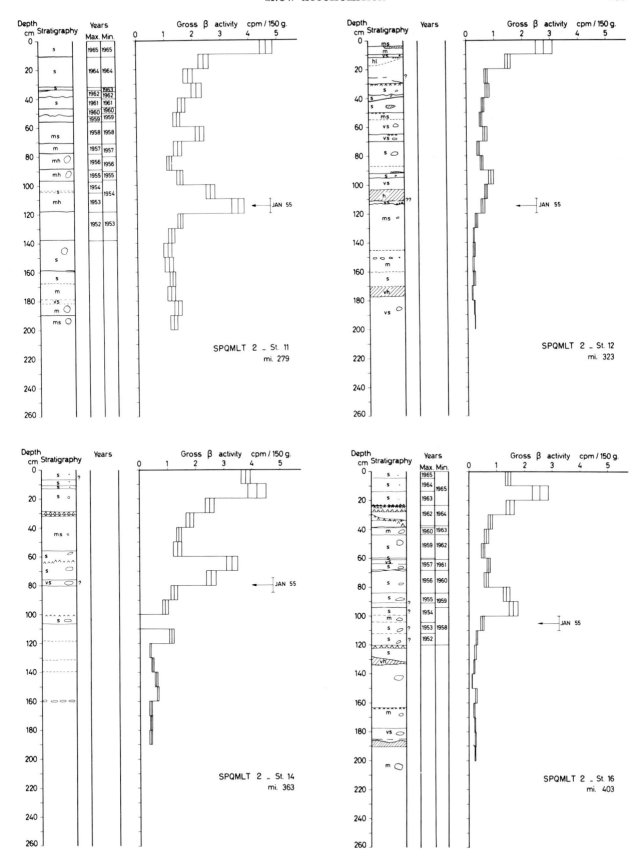

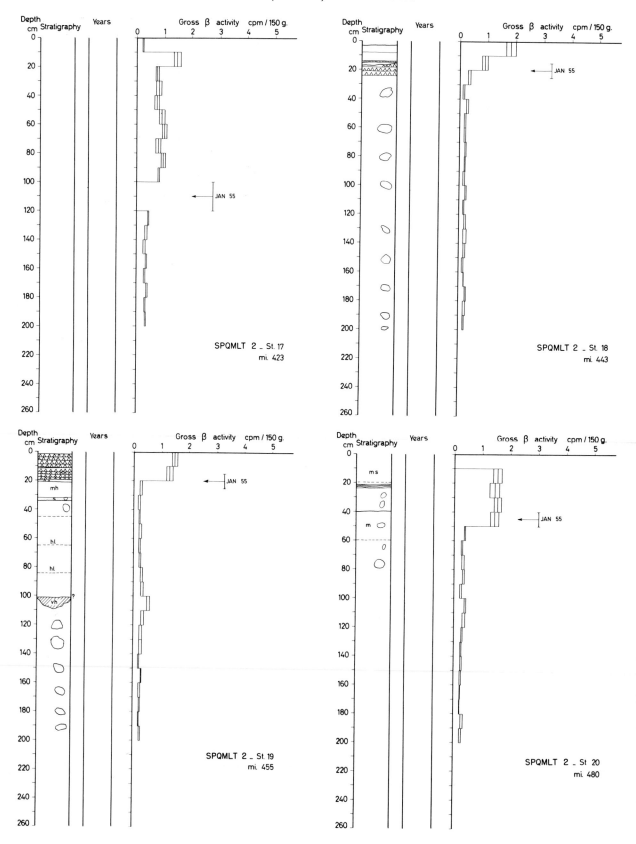

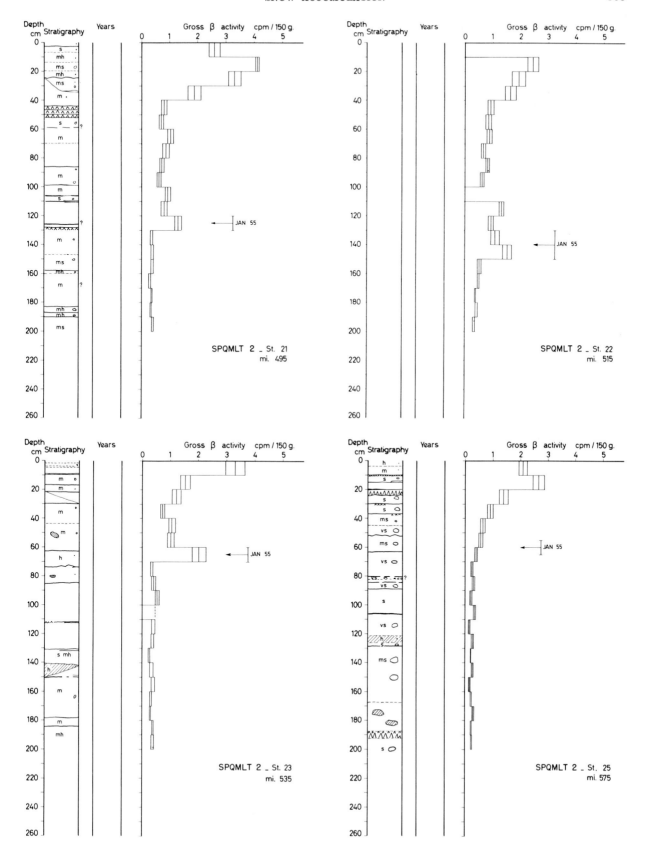

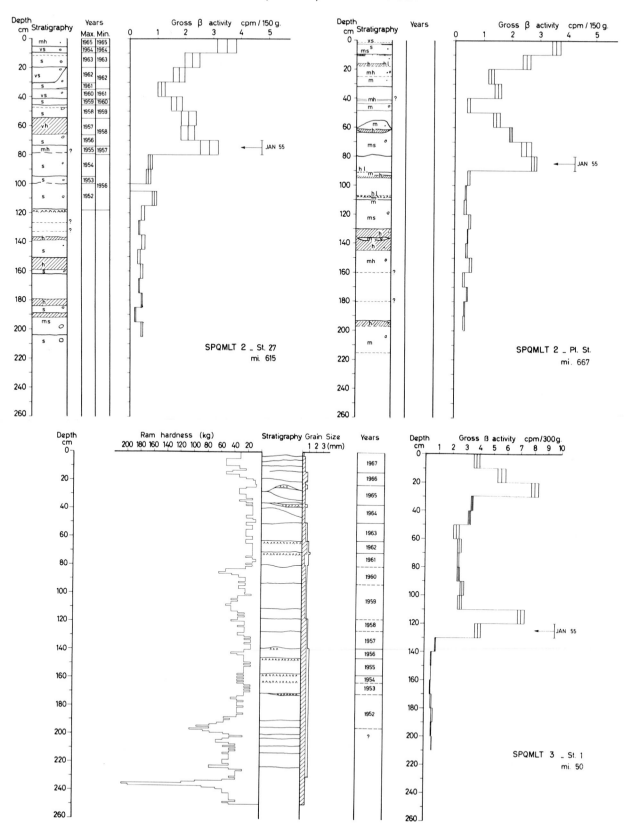

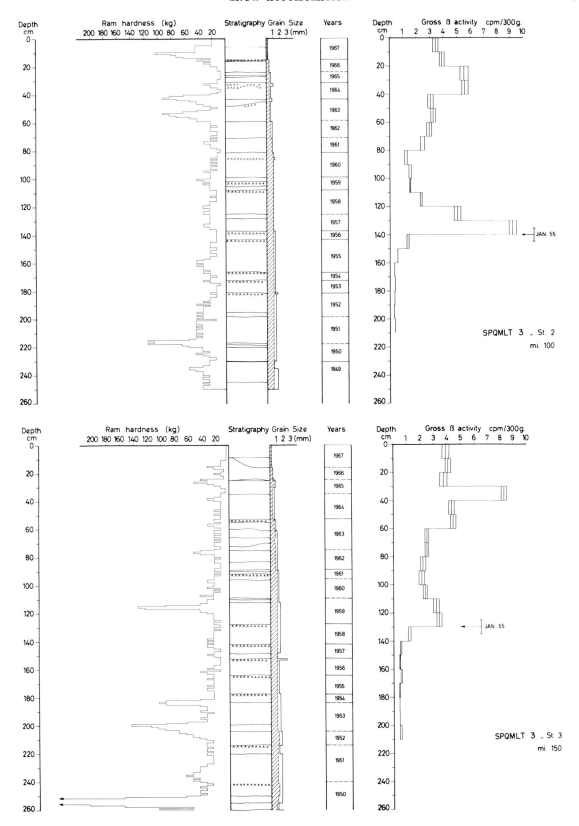

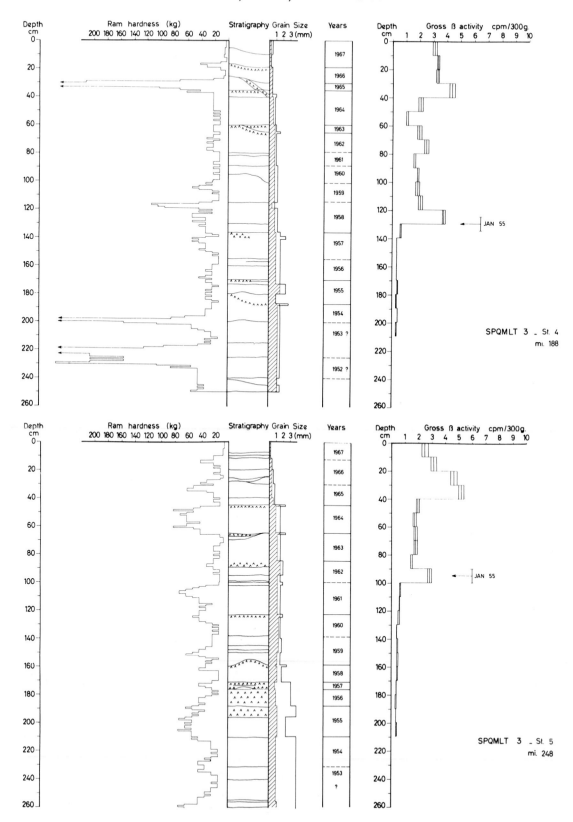

SNOW ACCUMULATION

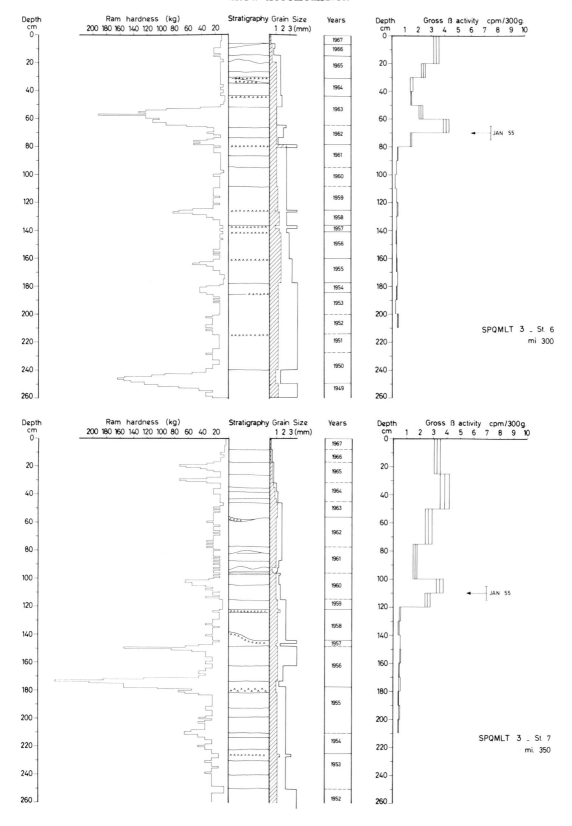

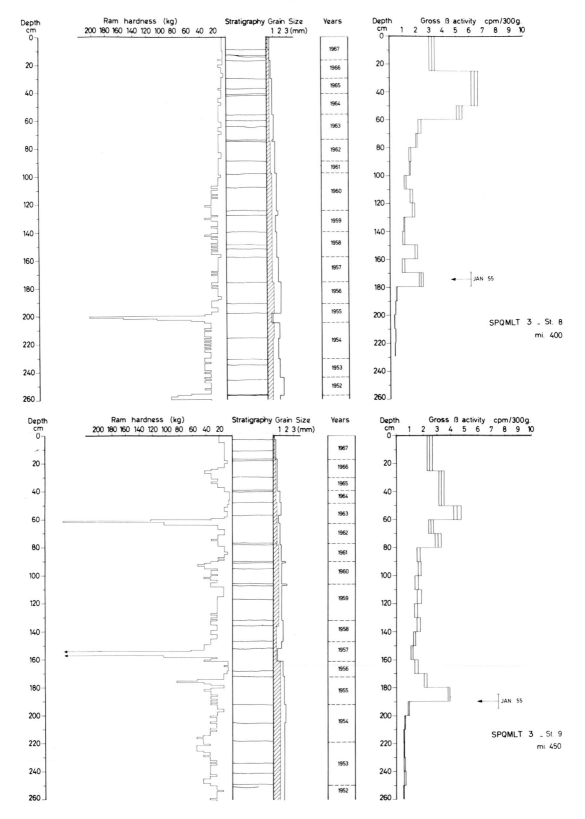

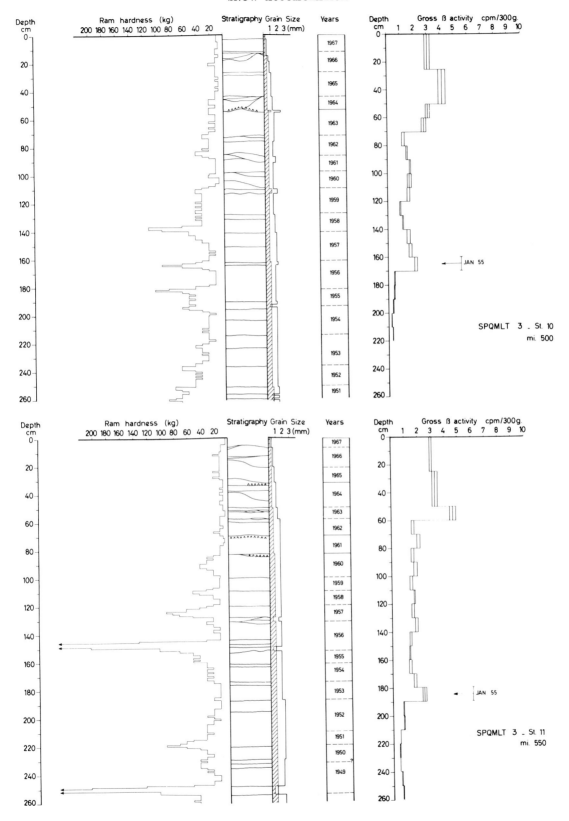

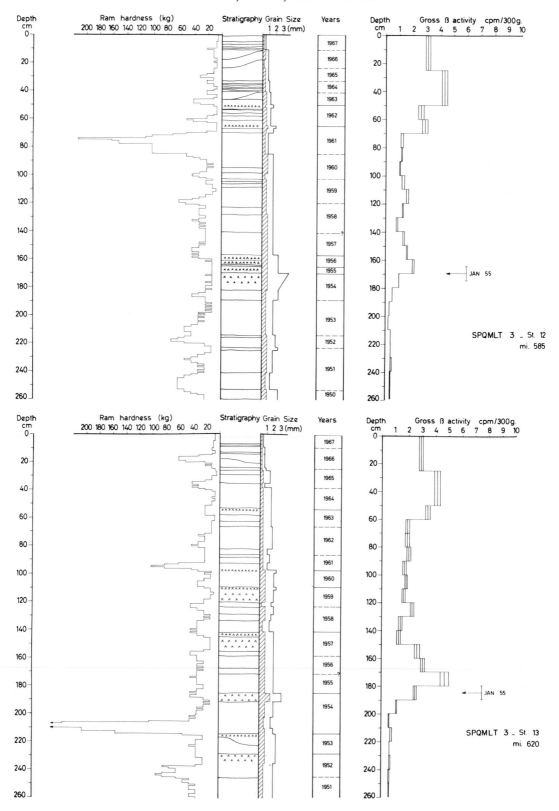

313

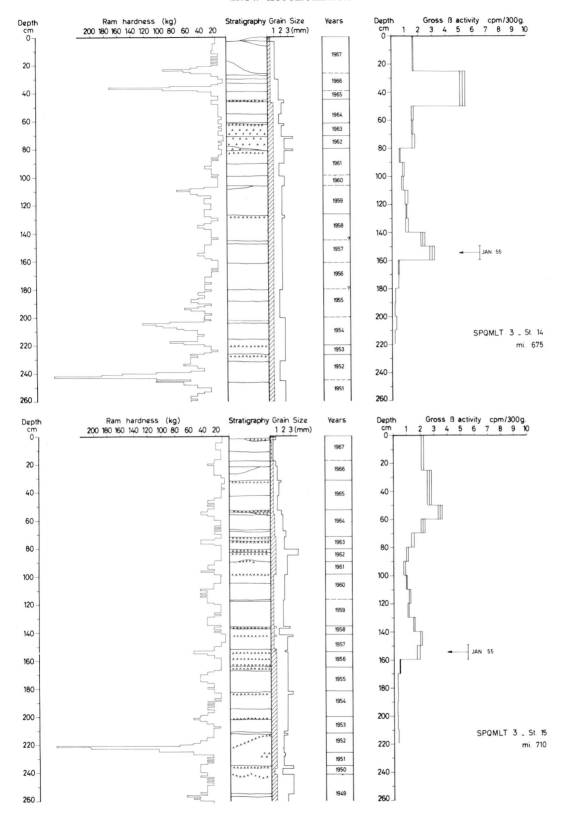

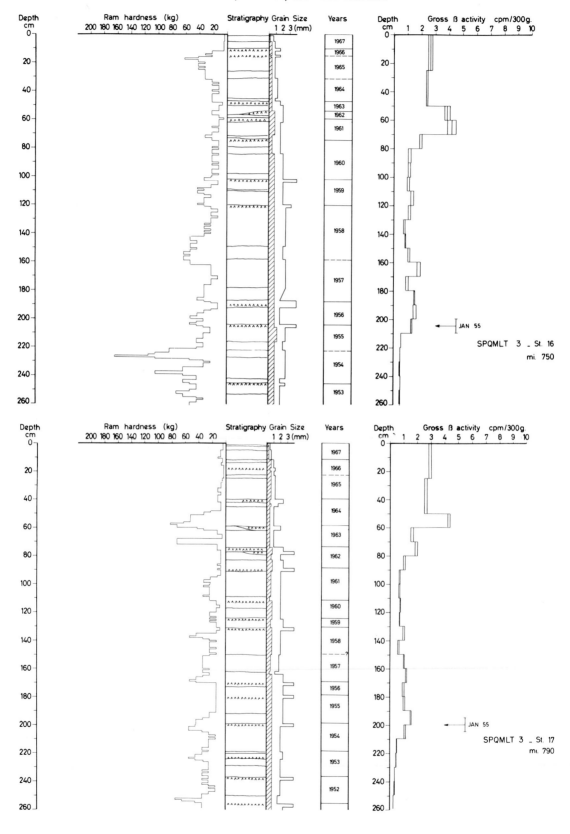

SNOW ACCUMULATION

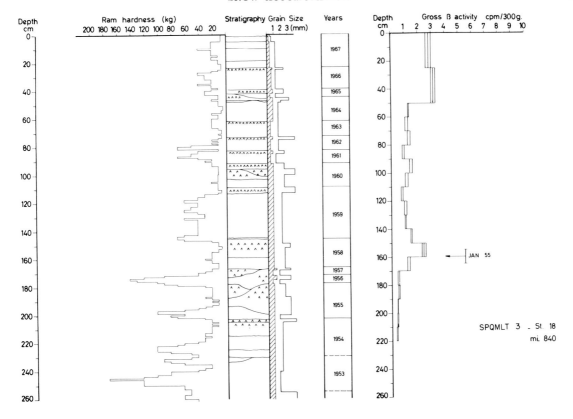

GLACIOLOGICAL STUDIES AT BYRD STATION, ANTARCTICA, 1963–1965

Richard L. Cameron

Institute of Polar Studies, Ohio State University, Columbus 43210

A series of snow studies initiated at new Byrd station in 1963 and continued through 1965 included a 100-stake accumulation net, two excavations in the firn, an elongated pit, and a deep pit. In 1964 and 1965, an accumulation rate of 10.3 g cm^{-2} yr^{-1} was recorded at the stakes; in the 14-meter-long pit, the accumulation rate for the period 1959 to 1963 was 13.7 g cm^{-2} yr^{-1}. On the wall of the long pit, 49% of the individual firn layers could be traced 14 meters, 31% could be traced 7 meters or more, and the remaining 20% were discontinuous. In the 10-meter-deep pit, vertical and horizontal strain rates were studied; the average vertical strain rate was 0.01032 yr^{-1}. Calculation of the accumulation rate, using Sorge's law and the measured vertical strain rates, gave 10.4 g cm^{-2} yr^{-1} for the period 1925 to 1964. Snow accumulation at Byrd station increased from 1948 to 1958 and has since been decreasing.

In January 1968 the first core hole to penetrate the bottom of the antarctic ice sheet was drilled at Byrd station through 2164 meters of ice. It was anticipated that the interpretation of the history of the firn and ice in the core would be based on the present conditions for snow accumulation at Byrd station and the areas up the flow line from the station.

In preparation for this deep drilling project, surface glaciological studies were initiated by the Institute of Polar Studies at Ohio State University and the Topographic Division of the U.S. Geological Survey. These studies included snow accumulation and the settling rate of the firn at Byrd station, and snow accumulation and surface strain up the flow line from the station. This report gives the results of the work at Byrd station for the period December 1963 to December 1965.

Snow accumulation was measured at 100 stakes set up over 1 km^2; an elongate pit was excavated to determine the snow accumulation over the recent past and to determine the feasibility of tracing specific firn layers for any considerable distance; and a 10-meter pit was excavated and instrumented to measure the settling of the firn and the rate of pit closure. The location of these sites relative to Byrd station is given in Figure 1.

SQUARE KILOMETER ACCUMULATION NET

Snow accumulation was measured at 100 bamboo poles spaced 100 meters apart in a net covering 1 km^2. The poles are topped with black flags and a saw mark was made on each pole two meters above the snow surface on December 17, 1963. Figure 2 is a topographic map of the square kilometer, showing a southeastward slope with a relief of more than 5 meters.

Four sets of measurements of the stake field are given in this paper, the most recent on December 17, 1965. The data are given in Table 1. Figure 3 is a plot of the cumulative accumulation. Using 0.35 g cm^{-3} as the snow density, the average annual accumulation for the period December 17, 1963, to December 17, 1965, is 10.3 g cm^{-2}. This can be considered the average annual accumulation rate for 1964 and 1965.

To examine the distribution of accumulation over this square kilometer, it was divided into quadrants, each of 25 poles (Figure 4), and the mean accumulation and the variance over the same two-year period have been computed (Table 2).

By an analysis of variance, it can be shown that quadrants I and IV have less than 5% probability of representing different populations. By contrast, quadrants II and III have a 90% probability of representing different populations. The accumulation is greater and has a lower variability where the wind is normal to the slope, as occurs in quadrant II. In much of the areas of quadrants I and IV, where the means and variances are very similar, the slope is normal to the prevailing wind. Quadrant III stands out as having a low mean accumulation and

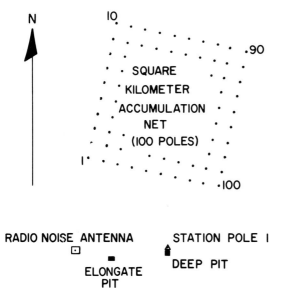

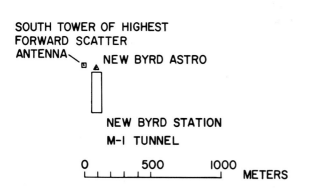

Fig. 1. Index map showing accumulation net, elongate and deep pits in relation to Byrd station.

of velocity or of stress on the surface would tend to keep the snow in motion and would erode the surface. *Schytt* [1958, p. 103] has shown a similar relationship of high wind speed and low accumulation on various slopes inland of Maudheim.

In a study of the distribution of snow accumulation near Byrd station, *Gow* [1965], using two accumulation lines 10 km long consisting of an east-west line with 21 poles and a north-south line with 22 poles, showed 'a strong topographic control on accumulation rates.' On his east-west line, in which there was a relief of about 30 meters and several marked troughs, there was more accumulation in the troughs. However, on the north-south line, with a relief of 60 meters, the high accumulation in two notable instances was on topographic highs or ridges, and immediately downwind of the highs there was a decrease in the accumulation rate. In the Wilkes station area, *Black and Budd* [1964] have also shown a decrease of accumulation rates downwind from the top of crests.

It can be concluded that a major local factor in the distribution of accumulation is the orientation of the slope relative to the prevailing winds.

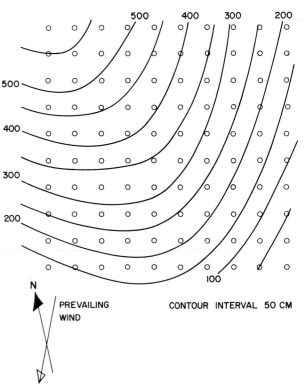

Fig. 2. The topography of the accumulation net.

a high variance, and here the slope is parallel to the prevailing wind.

These data suggest that the distribution of accumulation over the square kilometer is controlled by the directions of slope and prevailing wind. The low accumulation and high variability of quadrant III is probably caused by the increase in wind velocity as the air moves down a slope. This slight increase

TABLE 1. Snow Accumulation at the Square Kilometer for the Period December 17, 1963, to December 17, 1965*

Stake	Nov. 10, 1964	Δ	Feb. 2, 1965	Δ	Nov. 3, 1965	Δ	Dec. 17, 1965	Δ	ΣΔ
1	167.0	33.0	164.0	3.0	139.5	24.5	138.5	1.0	61.5
2	171.0	29.0	178.0	-7.0	152.5	25.5	145.0	7.5	55.0
3	190.0	10.0	174.0	16.0	151.0	23.0	152.0	-1.0	48.0
4	161.0	39.0	158.0	3.0	136.5	21.5	137.5	-1.0	62.5
5	178.0	22.0	171.0	7.0	148.0	23.0	142.5	5.5	57.5
6	171.0	29.0	159.0	12.0	134.0	25.0	130.0	4.0	70.0
7	179.5	20.5	173.0	6.5	148.5	24.5	150.5	-2.0	49.5
8	175.0	25.0	167.0	8.0	144.0	23.0	137.5	6.5	62.5
9	175.5	24.5	177.0	-1.5	142.5	34.5	144.5	-2.0	55.5
10	174.5	25.5	164.5	10.0	133.5	31.0	130.5	3.0	69.5
11	174.5	25.5	172.5	2.0	141.5	31.0	156.0	-14.5	44.0
12	186.5	13.5	177.0	9.5	147.5	29.5	148.0	-0.5	52.0
13	169.5	30.5	167.0	2.5	143.0	24.0	144.0	-1.0	56.0
14	174.0	26.0	172.0	2.0	148.5	23.5	145.5	3.0	54.5
15	170.0	30.0	170.0	0.0	147.5	22.5	149.0	-1.5	51.0
16	203.5	-3.5	193.5	10.0	141.0	52.5	139.5	1.5	57.5
17	187.0	13.0	181.5	5.5	146.5	35.0	144.5	2.0	55.5
18	162.0	38.0	156.5	5.5	140.5	16.0	131.0	9.5	68.5
19	184.0	16.0	174.5	9.5	153.5	21.0	154.0	-0.5	46.0
20	170.5	29.5	164.5	6.0	148.5	16.0	147.0	1.5	53.0
21	174.5	25.5	174.0	0.5	154.5	19.5	152.0	2.5	48.0
22	176.0	24.0	176.0	0.0	152.5	23.5	143.0	9.5	57.0
23	172.0	28.0	170.5	1.5	144.5	26.0	145.0	-0.5	55.0
24	181.0	19.0	177.5	3.5	156.5	21.0	152.5	4.0	47.5
25	179.5	20.5	172.5	7.0	147.0	25.5	144.5	2.5	55.5
26	166.5	33.5	167.5	-1.0	142.5	25.0	133.0	9.5	66.5
27	181.5	18.5	165.5	16.0	145.5	20.0	139.5	6.0	60.5
28	175.0	25.0	163.5	11.5	133.5	30.0	132.5	1.0	67.5
29	178.5	21.5	174.5	4.5	152.5	21.5	138.0	14.5	62.0
30	178.0	22.0	159.5	18.5	140.0	19.5	136.0	4.0	64.0
31	176.5	23.5	166.5	10.0	141.5	25.0	137.0	4.5	63.0
32	181.0	19.0	180.5	0.5	151.0	29.5	153.0	-2.0	47.0
33	177.5	22.5	174.5	3.0	131.5	43.0	135.5	-4.0	64.5
34	174.0	26.0	164.5	9.5	145.5	19.0	141.5	4.0	58.5
35	180.5	19.5	167.5	13.0	149.0	18.5	143.5	5.5	56.5
36	172.5	27.5	159.5	13.0	126.5	33.0	122.0	4.5	77.5
37	181.5	18.5	170.0	11.5	145.0	25.0	146.0	-1.0	54.0
38	173.5	26.5	166.0	7.5	142.0	24.0	136.5	5.5	63.5
39	179.0	21.0	175.5	3.5	153.5	22.0	151.5	2.0	48.5
40	183.5	16.5	177.5	6.0	155.0	22.5	153.0	2.0	47.0
41	176.0	24.0	195.5	4.5	173.5	22.0	182.0	-8.5	42.0
42	174.0	26.0	175.0	-1.0	155.0	20.0	148.5	6.5	51.5
43	190.0	10.0	180.0	10.0	161.0	19.0	161.0	0.0	39.0
44	176.5	23.5	174.0	2.5	141.5	32.5	137.5	4.0	62.5
45	186.5	13.5	176.0	10.5	158.5	17.5	154.5	4.0	45.5
46	171.0	29.0	158.5	12.5	139.5	19.0	134.0	5.5	66.0
47	181.5	18.5	181.0	0.5	150.0	31.0	146.0	4.0	54.0
48	180.0	20.0	171.0	9.0	144.5	26.5	142.0	2.5	58.0
49	162.5	37.5	161.0	1.5	139.5	21.5	136.5	3.0	63.5
50	168.0	32.0	161.5	6.5	139.5	22.0	139.0	0.5	61.0

TABLE 1. (continued)

Stake	Nov. 10, 1964	Δ	Feb. 2, 1965	Δ	Nov. 3, 1965	Δ	Dec. 17, 1965	Δ	ΣΔ
51	174.0	26.0	165.5	8.5	138.0	27.5	134.5	3.5	65.5
52	164.5	35.5	155.5	9.0	142.0	13.5	142.0	0.0	58.0
53	172.5	27.5	161.5	11.0	136.0	25.5	135.5	0.5	64.5
54	178.0	22.0	177.0	1.0	145.0	32.0	140.0	5.0	60.0
55	173.5	26.5	166.0	7.5	136.5	29.5	138.0	-1.5	62.0
56	177.0	23.0	176.0	1.0	136.0	40.0	131.5	4.5	68.5
57	179.0	21.0	173.5	5.5	148.0	25.5	148.5	-0.5	51.5
58	169.0	31.0	160.5	8.5	139.0	21.5	131.5	7.5	68.5
59	179.0	21.0	175.0	4.0	148.0	27.0	145.5	2.5	54.5
60	183.5	16.5	160.0	23.5	155.0	5.0	153.0	2.0	47.0
61	167.0	33.0	167.0	0.0	144.5	22.5	146.0	-1.5	54.0
62	180.5	19.5	177.0	3.5	146.0	31.0	147.0	-1.0	53.0
63	173.5	26.5	171.5	2.0	138.0	33.5	141.0	-3.0	59.0
64	167.0	33.0	163.5	3.5	128.5	35.0	128.0	0.5	72.0
65	183.0	17.0	171.0	12.0	145.0	26.0	135.0	10.0	65.0
66	163.0	37.0	160.5	2.5	138.0	22.5	125.0	13.0	75.0
67	179.5	20.5	167.5	12.0	138.0	29.5	134.0	4.0	66.0
68	190.5	9.5	178.0	12.5	137.5	40.5	135.0	2.5	65.0
69	176.0	24.0	169.0	7.0	143.5	25.5	139.0	4.5	61.0
70	175.0	25.0	173.5	1.5	143.0	30.5	136.5	6.5	63.5
71	173.0	27.0	170.5	2.5	138.5	32.0	134.0	4.5	66.0
72	162.0	38.0	161.5	0.5	132.0	29.5	129.5	2.5	70.5
73	176.0	24.0	175.5	0.5	138.0	37.5	137.0	1.0	63.0
74	181.5	18.5	166.5	15.0	140.5	26.0	140.0	0.5	60.0
75	182.0	18.0	176.0	6.0	139.5	36.5	139.0	0.5	61.0
76	178.5	21.5	177.5	1.0	146.5	31.0	147.0	-0.5	53.0
77	191.0	9.0	190.0	1.0	149.5	40.5	150.0	-0.5	50.0
78	178.5	21.5	164.5	14.0	150.0	14.5	143.5	6.5	56.5
79	180.5	19.5	170.0	10.5	145.0	25.0	147.0	-2.0	53.0
80	176.5	23.5	168.0	8.5	148.0	20.0	147.5	0.5	52.5
81	179.5	20.5	170.5	9.0	156.0	14.5	152.0	4.0	48.0
82	172.5	27.5	175.0	-2.5	143.5	31.5	141.0	2.5	59.0
83	173.5	26.5	165.0	8.5	141.0	24.0	140.0	1.0	60.0
84	177.0	23.0	167.0	10.0	142.0	25.0	138.5	3.5	61.5
85	176.0	24.0	159.5	16.5	135.5	24.0	133.5	2.0	66.5
86	173.5	26.5	171.5	2.0	140.5	31.0	138.5	2.0	61.5
87	172.0	28.0	170.5	1.5	139.0	31.5	136.5	2.5	63.5
88	165.0	35.0	164.5	0.5	139.0	25.5	137.0	2.0	63.0
89	165.0	35.0	167.5	-2.5	139.5	28.0	136.0	3.5	64.0
90	163.0	37.0	166.0	-3.0	136.5	29.5	126.5	10.0	73.5
91	160.5	39.5	164.0	-3.5	126.5	37.5	124.0	2.5	76.0
92	163.5	36.5	161.0	2.5	131.0	30.0	129.0	2.0	71.0
93	167.0	33.0	162.5	4.5	130.5	32.0	131.5	-1.5	68.0
94	177.5	22.5	172.5	5.0	143.5	29.0	141.0	2.5	59.0
95	167.5	32.5	169.5	-2.0	141.5	28.0	141.5	0.0	58.5
96	169.0	31.0	171.0	-3.0	134.5	36.5	136.0	-1.5	64.0
97	175.0	25.0	174.0	1.0	143.0	31.0	139.0	4.0	61.0
98	182.0	18.0	170.0	12.0	149.0	21.0	148.0	1.0	52.0
99	175.5	24.5	170.5	3.0	145.0	25.5	143.5	1.5	56.5
100	179.5	20.5	169.0	10.5	143.0	26.0	141.5	1.5	58.5

*The stakes were marked at 200 cm above the surface on December 17, 1963. The columns headed by a date give the measured distance, in centimeters, between the critical mark and the snow surface. The columns headed by Δ give the change in level of the surface (accumulation or ablation) between measurements. ΣΔ is the over-all accumulation of snow at the stake for the entire period.

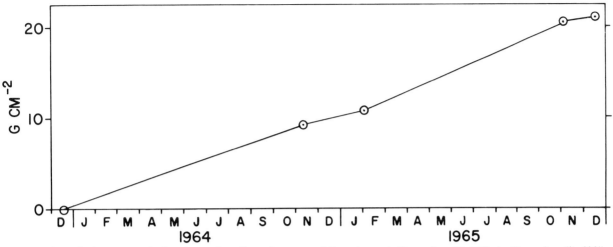

Fig. 3. Cumulative accumulation (g cm^{-2} yr^{-1}) at the square kilometer net, December 17, 1963, to December 17, 1965.

ELONGATE PIT

An elongate pit was excavated in the firn to study the stratigraphy over a short distance. The pit, oriented east-west, was 14 meters long, 2 meters wide, and 2 meters deep. The stratigraphy was recorded in detail at the midline of the north wall, individual layers were traced east and west of the midline to determine their continuity, and most of the pit wall was photographed. Density measurements were made at the midline and 3.5 meters away on either side.

The stratigraphy was displayed clearly on the pit wall by means of the following technique. After the pit had been made, the wall to be studied was cut smooth with a strong wire; a small trench, 1 meter deep and ½ meter wide, was dug parallel to and 50 cm behind the study wall; the pit was then completely covered with plywood. The smooth wall and the backlighting made the examination of the stratigraphy relatively easy (Figure 5).

The stratigraphy at the midline is given in Table 3. The relative hardness of the firn ranged from very soft, or loose, to very hard. Grain size varied from 0.5 to 4.0 mm. Depth hoar layers were very coarse, with grain size of 3.0 to 4.0 mm; winter snow was generally hard packed with a grain size of 1.0 mm or less.

The continuity of individual firn layers was determined by tracing the layer to the east and west

			I					II	
12.7	7.7	11.2	11.1	10.7	11.5	11.1	11.6	12.4	13.3
9.7	9.1	10.9	8.3	11.1	10.2	10.7	12.4	11.2	12.5
11.0	9.8	11.8	11.3	10.2	11.3	11.4	11.1	11.1	11.9
8.7	9.6	10.6	10.3	9.5	10.5	11.6	10.5	11.1	10.4
12.3	9.0	11.7	9.9	11.6	10.9	13.2	10.7	10.8	10.3
10.1	10.1	9.7	3.6	8.0	12.0	11.4	9.3	11.7	11.2
10.9	9.7	8.3	9.5	11.0	9.0	12.6	8.8	10.8	10.7
8.4	12.0	9.7	11.1	6.9	12.0	10.4	9.9	10.5	9.1
9.6	8.1	10.0	8.5	9.0	9.6	9.3	9.3	10.4	9.9
10.8	9.3	8.4	8.3	7.4	8.3	9.5	9.2	8.4	10.3
		III					IV		

Fig. 4. The mean annual accumulation (g cm^{-2} yr^{-1}) at each stake for 1964 and 1965. The net is divided into quadrants for analysis of the distribution of accumulation.

TABLE 2. Mean Accumulation and Variance over the Two-Year Period

Quadrant	Mean Accumulation, g cm^{-2} yr^{-1}	Variance, g cm^{-2} yr^{-1}
I	10.4	1.52
II	11.4	0.71
III	9.1	2.77
IV	10.1	1.35

Fig. 5. The elongate pit covered with plywood and showing the pit stratigraphy.

of the midline. In the midline stratigraphic section, there were 49 layers, and, of these, 49% were traced along the entire pit wall (14 meters), 31% were traced 7 meters or more, and 20% were discontinuous (Table 4). The layers varied in thickness along the length of the pit, the winter layers having the greatest variability and the most discontinuity. This is a result of the higher winds at Byrd station in the winter, which prevent an even accumulation of snow. The 5-meter-long section of the pit in Figure 6 shows the variability of the strata. This diagram has been drawn from a photographic mosaic. The fine-grained, dense snow is in highly irregular layers, and some of these are sastrugi. The most prominent coarse-grained layer occurs at a depth of 60 cm.

The tracing of layers for the correlation of snow studies over wide areas has been proposed by *Vickers* [1965]. In one instance, he traced an individual layer, on the basis of grain size, a distance of 200 miles with samples every 30 miles. The stratigraphic study in the elongate pit showed that nearly 50% of

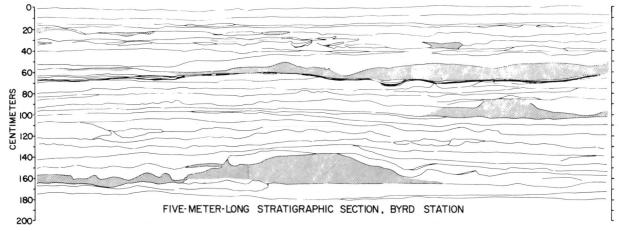

Fig. 6. A 5-meter-long section of the stratigraphy of the elongate pit, showing the variability of strata, the fine-grained dense firn (lined pattern), and the most prominent coarse-grained layer (solid pattern). Note the evenness of the top two layers, which were deposited during the summer with light winds. The fine-grained material is generally irregular and consists of sastrugi, a result of strong winds.

TABLE 3. Firn Stratigraphy at Middle of 14-Meter Pit, New Byrd Station, February 12, 1964

Layer	Depth, cm	Hardness*	Grain Size, mm	Remarks
1	0–2	M	<1.0	
2	2–9	S	0.5–1.0	
	9			Crust
3	9–13	S	1.0	
	13			Crust
4	13–17	M	1.0	
	17			Crust
5	17–19	VS	1.0–2.0	
	19			Crust
6	19–28	M	1.0	
7	28–31	VS	1.0	
8	31–33	VS	1.0	
9	33–34	M-H	<1.0	
	34			Layer 0.5 cm thick, continuous, very coarse; possible summer surface.
10	34–37	S	1.0	
	37			Crust
11	37–38	S	1.0	
12	38–53	M-S	1.0–2.0	
13	53–60	VH	1.0	Wind-packed, no real crust.
14	60–61	VS (coarse)	3.0–4.0	Vertical structure, very coarse; like depth hoar; summer surface (?).
15	61–65	loose	1.0–2.0	Loose
16	65–69	M	2.0	
	69			Crust
17	69–73	M	1.5–2.0	
18	73–77	M	1.0	
	77			Crust
19	77–80	M	2.0–3.0	
	80			Crust
20	80–95	VH	<1.0	Wind-packed
21	95–98	M-S	2.0	At about 95, very loose and soft.
	98			Crust
22	98–100	M-S	2.0–3.0	
	100			Crust; pinches out just to right.
23	100–104	M-S	2.0–3.0	
	104			Crust
24	104–112	M-S	2.0–3.0	
25	112–114	M-S	2.0	
	114			Layer of bonded grains, no definite crust.
26	114–118	M-H	1.0–2.0	
	118			Layer of bonded grains.
27	118–125	VH	1.0	
28	125–127	loose	2.0	
29	127–129	loose	1.0–2.0	
	129			Layer of bonded grains.

TABLE 3. (continued)

Layer	Depth, cm	Hardness*	Grain Size, mm	Remarks
30	129–133	M-S	2.0–3.0	
	133			Crust
31	133–138	M	2.0–3.0	
32	138–140	H	0.5–1.5	
	140			Crust
33	140–141	M	<1.5	
	141			Crust
34	141–145	M-S	2.0–3.0	
	145			Layer of bonded grains.
35	145–150	M-S	1.0–2.0	
	150			Crust
36	150–155	loose	2.0	Upper part is very loose.
	155			Crust
37	155–157	H	1.5–2.0	
	157			Thick crust!
38	157–163	M	1.0–3.0	
	163			Crust
39	163–171	VH	0.5–1.0	
	171			Crust; also at 171, 0.5-cm layer of coarse grains.
40	171–175	loose	2.0–3.0	
41	175–178	VH	1.5–2.0	
	178			Layer of bonded grains.
42	178–181	loose	2.0–3.0	
	181			Crust
43	181–182	loose	2.0–3.0	
	182			Layer of bonded grains.
44	182–185	M	2.0	
	185			Layer of bonded grains.
45	185–189	M-H	1.0–2.0	
46	189–190	loose	3.0	
	190			Crust
	194			Crust
47	194–200	M-S	2.0–3.0	No crust!
48	200–201	M-S	2.0–3.0	
	201			Crust
49	201–215			

*Hardness is represented as follows: S, soft; VS, very soft; M, medium; H, hard; VH, very hard; M-S, medium to soft; M-H, medium to hard.

the layers could be traced for 14 meters, and so it seems probable that especially well-defined layers could be traced beyond this.

The three density profiles taken along the pit at the midline and at 3.5 meters on either side (east and west) are plotted in Figure 7. From 75 cm depth to the pit bottom, the correspondence is good. The most prominent and consistent high-density layer is at 75 to 100 cm depth. Three more high-density layers show up well at lower levels. From the surface to 50 cm depth, however, correlation of the three density plots is practically impossible, although there is some similarity between the west and middle profiles.

TABLE 4. Continuity of Individual Layers in North Wall of 14-Meter-Long Pit

Layer	Depth, cm	Traced East	Traced West	Continuous*	Traced 7 Meters or More†	Discontinuous‡
1	0–2	Yes	Yes	X		
2	2–9	Yes	Yes	X		
3	9–13	Yes	Yes	X		
4	13–17	Yes	No		X	
5	17–19	Yes, for 3 m	Yes		X	
6	19–28	Yes	Yes	X		
7	28–31	Yes	Yes, but poorly defined	X		
8	31–33	No	No			X
9	33–34	Yes	Yes, but broken	X		
10	34–37	Yes, but poorly defined	Yes, but broken	X		
11	37–38	Yes, but broken	Yes, but broken	X		
12	38–53	Yes, but broken	Yes, but poorly defined	X		
13	53–60	Yes	Yes	X		
14	60–61	Yes	Yes	X		
15	61–65	Yes	Yes for 3–4 m		X	
16	65–69	Yes	Yes	X		
17	69–73	Yes	Yes	X		
18	73–77	Not clear	Not clear			X
19	77–80	Yes, but broken	Yes, for 3 m		X	
20	80–95	Yes	Yes	X		
21	95–98	Yes, but broken	Yes, but broken	X		
22	98–100	No	Yes		X	
23	100–104	Yes	Yes	X		
24	104–112	Yes	Yes	X		
25	112–114	Yes	Yes	X		
26	114–118	Yes, but broken	Yes, but broken	X		
27	118–125	Yes	Yes	X		
28	125–127	Yes	No		X	
29	127–129	No	No			X
30	129–133	Yes	Yes	X		
31	133–138	Yes, for 2 m	Yes, but broken		X	
32	138–140	Poorly defined	Poorly defined			X
33	140–141	No	No			X
34	141–145	Yes, for 5 m	Yes, but broken		X	
35	145–150	Yes, for 5 m	Yes, for 3 m		X	
36	150–155	Yes	Very poorly defined		X	
37	155–157	Yes, for 4 m	Yes, but broken		X	
38	157–163	Very poorly defined	Yes, for 2 m			X
39	163–171	Badly broken	Badly broken			X
40	171–175	Patchy	Yes		X	
41	175–178	Yes	Patchy		X	
42	178–181	Patchy	Patchy			X
43	181–182	Yes	Yes, for 3 m		X	
44	182–185	No	Very poorly defined			X
45	185–189	Yes	Patchy		X	
46	189–190	Yes	Yes	X		
47	190–194	Yes	Yes, but broken	X		
48	194–200	Yes	Yes	X		
49	200–201	No	No			X
				24 (49%)	15 (31%)	10 (20%)

*Layer traced continuously for entire 14 meters of pit wall.
†Layer traced at least 7 meters but less than 14 meters.
‡Layer discontinuous and poorly defined, so not traceable more than a few meters.

Table 5 gives the average densities for a number of intervals in the three profiles. It shows the rather wide variation from profile to profile for some of the intervals, such as 50–100 cm (west, 0.363 g cm^{-3}; middle, 0.394 g cm^{-3}; east, 0.382 g cm^{-3}), but a similarity for the over-all interval, 0–200 meters (west, 0.368 g cm^{-3}; middle, 0.371 g cm^{-3}; east, 0.368 g cm^{-3}).

Annual snow accumulation was determined at the pit midline, utilizing the stratigraphic descriptions and the density measurements. From the stratigraphic evidence alone (that is, sublimation layers and thick crusts), summer horizons were noted at depths of 34, 60, 157, and 189 cm. An increase in density, a slight increase in grain size, and some bonded grain layers indicate an additional summer horizon at 104 cm depth. Snow accumulation 3.5 meters east and west of the midline was determined using the snow density profiles of these sections and interpreting them with the midline section as a guide. In the west profile, summer horizons were at 27, 71, 93, 133, and 175 cm depth. In the east profile, summer horizons were at 39, 78, 108, 162, and 193 cm depth.

The annual accumulation determined at the three sections in the pit is given in Table 6. A change in accumulation of 12.7 to 14.4 g cm^{-2} yr^{-1} over a distance of only 7 meters is indicated. This is shown graphically on the three density profiles (Figure 7), where the summer horizon 1958–1959 is found at

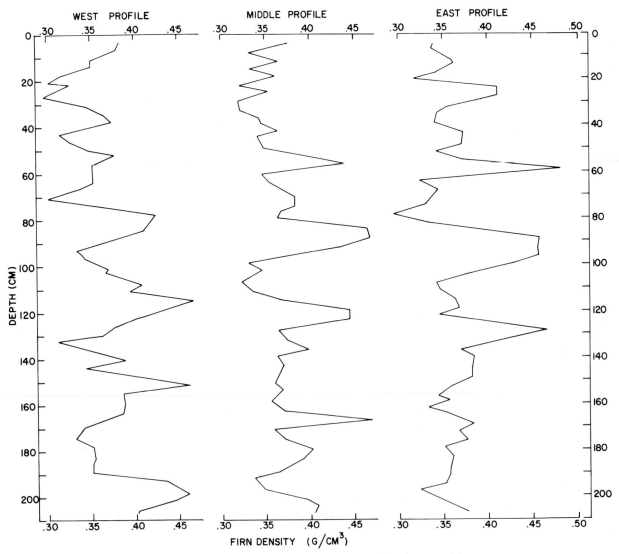

Fig. 7. Density profiles at three sections in the elongate pit; they are 3.5 meters apart.

TABLE 5. Firn Densities in Elongate Pit in Grams per Cubic Centimeter

Interval, cm	West	Mid-point	East	Avg.	West	Mid-point	East	Avg.	West	Mid-point	East	Avg.
0–50	0.341	0.344	0.357	0.347								
					0.353	0.367	0.369	0.363				
50–100	0.363	0.394	0.382	0.380								
									0.368	0.371	0.368	0.369
100–150	0.387	0.375	0.382	0.381								
					0.384	0.375	0.368	0.376				
150–200	0.381	0.375	0.354	0.370								

successively lower levels from west to east, 175, 189, and 198 cm.

SETTLING OF THE FIRN

To measure the settling of firn at Byrd station, a 10-meter pit was excavated during the period November 30 to December 4, 1963. Wooden dowels were set in the north wall of the pit every 25 cm for the first 3 meters and every 50 cm for the remaining 7 meters. A metal point was placed in the center of the head of each dowel, and distances between dowels were measured from metal point to metal point. The horizontal closure of the pit was measured between dowels on the opposing north-south and east-west walls, at the 3-, 5-, 7-, and 9-meter levels. The distances between dowels were measured with a 2-meter steel tape on December 8, 1963, November 11, 1964, and November 2, 1965.

The settling rate data are given in Table 7, and the strain rates are plotted in Figure 8. The average strain rate over the 10 meters during the first year was 0.01068, and that during the second year was 0.00996, averaging 0.01032 yr^{-1}.

The horizontal closure data are given in Table 8. The horizontal closure strain rate decreases markedly with depth, and there is considerable difference in strain rates at shallow depths, between north-south and east-west walls, the east-west closure strain rate being the higher.

ACCUMULATION FROM SORGE'S LAW

In areas where snow accumulates at a constant rate, where there is no melting, and the annual climatic cycle does not change appreciably in the number of years being considered, the firn densifies at a steady rate such that the depth-density curve remains time invariant. This relationship is known as Sorge's law

TABLE 6. Annual Snow Accumulation in Elongate Pit in Grams per Square Centimeter

Year	West	Midline	East	Mean
1963	9.2	11.6	13.8	11.5
1962	15.1	9.4	13.8	12.8
1961	8.3	17.1	11.8	12.4
1960	15.2	19.9	20.0	18.4
1959	15.6	12.4	12.8	13.6
Avg.	12.7	14.1	14.4	13.7

[*Bader*, 1954]. If the firn density and settling rate are known, then the accumulation can be determined as follows:

$$A = \frac{-vd^2}{dd/dh}$$

where A is the accumulation in g cm^{-2} yr^{-1}, v is the vertical strain rate per year, d is the average density of the interval, and dd/dh is the slope of depth-density curve.

A depth-density profile was measured from cores near the 10-meter pit on November 12, 1964 (Figure 9). The average density for the interval 0–987 cm is 0.477 g cm^{-3}, the slope of the curve is 2.25×10^{-4} g cm^{-3} cm^{-1}, and the vertical strain rate is 0.01032 yr^{-1}. This gives an accumulation rate of 10.4 g cm^{-2} yr^{-1}. The mass of snow in the interval, from surface to 987 cm depth, represents an accumulation of 39 years, or from 1925 to 1964.

For a similar period, 1925 to 1960 (35 years), Gow determined an accumulation rate of 14.0 g cm^{-2} yr^{-1} at old Byrd station, 9 km west of the new Byrd station. This difference in accumulation

TABLE 7. Settling of Firn in 10-Meter Pit

Peg	Crude Distance from Roof, cm	Measured Distance, cm, Dec. 8, 1963	Measured Distance, cm, Nov. 11, 1964	Δ	Measured Distance, cm, Nov. 2, 1965	Δ	Strain Rates, yr^{-1} 1st year (1964)	2nd year (1965)
Roof	0	25.7	23.6	2.1	22.7	0.9	0.088	0.038
1	25	23.8	23.4	0.4	22.8	0.6	0.021	0.026
2	50	25.9	25.6	0.3	25.1	0.5	0.013	0.021
3	75	24.8	24.7	0.1	24.6	0.1	0.004	0.004
4	100	23.9	23.6	0.3	23.3	0.3	0.013	0.013
5	125	25.4	25.4	0.3	24.7	0.4	0.013	0.016
6	150	25.7	25.6	0.1	25.1	0.5	0.004	0.016
7	175	23.7	23.4	0.3	23.2	0.2	0.014	0.009
8	200	26.0	25.6	0.4	25.4	0.2	0.017	0.008
9	225	24.7	24.5	0.2	24.2	0.3	0.009	0.012
10	250	23.3	22.8	0.5	22.5	0.3	0.023	0.013
11	275	28.2	28.0	0.2	27.8	0.2	0.008	0.007
12	300	48.5	47.9	0.6	47.2	0.2	0.013	0.004
13	350	49.2	48.7	0.5	48.2	0.5	0.011	0.010
14	400	51.4	51.0	0.4	50.4	0.6	0.008	0.012
15	450	51.2	50.7	0.5	50.2	0.5	0.011	0.010
16	500	50.4	50.2	0.2	49.8	0.4	0.004	0.008
17	550	44.6	44.1	0.5	43.8	0.3	0.012	0.007
18	600	54.6	54.4	0.2	54.2	0.2	0.004	0.004
19	650	47.6	47.1	0.5	46.8	0.3	0.011	0.006
20	700	51.6	51.1	0.5	50.6	0.5	0.010	0.010
21	750	40.3	40.2	0.1	40.1	0.1	0.003	0.003
22	800	58.3	58.1	0.2	57.7	0.4	0.004	0.007
23	850	51.4	51.3	0.1	51.1	0.2	0.002	0.004
24	900	51.8	51.6	0.2	51.2	0.4	0.004	0.008
25	950	45.8	45.6	0.2	45.1	0.5	0.005	0.011
26	1000							

Over-All Settling Date	Over-All Settling, cm	Strain Rate, yr^{-1}
Nov. 11, 1964	9.9	0.0107
Nov. 2, 1965	9.6	0.0099

of 3.6 g cm^{-2} yr^{-1} may be real and the result of different environments at old and new Byrd station.

SNOW ACCUMULATION NEAR BYRD STATION

The accumulation of snow in the vicinity of Byrd station has been determined by different methods over different periods of time. At old Byrd station, accumulation was measured from 1957 to 1961 at stakes, and snow pit and core studies have provided data for the interval 1893 to 1960. A list of the rates of annual accumulation, as determined by these studies, is given in Table 9. Gow, in his study of cores, showed that, for a 67-year period ending in 1960, the average annual accumulation was 14.4 g cm^{-2} yr^{-1}, and that the year-to-year accumulation can vary significantly (Figure 10). Also, his data show that from 1948 until 1958 there was a gradual increase in accumulation from 9.3 to 18.4 g cm^{-2} yr^{-1}; workers at Byrd station, such as Anderson, Long, and Pirrit and Doumani, did indeed record high accumulation (16.5, 17.8, and 21.8 g cm^{-2} yr^{-1}, respectively) during the International Geophysical Year.

The snow studies at new Byrd station all indicate generally low accumulation during the past few years. *Koerner* [1964] reported an average of 14.2 g cm^{-2} yr^{-1} from pits representing 1958 through 1962. *Gow* [1965] recorded an average of 11.0 g

$cm^{-2} yr^{-1}$ for two lines of stakes (21 and 22 poles) for the four years 1962 through 1965 (Table 10). The elongate pit study gave an average of 13.7 g $cm^{-2} yr^{-1}$ for the period 1959 through 1963, the square kilometer accumulation net gave an average of 10.3 g $cm^{-2} yr^{-1}$ for 1964 and 1965, and, from the deep pit study, Sorge's law gives an accumulation rate of 10.4 g $cm^{-2} yr^{-1}$. Along the first 12 km

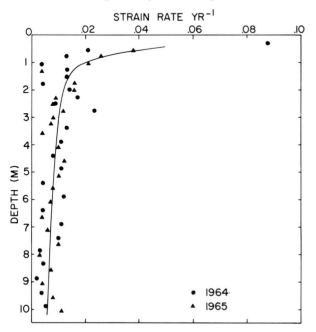

Fig. 8. Vertical strain rate in the deep pit for 1964 and 1965. The smooth curve is the mean of the two years.

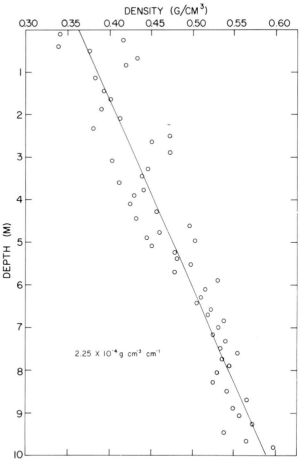

Fig. 9. Depth-density profile near the deep pit.

TABLE 8. Horizontal Closure in Centimeters of 10-Meter Pit

Depth and Direction	Dec. 8, 1963	Nov. 11, 1964	Δ	Strain Rate, yr^{-1}	Nov. 2, 1965	Δ	Strain Rate, yr^{-1}	2-Year Strain Rate, yr^{-1}
3 meters								
N-S	137.2	136.5	0.7	0.006	135.8	0.7	0.006	0.006
E-W	72.5	70.8	1.7	0.025	69.6	1.2	0.017	0.021
5 meters								
N-S	121.4	120.7	0.7	0.006	120.2	0.5	0.004	0.005
E-W	92.2	91.1	1.1	0.013	90.2	0.9	0.010	0.0115
7 meters								
N-S	126.7	126.2	0.5	0.004	125.8	0.4	0.003	0.0035
E-W	103.9	103.2	0.7	0.007	102.5	0.7	0.007	0.007
9 meters								
E-W	89.2	88.7	0.5	0.006	88.5	0.2	0.002	0.004

TABLE 9. Accumulation at Old Byrd Station

Period	No. of Stakes	Annual Accum., g cm^{-2}	Reference
March 1957 to Jan. 1958	7	16.5	*Anderson*, 1958
March 1957 to Dec. 1958	120	17.8	*Long*, 1961
Feb. 1959 to Oct. 1959	100	21.8	*Pirrit and Doumani*, 1960
Feb. 1959 to Nov. 1959	14	22.3	*Pirrit and Doumani*, 1960
Nov. 1960 to Oct. 1961	100	13.8	*Shimizu*, 1964
1893–1960	Pit and cores	14.4	*Gow*, 1961

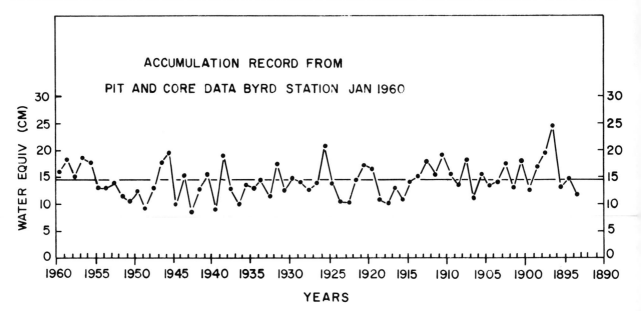

Fig. 10. Accumulation (g cm^{-2} yr^{-1}) at old Byrd station from pit and core studies [after *Gow*, 1961].

TABLE 10. Accumulation at New Byrd Station

Period	E-W Stakes (21), g cm^{-2} yr^{-1}	N-S Stakes (22), g cm^{-2} yr^{-1}
Feb. 1962 to Jan. 1963	11.0	10.4
Jan. 1963 to Jan. 1964	9.7	9.0
Jan. 1964 to Jan. 1965	13.1	12.6
Jan. 1965 to Jan. 1966	11.3	10.7
4-year average	11.3	10.7

From *Gow* [1965] and personal communication.

of the Byrd strain net to the northeast, the accumulation at 27 poles was 13.6 g cm^{-2} yr^{-1} in 1964, and, along the first 12 km of the Byrd–Whitmore Mountains traverse to the southeast, the accumulation at 4 poles, 1963 through 1965, was 14.5 g cm^{-2} yr^{-1} [*Brecher*, 1967].

Comparison of similar accumulation periods (Table 11) shows a fairly good agreement, consider-

TABLE 11. Snow Accumulation during Similar Time Intervals

Investigator	Method	Location	Period	Accumulation, g cm^{-2} yr^{-1}
Gow, 1961	Pit and core	Old Byrd	1925–1960	14.0
Koerner, 1964	Pit and core	New Byrd	1930–1961	14.4
Cameron, this paper	Sorge's Law	New Byrd	1925–1963	10.4
Cameron, this paper	100 stakes	New Byrd	1964–1965	10.3
Gow, 1965, and pers. commun.	43 stakes	New Byrd	1964–1965	12.9
Cameron, this paper	100 stakes and elongate pit	New Byrd	1962–1965	11.2
Gow, 1965, and pers. commun.	43 stakes	New Byrd	1962–1965	11.0

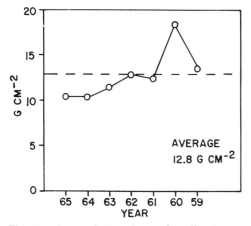

Fig. 11. Accumulation (g cm^{-2} yr^{-1}) at new Byrd station from the elongate pit and the square kilometer accumulation net.

ing the difference in measuring methods, numbers of stakes, and locations in and around old and new Byrd stations.

The accumulation near the Byrd stations varies from year to year; there is an indication of accumulation trends (increasing or decreasing); between 1948 and 1958, there was an increase in accumulation, and since then there has been a general decrease. A graph representing the accumulation as recorded in the elongate pit and at the stakes in the square kilometer net (accumulation years 1959 through 1965) shows the present downward trend (Figure 11).

CONCLUSIONS

The square kilometer accumulation net recorded an accumulation rate of 10.3 g cm^{-2} yr^{-1} for 1964 and 1965. The distribution of accumulation over the net showed that the accumulation is controlled by the direction of slope and the prevailing wind such that there is a decrease in accumulation downwind of crests.

The elongate pit revealed an average accumulation of 13.7 g cm^{-2} yr^{-1} for the period 1959 through 1963. These data and the 100-stake data for 1964 and 1965 give an accumulation rate of 12.8 g cm^{-2} yr^{-1}. Tracing of the 49 individual firn layers along the 14-meter pit wall showed that nearly 50% could be followed the entire length.

In the deep pit, the average vertical strain rate from the surface to 10 meters was 0.01032 yr^{-1}. Utilizing the vertical strain rate in a calculation of Sorge's law, an accumulation rate of 10.4 g cm^{-2} yr^{-1} was obtained for the period 1925 to 1964.

Snow accumulation in the vicinity of Byrd station shows increasing and decreasing trends. Accumulation increased from 1948 to 1958, and has been decreasing since then.

Acknowledgments. Mr. Arthur Rundle ably assisted the author in the field work during the 1963–1964 season. In addition, Messrs. Helmut Jaron and Robert Rutford and Dr. Chester Pierce aided in the excavation of the deep pit and the elongate pit. The accumulation stakes have been remeasured by Messrs. James Gliozzi and Henry Brecher of the Institute of Polar Studies. The author is grateful to Mr. Michael O'Kelley of the Institute of Polar Studies for helpful discussions on the treatment of the data. The manuscript has been critically read by Drs. Arthur Mirsky and Colin Bull of the Institute of Polar Studies, and by Mr. Anthony Gow of the U.S. Army Cold Regions Research and Engineering Laboratory.

This work has been financed by National Science Foundation grant GA-50 awarded to Ohio State University Research Foundation.

Contribution 184 from the Institute of Polar Studies, Ohio State University.

REFERENCES

Anderson, V. H., Byrd station glaciological data 1957–58, *Ohio State Univ. Res. Found. Rep. 825-1,* part 2, December 1958.

Bader, H., Sorge's law of densification of snow on high polar glaciers, *J. Glaciol., 2,* 319–323, 1954.

Black, H. P., and W. Budd, Accumulation in the region of Wilkes, Wilkes Land, Antarctica, *J. Glaciol., 5,* 3–15, 1964.

Brecher, H. H., Accumulation between Mount Chapman and "Byrd" station, Antarctica, *J. Glaciol., 6,* 573–577, 1967.

Gow, A. J., Drill-hole measurements and snow studies at Byrd station, Antarctica, *U. S. Army Corps Eng. SIPRE Tech. Rep. 78,* 12 pp., 1961.

Gow, A. J., Snow studies in Antarctica, *U. S. Army Corps Eng. SIPRE Res. Rep. 177,* 20 pp., 1965.

Koerner, R. M., Firn stratigraphy studies on the Byrd–Whitmore Mountains traverse, 1962–1963, *Antarctic Snow and Ice Studies, Antarctic Res. Ser.,* vol. 2, pp. 219–236, AGU, Washington, D. C., 1964.

Long, W. E., Marie Byrd station and traverse glaciological data, 1958–1959, *Ohio State Univ. Res. Found. Rep. 825-2,* part 11, January 1961.

Pirrit, J., and G. A. Doumani, Glaciology, Byrd station and Marie Byrd Land traverse, 1959–1960, *Ohio State Univ. Res. Found. Rep. 968-2,* November 1960.

Schytt, V., *Glaciology 2, A,* Snow studies at Maudheim, *B,* Snow studies inland, *C,* The inner structure of the ice shelf at Maudheim as shown by core drilling, Norwegian Polar Institute, Oslo, 151 pp., 1958.

Shimizu, H., Glaciological studies in West Antarctica, 1960–1962, *Antarctic Snow and Ice Studies, Antarctic Res. Ser.,* vol. 2, pp. 37–64, AGU, Washington, D. C., 1964.

Vickers, W. A., A study of ice accumulation and tropospheric circulation, Ph.D. thesis, McGill University, Montreal, 1965.

STRATIGRAPHIC STUDIES IN THE SNOW AT BYRD STATION, ANTARCTICA, COMPARED WITH SIMILAR STUDIES IN GREENLAND[1]

Carl S. Benson

Geophysical Institute, University of Alaska, College 99701

The stratification exposed on the long tunnel walls during construction of new Byrd station, Antarctica, was studied in the 1961–1962 austral summer. The well-exposed sequences of strata showed good lateral continuity; annual increments of accumulation were in the range of 17 ± 3 g cm^{-2} yr^{-1}. The rate of accumulation appears to have changed from about 20 g cm^{-2} yr^{-1} between 1947 and 1954 to about 14 g cm^{-2} yr^{-1} from 1955 to 1961. The average accumulation rate at new Byrd was 1.18 times greater than that determined at old Byrd by A. J. Gow. The differences in accumulation at these two stations, which are 10 km apart, are attributed to topographic features in the Byrd station area rather than to differences in the amount of snowfall. It is demonstrated how a migrating surface topography can produce a variable rate of accumulation at a given station even though the rate of snowfall remains constant. This is an important complication in the interpretation of core obtained from deep drilling. The facies parameters at Byrd station are compared with parameters at six stations on the Greenland ice sheet that have comparable mean annual temperatures and rates of accumulation. The annual sequences of strata are similar, but the absolute density values are higher at Byrd station. The similarities and differences in the upper 6 meters of snow strata, at the Antarctic and Greenland stations, are interpreted by analyzing the differences in wind action and range of temperature. Byrd station is windier and has a smaller temperature range; these differences favor the development of harder, higher-density snow at Byrd Station.

The field investigations described in this paper were carried out at Byrd station between December 13 and 29, 1961, when the construction of new Byrd station was underway. New Byrd and old Byrd stations are 10 km apart at about 80°S, 120°W and at an elevation of 1500 meters above sea level. The mean annual temperature is −28°C as determined by repeated measurements made 8 and 16 meters below the snow surface [*Anderson*, 1958, pp. 69–72], and as determined from U.S. Weather Bureau surface data.

During construction, a network of tunnels was made (Figure 1). Buildings subsequently constructed inside the tunnels were protected from temperature extremes, inconveniences from wind-drifted snow, and stresses associated with densification of snow in this area of continuous accumulation. The long, deep tunnel cuts provided an unusually good opportunity for observing the continuity of snow strata.

Sequences of snow layers were observable on all clean tunnel walls. The stratigraphy was emphasized in some areas by heat and smoke resulting from construction. This was especially evident in the main entrance of tunnel L-3 (Figure 1), where the garage and several heated workshops were located.

A site was selected in tunnel L-9 for detailed measurements. During the time of these studies, it was one of the few areas that was free of heavy traffic and construction activities. After the stratigraphic studies were completed, this area was further excavated for construction of buildings 20 and 34 (Figures 1 and 2).

The upper snow layers, from 0 to 380 cm, were studied at point 2 on the west-facing wall of a fresh cut made on December 21, 1961, for building 34. Undisturbed snow was well exposed from 306 to 625 cm below the snow surface on the test wall of tunnel L-9 (Figure 2). (Above about 306 cm, the snow

---

[1] The studies described in this paper were carried out while the author was a Visiting Research Associate at the Institute of Polar Studies, The Ohio State University, Columbus, Ohio.

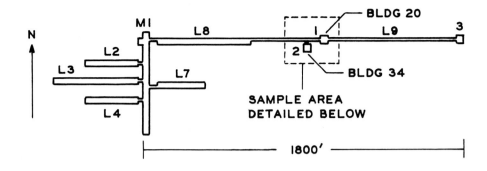

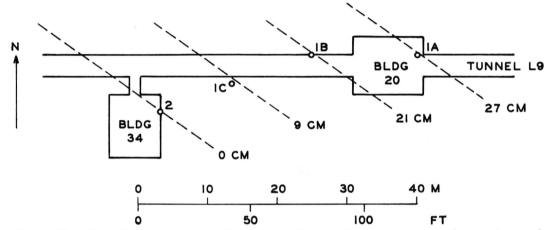

Fig. 1. Map of new Byrd station, Antarctica, and detailed map of the snow stratigraphy area in tunnel L-9. The dashed contour lines (relative to point 2) running through points $1A$, $1B$, $1C$, and 2 represent the 1961 snow surface topography as determined from a topographic map made by U.S. Navy surveyors.

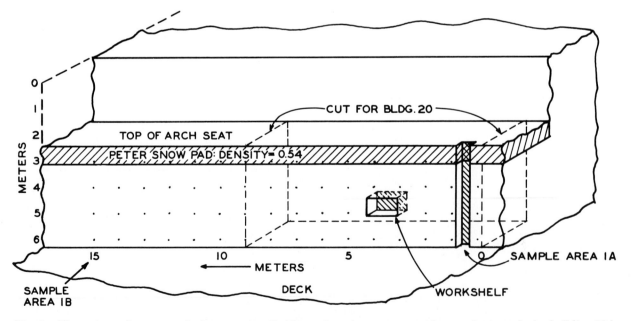

Fig. 2. The main study area on the L-9 tunnel wall. The region where an excavation was later made for building 20 is indicated. The points on the wall represent the locations of welding rods placed at 1-meter intervals.

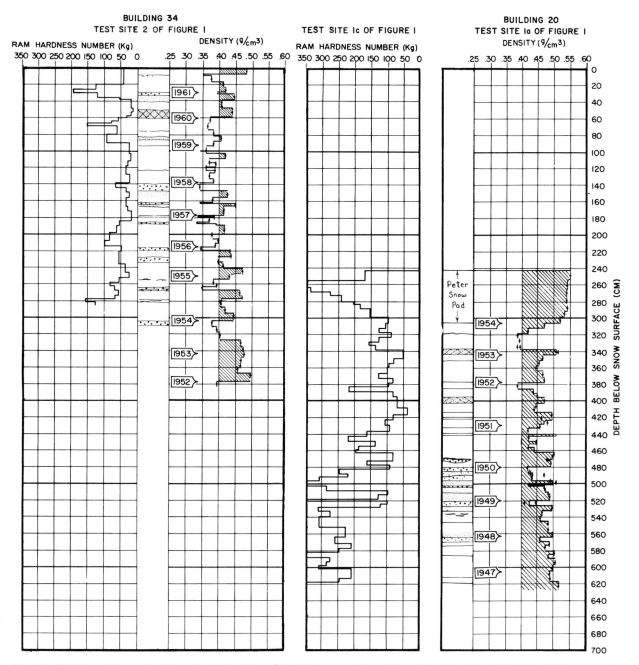

Fig. 3. Stratigraphic profiles from test sites 1A and 2. A Rammsonde profile made at test site 1C is plotted between the profiles at points 1A and 2 which were correlated by leveling between them, and by matching the prominent low-density layer that occurs near the 320-cm depth. The density profile is cross hatched where it exceeds 0.40 g cm^{-3}. The stratigraphic positions of March 1947 to 1961 (from estimate A, Table 3) are indicated.

strata in tunnel L-9 had been disturbed by the construction activities before December 13; in particular, the Peter Snow Miller had prepared a hard, compacted layer that formed the top of trench L-9.) This undisturbed snow was studied in detail at the place labeled 1A in Figures 1 and 2. By combining measurements at points 1A and 2, a complete stratigraphic section extending from the December 1961 snow surface to a depth of 625 cm was obtained. The data from these two points are graphically displayed in Figure 3. The two parts of this stratigraphic section are 27.5 meters apart, and the original snow surface elevation was 27 cm higher at point 1A than at point 2. The separate stratigraphic sections were correlated by leveling between points 1A and 2 and by matching the low-density layers that appear just below the 3-meter depth in each section.

The depth-density profiles in Figure 3 were integrated by summing 20-cm-depth increments and omitting the high-density Peter Snow Pad (Table 1) to obtain the depth-load curve in Figure 4. The

TABLE 1. Depth-Load Data from Integrated Depth-Density Data

Δx, cm	0–380 cm Depth			320–620 cm Depth		
	$\Delta\sigma$, g cm^{-2} (or cm H$_2$O)	$\Sigma\sigma$, g cm^{-2} (or cm H$_2$O)	Avg. ρ, g cm^{-3}	$\Delta\sigma$, g cm^{-2} (or cm H$_2$O)	$\Sigma\sigma$, g cm^{-2} (or cm H$_2$O)	Avg. ρ, g cm^{-3}
0–20	8.40	8.40				
20–40	8.48	16.88				
40–60	8.52	25.40	0.404			0.404
60–80	7.41	32.81				
80–100	7.63	40.44				
100–120	7.80	48.24				
120–140	7.44	55.68				
140–160	7.64	63.32	0.387			0.387
160–180	8.00	71.32				
180–200	7.85	79.17				
200–220	7.56	86.73				
220–240	8.30	95.03				
240–260	8.62	103.65	0.412			0.412
260–280	8.37	112.02				
280–300	8.33	120.35				
300–320	7.99	128.34 . . . . $\longrightarrow$		128.34		
320–340	8.79	137.13		8.07	136.41	
340–360	9.38	146.51	0.432	9.38	145.79	0.431
360–380	9.39	155.90		9.08	154.87	
380–400				8.61	163.48	
400–420				9.26	172.74	
420–440				9.08	181.82	
440–460				8.74	190.56	0.458
460–480				9.76	200.32	
480–500				8.98	209.30	
500–520				9.58	218.88	
520–540				9.30	228.18	
540–560				9.55	237.73	0.479
560–580				9.54	247.27	
580–600				9.90	257.17	
600–620				9.90	267.07	
620–640						
640–660						
660–680						
680–700						

These data were obtained by integrating depth-density profiles (Figure 3). Two profiles were used to achieve a total depth span of 0 to 620 cm below the December 1961 snow surface. Density data spanning the depth range from 0 to 380 cm are from the east wall of the excavation for building 34 (point 2 of Figure 1). Data spanning 320 to 620 cm are from the north wall of the excavation near building 20 (point 1A of Figure 1).

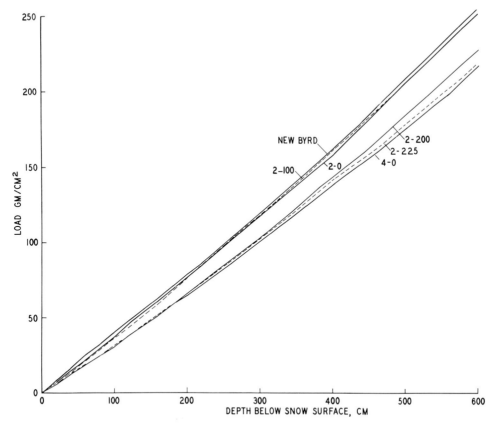

Fig. 4. The depth-load curve from Table 1, plotted together with curves from the selected Greenland stations identified in Tables 6 and 7 (station 2-0 has the same location as Camp Century, Table 11).

curves from the Greenland stations (2-0, 2-100, 2-200, 2-225, and 4-0) shown in Figure 4 are discussed below in the section Facies Parameters.

The questions considered in this paper concern (1) the detailed continuity of strata in the long continuous exposures at Byrd station, (2) the nature of annual units of strata and their water equivalent, and (3) the comparison of facies parameters measured at Byrd station on the Antarctic ice sheet with those measured at similar stations on the Greenland ice sheet.

The annual accumulation determined in this study was slightly higher than that determined by Gow [1961] from a 19-meter core at old Byrd station. An attempt to explain this in terms of topographic variations led to an extension of previous work. In particular, it is demonstrated that a migrating surface topography can produce a variable rate of accumulation at a fixed point, even though the rate of snowfall remains constant. The effects of topographic variations on a smaller scale were investigated by examining the variability of data from the stake farm at Byrd station. It was found that such data reveal an annual variation in sastrugi development.

CONTINUITY OF STRATA

Observations on the Tunnel Walls

Stratigraphic continuity was measured on the long, well-exposed north wall of tunnel L-9 by extending the observations westward from point 1A of Figure 2 where density samples were obtained. A coordinate system was established on the test wall by placing welding rods at 1-meter intervals horizontally and vertically. The stratigraphy was exposed by careful cleaning and is sketched in Figure 5.

Lensing of individual layers is common, especially hard wind-packed layers, as can be seen clearly in Figure 6. However, this lensing does not hopelessly confuse the stratigraphy; indeed the continuity of some individual layers is good. In particular, the strata just below 3 meters and immediately

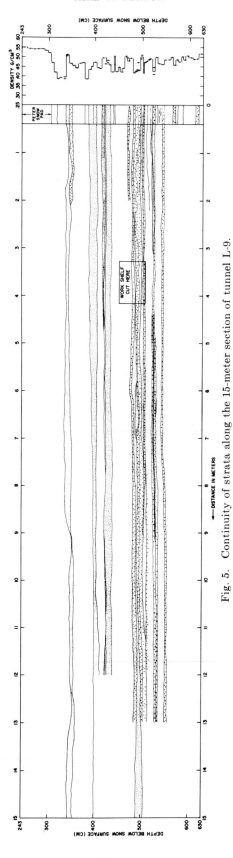

Fig. 5. Continuity of strata along the 15-meter section of tunnel L-9.

Fig. 6a. The study area (Figures 1 and 2).

Fig. 6b. The first 6 meters of the study area with the layering accentuated by sunlight.

Fig. 6c. Detailed layered structure in the first 4 meters of the study area.

Fig. 6d. Density samples cut from the pit wall were weighed on a triple-beam balance mounted in the shelf shown here.

above and below 4 meters were easily traced along the wall of tunnel L-9 and proved useful in correlating the density profiles measured at points 1A and 2.

Stratigraphic Expressions of Surface Irregularities

The snow surface is rarely a smooth plane, and the variations of thickness in individual layers revealed in pit walls are expressions of buried surface roughness. Surface relief is developed to the maximum extent by wind action in the form of sastrugi, and the greatest variation in buried strata is accordingly found in the hard wind-packed layers. However, the extreme surface roughness represented by sastrugi is nearly always smoothed to some extent by erosion processes before burial. The lensing hard layers that intercept 1A at 338–344, 385–403, and 490–500 cm, and are traced across the wall in Figure 5, are interpreted as the remains of winter sastrugi layers that have been subdued by surface erosion during the summer.

Erosion and deposition occur simultaneously on the surface of an ice sheet [*Benson*, 1962, Figs. 7a–7h] and lead to the observation that, although the surface is frequently rough, the strata exposed in pit walls tend to be horizontal. The maximum surface relief occurs near the end of winter, and the minimum occurs near the end of summer. *Gow* [1965] described the action of sublimation and deflation in eroding sastrugi forms during summer at the South Pole. *Orheim* [1968] discussed the same

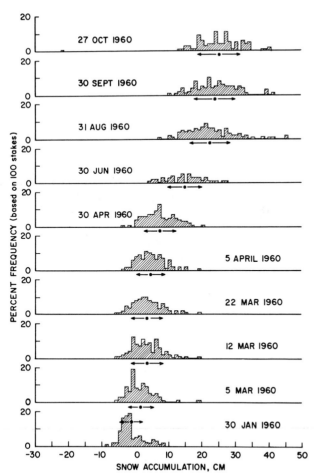

Fig. 7. Histograms of accumulation measurements made in the Byrd station stake farm during 1960 by P. Parks (personal communication, 1962). The average value of each set of measurements is represented by a dot, and the standard deviation is shown by the tips of the arrowheads.

Surface Irregularities at Byrd Station

The surface irregularity at Byrd station is illustrated by data from the 1 km² stake farm set out by Mr. George Toney during the 1957–1958 summer. One hundred bamboo poles were set in a square grid, 1 km on a side. Some of the accumulation measurements made in the stake farm since January 29, 1958, were summarized by *Long* [1961, p. 13], and several contour maps of the accumulation patterns were presented by *Shimizu* [1964]. The microscale surface irregularity at Byrd station, and the way it changes with storms, was illustrated by contour maps made by *Long* [1961, pp. 12–17] in a square area, 100 meters on a side, which was set aside for this purpose in a protected glaciological study area near the stake farm.

P. Parks (personal communication, 1962) established a reference mark on each stake on December 27, 1959, and made 10 complete sets of measurements of the snow surface relative to the stakes between January 30 and October 27, 1960. His data, summarized in Figure 7 and Table 2, show that the maximum irregularity of the snow surface occurred late in winter (standard deviation of 6.8 cm on August 31) and that the minimum irregularity occurred in summer (standard deviation of 3.3 cm on January 30). This seasonal variation in surface roughness agrees well with the interpretation of summer erosion of sastrugi before, or concurrent with, burial.

The surface irregularities exert a microscale con-

processes at the Plateau station and included measurements from artificial snow features. The heat and mass balance of horizontal and inclined snow surfaces of sastrugi dimensions were measured at Plateau station by *Weller* [1969]. He found significant differences in the amount of radiation absorbed by the various faces of sastrugi forms to be the basic course of the observed preferential erosion which tends to level the surface. I have repeatedly been impressed by this horizontality of strata in the study of nearly 200 pits in Greenland, Antarctica, and in high-altitude snow fields of Alaska and, after comparing notes with Gow, Giovinetto, and Weller on this point, am convinced that this is generally observed, even in areas of pronounced surface roughness.

TABLE 2. Accumulation Data Obtained from the Byrd Station Stake Farm during 1960

(The numbers represent the distance in centimeters from the snow surface to a reference datum marked on each of the 100 stakes on December 27, 1959.)

No.	Date 1960	Avg. A	Range	Std. Dev. σ	σ/A
1	Jan. 30	−1.1	18	3.3	−3.00
2	March 5	1.6	26	3.9	2.44
3	March 12	3.6	26	4.6	1.28
4	March 22	3.7	26	4.5	1.22
5	April 5	4.8	24	4.2	0.875
6	April 30	7.5	25	4.6	0.614
7	June 30	15.1	23	6.1	0.404
8	Aug. 31	22.7	38	6.8	0.299
9	Sept. 30	24.1	32	6.0	0.249
10	Oct. 27	25.4	27	6.2	0.244

*The anomalous reading of −21.8 at stake 116 was omitted.

trol on the distribution of accumulation. One stake may record a large increment of accumulation over a given time span because it is near the crest of a dune or sastrugi ridge at the time of measurement. Conversely, a stake in, or near, the bottom of a trough will record a smaller increment of accumulation. The next storm may change this pattern by partly filling in the troughs and eroding the crests.

The usefulness of accumulation measurements made on a single stake has long been open to question, especially when the measurements span a short period of time. The purpose of a stake farm, of course, is to provide a set of values that can be examined statistically to overcome local irregularities. It is hoped that, as time goes on, the microrelief will vary in such a way as to make an individual stake record more useful. That this does sometimes happen is demonstrated in Figure 8 by the data from stakes 30 and 93; on August 31 their records differed by 30 cm, but on September 30 the difference was only 2.2 cm. A comparison of individual stake records with the average value in the farm can be obtained by plotting the ratio of standard deviation to average accumulation against the accumulation for each set of measurements. These data are listed in Table 2 and are plotted in Figure 9. From Figure 9 we see that, at Byrd station, the standard deviation becomes less than 20% of the average accumulation value after 25 cm of snow accumulation occurs. This is true even during late winter, when the standard deviation is at its maximum value (Table 2).

Sequences of Strata, Lensing Layers, and Annual Units

The interpretation of snow strata is based more on the recognition of similar layered sequences than on the positive identification of specific layers. The

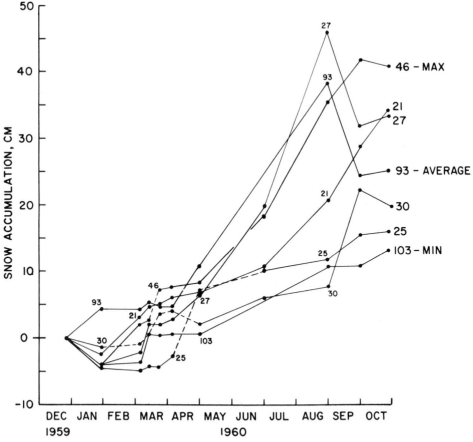

Fig. 8. Accumulation of snow on several selected stakes in the Byrd station stake farm during 1960. The maximum, minimum, and average values for the entire set of 100 stakes over the 10-month period are shown.

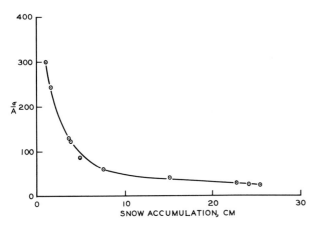

Fig. 9. The ratio of standard deviation to accumulation plotted against accumulation for the 1960 stake farm data at Byrd station (see Table 2).

stratigraphic sequence represents a response to changes in the environment of deposition and diagenesis. Since these changes occur over an annual cycle, similar sequences are produced each year. The possibility that lensing layers, resulting from surface irregularities, may destroy the annual stratigraphic sequence has been a point of concern to some writers [Crary, 1963]. Gow's observation at the South Pole is an excellent demonstration that a useful stratigraphy can exist in spite of pronounced surface irregularities.

In the present study, the details of lensing layers in the tunnels (Figures 5 and 6) did not interfere with interpretation of the major stratigraphic sequences that represent annual layers. The specific effect of lensing layers on the correlation of strata over short distances was investigated in Greenland by making Rammsonde profiles at intervals of 0.1 mile (0.16 km) on four 1-mile-long (1.6-km-long) courses. The annual accumulation on these four courses ranged between 16 and 40 cm water equivalent, and the mean annual temperature ranged between $-22°C$ and $-31°C$. The depth of individual maximums or minimums varied over a range of 20 to 40 cm, which is negligible compared with the horizontal distance of about 10^5 cm [Benson, 1959, Fig. 24]. The continuity of maximums or minimums on the ram profiles was best where melt was negligible, where at least some of the individual layers were 10 or more cm thick, and where density variations were pronounced. Although such irregularities exist, the continuity of sequences of strata has been followed continuously by pit studies spaced 10 to 25 miles (16 to 40 km) apart and by Rammsonde profiles spaced 5 miles (8 km) apart for over a thousand miles (1600 km) of continuous traverse on the Greenland ice sheet.

Exceptionally well developed sastrugi may persist for several years in some places, as Crary [1963, pp. 36–37] has pointed out. If such a 'sastrugi field' exists persistently in a particular area, it may be that sequences of strata, if identifiable at all in such places, do not record annual cycles. The strata may consist exclusively of lensing layers or may even be so homogenized by continued wind action that no useful stratigraphic record exists. However, in areas where sequences of snow strata do record annual cycles, lensing in itself does not destroy stratigraphic continuity. Sometimes the lensing of hard wind slabs is useful because the lenticularity may, in itself, provide a basis for correlation if a zone of wind slabs can be recognized in the stratigraphic sequence over a traverse. They are good stratigraphic horizons, because conditions leading to erosion with sastrugi and dune formation on the snow surface, with or without new deposition, are widespread. Sometimes the degree of induration of hard layers, which have been correlated for 100 miles (160 km) or more in Greenland, varies laterally with bona fide wind slabs representing local extremes. Individual wind slabs at a given location are preserved for several years, as Benson [1962, p. 28] has demonstrated.

In summary, there is abundant evidence that useful stratigraphic sequences exist in perennial snow deposits. The long continuous exposures available on the tunnel walls at Byrd station provide a unique opportunity to observe snow stratigraphy. In general, an increase in the horizontal exposure of snow strata improves one's ability to interpret it [Benson, 1962; Koerner, 1964]. Thus, the walls of a pit are easier to interpret than a core, and the long, smooth tunnel walls (Figure 6) are the best exposures of snow stratigraphy available.

ANNUAL UNITS

The most important goal in analyzing snow stratigraphy is the determination of annual units. If annual units can be determined in the snow strata, the annual accumulation can be calculated. The water equivalent of the snow in each annual unit is determined by integrating the depth-density profile over the depth interval involved. The annual accumula-

Selection of a Reference Datum in the Annual Stratigraphic Sequence

As was indicated above, the interpretation of snow strata is based on identifying similar layered sequences rather than on the absolute identification and correlation of specific layers. It is hoped that the annual sequence of strata will include a reference datum that separates one year's accumulation from the next. Ideally, such a reference datum should form within a short time interval and should be recognizable in all facies.

The ideal depth-density profile is a saw-toothed curve with density increasing gradually through the winter strata and decreasing abruptly at the fall discontinuity. This ideal relationship is sometimes developed to the extent that it is easy to identify [*Benson*, 1962]. Idealized drawings of the relation between summer and winter strata, with special reference to grain size, have been presented by *Shimizu* [1964, p. 48].

There are, of course, departures from the ideal profiles. The relationship varies from year to year at a given site, and from one site to the next across the several glacier facies. However, annual units have been correlated along 1100 miles (1770 km) of traverse, spanning the wetted, percolation, and dry-snow facies on the Greenland ice sheet. In general, summer strata are coarser grained and have lower density and hardness values than winter layers. Also, there is usually more variability in summer layers; coarse-grained, loose layers alternate with finer-grained, high-density layers or even wind slabs of variable thickness. Winter layers are generally more homogeneous, having higher density and finer grain size than summer layers [*Benson*, 1962, p. 30].

Stratigraphic Analysis of the Data from New Byrd Station

Three estimates, A, B, and C, of the locations of fall discontinuities in the stratigraphic data from new Byrd station, plotted in Figure 3, are listed in Table 3. The depth span of slightly more than 6 meters includes 15, 14, or 13 years according to estimate A, B, or C, respectively. Estimate A is considered to be the best. The three estimates are listed here to demonstrate the effects produced by errors in identify-

TABLE 3. Stratigraphically Determined Depths

		Estimate A			Estimate B			Estimate C		
No.	Year (March)	Depth z, cm	Load σ, g cm^{-2}	Load Increment $\Delta\sigma$, g cm^{-2}	Depth z, cm	Load σ, g cm^{-2}	Load Increment $\Delta\sigma$, g cm^{-2}	Depth z, cm	Load σ, g cm^{-2}	Load Increment $\Delta\sigma$, g cm^{-2}
0	Dec. 1961	0	0	0	0	0	0	0	0	0
1	1961	30	12.5	12.5	30	12.5	12.5	40	16.9	16.9
2	1960	60	25.5	13.0	60	25.5	13.0	90	36.5	19.6
3	1959	94	38.0	12.5	94	38.0	12.5	139	55.5	19.0
4	1958	139	55.5	17.5	139	55.5	17.5	178	71.0	15.5
5	1957	178	71.0	15.5	178	71.0	15.5	205	81.0	10.0
6	1956	216	85.0	14.0	216	85.0	14.0	250	99.5	18.5
7	1955	250	99.5	14.5	250	99.5	14.5	305	122.5	23.0
8	1954	305	122.5	23.0	305	122.5	23.0	378	154.0	31.5
9	1953	344	139.0	16.5	378	154.0	31.5	430	177.0	23.0
10	1952	378	154.0	15.0	430	177.0	23.0	480	200.3	23.3
11	1951	430	177.0	23.0	480	200.3	23.3	521	219.5	19.2
12	1950	480	200.3	23.3	521	219.5	19.2	564	239.8	20.3
13	1949	521	219.5	19.2	564	239.8	20.3	605	259.5	19.7
14	1948	564	239.8	20.3	605	259.5	19.7			
15	1947	605	259.5	19.7						
Avg.				17.3			18.6			19.9

Three estimates (A, B, and C) of the depth to a reference datum were made. The integrated load values of Table 1 as read from the depth-load curve in Figure 4 are listed next to the depth estimate and are labeled σ. The load increment from one reference level to the next is labeled $\Delta\sigma$.

ing annual units. Estimate B differs from A by omitting the level identified at 344 cm. Estimate C identifies two annual units in the top meter, whereas A and B both identify three; C also differs from A and B in that it defines a fall discontinuity at 205 cm instead of at 216 cm as in A and B. The three estimates are identical for depths below 378 cm. Some of the fall discontinuities are especially well developed and appear in all attempts to date the strata; the best ones, in rough order of quality, occur at the following depths: 480, 521, 430, 378, 564, 250, 216–205, 139, and 178 cm.

The depths of annual units listed for estimate A are plotted in Figures 3 and 10. The two depth-time curves drawn in Figure 10 were calculated by assuming constant accumulation rates to obtain sequences of load values; the depth-load curve (Figure 4) was then consulted to obtain the corresponding sequences of depth values (Table 4). The range of annual increment values involved in estimate A is 12.5 to 23.3 cm water equivalent. However, the curves drawn for constant rates of 14 and 20 g cm^{-2} yr^{-1} bracket the data points, and a rate of 17 ± 3 g cm^{-2} yr^{-1} appears to be a good representative value. This is slightly higher than the value of 14 to 15 g cm^{-2} yr^{-1} obtained by *Gow* [1961] in analyzing cores from a 19-meter auger hole at old Byrd station.

The stratigraphic interpretation represented by estimate A indicates decreasing accumulation from 1947 to 1961, and, rather than decreasing steadily, the accumulation values seem to change abruptly

TABLE 4. Calculated Depths Assuming Constant Accumulation

(Depth values, corresponding to the constant load increments, were read from depth-load curve of Figure 4.)

No.	Year	$A = 14$ g cm^{-2} yr^{-1}		$A = 20$ g cm^{-2} yr^{-1}	
		σ, g cm^{-2}	z, cm	σ, g cm^{-2}	z, cm
1	1961	14	33	20	48
2	1960	28	67	40	99
3	1959	42	104	60	151
4	1958	56	141	80	202
5	1957	70	176	100	251
6	1956	84	213	120	300
7	1955	98	246	140	346
8	1954	112	280	160	392
9	1953	126	314	180	436
10	1952	140	346	200	479
11	1951	154	377	220	522
12	1950	168	410	240	565
13	1949	182	440	260	605
14	1948	196	470		
15	1947	210	501		
16	1946	224	530		
17	1945	238	560		
18	1944	252	590		
19	1943	266	618		

between 1954 and 1955. This relationship is more strongly suggested in Figure 12 than in Tables 3 and 4. Between 1947 and 1954, the depth-time points are roughly parallel to the curve calculated for 20 g cm^{-2} yr^{-1}, whereas they parallel the 14 g cm^{-2} yr^{-1} curve from 1955 to 1961.

Measurements from the stake farm at old Byrd station provide a check on the stratigraphic interpretation. Average values measured over a four-year span are plotted in Figure 11. The measurements were not continuous, and so the values for each year are plotted side by side. The rates of snow accumulation can be used to estimate the accumulation in terms of water equivalent per year. To do this it is necessary (1) to assume that the monthly rates can be used to calculate annual accumulation if data are available for less than the full 12 months, and (2) to select a representative depth-load curve to convert from snow depth to water equivalent. The depth-load curves in the top meter of snow vary only slightly from place to place in the Byrd station area, and a representative curve can be used. Proceeding in this manner, the rates of accumulation calculated from the data in Figure 11 are listed in Table 5.

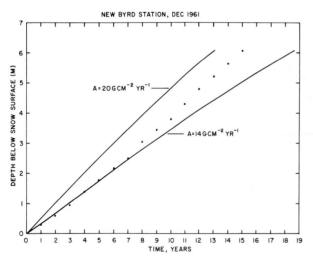

Fig. 10. Stratigraphically determined depths of the fall discontinuities according to estimate A in Table 3.

It is clear from Table 5 and Figure 11 that the rate varies within a given year and from one year to the next. For example, the average rate of 4 cm/month in 1958 includes the low value of 2.5 cm/month from July to December. This is similar to the nearly constant value of 3 cm/month measured over a 9-month span in 1960. Also, the highest value of 25.3 g cm^{-2} yr^{-1} obtained in 1959 is followed by the lowest value of 14.6 g cm^{-2} yr^{-1} in 1960. According to *Gow and Rowland* [1965], the accumulation values at Byrd station remained low during 1962, 1963, and 1964. However, regardless of the variability, the accumulation values calculated from the stake data are in reasonably good agreement with the values obtained from the stratigraphic analysis A.

TOPOGRAPHIC EFFECTS ON ACCUMULATION

The apparent change in rate of accumulation may represent a real change in the rate of snowfall, or it may be, at least in part, due to topographic effects. The topographic features in question are ridges, troughs, and step-like descents along slopes [*Swithinbank*, 1958; *Benson*, 1959, pp. 15–16; *Bader*, 1961, p. 14; *Black and Budd*, 1964, pp. 11–15; *Gow and Rowland*, 1965]. Although the forms are irregular, we can characterize their dimensions by 'wavelengths' of several kilometers and amplitudes of about 10 meters. They are not visible from the air, in contrast to the smaller waves associated with the cyclonic blizzards reported by *Dolgushin* [1961, p. 67]. Though the latter waves clearly migrate with the wind, there is evidence that the larger features (of concern to us) migrate against the wind [*Black and Budd*, 1964]. There are two important points about these larger features: (1) they control the local distribution of accumulation, there being larger amounts in the troughs than on the crests; and (2) migration of these features along the surface, at a rate unrelated to the flow of the glacier itself, would cause a variable rate of accumulation at a fixed point such as the place where a deep core might be obtained. This is a source of complication to anyone attempting to interpret deep cores [Benson, in *Bader*, 1962, Appendix B, p. 1].

In order to illustrate the potential problems involved, let us assume for simplicity that: (1) the rate of snowfall has remained constant in an area of several hundred square kilometers which includes both new and old Byrd stations; (2) the stratigraphic record from 1947 to 1961 is correctly interpreted; and (3) the variability is due solely to the

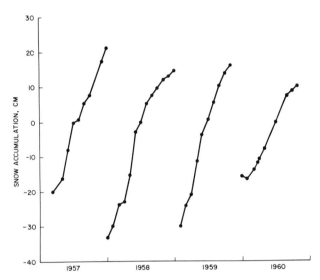

Fig. 11. Average accumulation of snow measured on 100 stakes over a 4-year period in the Byrd station stake farm. For convenience, the data from each year are plotted side by side with an arbitrary zero reference placed at mid-year. The data for 1957 were taken from *Anderson* [1958, pp. 65–68]; for 1958, from *Long* [1961, p. 13]; for 1959, from *Pirrit and Doumani* [1961, p. 9]; for 1960, from P. Parks (personal communication, 1962).

TABLE 5. Accumulation Calculated from Stake Data of Figure 11

Year	Time Interval Covered by Data, months	Avg. Rate Snow Accum. over Obs. Period, cm mo^{-1}	Extrap. Ann. Rate, cm yr^{-1}	Cal. Ann. Accum. in Water Equiv., g cm^{-2} yr^{-1}
1957	9.5	4.4	53	22.7
1958	12	4	48	20
1959	9	5	60	25.3
1960*	9	3	37	14.6

*The 1960 data are taken over a nine-month span to avoid the one month of ablation at the beginning of the record. If it is included, the time span is 10 months and the rate is 2.5 cm mo^{-1} (like the last half of 1958) rather than 3 cm mo^{-1}.

effects of mobile topographic features. In the light of these assumptions, Figure 10 would indicate that the local topographic setting changed in such a way that the rate of accumulation from 1955 to 1961 was about ⅓ less than it was from 1947 to 1954 (14 as compared to 20 g cm^{-2} yr^{-1}). *Gow*'s [1961] analysis of the 19-meter core from old Byrd, 10 km away from new Byrd, shows a similar variation for approximately the same time span. Furthermore, the variations at old and new Byrd stations are in the opposite directions; *Gow*'s [1961, Table IV] interpretation shows ⅓ more accumulation between 1954 and 1959 than between 1947 and 1954 (17 as compared to 12 g cm^{-2} yr^{-1}). According to our simplifying assumption of a constant rate of snowfall in the area encompassing both stations, we would conclude that the topographic settings at old and new Byrd stations changed in opposite ways.

Two questions come to mind: (1) how much variation in accumulation can be caused by topography; and (2) can this topography migrate and thereby produce variable accumulation at a specified point?

There are two sources of data that contribute to answering the first question. First, during the construction of new Byrd station, a topographic map was made (the contours on Figure 1 were derived from it) that showed the new station to lie in a broad depression (D. DeVicq, personal communication, 1961). This observation is compatible with the interpretation that the average accumulation rate for 15 years was 1.18 times greater at new than at old Byrd station. The 15-year time spans are roughly the same; specifically, the data are:

| 1945–1959 | Old Byrd | 14.7 g cm^{-2} yr^{-1} | *Gow* [1961, Table IV] |
| 1947–1961 | New Byrd | 17.3 | This paper, Table 3 and Figure 10 |

Second, *Gow and Rowland* [1965] established a base line on February 16, 1962, between new and old Byrd stations and measured accumulation at 0.5-km intervals along it in January 1963, January 1964, and January 1965. Maximum accumulation values along both lines (one parallel and one perpendicular to the prevailing winds) were twice as great as the minimums. Their measured variations in accumulation for three years along the 10-km line were directly related to topography. This makes it seem reasonable that the variation by a factor of 1.18, determined from 15 years of snow strata, at the ends of the line could also be due to topography.

The question about migration rates of the surface topography cannot yet be answered by direct observation. However, an interesting approach to this problem was made by *Black and Budd* [1964]. They observed, as did *Gow and Rowland* [1965], that maximum changes in rates of accumulation are associated with maximum changes in slope. An implication of this is that accumulation maximums and minimums do not coincide with corresponding minimums and maximums of the surface topography. Black and Budd argued that this will prevent the depressions from being filled in; instead, it will cause the topographic features to migrate. Furthermore, they show that the direction of this migration will be opposite to the direction of the wind. After making several simplifying assumptions, they arrived at a descriptive equation for the waveforms and calculated an average 'wave' velocity of 27 m yr^{-1} near Wilkes station. When the same equation is used with *Gow and Rowland*'s [1965] data, one calculates an average velocity of 13 m yr^{-1}. However, if one calculates the rate of motion for specific features, i.e., at the 2- and 6-km positions of *Gow and Rowland*'s Figure 1 [1965, p. 844], one arrives at rates of 11 and 41 m yr^{-1}, respectively. This introduces the complicating concept of individual features migrating at different rates.

Finally, it is of interest to consider how long it would take for such migratory features to move enough to affect the accumulation rate at a given station. If it is assumed that a feature would have to move 2 km, and if we use a range of rates from 10 to 40 km/yr, then the time required for accumulation at a given point to change from a maximum to a minimum value is of the order of 50 to 100 years.

FACIES PARAMETERS

It is useful to stress the similarities between the polar regions and to look for specific causes and effects to explain their differences. In this connection, it is interesting to compare conditions at selected points on the Greenland and Antarctic ice sheets. The facies classification of glaciers [*Benson*, 1962, 1967] provides a frame of reference, and the parameters that can be used to select points for comparison are the mean annual temperature and

TABLE 6. Byrd Station, Antarctica, Compared with Stations in Northwest Greenland
That Have Similar Rates of Accumulation and Mean Annual Temperatures

Station	Latitude	Longitude	Elevation, meters	Avg. Rate Accum., cm water equiv./year	Mean Ann. Temp., °C
Byrd	80°S	120°W	1500	17.3	-28
Greenland*					
Station 4-50	76°19'N	45°06'W	2720	17.5	-31
Station 4-25	76°38'N	45°42'W	2674	17.5	-31
Station 4-0	76°58'N	46°59'W	2616	16.5	-31
Station 2-225	77°04'N	48°01'W	2536	18.5	-31
Station 2-200	77°10'N	49°46'W	2460	22	-29
Station 2-175	77°03'N	51°20'W	2390	24	-28

*Traverse stations [Benson, 1962].

the rate of accumulation. Stations that can be considered are listed in Table 6.

Several Greenland stations are listed to indicate a range of values close to those measured at Byrd station. Where the temperatures are in close agreement, the Greenland accumulation values are slightly higher, and where the accumulation values are in close agreement, the Greenland temperatures are slightly lower. These stations with comparable rates of accumulation and mean annual temperatures have similar stratigraphic features, especially in the annual variation of snow density values, in the top 10 meters of snow, as should be expected. However, the range of density values is greater at Byrd station than at the Greenland stations (Figure 4). This is most easily demonstrated by numerically comparing the top 5 meters of snow; this depth includes several (about ten) annual units at each station, so it is meaningful in terms of facies. The comparison is made in Table 7.

The difference between ranges of density values measured at Byrd station and at Greenland stations with comparable temperatures and rates of accumulation is significant. Indeed, the Byrd station density and load values, considered by themselves, would not even fit into the dry-snow facies on the west slope of the Greenland ice sheet. The 5-meter load values in the dry-snow facies of the Greenland ice sheet are invariably less than 200 g cm^{-2}; they exceed 200 g cm^{-2} a little below the dry-snow line, and increase gradually through the percolation facies to about 225 g cm^{-2} near the saturation line [Benson, 1962, Fig. 46]. The snow at Byrd station clearly does not fit into the percolation facies because of the lack of melt evidence and the low mean

TABLE 7. Comparison of Snow Density and Load in the Top 5 Meters at the Stations Listed in Table 6

Station	Avg. Dens. in Top 5 m Snow, g cm^{-3}	Load 5 m Below Snow Surface, g cm^{-2}
Byrd	0.418	209
Greenland		
Station 4-50	0.356	178
Station 4-25	0.356	178
Station 4-0	0.350	175
Station 2-225	0.356	178
Station 2-200	0.370	185
Station 2-175	0.372	186

annual temperature. Therefore, the different ranges of density and load values expressed in Figure 4 and Table 7 require an explanation. Two factors seem to be involved: wind action and range of temperature. These factors varied only slightly from place to place on the west slope of the Greenland ice sheet, where the facies were originally defined, and consequently they were not treated as variables. However, they vary significantly between Byrd station and the Greenland stations considered here. Byrd station is windier and has a smaller temperature range; these differences favor the development of denser, harder snow at Byrd.

Range of Temperature

The average temperature of the warmest month exceeds the mean annual temperature by 15°C at Byrd station (Table 8). The average temperature of the warmest month at the specified Greenland stations exceeds the mean annual value by several de-

TABLE 8. Average Monthly Air Temperatures in Minus Degrees Centigrade

Byrd Station, Antarctica*

Month	1958	1959	1960	1961	1962	1963	1964	Avg.
Jan.	15	17	16	13	15	14	17	15.3
Feb.	15	18	24	19	23	21	22	20.3
March	28	25	32	29	28	31	25	28.2
April	30	23	30	24	37	29	31	29.1
May	27	33	33	36	29	34	31	31.8
June	41	33	31	34	41	28	32	34.2
July	41	34	34	36	36	40	32	36.1
Aug.	42	39	41	37	34	35	41	38.4
Sept.	40	36	34	33	43	37	35	36.8
Oct.	30	34	27	37	33	27	31	31.2
Nov.	22	23	24	22	22	19	19	21.6
Dec.	17	14	14	16	15	17	14	15.3
Avg.	29	28	28	28	29	28	27	28.1

	Northice, Greenland†				Site 2, Greenland‡				
Month	1952	1953	1954	Avg.	1953	1954	1955	1956	Avg.
Jan.		40.6	41.2	40.9		31.7	36.1	36.8	34.8
Feb.		46.3	37.2	41.8		30.6	31.1	35.7	32.4
March		44.7	41.8	43.2		33.9	34.4	35.2	34.5
April		31.5	33.1	32.3		23.9	30.7	26.7	27.1
May		22.6	20.4	21.5		16.1	16.4	19.5	17.3
June		12.4	13.2	12.8		8.9	8.3		8.6
July		10.2		10.2		6.1	8.2		7.2
Aug.		13.2		13.2	8.3	7.2	12.0		9.2
Sept.		26.1		26.1	20.0	Incomp.	18.9		19.4
Oct.		36.5		36.5	28.9	25.6	18.2		24.2
Nov.	39.2	40.8		40.0	35.0	27.8	31.7		31.5
Dec.	40.3	44.2		42.2	41.7	38.3	36.8		38.9
Avg.		30.8		31.8		23.6			23.8

*U.S. Weather Bureau data.

†Values were obtained by averaging the four synoptic readings (made at 0000, 0600, 1200, and 1800 GMT) tabulated by *Hamilton and Rollitt* [1957].

‡Station 2-100 [*Benson*, 1962, p. 45].

grees more than this. Although there are no records from these particular stations, there are records (Table 8) from stations 100 miles (160 km) west and 150 miles (240 km) northeast of them; namely, site 2 [*Benson*, 1962, pp. 44–45] and Northice of the British North Greenland Expedition [*Hamilton and Rollitt*, 1957, pp. 43–44]. The average temperature of the warmest month departs from the mean annual value by 17.7°C at site 2 and by 21.6°C at Northice. The data from the three stations have the same general appearance when graphed to show deviations from the mean annual temperature (Figure 12). The winter minimum is broader and departs less from the mean value than does the sharper summer peak. Data from Eismitte, at 71°N, 40°W, during 1930–1931 [*Sorge*, 1935] agree in phase with

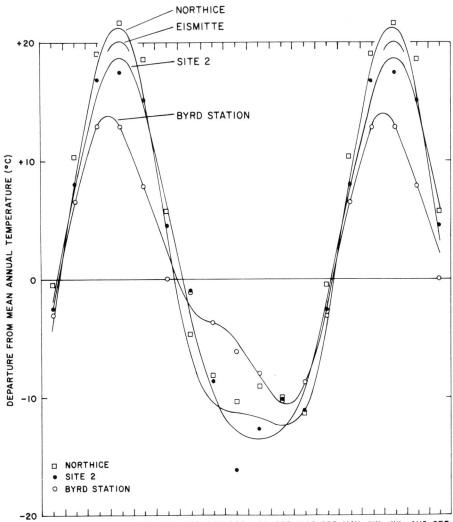

Fig. 12. Mean monthly temperature data for Byrd station, Antarctica, and site 2, Eismitte, and Northice in Greenland. The data are plotted as departures from the mean annual values, which are as follows: Byrd station, −28.1°C; Site 2 (station 2-100), −24.4°C; Eismitte, −31.0°C; Northice, −31.8°C.

the data from site 2, but the amplitude at Eismitte is about 10% greater [*Benson*, 1962, pp. 46–52 and Fig. 36]. Only the peak of the Eismitte curve is shown in Figure 12 between the curves for site 2 and Northice.

The curves drawn in Figure 12 were calculated from the first two terms of a Fourier series, which, in each case, gives a good representation of the data [*Benson*, 1962, pp. 44–46]. In addition to providing smooth curves that are easy to compare, the series yields numerical parameters that are useful for comparison. Since only the first two terms of the series are used, the temperature T can be expressed as a function of time t as follows:

$$T = f(t) = \sum_{n=0}^{2} a_n \cos nt + b_m \sin nt$$
$$= a_0 + a_1 \cos t + b_1 \sin t + a_2 \cos 2t \quad (1)$$
$$+ b_2 \sin 2t$$

TABLE 9. Fourier Coefficients for Equations 1 and 2 Calculated from the
Data in Table 8 and Used to Construct the Curves in Figure 12

Coeff.	Northice	Eismitte*	Site 2*	Byrd Station
a_0	$-31.8°C$	$-31.0°C$	$-24.4°C$	$-28.1°C$
a_1	-16.45	-17.46	-16.1	-10.37
b_1	0.97	1.63	0.55	-0.05
a_2	4.68	2.6	2.47	2.33
b_2	-1.12	-1.6	-0.20	-3.00
r_1	± 16.48	± 17.54	± 16.11	± 10.37
r_2	± 4.82	± 3.05	± 2.48	± 3.80
P	$-3°22'$	$-5°20'$	$-1°58'$	$0°17'$
Q	$-13°31'$	$-31°35'$	$-4°34'$	$-52°10'$

*Both the Eismitte and Site 2 data were corrected to give snow surface temperature. The difference between these values and the air temperature values measured in the instrument shelter is slight; it amounts to differences of 0.74°C and 0.6°C in the mean annual values at Eismitte and Site 2, respectively [Benson, 1962, p. 45]. Therefore, these corrections do not adversely affect the comparison of the four stations.

TABLE 10. Dates of Maximum Values for the
Harmonic Temperature Functions

Station	Annual Wave	Semiannual Wave	
Northice	July 12	July 8	Jan. 8
Eismitte	July 10	July 1	Jan. 1
Site 2	July 12	July 13	Jan. 13
Byrd station	Jan. 15	Dec. 19	June 19

This equation can also be written as a series involving only sine or cosine terms. In the latter case, it becomes:

$$f(t) = a_0 + r_1 \cos(t - P) + r_2 \cos(2t - Q) \quad (2)$$

where

$$r_1 = \pm(a_1^2 + b_1^2)^{1/2} \quad r_2 = \pm(a_2^2 + b_2^2)^{1/2}$$

$$P = \tan^{-1} b_1/a_1 \quad Q = \tan^{-1} b_2/a_2$$

The numerical values of the parameters are listed in Table 9. The parameters were determined by slightly idealizing the situation and using a 360-day year so that each month consists of 30 days (or 30 degrees). The time $t = 0$ occurs on January 15 for the Greenland stations and on July 15 for the Antarctic station. Thus the phase angles P and Q, expressed as degrees of arc (or days), indicate maximum values for the annual and semi-annual waves as shown in Table 10.

Departures from the mean for the Greenland stations increase with distance from the coast and with elevation. This may be interpreted as an increase in continentality. If we consider the amplitude of the annual temperature cycle as an index of continentality, it would follow that Byrd station is in a more maritime climatic setting than are the specific Greenland stations considered here. The more maritime nature of the climate at Byrd station is consistent with the generalized storm tracks across this part of West Antarctica shown by Astapenko [1960, pp. 52–56].

The lower summer temperatures at Byrd station cause the fall temperature gradient to occur in a lower range. Because vapor pressure decreases with decreasing temperature, there is less water vapor available for transport within the snow at Byrd station during fall. The result is less depth-hoar development and, therefore, higher-density snow at Byrd station than at the north Greenland stations, even though the mean annual temperatures are approximately the same at all the stations. The shifting of the temperature gradient to lower ranges with increased elevation along the traverse at 77°N in Greenland also produces a decrease in the amount of depth hoar in the fall discontinuity layer [Benson, 1962, p. 32].

Wind Action

The average wind speed at Byrd station is at least 2 knots greater than at the selected Greenland sta-

TABLE 11. Average Monthly Wind Speed in Knots

	Byrd Station, Antarctica*						
Month	1957	1958	1959	1960	1961	1962	1963
Jan.	10.5	13.1	12.2	9.9	13.1	11.7	12.0
Feb.	12.5	18.7	13.6	12.7	12.3	13.4	16.3
March	17.2	14.0	15.6	17.0	13.5	17.2	14.3
April	17.3	14.9	14.1	14.5	20.5	21.6	19.1
May	16.6	19.5	15.4	20.1	17.2	23.2	19.7
June	15.5	12.3	17.2	21.3	15.7	21.6	21.4
July	19.0	15.6	16.3	20.0	15.7	23.3	21.0
Aug.	23.5	15.9	18.6	15.7	22.2	21.1	17.3
Sept.	22.3	19.0	18.3	21.0	19.7	16.5	20.3
Oct.	15.0	17.6	14.9	19.8	14.4	20.3	18.6
Nov.	16.6	15.8	14.4	14.1	14.0	14.5	16.4
Dec.	13.5	12.4	11.5	13.5	12.0	14.0	8.8
Yearly Avg.	16.6	15.7	15.2	16.6	15.9	18.2	17.1

Average of 7 years' data, 16.5 knots

	Northice, Greenland†			Camp Century, Greenland‡			
Month	1952	1953	1954	1960	1961	1962	1963
Jan.		13.6	17.8		10	14	17
Feb.		13.7	18.2		10	15	14
March		14.2	17.1		11	13	12
April		15.0	15.2		11	11	12
May		13.7	10.0		10	13	11
June		11.1	9.1		9	11	8
July		13.7			10	10	12
Aug.		15.3			12	13	9
Sept.		13.8			11	11	13
Oct.		14.8		11	13	12	14
Nov.	17.8	15.2		12	13	11	
Dec.	14.5	15.5		11	14	12	
Avg.		14.1			11	12	

Average of 20 months' data, 14.5 knots   Average of 3 years' data, 12 knots

*U.S. Weather Bureau Data.
†Hamilton and Rollitt [1957].
‡Bates [1965]. Camp Century is near station 2-0 [Benson, 1962]; climatologically, it is similar to site 2 (station 2-100, according to Haywood and Holleyman [1961].

tions (Table 11), producing more wind packing at Byrd station. The average wind speed at the selected Greenland stations probably lies within the range of 12 and 14.5 knots determined at Camp Century and Northice, respectively. The Byrd station average of 16.5 knots represents a significant increase in terms of the effect of wind on the snow.

This is illustrated by the observation of *Diamond and Gerdel* [1957, p. 1] at site 2:

> When wind speeds were below 15 knots, blowing snow was reported about 4–7% of the time. When wind speed was between 15–19 knots, blowing snow occurred 28–32% of the time. At 20–24 knots, blowing snow was recorded 54–67% of the time, and for periods when the

wind speed was in excess of 30 knots, blowing snow occurred 79–92% of the time.

The data from Northice and Camp Century, which provide incomplete coverage of only 3 or 4 years, are not as satisfactory as the long-term U.S. Weather Bureau records at Byrd station. Nevertheless, it is clear that Byrd station is windier. Although this is well illustrated in Table 11, it can be emphasized by comparing the maximum values. The maximum monthly average value of 19 knots listed for Camp Century was exceeded during 35 months, or 41.7% of the Byrd station record. The corresponding maximum of 18.2 knots at Northice was exceeded during 27 months, or 32.2% of the Byrd station record. Also, the average monthly wind speed never exceeded 20 knots in the records from Camp Century or Northice, but they exceeded 20 knots during 17 months, or 20.2% of the records from Byrd station.

It seems clear that both temperature range and wind action vary in such a way as to reinforce each other in causing the observed differences in density between Byrd station and the Greenland stations with comparable mean annual temperatures and rates of accumulation.

Acknowledgments. The hospitality and assistance provided by Lt. David De Vicq, commander of U.S. Navy MCB-1 at new Byrd station, and by Mr. Kendall Moulton, U.S. Antarctic Research Program representative at old Byrd, are gratefully acknowledged. Thanks are also due Mr. Perry Parks for providing his unpublished data from Byrd station. I am especially pleased to express my thanks to Dr. R. P. Goldthwait and Dr. Colin Bull of the Institute of Polar Studies, Ohio State University, for providing the opportunity to make this study and for their patience and encouragement during the time involved in completing it, and to the National Science Foundation and Ohio State University, whose support made it possible. Assistance in preparation of drawings and typescript for this paper was provided by a grant from Dr. Terris Moore, former president of the University of Alaska. Dr. John Mercer, Mr. Arthur S. Rundle, and Dr. Colin Bull made valuable comments on an early draft of the manuscript.

Contribution 123, Institute of Polar Studies, Ohio State University.

REFERENCES

Anderson, V. H., USNC-IGY Antarctic glaciological data, field work 1957 and 1958, *Rep. 825-1,* part 2, Ohio State Univ. Res. Found., Columbus, 269 pp., 1958.

Astapenko, P. D., Atmospheric processes in the high latitudes of the southern hemisphere, translated from Russian by Israel Program for Scientific Translations, 1964, 286 pp., 1960.

Bader, H., The Greenland ice sheet, *Cold Regions Sci. Eng.,* part 1, sect. B2, Cold Regions Res. Eng. Lab., Hanover, N.H., 18 pp., 1961.

Bader, H., Scope, problems and potential value of deep core drilling in ice sheets, *Spec. Rep. 58,* Appendices A and B, Cold Regions Res. Eng. Lab., Hanover, N.H., 6 pp., 1961.

Bates, R. E., Camp Century wind and temperature summaries from October 1960 through October 1963 continuous, unpublished summaries available in the files at the Cold Regions Res. Eng. Lab., Hanover, N.H., 1965.

Benson, C. S., Physical investigations on the snow and firn of northwest Greenland during 1952, 1953, and 1954, *Res. Rep. 26,* Snow, Ice, Permafrost Res. Estab., Wilmette, Ill., 62 pp., 1959.

Benson, C. S., Stratigraphic studies in the snow and firn of the Greenland ice sheet, *Res. Rep. 70,* Appendices A–C, Snow, Ice, Permafrost Res. Estab., Wilmette, Ill., 93 pp. (summarized in *Folio Geograph. Danica, 9,* 13–35, 1961), 1962.

Benson, C. S., Polar regions snow cover, in *Physics of Snow and Ice,* Proceedings of the Sapporo Conference, 1966, part 2, pp. 1039–1063, Inst. Low Temp. Science, Hokkaido Univ., Sapporo, Japan, 1967.

Black, H. P., and W. Budd, Accumulation in the region of Wilkes, Wilkes Land, Antarctica, *J. Glaciol., 5*(37), 3–15, 1964.

Crary, A. P., Results of United States Traverse in East Antarctica, 1958–1961, *I.G.Y. Glaciol. Rep. 7, World Data Center A, Glaciology,* American Geographical Society, New York, 144 pp., 1963.

Diamond, M., and R. W. Gerdel, Occurrence of blowing snow on the Greenland ice cap, *Res. Rep. 25,* Snow, Ice, Permafrost Res. Estab., Wilmette, Ill., 1957.

Dolgushin, L. D., Zones of snow accumulation in eastern Antarctica, International Association of Scientific Hydrology, I.U.G.G., General Assembly of Helsinki, Symposium on Antarctic Glaciology, *I.A.S.H. Publ. 55,* pp. 63–70, 1961.

Gow, A. J., Drill hole measurements and snow studies at Byrd station, Antarctica, *Tech. Rep. 78,* Cold Regions Res. Eng. Lab., Hanover, N.H., 1961.

Gow, A. J., On the accumulation and seasonal stratification of snow at the South Pole, *J. Glaciol., 5*(40), 467–477, 1965.

Gow, A. J., and R. Rowland, On the relationship of snow accumulation to surface topography at 'Byrd Station,' Antarctica, *J. Glaciol., 5*(42), 843–847, 1965.

Hamilton, R. A., and G. Rollitt, British North Greenland Expedition 1952–54, Climatological Tables for the Site of the Expedition's Base at Britannia Sø and the Station on the Inland-Ice 'Northice,' *Meddelelser om Grønland, 158*(2), 1957.

Haywood, L. J., and J. B. Holleyman, Climatological means and extremes on the Greenland ice sheet, *Res. Rep. 78,* Cold Regions Res. Eng. Lab., Hanover, N.H., 1961.

Koerner, R. M., Firn stratigraphy studies on the Byrd-Whitmore mountains traverse, 1962–1963: *Antarctic Snow* and Ice Studies, Antarctic Res. Ser. vol. 2, pp. 219–237, AGU, Washington, D.C., 1964.

Long, W. E., Glaciology, Byrd station and Marie Byrd Land traverse, 1958–1959, *Rep. 825-2,* part 11, Appendices 1–2, Ohio State Univ. Res. Found. 20 pp., Columbus, 1961.

Orheim, O., Surface Snow Metamorphosis on the Antarctic Plateau, *Norsk Polarinstitutt Årbok 1966,* pp. 84–91, Oslo, Norway, 1968.

Pirrit, J., and G. A. Doumani, Glaciology, Byrd station and Marie Byrd Land traverse, 1959–1960, *Rep. 968-2,* Appendices 1–4, Ohio State Univ. Res. Found. 11 pp., Columbus, 1961.

Shimizu, H., Glaciological studies in West Antarctica, 1960–1962, *Antarctic Snow and Ice Studies, Antarctic Res. Ser.,* vol. 2, pp. 37–64, AGU, Washington, D.C., 1964.

Sorge, E., Glaziologische Untersuchungen in Eismitte (Glaciological investigations in Eismitte), in *Wissenschaftliche Ergebnisse der Deutschen Grönland Expedition Alfred Wegener 1929 and 1930–1931,* vol. 3 (text in German), F. A. Brockhaus, Leipzig, 1935.

Swithinbank, C. W. M., *Norwegian-British-Swedish Antarctic Expedition, 1949–52, Sci. Results 3,* Glaciology 1, parts D and E, Norsk Polarinstitutt, Oslo University Press, Oslo, 1958.

Weller, G. The heat and mass balance of snow dunes on the Central Antarctic Plateau, *J. Glaciol.,* 8(53), 277–284, 1969.

INVESTIGATION OF PARTICULATE MATTER IN ANTARCTIC FIRN

Wayne L. Hamilton and M. E. O'Kelley

Institute of Polar Studies, Ohio State University, Columbus 43210

Frequency-size distribution data for water-insoluble microparticles in firn from C site, at the base of the Antarctic Peninsula, and from the South Pole are discussed and compared with measurements from other locations with the following results: (1) At C site, the recent annual mass accretion rate of particulate material (about 2.5 μg cm^{-2} yr^{-1}) appears to be relatively constant and independent of the annual snow accumulation. (2) In general, higher density firn at the South Pole contains a higher concentration of particulate material than the lower density layers. The recent annual mass accretion rate of dust is about 0.5 μg cm^{-2} yr^{-1} at Pole station. (3) Meltwater specific electrical conductance values for South Pole firn range about 1.1 to 2.8 μmhos cm^{-1}. Measured values correspond to salt concentrations (as NaCl) of between 75 and 500 ppb. The average dust concentration at Pole station is only 24 ppb. (4) Annual accretion of particulate material and particle concentration in firn are related to elevation at the base of the Antarctic Peninsula and in the interior of the continent by inverse exponential functions. (5) Precipitation at Byrd station was about 14 times dirtier in 1959 than in the years ca. 600 A.D., and the cleanest antarctic surface snow so far measured is about 2.6 times dirtier than the 600 A.D. ice. A recent large increase in atmospheric turbidity is indicated.

Earlier investigations of microparticles in polar glacier ice and firn have been reported by *Marshall* [1962], *Taylor and Gliozzi* [1964], and *Bader et al.* [1965]. These investigators used the Coulter Counter, an electrical sensing zone size analyzer, to count micron-size particles as a function of size in melted samples. They determined that the volume concentration of particles varied with depth in the ice sheet. *Hamilton* [1967] and *Hamilton and Langway* [1967] established that this variation reflects the seasonally changing dust content of precipitation, and that several hundred years ago late winter–early spring precipitation contained more dust than that falling at other times of the year in north central Greenland. Here data are presented for precipitation that fell in the past three decades in Antarctica.

For ease of comparison with earlier papers, the volume concentration of particles in sample meltwater is given in mk units. The volume concentration in parts per billion is approximately 5 mk. The absolute value of the slope of the log-log cumulative frequency-size distribution between about 0.6 and 1.5 μm diameter is given by m, and k is the number of particles per μl larger than 1 μm^3. The derivation of mk is given by *Bader et al.* [1965].

A detailed description of the mechanism of deposition of atmospheric dust on polar ice sheets has been given by *Hamilton* [1969], and the current report is a link between earlier (pre-1966) investigations and that account.

The sampling sites discussed below are shown on the map in Figure 1.

C SITE, ANTARCTIC PENINSULA

The firn core taken at C site (73°W, 75°S) spans the depth interval from 1.01 to 7.91 meters relative to the 1961–1962 summer surface. It was cut into small segments several centimeters long, which were trimmed to remove possible contamination, melted, and analyzed in the clean room laboratory at the Institute of Polar Studies.

Values of mk versus water equivalent depth (sample length $\times$ sample density) are plotted with the firn stratigraphy in Figure 2. Zero depth in the figure corresponds to 1.01-meter depth in the firn, because density was not measured above 1.01 meters. The hoar layers indicated in the figure are

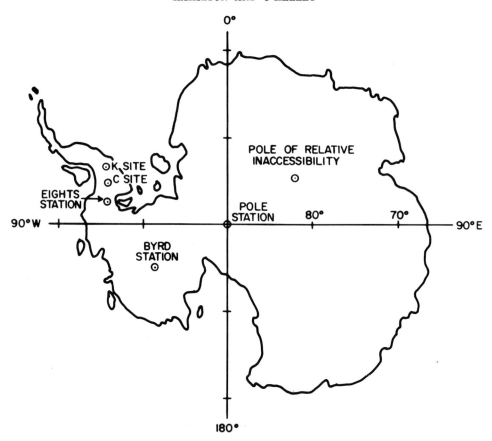

Fig. 1. Map of Antarctica showing location of Pole station, Byrd station, Eights station, C site, K site, and Pole of Relative Inaccessibility.

zones of coarse crystals developed on and within the snow, and the crusts are caused both by wind packing and solar radiation.

The mk-depth profile has been subjected to the 'runs test' of randomness to examine the possibility of random variation. The null hypothesis of randomness was rejected at a significance level of 0.05.

Peaks in the mk profile have been selected by inspection to determine the wavelength, in centimeters water equivalent, of dust concentration. Criteria for selecting the peaks were that a peak should rise above the base line at least 6 mk units and should consist of at least two relatively dirty samples set off from neighboring peaks by much cleaner samples. The method of selection is subjective, but in the absence of more field data it is believed to be a satisfactory approach. Cycles are represented by dashed lines in Figure 2.

Depending on whether or not the questionable peak at 221 cm is counted, the average wavelength of cycles of dust concentration is either 22.8 or 24.4 cm water equivalent. *Shimizu* [1964] measured annual snow accumulation in the upper 2 meters of firn at C site by conventional stratigraphic techniques and reported a value of 25.1 cm water equivalent per year. Our reinterpretation of his field data yields a somewhat lower value: 23.3 cm water. If we accept Shimizu's result, the conclusion is that there is an annual cycle of dust accumulation at C site. This interpretation differs from that of *Gliozzi* [1966], who observed two cycles per year in the mk profile in firn from nearby Antarctic Peninsula sites.

There is no obvious correlation between mk and stratigraphic features, so we are unable to say that peaks correspond to a certain season's snow accumulation.

SOUTH POLE PIT

More than 100 firn samples were collected in vertical profile from the wall of a 2-meter pit excavated at an accumulation stake near Pole station during the austral summer 1964–1965. Samples were cut from the firn with a 1-inch-diameter stainless steel cylinder, placed in clean plastic bottles, and shipped frozen to the Institute of Polar Studies.

In the middle part of the profile (Figure 3), the sampling was nearly continuous. In the upper and lower parts, the samples were spaced 3 to 4 cm apart. A number of duplicate samples were taken in the upper half of the pit, horizontally adjacent to the main sample profile.

In the laboratory, samples were selected for analysis in random order. The particle concentration data for all the samples are plotted in Figure 3, along with density data, accumulation stake chronology, and the levels of δO^{18} (oxygen isotope) 'summers' [*Picciotto*, 1967] in the same wall of the same pit.

The South Pole dust concentration data have also been investigated by the 'runs test' of randomness, and the null hypothesis of randomness was rejected at a significance level of less than 0.05. There were fewer runs (more peaks) than would be expected in a random population.

Figure 3 shows that at the South Pole there is a general correspondence between dirty layers and high firn density. The mk peaks at 68, 80, and 138 cm are notable exceptions. *Giovinetto* [1960] found that South Pole firn density is usually higher in winter layers; therefore it may be that precipitation falling during the coldest months is dirtier than that occurring at other times of the year at the South Pole. This seasonal relationship is similar to that observed in Greenland ice by *Hamilton and Langway* [1967]. The absence of an obvious correlation between oxygen isotope and microparticle stratigraphy at Pole station may be due to post-depositional modification of the snow, either by wind mixing of the surface layer or by vapor transfer within the layers.

The South Pole pit data do not appear to conform to a smoothly cyclical pattern such as that observed elsewhere. The irregularities in the mk profile may be due in part to deflation and redeposition of snow by wind. The fact that the upper and lower parts of the profile appear to have better-developed cycles than the middle part suggests that the pattern may be related to the length of the sampling interval.

DISCUSSION

Annual Mass Accretion of Dust

The snow stake accumulation measurements at Pole station permit determination of the volume of an-

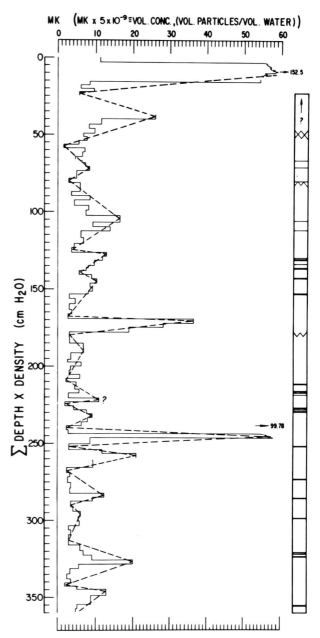

Fig. 2. C site, Antarctic Peninsula. The mk profile goes from 1.01 to 7.91 meters depth relative to the 1961-1962 summer surface. Depth is given in centimeters water equivalent. Concentration cycles are represented by dashed line. Hoar layers and crusts are represented in the stratigraphy column by zig-zag and straight lines, respectively.

nual particle deposition. About 0.5 μg of dust (assumed density 2.5 g cm^{-3}) accumulates annually on 1 cm^2 of the surface at the South Pole.

In Figure 4, values of integrated average mk for annual layers at C site (determined from the 12

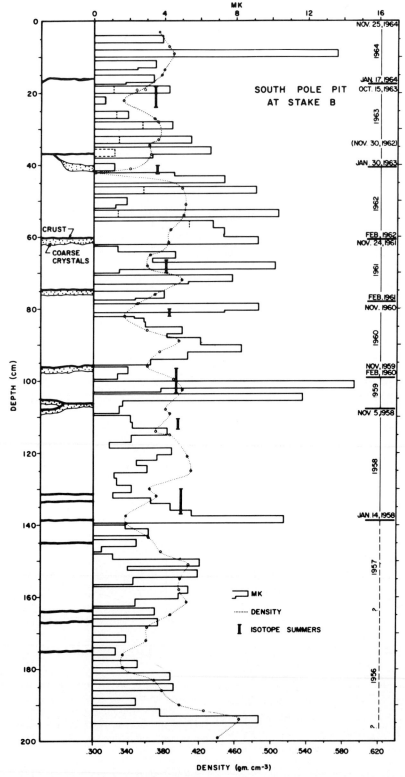

Fig. 3. South Pole pit at accumulation stake B. The upper and lower abscissas represent, respectively, *mk* and density. *Picciotto*'s [1967] isotope summers and accumulation stake chronology are shown.

complete trough to trough cycles represented by the dashed line in Figure 2) are plotted against annual snow accumulation (g cm^{-2} yr^{-1}). The calculated, average mass accretion rate of dust is about 2.5 μg cm^{-2} yr^{-1}. The solid curve in the figure represents the model for which a constant annual deposit of particulate matter (2.5 μg) is mixed with snow deposition, which varies in amount from year to year. The particle concentration data are in rough agreement with the model illustrated by the curve.

South Pole Meltwater Conductance

Meltwater conductance is a measure of the soluble salt concentration in the firn [*Gow*, 1968a], and it has been shown by *Murozumi et al.* [1969] that major ionic species in polar precipitation are present in approximately marine salt abundance ratios. It follows that variations of conductance values in melted samples from a vertical profile through very cold firn provide a measure of the changing contribution of salt spray deposited with snow.

In Figure 5, South Pole firn meltwater conductance measurements are illustrated. Control samples of deionized, filtered laboratory water gave better than 1% repeatability at the 1.8 μmho cm^{-1} level. Measured values have been temperature-corrected to 20°C, and as much as 0.8 μmho cm^{-1} may be due to CO_2 dissolved in the water.

The samples were selected for analysis in random order; therefore, the trends in the profile are certainly real. The conductance values seem to conform to two wave shapes; one cycling between about 1.1 and 1.9 μmho cm^{-1} over a time interval of years, and the other superimposed on the first and varying over an interval of only a few months. It seems possible that the long-period variation is re-

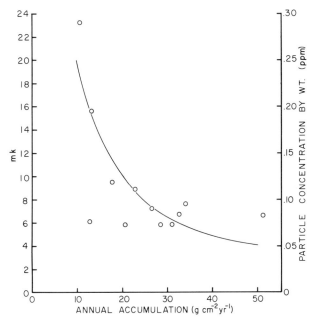

Fig. 4. Integrated average mk versus microparticle concentration wavelength (assumed to be annual water accumulation) on C site mk profile. Curve represents locally constant annual fallout model.

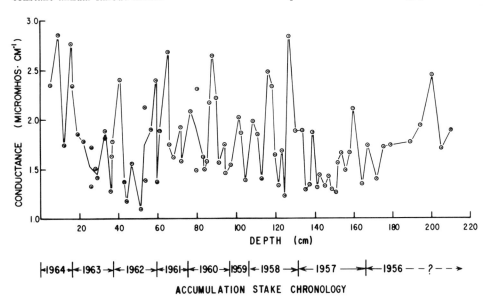

Fig. 5. Meltwater specific electrical conductance, South Pole accumulation stake B.

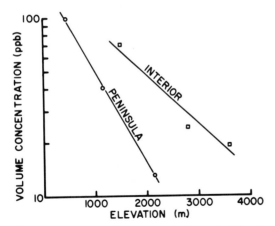

Fig. 6a. Average dust concentration (ppb) at Antarctic sites over the past several decades versus site elevation. Antarctic Peninsula and interior sites are represented by circles and squares, respectively. K and Eights station data are from *Gliozzi* [1966].

lated to changing intensity of meridional circulation or low-level tropospheric turbulence. The short-period variation may indicate the arrival of a succession of frontal storms.

Assuming that the samples were CO_2-saturated, the extreme measured values correspond to NaCl concentrations of between 75 and 500 ppb. Salt is much more abundant than dust, as *Murozumi et al.* [1969] showed for Byrd station firn.

Particle Accumulation versus Elevation

Values of particle concentration and annual particle accumulation are plotted in Figures 6a and b for C site and the South Pole, as well as two other sites at the base of the Antarctic Peninsula (Eights station and K site) and two additional sites in the interior (Byrd station and the Pole of Relative Inaccessibility). The Byrd station value is from the 1959 layer (Hamilton, unpublished). The dust concentration for the Pole of Relative Inaccessibility is from *Hamilton* [1969].

The Eights station and K site values of particle concentration were calculated from the original data of *Gliozzi* [1966]. Annual snow accumulation values used in deriving the dust accumulation values in Figure 6b are presented in Table 1. Two or more values were used at each peninsula site, because the uncertainty of the snow accumulation measurements is much greater there than in the interior.

Figure 6a shows that there is an inverse, exponen-

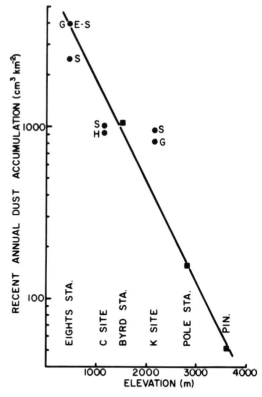

Fig. 6b. Average annual volume of dirt accretion (cm^3 km^{-2}) over the past several decades versus site elevation. Antarctic Peninsula and interior sites indicated by circles and squares, respectively. K and Eights station microparticle data are from *Gliozzi* [1966]. The snow accumulation data used to calculate plotted values of annual dust accretion (indicated by letters beside plotted points) are given in Table 1.

tial relationship between dust concentration and elevation, both at the peninsula sites and in the interior. It is not surprising that the values for peninsula and interior sites fit two different exponential functions; the precipitation regimes are quite different. Temperature, surface slope, and annual snow accumulation are all greater near the coast. The measurements show a greater dust concentration in the interior than on the peninsula for any given elevation. This suggests that toward the interior the air may become somewhat deficient in moisture relative to particulate material. Perhaps dust is being added to the interior by some process not associated with frontal storms moving in from the coast, for example subsidence in the upper atmosphere.

The annual dust deposition values plotted in Figure 6b also seem to be related to elevation by an

TABLE 1. Snow Accumulation for the Sampled Interval at Various Antarctic Sites

Site	Accumulation g cm^{-2} yr^{-1}	Reference	Remarks
Eights station	37.4–42.1	*Gliozzi*, 1966	2 microparticle cycles per year
	40	*Epstein and Sharp*, 1967	1 δO^{18} cycle per year
	25.1	*Shimizu*, 1964	Pit-wall stratigraphy
C site	24.4	This paper	1 microparticle cycle per year
	25.1	*Shimizu*, 1964	Pit-wall stratigraphy
K site	56.3–63.3	*Gliozzi*, 1966	2 microparticle cycles per year
	73.4	*Shimizu*, 1964	Pit-wall stratigraphy
Byrd station	15.5	*Gow*, 1968b	Pit-wall stratigraphy
Pole station	6.5	*Picciotto*, 1967	Fission-product activity
Pole of Relative Inaccessibility	2.7	*Picciotto*, 1967	Fission-product activity

inverse exponential function. At sites where annual snow accumulation is well known (Eights station, Byrd station, Pole station, and the Pole of Relative Inaccessibility), values fall on or very close to the line drawn in the figure. At C site and K site, however, there is not enough confidence in the stratigraphy to be able to say whether or not the scatter of dust accumulation values away from the line is representative.

Increasing Dust Deposition at Byrd Station

Figure 7 shows the increase in precipitation dirtiness (mk) at Byrd station over the past fourteen centuries. The recent value is from the 1959 layer. Firn from the 1960 level and above is more than twice as dirty as the 1959 layer, but it was not used in this analysis because it may contain station contamination. The values represented by squares and circles are from *Bader et al.* [1965] and *Hamilton* [1967], respectively. The sample ages have been calculated from the stratigraphy of *Gow* [1968b] and may be subject to some reinterpretation in view of more recent estimates of ice chronology [e.g., *Epstein et al.*, 1970].

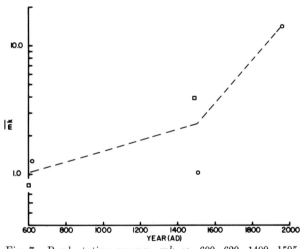

Fig. 7. Byrd station average mk ca. 600, 620, 1490, 1505, and 1959 A.D. The 1959 value is from this laboratory (Hamilton, unpublished). Values from *Bader et al.* [1965] are shown by squares, and those given by *Hamilton* [1967] are shown by circles.

These data strongly suggest a recent large in-

TABLE 2. Concentrations of Dust in Recent Antarctic Firn

Sampling Site	Depth, meters	Dust Vol. Conc., ppb
Eights station, 1961–1962	4–22	98
C site, 1961–1962	1–8	40
K site, 1961–1962	3–24	13
South Pole, 1964–1965	0–2	24
Byrd station, 1963–1964	1.4–1.85	70
Pole of Relative Inaccessibility, 1965–1966	0–1	19

crease in the amount of dust deposition in Antarctica. Nowhere have we found, near the surface, snow strata with average dust concentrations as low as those found at 250 meters depth (~600 A.D.) beneath Byrd station. Those sample sequences contained only about 5 ppb dust, compared with 13 ppb at K site, the cleanest surface site.

Such large time (depth) variations in dust content can provide valuable information on changing atmospheric turbidity, and discrete dirty layers can serve as useful index horizons for tracing flow lines within the ice sheet.

Tabulated Dust Concentrations

Table 2 gives average volume concentrations (ppb) of dust in recent antarctic firn. Depths are relative to the surface at the sampling time.

Acknowledgments. We thank James Gliozzi for collecting the firn samples at the South Pole, Roy M. Koerner for valuable discussions, and Colin Bull and Arthur Mirsky for their constructive review of the manuscript.

This study was financed by NSF grant GA-125 awarded to the Ohio State University Research Foundation with additional support from the Ohio State University.

Contribution 91 from the Institute of Polar Studies, Ohio State University.

REFERENCES

Bader, H., W. L. Hamilton, and P. L. Brown, Measurement of natural particulate fallout onto high polar ice sheets, 1, Laboratory techniques and first results, *Res. Rep. 139*, Cold Regions Res. Eng. Lab., Hanover, N.H., 1965.

Epstein, S., and R. P. Sharp, Oxygen- and hydrogen-isotope variations in a firn core, Eights station, Western Antarctica, *J. Geophys. Res.*, 72(22), 5595–5598, 1967.

Epstein, S., R. P. Sharp, and A. J. Gow, Antarctic ice sheet: Stable isotope analyses of Byrd station cores and inter hemispheric climatic implications, *Science*, 168(3939), 1570–1572, 1970.

Giovinetto, M. B., Glaciology report for 1958, South Pole station, *Ohio State Univ. Res. Found. Rep. 825-2-Part 4*, 1960.

Gliozzi, J., Size distribution analysis of microparticles in two Antarctic firn cores, *J. Geophys. Res.*, 71(8), 1993–1998, 1966.

Gow, A. J., Electrolytic conductivity of snow and glacier ice from Antarctica and Greenland, *Res. Rep. 248*, Cold Regions Res. Eng. Lab., Hanover, N.H., 1968a.

Gow, A. J., Deep core studies of the accumulation and densification of snow at Byrd station and Little America V, Antarctica, *Res. Rep. 197*, Cold Regions Res. Eng. Lab., Hanover, N.H., 1968b.

Hamilton, W. L., Measurement of natural particulate fallout onto high polar ice sheets, 2, Antarctic and Greenland cores, *Res. Rep. 139*, Cold Regions Res. Eng. Lab., Hanover, N.H., 1967.

Hamilton, W. L., Microparticle deposition on polar ice sheets, *Ohio State Univ. Inst. Polar Studies Rep. 29*, 1969.

Hamilton, W. L., and C. C. Langway, Jr., A stratigraphic correlation of microparticle concentrations with oxygen isotope ratios in 700 year old Greenland ice, *Earth Planet. Sci. Lett.*, 3(4), 363–366, 1967.

Marshall, E. W., The stratigraphic distribution of particulate matter in the firn at Byrd station, Antarctica, in *Antarctic Research, Geophys. Monograph 7*, pp. 185–196, AGU, Washington, D.C., 1962.

Murozumi, M., Tsaihwa J. Chow, and C. Patterson, Chemical concentrations of pollutant lead aerosols, terrestrial dusts, and sea salts in Greenland and Antarctic snow strata, *Geochim. Cosmochim. Acta*, 33(10), 1247–1294, 1969.

Picciotto, E. E., Geochemical investigations of snow and firn samples from East Antarctica, *Antarctic J. U.S.*, 2(6), 236–240, 1967.

Shimizu, H., Glaciological studies in West Antarctica 1960–1962, in *Antarctic Snow and Ice Studies, Antarctic Res. Ser.*, vol. 2, edited by M. Mellor, pp. 37–64, AGU, Washington, D.C., 1964.

Taylor, L. D., and J. Gliozzi, Distribution of particulate matter in a firn core from Eights station, Antarctica, in *Antarctic Snow and Ice Studies, Antarctic Res. Ser.*, vol. 2, edited by M. Mellor, pp. 267–277, AGU, Washington, D.C., 1964.

GLACIAL GEOLOGY OF THE VICTORIA VALLEY SYSTEM, SOUTHERN VICTORIA LAND, ANTARCTICA

PARKER E. CALKIN

Department of Geological Sciences, State University of New York, Buffalo 14214

In southern Victoria Land, Antarctica, the inland ice plateau is bounded by a north-south mountain range through which outlet glaciers have carved valleys. Most of these valleys are still ice filled, but an amelioration of climate has caused the glaciers to retreat from some, including the five valleys that constitute the Victoria Valley system. Two major glaciations are recorded in the Victoria Valley system, but they may have been preceded by others. The first distinguishable glaciation, the Insel glaciation, was an eastward flow of ice from the inland plateau through the valleys to the coast. The Insel drift includes very silty till and erratic pebbles and cobbles on mesas 300 to 600 meters above the valley floors. The till lacks morainal topography, and upstanding boulders are rare. During the recessional phase of the Insel glaciation, deep meltwater channels were cut. Since the end of the glaciation, the shapes of the major valleys have not changed significantly. The second, or Victoria, glaciation was marked by strong invasions from local ice fields and from the coast, and weaker invasions from the inland ice plateau. This glaciation, which probably began more than 30,000 years B.P., is subdivided into three episodes. The Bull drift episode included the most extensive glaciers. At the maximum, the area was invaded by at least six glacier tongues that extended up to 20 km beyond their present positions, nearly filling the valley system. Two large end moraines are well preserved, but most of the morainal topography is now subdued. During the subsequent Vida drift episode, the regimen was more vigorous. The retreat of the glaciers from their maximums of the Bull drift episode stopped about 10 km from their present positions. The glaciers readvanced locally. Large outwash fans and kames formed at the borders of proglacial lakes. With continued retreat, thick ground and end moraines were deposited. These moraines are moderately well preserved and hummocky, standing several meters above adjacent deposits of the Bull drift episode. Upstanding boulders are much more plentiful than on the older drifts, but are cavernously weathered. Vida till is very sandy. During the Packard drift episode, which continues to the present, the glacier regimen has been less vigorous. Most of the deposits represent a slow regular retreat of the glaciers to their present positions. The Packard drift occurs largely as ground moraine, but includes areas of kame and kettle topography, and very bouldery ablation moraine still ice-cored. In most areas, there is no sharp break in weathering between the Packard and Vida deposits, but the Packard are generally fresher than the Vida deposits.

Almost all of the antarctic continent is ice-covered, but around the margins are small ice-free areas exposed by recession of local glaciers and the main ice sheet. The largest single ice-free area is in the rugged mountain and valley area of southern Victoria Land (Figure 1) west of McMurdo Sound. This report concerns part of this ice-free area, a region of 45 km east-west by 28 km north-south, comprising five interconnected valleys referred to here as the Victoria Valley system (Figures 2 and 3).

Studies of land forms and surficial deposits were undertaken there in the hope of gaining knowledge of the glacial history of Antarctica as displayed by fluctuations of the ice at the edge of the continent.

Most of the field work was done in the 1960–1961 and 1961–1962 seasons. A brief period was also spent in the area in 1968–1969. Mapping initially undertaken by sketching, ground photography, and on oblique air photographs, was later transferred to U.S. Navy vertical photographs, and in turn to a U.S.G.S. topographic map enlarged to a scale of 1:50,000 (Figure 2).

Some of the unnamed geographic features of the valley system have been given names to simplify this report and the maps. These names include:

Bullseye Lake
Balham Lake
Webb Lake

Victoria Upper Lake
Haselton ice fall
Webb ice fall
McKelvey moraine
Bullseye moraine
Orestes Valley

DESCRIPTION OF AREA

The inland ice of East Antarctica is bounded along the west side of the Ross Sea and Ross ice shelf by a mountain range that extends along 160°E longitude from 67° to 85°S latitude. The Ross Sea part is known as Victoria Land. Here outlet glaciers from the inland ice have shaped east-west valleys that are still occupied by large glaciers, but in the Victoria Valley system and Wright Valley glaciers have disappeared, leaving an area of 4000 km² almost free of ice (Figure 3).

The Victoria Valley system is bounded on the west by the inland ice sheet, on the north and south by mountain ranges whose peaks reach from 100 meters in the east up to 2500 meters above sea level in the west (Figure 2), and on the east by Wilson Piedmont glacier.

The system consists of five main interconnecting valleys lying at 400 to 1000 meters above sea level. Of the 246 km² of valley floor, more than 75% is mantled by glacial and colluvial deposits. Four large glacier tongues flow into the valley system from the north or east.

Under a cold desert climate, meltwater is meager and streams may flow from these glaciers a few weeks a year. Several large, perennially frozen lakes exist in the valley system. Most of these, including the 5-km-long Lake Vida (Figure 3) [*Calkin and Bull*, 1967] are probably frozen to their bottoms. In addition, there are many small ephemeral saline or brackish ponds.

The Victoria Valley system is connected southward to the long and narrow ice-free Wright Valley, whose floor is between 100 and 700 meters lower.

Inland Ice Plateau and High Rock Thresholds

Immediately above the west end of the Victoria system, the nunataks of Mistake Peak, Shapeless Mountain, and Mount Bastion rise 100 to 300 meters above the bordering inland ice plateau (Figure 1). Results of the Victoria Land traverse [*Crary*, 1963] suggest that the ice plateau surface may rise slightly from about 2200 meters at the valley heads (near 160°E) to 2300 to 2400 meters some 45 km inland. To the west the surface falls by the same amount before rising steadily to the interior of East Antarctica at over 3600 meters. Thus the only ice that can flow eastward through much of these Victoria Land mountain ranges accumulated between about 158°E and 160°E. Furthermore, plateau ice surface elevations in this part of East Antarctica indicate that the general ice flow appears to diverge to either side of a line extending west from the McMurdo Sound area [*Crary*, 1963, p. 22; *Giovinetto*, 1964].

The preferential starvation of the glaciers feeding into the valleys of the Victoria system may thus be related to the thickness and subglacial topography of the margin of the inland ice. At the west ends of the ice-free valleys are high bedrock thresholds (Figures 4 and 5) which were overridden only at times when the surface of the inland ice was higher than at present.

Evidence that the bedrock surface below the inland ice plateau decreases westward from the Victoria Valley system is suggested by the absence of nunataks projecting above the ice surface farther west [*Bull*, 1960]. A westward decrease is now clearly proved by gravity and seismic studies [*Crary*, 1963] in the Skelton glacier area (Figure 1), and by results of seven radio echo sounding traverses made in the area between the Skelton and Mackay glaciers by scientists of Scott Polar Research Institute, Cambridge. Southeastward down the Skelton and Taylor glaciers from the inland ice plateau, between longitude 157° and 160°E, the subglacial surface suddenly rises from a 1800-meter depth to 200 meters as these glaciers apparently thin over high subglacial bedrock thresholds. The large icefalls between 160° and 162°E longitude on the Ferrar and Mackay glaciers indicate the presence of subglacial thresholds, but these glaciers are still able to reach the sea; similarly, the Skelton reaches the Ross ice shelf, although little nourishment now comes from the inland ice [*Wilson and Crary*, 1961, p. 877; *Crary*, 1966, p. 912]. However, the lower Taylor glacier is nearly cut off from the inland ice plateau and now terminates 30 km from the coast. Any further lowering of the plateau surface would probably cause this glacier to stagnate and retreat to the bedrock threshold [*Bull et al.*, 1962, p. 72]. In such an event, the western end of

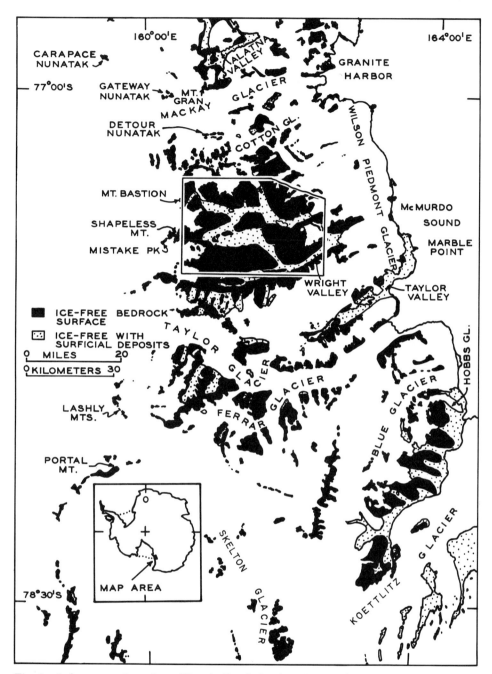

Fig. 1. Index map of southern Victoria Land showing the location of the Victoria Valley system and area of Figure 2.

Taylor Valley would resemble that of Wright Valley and the Victoria Valley system.

Bedrock Geology

Much of the following discussion and nomenclature is taken from the work of *McKelvey and Webb* [1962] and *Allen and Gibson* [1962]. The rock stratigraphic names given below in parentheses are the equivalent terms applied by *Gunn and Warren* [1962] in southern Victoria Land.

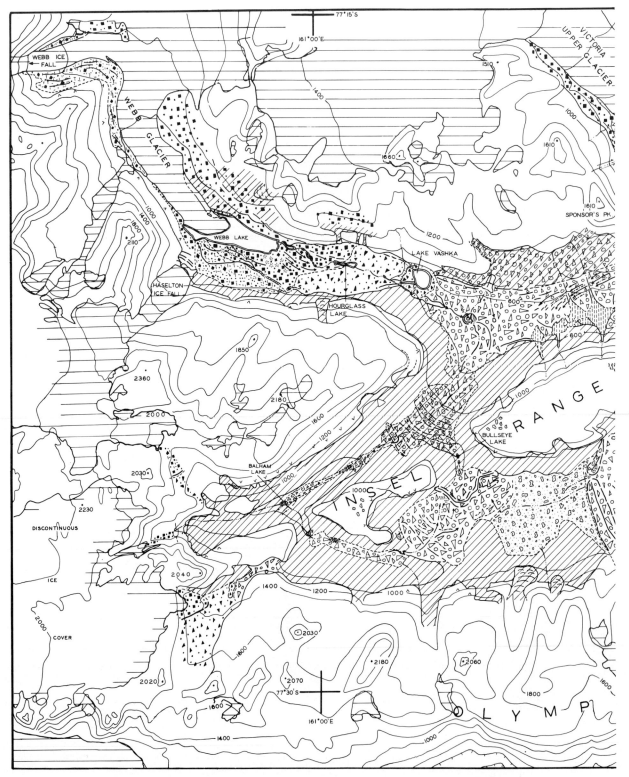

Fig. 2a. Glacial geology of the Victoria Valley system (see legend, page 368–369).

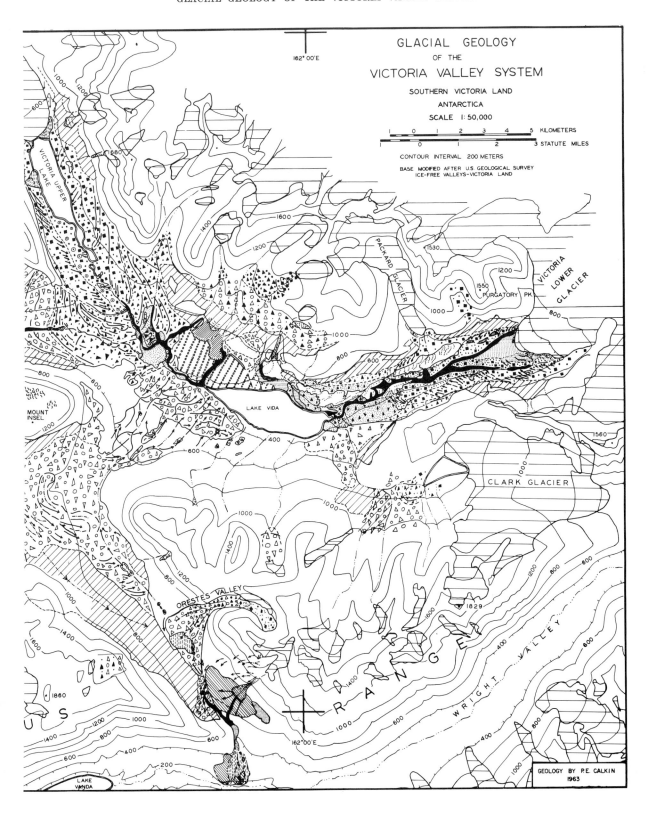

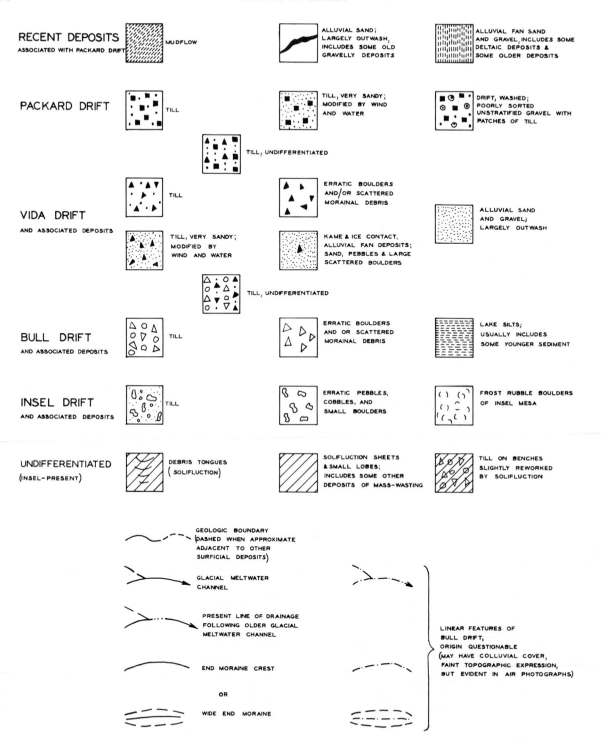

Fig. 2b. Legend for map of Victoria Valley system (pages 366–367).

The Victoria Valley system is underlain by igneous, metamorphic, and sedimentary rocks (Figure 6) of Cambrian or Late Precambrian to Mesozoic age. The basement complex, cropping out over the eastern half of the Victoria Valley system, consists of the folded metasediments of the Asgard forma-

GLACIAL GEOLOGY OF THE VICTORIA VALLEY SYSTEM

 BARCHAN SAND DUNES (DOTS OUTLINE CRESTS)

 WHALEBACK-SHAPED MANTLES (DOTS OUTLINE CRESTS)

 SAND SHEET DEPOSITS

RECENT DEPOSITS ASSOCIATED WITH PACKARD DRIFT

 ICE-CORED MORAINE; VERY BOULDERY IN BARWICK V., VERY SANDY IN LOWER VICTORIA V.

 ALLUVIAL SAND & GRAVEL; OUTWASH NOT ASSOCIATED WITH PRESENT DRAINAGE

PACKARD DRIFT

 ALLUVIAL FAN SAND AND GRAVEL

 DEBRIS LOBE; UNSORTED AND SANDY

VIDA DRIFT AND ASSOCIATED DEPOSITS

 DEBRIS FAN: UNSORTED, SILT, TILL-LIKE DEPOSIT, POCKETS OF STRATIFIED SAND & SILT (DOTS INDICATE MORE SANDY DEPOSITS)

BULL DRIFT AND ASSOCIATED DEPOSITS

INSEL DRIFT AND ASSOCIATED DEPOSITS

UNDIFFERENTIATED (INSEL–PRESENT)

◯ LAKE OR DRY LAKE BED

⋯⋯ KETTLE PONDS OF ICE-CORED MORAINE, BARWICK V.

S SALINE WATER OR SALT DEPOSIT

⋋⋌ KETTLE HOLE AND KNOB-KETTLE TOPOGRAPHY

v v v TALUS—IN ORDER OF TENS OF METERS THICK, ACTIVE & INACTIVE

⁀⁀⁀ BORDERING SCARP

tion (Koettlitz marble) associated with or cut by younger rocks of the Granite Harbor intrusive complex [*Gunn and Warren*, 1962]. These basement rocks, consisting of white granular marbles, paragneisses, granulites, and quartzofeldspathic schists, were subjected to strong folding during the Early Paleozoic (?) Ross orogeny [*Gunn and Warren*, 1962]. They strike northwest and dip 45° to 90° southwest.

The oldest formation of the Granite Harbor in-

Fig. 3. Aerial view from 15,000 feet (4500 meters) looking southwest into the Victoria Valley system. Symbols are: *d,* barchan dunes; *dm,* whaleback sand mantles. U.S. Navy photograph, TMA-353, no. 197, F 31, December 19, 1957.

trusive complex is the Olympus granite gneiss. This pretectonic rock forms a transitional zone between the older metasediments and the younger syntectonic Dais granite (Larsen granodiorite). The widespread, post-tectonic pink Vida granite (Irizar granite) cuts across these rocks and in turn is cut by dikes of the Vanda lamprophyre and porphyry. A recent, detailed study of the granitic rocks in Victoria Valley has been undertaken by *Fikkan* [1968].

The basement complex is truncated by an erosion surface known as the Kukri peneplain [*Debenham,* 1921*b,* p. 105]. West of Victoria Valley and Bull Pass, this surface is overlain by Beacon rocks [*Mirsky,* 1964; *Mirsky et al.,* 1965] of Upper Paleozoic to Mesozoic age. In this area, the Beacon rocks (Beacon group, *Allen and Gibson* [1962]; Beacon sandstone, *Gunn and Warren* [1962]) consist of a sequence up to 1500 meters thick in which white, cross-bedded quartzose sandstone predominates. The Beacon rocks dip 3° to 5° west.

Very large sills and associated dikes of the Ferrar dolerites of Jurassic-Cretaceous age [*McDougall,* 1963; *Evernden and Richards,* 1962; *Hamilton,* 1965] intrude the Beacon rocks and the underlying basement complex. Three main bodies are recognized. Sill *a* (basement sill) is a uniform sheet, 427 meters thick, that cuts across basement structures some 450 meters below the peneplain surface and dips 3° southwest. Sill *b* (peneplain sill) has been intruded along the contact of the Kukri peneplain and Beacon rocks. Several thinner sills, sill *c* of Wright Valley, linked in places by dikes, cut the Beacon rocks.

Fig. 4. Aerial view from 20,000 feet (6100 meters) looking southwest along McKelvey Valley (left) and Balham Valley (right). Symbols are: *ID*, Insel drift; *BD*, Bull drift; *s*, solifluction deposits. U.S. Navy photograph, TMA-540, no. 234, F 33, November 7, 1959.

The present structure and much of the basic relief of this area are believed to have been formed by block-faulting during the Victoria 'orogeny' [*Gunn and Warren*, 1962, p. 56] of Tertiary and Quaternary times.

Glacial Sculpture

In cross section, three important divisions of the valley area can be seen: (1) the deep valleys themselves; (2) glacial shoulders or benches and structural terraces such as those that form the upper surface of the Insel Range; and (3) the network of cirques and cirque-headed alpine valleys cut in the neighboring ranges.

Valleys. The valleys, from 10 to 27 km long and averaging more than 2 km wide at the bottoms, are arranged in a branching pattern with trends of either east-northeast perpendicular to the coastline, or southeast. Such parallel trends are also clearly evident in the major glacial valleys between the Ferrar and Mawson glaciers [*Allen and Gibson*, 1962, p. 241], which parallel and sometimes follow faults mapped by *Gunn and Warren* [1962, pp. 57–58] and *Angino et al.* [1962, pp. 1555–1556]. No faults have been recognized along the valleys of the Victoria system, but contact relations (Figure 6) may indicate that some valleys were initiated by streams which selectively cut parallel to structural or lithologic weaknesses.

The valleys exhibit broad U shapes that are probably related in part to the resistance of dolerite sills to vertical abrasion. *Gunn and Warren* [1962, p. 60]

Fig. 5. Aerial view from 20,000 feet (6100 meters) looking southwest into western Barwick Valley. Symbols are: *BD,* Bull drift; *VD,* Vida drift; *PD,* Packard drift; *ic,* ice-cored moraine; *kk,* knob and kettle topography. U.S. Navy photograph, TMA-540, no. 288, F33, November 7, 1959.

note that the valleys contrast with the much deeper and narrower glacial troughs of New Zealand. The valley floors, consistently 700 to 1200 meters below adjacent peaks, increase in width from about 1.5 km at their upper ends to about 4 km where they join. Valley floors of the Victoria system have a maximum elevation of 1000 meters. Upper and lower Victoria, Barwick, eastern McKelvey, and Balham valleys slope irregularly down to the shallow interior basin of Lake Vida at 390 meters.

Glacial benches. The uniform trough wall is interrupted on the north side of Barwick and Victoria valleys by three prominent sloping benches or glacial shoulders up to 2 km wide between 600 and 1000 meters elevation. These benches rise upward from the top of a dolerite sill, 100 to 200 meters above the main valley floor, to near accordant floors of cirques. These structurally controlled benches may be a result, in part, of the joining together of adjacent trough or cirque floors of the bordering valleys by the abrasion of large trunk glaciers. More narrow benches occur in lower Victoria Valley at 800 to 1000 meters elevation (Figure 7).

The most conspicuous high, flat surfaces are the two mesas of the Insel Range, formed by the eroded and weathered top of dolerite sill *b* (Figure 8).

Cirques. The walls above the valley floors in the Victoria Valley system contain more than 60 cirques or cirque-headed alpine valleys. Less than 15% are occupied by large ice masses, and most of these are at the far eastern end of the valley system (Figure 3). Most of the ice-free cirques reflect a strong

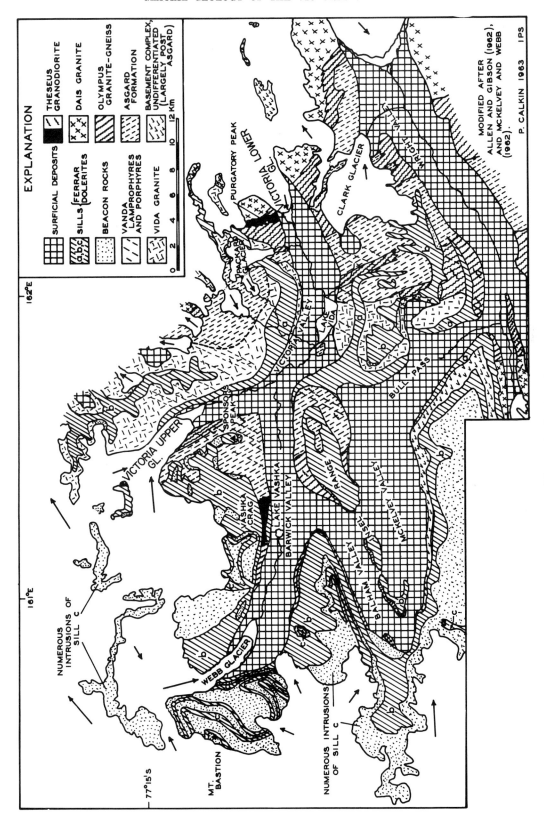

Fig. 6. Bedrock geology.

Fig. 7. Aerial view from 20,000 feet (6100 meters) looking northeast into Victoria Valley. Symbols are: *BD*, Bull drift; *VD*, Vida drift; *PD*, Packard drift; *dl*, debris lobe; dashed line, 20-meter contour above present surface of Lake Vida (390 meters) and possible shore line of the lake during Vida drift episode. Control points: *a*, 14-meter east base of outwash fan; *b*, 22 meters (2-meter-high ridge) on alluvial fan; *c*, 21-meter base of hummocky area on alluvial fan; *d*, 22-meter base of debris lobe and break to 5° slope; *f*, 20-meter upper third of kamefan, where algal peat sample was taken. Elevations locate some glacial benches. U.S. Navy photograph, TMA-540, no. 232, F 31, November 7, 1959.

change in climate and probably a reduction in accumulation since the time of active formation. They show marked differences in elevation, shape, and degree of development which are largely related to lithology and exposure.

One of the most marked characteristics of the cirques is the predominantly north-northeast or south-southwest orientation. More than 53% of the cirques face northeast. The troughs beyond these cirques average more than 1500 meters in length. Because the rocks dip to the west, the preferred orientation and differences in length of the cirques are not structurally controlled. These phenomena are associated with (1) the increased freeze-thaw action and greater erosion in the north-facing cirques; (2) the action of the strong winds from the southwest quadrant; and (3) perhaps greater accumulation on the north-facing slopes from the dominant northeasterly precipitation-bearing summer storms. Northeast-facing slopes trap snow from both northeast and southwest winds, whereas the stronger south or west winds cause the snow to accumulate in long drifts of this orientation.

Elevations at the bases of the headwalls of the

Fig. 8. Aerial view from 20,000 feet (6100 meters) looking northeast into the Victoria Valley system. Elevations indicate some glacial benches. U.S. Navy photograph, TMA-542, no. 241, F 31, November 7, 1959.

cirques show (1) a decrease in elevation eastward, especially in western Barwick and McKelvey valleys; and (2) a difference in elevation between the lower-lying, southwest-facing cirques on the north valley walls and the higher-lying, generally northeast-facing cirques of the south walls. The significance of the eastward decrease in elevations is difficult to determine, since both the valley floors and general summit levels show similar trends. The descending elevations may control the height to which cirques can be cut, or, if glacial erosion has gone on long enough, the summits may have been lowered to follow the highest level of cirque formation. Summertime meteorological observations show that the western parts of these valleys are warmer and drier than the eastern parts. In addition, the more common occurrence in the west of dark rocks with low albedo might explain an eastward descent of the orographic snowline. Furthermore, the snow comes from the northeast.

Climate

The Victoria Valley system lies within an arid zone in which low precipitation, low temperatures, and high winds are characteristic. Mean temperatures at McMurdo station, Ross Island, the nearest permanent weather station, vary from 4.2°C in January to −26.6°C in August [*Loewe*, 1963]. Mean summer temperatures in the Victoria system may be as much as 4°C higher than at McMurdo [*Bull*, 1966], the mean annual temperature being near −20°C [*Nichols*, 1964b].

Figure 9 shows some climatic observations for December and January 1959–1960 and for the

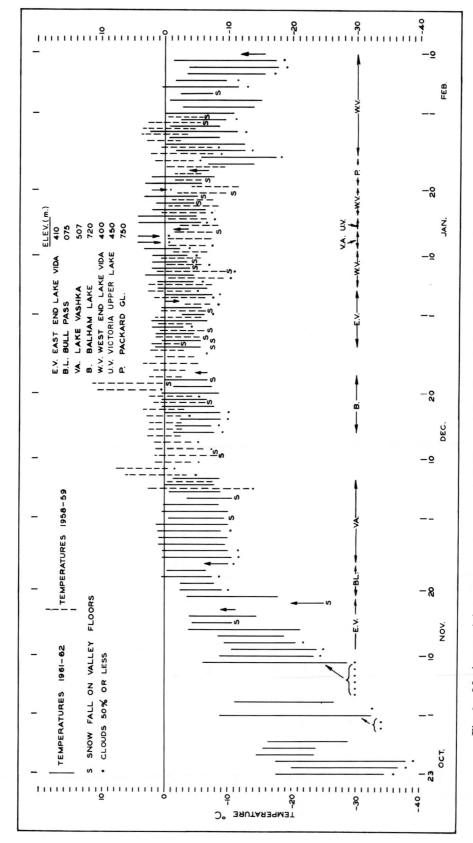

Fig. 9. Maximum-minimum temperatures, precipitation, and cloudiness during the 1958–1959 and 1961–1962 field seasons.

1961–1962 field season. Although snow fell during the field season on some areas of the valley floor, it rarely exceeded 6 to 10 cm, and rarely remained on the ground longer than a day. Observations of spring snow depths and stake measurements on the Packard glacier indicate that winter precipitation may be of similar order of magnitude.

Winds of the Victoria system are predominantly either from the east or from the southwestern quadrant. The easterly winds from McMurdo Sound are the most constant and predominant winds in Victoria Valley east of Lake Vida. During some 68 days after November 9, 1961, in which observations were made in the Lake Vida area, the wind blew almost constantly from the east and averaged more than 8 knots. However, west of Victoria Valley, the southwest or westerly katabatic winds off the inland ice plateau were nearly as common as the easterlies. These katabatics were more sporadic but also very much stronger; during a few days in February, gusty southwesterly winds up to 45 knots and steady winds of 20 to 25 knots were recorded as far east as Lake Vida.

The easterly winds carry more moisture and are slightly colder than the westerly winds, which are heated adiabatically in their rapid descent from the plateau. Winter snow remains in the eastern end of the valley system, though it may be absent on the valley floors to the west. In addition, as in the adjacent valleys, summer snowfall is heavier in the eastern end of the Victoria system.

An important climatic factor in the ice-free nature of the Victoria system is the strong positive radiation balance. *Bull* [1966] estimates that there is a net gain of 29,000 cal/cm^2 for ice-free areas compared with a loss of 4000 cal/cm^{-2} for permanently snow-covered areas of the adjacent Wright Valley area.

Under such a positive radiation balance and low accumulation, the size of the ice-free area tends to increase continuously. The low albedo of the rock causes an increase in local summer air temperature, which in turn causes a reduction in relative humidity and increase in ablation.

Summary of Recent Glacial Action

Four major local glaciers enter the valleys of the Victoria system (Figure 2). The largest are the Victoria Upper glacier and Victoria Lower glacier at opposite ends of Victoria Valley. The Upper glacier (Figure 10) occupies an area of about 80 km^2, more than two-thirds of which is made up of névé fields above 1000 meters elevation. The Victoria Lower glacier (Figure 3), a west-flowing tongue of the Wilson Piedmont glacier, is much smaller. The Webb glacier (Figure 5) occupies the inland end of

Fig. 10. Terminus of Victoria Upper glacier. View north.

Barwick Valley, and a smaller, alpine glacier, the Packard glacier (Figure 3), lies on the north side of lower Victoria Valley.

Five tongues of the inland ice occur at the western margin of the valley system. These are the Webb and Haselton icefalls (Figure 5), two tongues at the head of Balham Valley, and a glacier tongue at the inland end of McKelvey Valley (Figure 4). Except for the Webb icefall, these tongues are nearly cut off from the inland ice and are slowly wasting away.

Studies made in 1961–1962 [*Calkin*, 1964*b*] and 1966 indicate that, although some ice still comes over the Webb ice fall from the inland ice plateau, strong, dry katabatic winds cause the tongue of the Webb glacier (at about 775 meters) to be lowered as much as 6 cm in two months of the summer; the tongue itself is now stagnant. Net ablation is slightly less on the terminus of the Victoria Lower glacier (about 1 g/cm^2 for November through January 1961–1962), which is exposed to moisture-bearing winds, but equal or greater ablation occurs below 800 meters on the Packard glacier; the surface ablation of the Victoria Upper glacier has not been measured. Snow pit studies on the Packard glacier suggest that during 1961 there was a net accumulation of snow over most of the glacier. However, measurements made within the ablation zone (below 100 meters) in 1966 and over the whole glacier in December of 1968 showed a general decrease in the area of accumulation since 1962.

Surveys show that near their termini the Victoria Lower and Upper glaciers may be moving more than 1.5 m/yr. According to surveys made in 1962 and 1966, the Packard glacier has a similar regimen, movement being of the order of 1 m/yr over most of the glacier and the maximum being 2 m/yr at the firn limit.

The fronts of the Victoria Lower and Upper glaciers, of the Packard glacier, and of the tongues from the inland ice have probably not retreated significantly in recent years. This is suggested by the occurrence of active, well-developed contraction polygons a few meters from the fronts, and by the occurrence of some cavernously weathered boulders a few tens of meters away. Many of these boulders appear to have weathered in place. The possibility that these glaciers have advanced recently cannot be disproved at this time.

Recent studies by A. T. Wilson (written communication, 1966) may suggest that glaciers of the Ross Sea region are experiencing an advance that started 150 years ago. In Wright Valley, a few alpine glaciers have extended beyond their terminal ice-cored moraines [*Calkin et al.*, 1970]. However, the very recent general inactivity of glacier fronts in the surrounding lowland areas of McMurdo Sound have been documented by *Péwé* [1962; see also *Péwé and Church*, 1962]. *Péwé* [1962, p. 93] noted that 'comparison of the positions of existing glacier fronts with positions on photographs of 50 years ago shows no movement of the front or appreciable thickening or thinning of the glaciers during the last 50 years.'

Little morainal material is being deposited at any of the glacier termini in the Victoria Valley system.

GLACIAL GEOLOGY BACKGROUND

Previous Work

Mercer [1962, p. 7] has summarized the literature on glacier variations in Antarctica, including the southern Victoria Land area. He stated that observations in the Antarctic since 1874 have given the following broad picture of glacier variations: (*a*) everywhere the ice cover has been greater than at present; (*b*) in the subantarctic islands and much of Antarctic Peninsula, ice shelves and many glaciers have been receding during recent decades, probably because of rising temperatures; (*c*) in Antarctica, except for Antarctic Peninsula, the ice margins are either stationary or receding very slowly; (*d*) the antarctic ice sheet as a whole may have a positive regimen.

Evidence of the formerly more extensive ice cover in the area of McMurdo Sound and southern Victoria Land and the hypothesis of an earlier 'flood glaciation epoch' were advanced by geologists of the early British expeditions [*Scott*, 1905, p. 360; *David and Priestley*, 1914, pp. 287–289; *Wright and Priestley*, 1922, p. 438]. *Nichols* [1953, 1960] and *Mercer* [1962] give more complete lists and summaries. In their discussion, these early workers did not mention more than one advance, nor did they attempt to correlate this 'glacial flood epoch' in Victoria Land with that recorded at other places in Antarctica [*Péwé*, 1960, p. 498].

In 1946–1947, *Hough* [1950] secured deep-sea

cores from the mouth of the Ross Sea that were believed to span a period of about one million years. Later *Thomas* [1960] considered the stability of the Ross ice shelf in this area from the study of two ocean-bottom cores. However, the first terrestrial study and mention of multiple glaciation in this region was made by Péwé in 1957–1958 [*Péwé*, 1958, 1960]. His examination of the glacial deposits of the ice-free areas of McMurdo Sound and particularly in Taylor Valley revealed [1960, p. 498] 'at least four major fluctuations of the ice cap, each successively less extensive than the former.' These he called the McMurdo, Taylor, Fryxell, and Koettlitz glaciations.

During expeditions of the Victoria University of Wellington (New Zealand) into the Wright Valley and Victoria system, reconnaissance studies of the glacial deposits and land forms were made by *Bull et al.* [1962]. They recognized evidence of four glaciations (first through fourth) in the area.

During the summers of 1959–1960 and 1960–1961, more detailed studies were made by R. L. Nichols and associates of the Tufts University Antarctic Expeditions in the eastern two-thirds of Wright Valley. *Nichols* [1961, 1962, 1964a, 1966, 1971] also recognized four glaciations, which he referred to as Vanda, Pecten, Loop, and Trilogy glaciations.

Before the work in the Victoria system in 1960–1961, the author had made a brief study of the deposits in the Mount Gran area and the adjacent ice-free Alatna Valley. Two major glaciations (as defined in this section) are distinguished in this area [*Calkin*, 1964a].

Studies undertaken by Denton and Armstrong in the Taylor Valley and adjacent areas during 1967–1968 and 1968–1969 have shown that the oldest drifts in the ice-free valleys bordering McMurdo Sound may be from 2 to 4 m.y. old. In addition, these studies have allowed differentiation of at least five episodes of advance and retreat of the Taylor glacier (from youngest to oldest: Taylor glaciations 1, 2, 3, 4, 5, etc.), three for the alpine glaciers (Alpine 1, 2, 3) and four glaciations from the Ross Sea. Field work in Wright Valley in the 1968–1969 season by *Calkin et al.* [1970] has shown a similar set of glaciations: Wright Upper glaciations 1, 2, 3, and 4; earlier, valley-cutting glaciations, Alpine 1, 2, and 3; Wright Lower glaciation; and the Trilogy, Loop, and Pecten glaciations designated by *Nichols* [1961, 1962, 1963, 1964a, b, 1965, 1966, 1971].

Succession

At least two major glaciations, defined on the basis of deposits and land forms, are recorded in the Victoria Valley system. The first, the Insel glaciation, represents the strongest invasion of ice from the inland ice plateau and may have been preceded by one or more still earlier glaciations in this area. The second, the Victoria glaciation, was marked by many reversals in direction of ice flow, with a strong invasion of ice from local ice fields and from the coastal area. Weaker invasions from the inland ice plateau have been correlated with episodes of this last glaciation. However, because of the possibility that the latter are related to surges of the inland ice, this correlation is tentative. The Victoria glaciation has been subdivided into three 'episodes,' which involve glacial advances, stillstands, and retreats clearly subsidiary to and within the major advance and retreat comprising the Victoria glaciation. The last of these episodes extends to the present.

Methods

The deposits that define and give their names to the glaciation and glacial episodes, i.e., Insel, Bull, Vida, and Packard drifts, are often differentiated of necessity by inferred geologic history or by primary or secondary surface form (erosional morphology). The lithologic descriptions of the drifts are based almost entirely on their characteristics within the zone of weathering and frost action.

A few of the more useful techniques of field differentiation are discussed below.

Cavernous weathering. The cavernous weathering of boulders (Figure 11 and Table 1) has been used to determine the relative ages of moraines in the Victoria Valley system [*Calkin and Cailleux*, 1962].

An area on each moraine was selected that appeared to represent the average weathering. In each area, about 50 boulders of standard or larger size (Table 1) were measured. The height and length of each boulder and the number and depth of every hollow within the boulder deeper than 9 cm were recorded.

This method was used with most success on

even-textured rock groups such as gneisses with low percentages of platy minerals.

Surficial boulder frequency. The procedure for obtaining the frequency of surficial boulders and determining percentages of the major rock types was as follows: 300 meters were paced off, when possible in a straight line of arbitrary direction. Over this course, every upstanding boulder more than 20 cm high within arm's reach was counted and its lithology was noted.

The frequency depends on original number and lithology of the boulders. However, it also indicates the degree of weathering and erosion and the relative age of the deposit. Table 2 shows some of the average frequencies on adjacent glacial deposits. Averages of each of the four columns show a general decrease from left to right, that is, from recent to ancient. This is best shown by comparing deposits of one glacier from its terminus down valley; for example 156, 77, and 36 for the north side of lower Victoria Valley.

Lithologic counts of pebbles and boulders. Lithologic counts of the very large pebbles were made throughout the valley area. The pebbles were taken from the surface; it was not possible to dig to the unweathered zone.

Two size groups of surface boulders were distinguished for the lithologic counts; those between 20 and 70 cm in vertical and horizontal dimensions and those larger. The frequencies of boulders of selected rock types in each size group are shown on the sample location map (Figure 12).

Weathering and texture of deposits within the ice-free zone. One composite sample was taken at each locality shown in Figure 12. The upper and lower limit of the sample varied between 10 and 60 cm, depending on the thickness of the surficial 'ice-free zone' (this may be the dry or moist zone above ice-cemented material and may or may not coincide with the active layer; it is generally less than 60 cm and frequently less than 30 cm thick). At selected localities, two samples were taken, one at depth and one 10 cm below the surface.

In the laboratory, samples of the material less than 2 mm in diameter were dispersed and sieved, and the results were plotted and analyzed (Table 3)

Fig. 11. Cavernously weathered boulder of gneiss in the Bull drift of Clark valley.

TABLE 1. Summary of Cavernous Weathering Data for Granitic Boulders Higher Than 30 Centimeters and Larger Than 70 Centimeters*

		S Side Lower Victoria Val.			E Side Upper Victoria Val.				Lake Vashka to Webb Glacier (Barwick Val.)			Below Orestes Val., B	Solifluction, SE McKelvey Val.	Wright Val., V?
		P	V	B	P	P†	V	B‡	P	V	V			
Boulders, total		12	56	86	36	51	60	100	28	20	60	83	88	44
Lower than	50 cm	5	52	68	28	63	62	100	19	17	40	57	90	43
Higher than	50 cm	17	60	100	28	33	59	100	39	22	69	87	87	45
Shorter than	120 cm	6	43	74	38	50	60	100	32	22	42	82	75	50
Longer than	120 cm	2	70	96	33	53	60	100	21	17	66	88	95	36
Avg. no. hollows per boulders with hollows		1.0	2.3	2.5	1.3	1.4	2.6	3.4	1.8	1.3	2.4	2.3	2.5	2.7
% hollows 20 cm deep on boulders 50 cm high, 120 cm long		0	58	62	0	22	37	42	2.5	0	13	47	33	3.1

*Relative ages: P, Packard; V, Vida; B, Bull.
†Nearer glacier.
‡Boulders are on far eastern end of ground moraine and are not typical of whole moraine, where second stage cavernous weathering is more typical.

TABLE 2. Frequency of Upstanding Boulders on Moraines

Deposits of Victoria Glaciation												Deposits of Insel Glaciation			
Packard				Vida				Bull							
Location*	C†	Avg.‡	Range	Location*	C†	Avg.‡	Range	Location*	C†	Avg.‡	Range	Location*	C†	Avg.‡	Range
McKelvey Valley															
M1	83	249		M3	36	36		M5–7	211	70	41–86	M15–18	179	24	16–31
Balham Valley				B1,2	137	154	146–162	B3–6 §	198	50	38–55				
								M11,12A 14 §	185	61	54–75				
								M9,10 12,13	124	33	21–41				
Barwick Valley															
BW1,2	222	111	99–123	→BW4,5,8, 10–12	570	95	43–157	→BW13–17	182	32	13–50				
Upper Victoria Valley															
VU1,3,4,5,8–10	791	195	71–349	→VU6,7, 14,15	185	46	31–71	→VU16,20	46	23	15–31				
								VU22,24	88	27	12–52				
								M19–29	453	33	18–51				
								VU36,38§	101	48	47–48				
Lower Victoria Valley															
VL1	156	156		→V12	77	77		→VL4	36	36					
VL5,7	72	32	10–53	VL11	39	39		VL13	41	20					
VL9	41	41													

*See Figure 12.
†Number of boulders counted.
‡Upstanding boulders per 300-meter traverse (arms width); i.e., per approximately 600 m².
§End moraine, predetermined traverse direction.
Arrows connect contiguous deposits.
Relative age and distance from glacier increase from left to right.

according to *Inman*'s method [1952]. Of 140 samples analyzed, 30 were examined for clay content by the pipette method.

The oldest tills (in surface ice-free zone) are usually much higher in the silt-clay fraction than the fresher deposits. The percentage of clay itself appeared to correlate only very roughly with the percentage of silt clay and the weathering of the surface boulders. The sorting is also slightly poorer in the older deposits.

These relationships of texture and sorting are unusually well displayed in the surface of the ground moraine of the Victoria Upper glacier, where general decrease in mean grain size, increase in the silt-clay fraction, and decrease in sorting can be seen with increasing distance from the present glacier terminus (Figure 13). A similar increase of fines with age of drifts was carefully measured by *Ugolini* [1964; *Ugolini and Bull*, 1965] in the adjacent Wright Valley. This variation may be related in part to a change in mode of deposition and general state of the glaciers, to progressive weathering with time, or to both. As *Ugolini and Bull* [1965, pp. 254, 264] noted, had the former glaciers been thicker, or the mean air temperature higher, or both, basal melting might have occurred. The resulting greater abrading power of these glaciers, together with deposition by 'lodgment' from the basal ice, might explain the differences in amount of fine material in the tills. The more recent thin, cold glaciers leave only partly sorted, sandy 'ablation' till, and outwash streams at present do not carry rock flour.

Some evidence for a lodgment origin for the oldest glacial deposits is suggested by the shape of McKelvey and Bullseye end moraines discussed later. In addition, striated boulders, suggesting subglacial transport, are plentiful in at least one locality of very old (Bull) till. They are also present in young (Vida and Packard) tills west of Lake Vashka, which have an exceptionally fine texture compared to other deposits of the same age. Another line of reasoning is that, had the silty deposits ever been as sandy as those of the present, polygons would probably have formed. However, fossil polygons were not recognized over much of the older silty till.

Much evidence may also be gathered to support the second (weathering) hypothesis. The rocks carried and deposited by the glaciers in the Victoria Valley system, with the exception of some hornfelses and schists most common west of Lake Vashka,

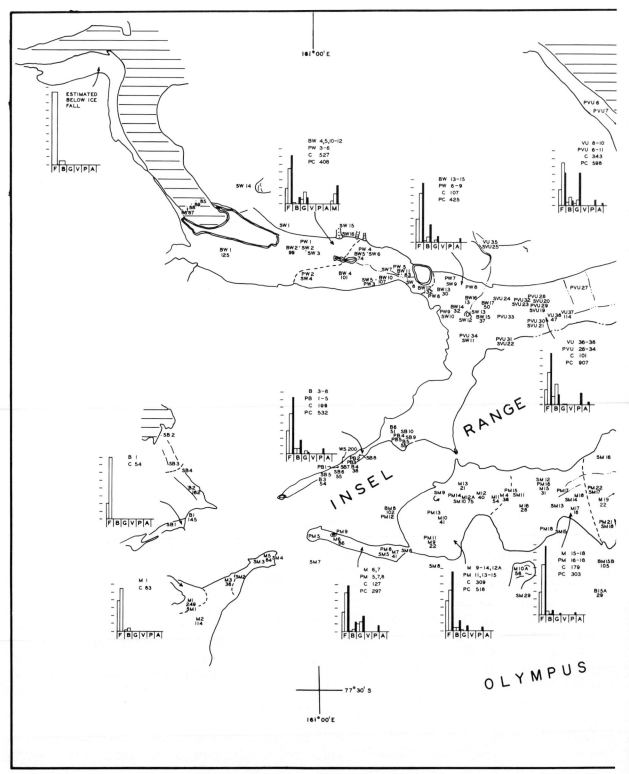

Fig. 12. Drift in the Victoria Valley system.

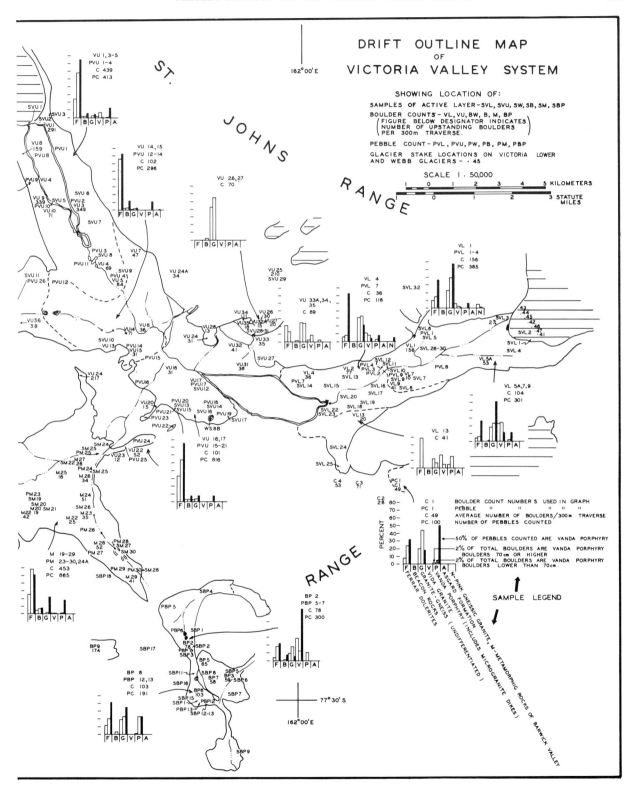

TABLE 3. Some Average Textural Characteristics of Till Particles, Less Than 2 mm Size, from Ice-Free Zone*

Sample (See Fig. 12)	Deposit†	No. Samples	% Silt-Clay Avg.	% Silt-Clay Range	Phi Mean Diameter‡ Avg.	Phi Mean Diameter‡ Range	Phi Deviation‡ (sorting) Avg.	Phi Deviation‡ (sorting) Range
McKelvey Valley								
SM1	P	1	7.0		1.1		1.4	
SM2–6	B	5	19.0	9–29	2.3	1.6–2.7	1.6	1.2–1.9
SM12,13,14B,15	I	4	26.8	7–43	2.5	1.7–2.9	1.5	1.0–1.8
Balham Valley								
SB1–4	V-P	4	5.0	2–9	0.8	0.7–1.1	1.1	0.9–1.5
SB^{E}5,6,8–10	B	7	16.4	3–55	1.9	1.9–3.5	1.1	0.8–1.4
SM10,11								
Barwick Valley								
SW1B,2,4–6,14	V-P	6	17.0	5–36	2.1	1.4–2.6	1.5	1.0–2.2
SW9,10	B	2	38.5	28–49	2.7	2.4–2.9	1.9	1.8–1.9
Upper Victoria Valley								
SVU3,6,8,9	P	4	2.8	1–6	1.6	1.6–1.7	0.8	0.6–1.1
SVU10,12B–18	B	8	26.4	20–36	2.5	2.2–2.9	1.8	1.4–2.2
SM19,21–28	B	9	28.3	11–60	2.4	1.2–3.2	1.8	1.5–2.0
SVU11,20–21,23,24	B	5	23.8	5–48	2.3	1.6–2.7	1.3	0.7–1.9
Lower Victoria Valley								
SVL3–10	P	8	4.4	1–10	2.0	1.7–2.2	1.0	0.6–1.3
SVL16–20	V	5	4.8	1–8	1.6	1.5–1.7	1.4	1.1–1.6
Victoria Upper, Lower and Packard glaciers								
SVU1,SVL1,2,32 (superglacial moraine)	Pres.	4	3.6	1–10	1.9	1.1–2.6	1.0	0.8–1.3

*As far as possible, definitely recognized lake or solifluction deposits have been eliminated from averages. Deposits are largely ground moraine, except where noted as single end moraine samples (^{E}). Sampling sites are grouped according to major depositing ice source and are listed in order of increasing distance from valley head and ice source.
†I, Insel drift (oldest); B, Bull drift; V, Vida drift; P, Packard drift (youngest).
‡Based on 5 sand fractions plus silt-clay fraction.

Phi Units	Wentworth Equivalent	Phi Deviation (Sorting) Scale
0–1	Very coarse sand	0.50–0.75, well sorted
1–2	Coarse sand	0.75–1.50, moderately sorted
2–3	Medium sand	1.50–2.00, poorly sorted
3–4	Fine sand	
4–5	Very fine sand	

are granular types which probably would not readily produce large quantities of rock flour during glacier transport.

Mechanical analysis of boulders of more than 30 cm diameter, disintegrated or decomposed in situ in old, silty till, is as follows:

Lithology	2-mm	Sand	Silt Clay
Medium-grained granite gneiss	80%	18%	2%
Green hornfels-schist	53	36	11
Coarse-grained dolerite			
A	55	41	4
B	33	62	5

It seems certain that a significant percentage of this fine material may originate from in situ disintegration or decomposition of the same volume of glacially deposited pebble or sand material so common in the younger moraines. The poorer sorting in the older deposits may also be a result of this in situ weathering. *Ugolini* [1964] has clearly shown that time has been an important factor in soil development in the eastern end of Wright Valley. He has shown that, among other weathering processes, orthoclase was probably altered to illite clay and secondary chlorite decomposed in the surface horizons.

If truly unweathered samples could be taken from below the ice-cemented zone in the silty deposits,

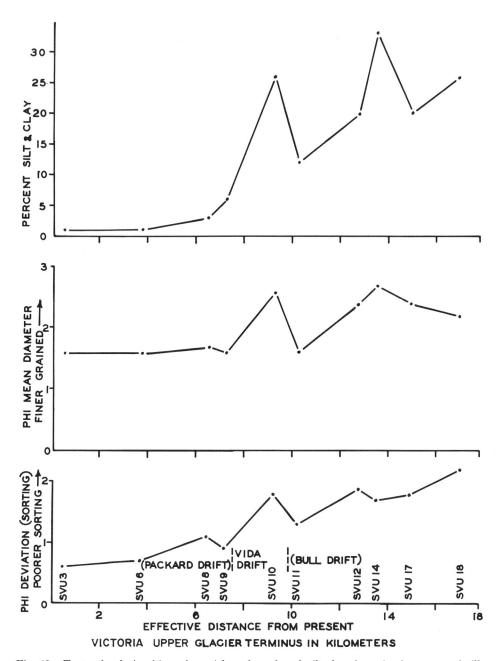

Fig. 13. Textural relationships of particles of sand and silt-clay from ice-free zone of till deposited by Victoria Upper glacier. See Figure 12 for sample locations.

their similarity or dissimilarity with samples from the ice-free zone above might shed considerable light on the problem. However, the weathering relations mentioned above, together with field observations, suggest that at least a part of the textural differences from old to young tills may be due to weathering, influenced by time and climate (Table 3).

Patterned ground. Assuming that increasing percentages of silt-clay reflect increasing age of the deposits, their texture and age may be inferred from the presence and degree of development of soil pat-

Fig. 14. Polygon and furrows in ice-cored moraine, western Barwick Valley.

terns. The use of patterned ground in age determinations in southern Victoria Land has been studied extensively by *Black and Berg* [1963, 1964; *Berg and Black*, 1966].

Thermal contraction polygons have formed in nearly all the sandier deposits of the Victoria Valley system. They are similar to the typical ice-wedge polygon of the Arctic [*Lachenbruch*, 1962], but because of the arid conditions they frequently consist of varying amounts of sandy or stony debris in place of the ice wedge [*Péwé*, 1959]. Polygons of this type are best developed in ground with the higher coefficients of expansion. The coefficients of expansion are in turn increased by the presence of moisture or ice in the ground. In general, icy permafrost, ground ice, and polygons in this area are best developed in the sandy or coarser deposits that contain less than 11% silt-clay in the 2-mm or finer fraction.

The net loss of soil moisture with time through vapor transfer [*Black and Berg*, 1963] and liquid capillary conduction with corresponding lowering of the underlying ice-cemented layer is well verified in the ice-free valleys here. However, the deficiency of moisture in the older, silty drift may also be due to the much greater capillarity and reduced permeability of the finer deposits. The capillarity at 0°C for even coarse silt is 30 to 100 cm, while it is 3 to 10 meters for fine silt. On the other hand, permeability for coarse silt is 38 to 3 cm/hour for fine silt [*Beskow*, 1935, p. 86]. In this climate, where potential evaporation greatly exceeds precipitation, moisture has been slowly removed from the silty deposits, but their low permeability has prevented their regaining moisture during periods when snow is melting.

The occurrence of permafrost and patterned ground in coarse-grained deposits, and their absence in silt and clay, has in the past been considered a striking anomaly. However, the same phenomena occur in the arid region of North Greenland [*Davies*, 1960].

Péwé [1962] suggested that after initial development of polygons there is, with time, an enlargement of the contraction crack by filling with ice, sand, or larger particles. Therefore, the furrows produced at the surface by widening of the wedge should also grow larger with time. Assuming that growth rates measured in Victoria Land over the past few years are representative, *Berg and Black* [1966] have concluded that a considerable part of the deglaciation of Victoria Land has occurred in late Wisconsin to recent time (a conclusion generally denied by the work of *Nichols* [1971] and *Denton et al.* [1970], and it is clear from observations in the Victoria Valley system that this criterion of age must be used with considerable caution. Furrows do not always become wider with increasing distance from the present glacier fronts. In keeping with the control imposed by the coefficient of thermal expansion, the best-developed polygons with the deepest and widest furrows in the Victoria Valley system occur in the coarse, blocky, ice-cored moraine of western Barwick Valley (Figure 14). Evidences for periods of more moisture, as well as more aridity, in the past are well documented in this report and in many others [*Ugolini and Bull*, 1965], and variations in rate of wedge growth must have been considerable.

Nomenclature

In this report, the term 'glaciation' has been used to denote an event during which extensive glaciers developed, attained a maximum extent, and receded. The definition of 'glaciation' used strictly as a geologic-climatic unit for Quaternary [*American Commission on Stratigraphic Nomenclature*, 1961] has been purposely avoided, because one or both of the glaciations defined in this paper could have been in some part due to glacial surging [*Wilson*, 1964a; *Calkin et al.*, 1970].

Fundamentally the definition used in this paper is

the one followed by Péwé, Nichols, *Denton et al.* [1970], and Bull et al. An interglaciation is defined as 'an episode during which the climate was incompatible with the wide extent of glaciers that characterize a glaciation' [*American Commission on Stratigraphic Nomenclature*, 1961, p. 660]. In this sense, no formal interglaciations in the McMurdo Sound area of Antarctica have been distinguished by previous authors and none is distinguished at this time. *Nichols* [1965, p. 447] has reviewed evidence proving that there has been little deglaciation in Antarctica since the onset of the Pleistocene. The lack of true interglaciations in Antarctica such as have occurred in the northern hemisphere has also been noted by *Różycki* [1961, p. 278]. It is probable that glaciers have never left the southern Victoria Land area since their initiation after uplift of the Transantarctic Mountains some time in the Tertiary [*Mercer*, 1968]. In addition, there is no depositional evidence in the Victoria Valley system for an interglaciation as defined above.

The term 'episode' has been used for subdivision of the Victoria glaciation, because the author felt it was useful to correlate events within the Victoria Valley system on the basis of weathering and similarity of sequence which involved contemporaneous stillstand, advance, and retreat in the divergent parts of the Valley system. Further work may show that some or all of these episodes deserve the status of glaciations as defined above.

EARLY HISTORY OF GLACIATION

This section considers the period from the initiation of local glaciation up to, but not including, the retreat from the last major advance into the valley system from the inland ice plateau (Insel glaciation). The interpretation presented is based largely on glacial sculpture.

Preglacial Topography

Because no direct evidence is available, it is not certain whether the directions of the valleys in the Victoria system follow a preglacial drainage pattern, are controlled by a fault pattern, or perhaps have some other control [*Nichols*, 1966, p. 7]. The fact that upper Victoria Valley and Bull Pass follow geologic contacts suggests that the glaciers may have enlarged preglacial, structurally oriented stream valleys [*David and Priestley*, 1914, p. 201; *Taylor*, 1922, p. 180]. *Nichols* [1961] implies a stream ancestry for the Wright Valley in saying that the oldest glaciers produced a reversal of drainage. On the other hand, *Gunn and Warren* [1962, p. 60] note: 'It seems unlikely that large rivers have existed in Victoria Land even in interglacial periods, and it is probable that the present topography has been produced entirely by glacial dissection of block-faulted mountain ranges.' More recently, *Calkin* [1964a] and *Smith* [1965] have called attention to the labyrinthine complexes of bedrock channels at the heads of Alatna and Wright valleys, which indicate torrential stream activity of probable local interglacial times.

Local Initiation of Glaciation

In southern Victoria Land, the main source of moisture has probably always been from the northeast. With a deterioration of climate, glaciation probably started in the mountainous coastal areas rather than in the drier and lower inland areas west of the mountains. The time of this onset is not determined; according to *Mercer* [1968; *Minshew and Mercer*, 1971] it may have followed closely the uplift of the Transantarctic Mountains some time in the Tertiary. *Rutford et al.* [1968] consider that a striated bedrock surface and tillite of Miocene or Pliocene age in the Jones Mountains, Antarctica, mark an early stage in the formation of the recent ice sheet.

Paleomagnetic study of deep-sea cores from the Antarctic region, by *Opdyke et al.* [1966], indicates that ice-rafted glacial detritus was being deposited here at least 2.5 m.y. ago. More recent work [*Bandy and Casey*, 1969] indicates that the first appearance of glacial deposits in most cores occurs well into the Gilbert reversed magnetic epoch, or about 4.4 m.y. ago. Potassium-argon ages from basaltic cones that erupted on the floors of Wright Valley [*Nichols*, 1971] and Taylor Valley [*Denton et al.*, 1970] indicate that these valleys were cut by glaciers more than 3.5 to 3.7×10^6 years ago.

Theories about the evolution of the inland ice have been advanced by *Taylor* [1922, p. 180] and *Bull et al.* [1962, p. 72]. A coalescence of the glaciers draining toward the low interior, perhaps accompanied by increased snowfall, caused a thin but growing ice sheet to form. By direct accumulation in the interior, this ice sheet attained a thickness sufficient to cause reversal of ice flow in the east-west valleys.

Taylor [1922, pp. 174–186] considered that the form of the outlet valleys, such as the Taylor and Ferrar valleys, is due initially to headward erosion of cirque glaciers subsequently modified and enlarged by the overflow of ice from the inland plateau. However, *Gunn and Warren* [1962, pp. 60–63] rightly challenge this thesis, noting that high-level moraines are continuous down-valley from the present glaciers at the heads of ice-free valleys, so that these are remnants of former outlet glaciers. The cirque form is merely due to selective erosion of parts of the boundary scarp by overflow of the plateau glacier.

Within the Victoria Valley system, the three valleys opening onto Barwick Valley and the Webb glacier show the reshaping effect of ice from the inland plateau on cirques (Figure 5). The Webb and Victoria Upper glaciers are now small and relatively inactive, and are fed from their own névé fields, but, even before the inundation of the area by ice flowing from the inland plateau, the western parts of the valleys were occupied by larger glaciers than at present. Both glaciers head in areas occupied by large cirques, many of which are now nearly free of ice.

Invasion of the Inland Ice

After the initial glaciation of the coastal mountainous area and the development of cirques, small névé fields, and valley glaciers like the Webb and Victoria Upper glaciers, the area was inundated by ice flowing from the inland plateau. The quantity of ice and its modifications of the valley forms vary from pass to pass. In the Balham and McKelvey valleys, no cirque-like form had been produced in the initial stages, and their courses, normal to the coastline, are more typical of outlet glaciers of this region than the courses of either the Webb glacier or the upper Victoria Valley. The deep basins at the heads of Balham and McKelvey valleys (Figure 3) are not found in any of the ice-free cirque valleys of the Victoria Valley system, but are typical of the heads of outlet valleys such as Wright and Alatna. In addition, valley floors that rise steeply westward to the plateau through narrow stepped troughs as in Balham Valley, or more abruptly over wide dolerite steps leading to the high plateau as in McKelvey Valley, are common for the larger ice-free outlet valleys of Victoria Land.

The similarities of the stepped valleys above to such resistant bedrock scarps or benches beneath some active outlet glaciers which reach into or through the Victoria Land ranges have been cited previously. All of these, including Balham and McKelvey valleys, appear to be the end product of inland ice overflowing bedrock thresholds and consequent erosion through sandstone or basement rocks intruded by thick dolerite sills or both. The processes and resulting forms are clearly described by *Gunn and Warren* [1962, pp. 60–63] and by *Bull et al.* [1962]. They consider that back-cutting occurred, which Gunn describes as a process similar to plunge-pool erosion and the resultant formation of 'pseudo-cirques' and benches on the dolerite sills. These forms are a consequence of the columnar-jointed dolerite that more strongly resists vertical erosion than the intruded rocks, but that, once penetrated, is more easily worn back by lateral corrasion.

During maximum ice inundation, some true cirque-cut topography was modified and perhaps erased. With continued invasion of ice, cutting back of the plateau border scarp continued; Balham and McKelvey valleys were formed or enlarged or both, and the Insel Range mesas were formed or accentuated.

When the retreating scarp line intersected the unusually resistant series of dolerite sills (sill *c*) within the Beacon rocks, the headward enlargement of these valleys was impeded. The consequent prolonged overflow of ice at this location facilitated excavation of the deep bedrock hollows at the heads of Balham and McKelvey valleys.

While McKelvey and Balham valleys were being cut, the Webb glacier, substantially enlarged by glacier tongues from the inland ice, expanded southeastward, merging with the undivided ice stream in the early Balham and McKelvey valleys and with the Victoria Upper glacier. Some of the McKelvey ice stream probably broke through a col ridge at the head of Bull Pass to form a tributary of the outlet glacier occupying the Wright Valley.

High-level valley walls and benches are continuous from upper Victoria through lower Victoria and Clark valleys (Figure 8). This is consistent with the theory that a continuous stream of ice from local ice fields and the inland ice plateau flowed through the valley system. Truncated spurs between alpine valleys on the southwestern margin of the Clark Valley preserve an indication of the earliest southeastward-moving outlet glacier.

INSEL GLACIATION

Insel Drift and Associated Deposits

The sequence of cirque and alpine valley cutting and inundation by inland ice may have been repeated many times, but any evidence of early glaciations has been obliterated by later ones. Therefore, the oldest event that can be definitely recognized is the last inundation by the inland ice, here called the Insel glaciation. This is represented by two contrasting glacial deposits, one on the Insel Range and the other 200 to 300 meters below at the eastern end of the McKelvey Valley, and by a succession of glacial channels cut in the bedrock southwest of Lake Vida.

Insel Range. The oldest of the glacial deposits recognized in the Victoria Valley system consists of a small number of individual boulders, cobbles, and pebbles that rest on the weathered dolerite sill cap of the Insel Range mesas, 200 to 600 meters above the adjacent valley floors. The most profuse accumulation occurs within the low, shallow saddle at an elevation of about 1200 meters (400 meters above the valley floor) on the westernmost mesa (Figure 4). Here the deposits consist of a few 20- to 50-cm boulders of Beacon rocks (quartzite), a few cobbles of granitic rock, one of which is Dais granite, and scattered small pebbles of quartz and quartzite with a few dike rocks of Vanda porphyry.

Most of these erratics have probably been derived from exposures within the valley system (Figure 6). A few other stones of sandstone, quartzite, and basalt were found over this mesa, but their derivation is uncertain.

The position, paucity, lithology, and form of these deposits suggest great antiquity. They are spread over an area of 20 km², high above the valley floor on a surface untouched by later glaciations. They occur largely as individual pebbles and cobbles; only the most resistant rock types still exist as boulders. The present form of most erratics is due entirely to wind abrasion.

Sheets and irregular accumulations of frost rubble of dolerite are localized in nivation hollows and small cirques at the margins of the mesas of the Insel Range (Figure 15). A basalt dike that transects the Mount Insel mesa stands up to one meter above the only slightly less resistant dolerite (diabasic) cap rock.

McKelvey Valley. Insel drift on the eastern floor of McKelvey Valley consists of a till sheet (Figure 16), probably many meters thick, upon

Fig. 15. Frost rubble of weathered dolerite on the Insel Range. View southeast. The white map case in the center is 25 cm square.

Fig. 16. Till sheet of Insel drift. View southeast from near Bullseye moraine in McKelvey Valley.

which a poor but distinguishable weathering profile has developed. This deposit grades rapidly into the McKelvey and Bullseye (end) moraines on the east (Figure 4) and associated deposits correlated with the Bull drift episode on the west.

Much of the Insel drift has been covered by solifluction sheets from the valley walls, so that the unaltered exposures are narrow and limited. No morainal form remains, and frost polygons occur only near the borders of the solifluction sheets.

Most of the boulders on the surface have been reduced to ground level by weathering and erosion: they are much rarer than on the younger surfaces (Table 2). More than 90% of the recognizable boulder remnants and pebbles consist of the most resistant medium- to fine-grained dolerite with a marked deficiency of the more easily weathered sandstone common in surrounding outcrops and younger drifts. That these deposits were transported by glaciers invading from the west (inland ice plateau) is suggested by the presence of some Dais granite boulders, similar to that cropping out to the west, and the deficiency of boulders and pebbles of porphyry common in outcrops to the east [*Calkin*, 1964b].

Within the top 60 cm of the till, most of the boulders have disintegrated to cobble or smaller size. The sorting of the fine material is moderate to poor, but the material is uniformly silty and loose. Analyses show that there is from 27 to 43% silt-clay, of which 4 to 11% is clay (Table 3 and Figures 2 and 12).

A shallow and poorly developed weathering profile, defined by oxidation and salt accumulation, occurs in these deposits. Below the 2- to 8-cm pebble and coarse sand lag deposit, a typical profile consists of an upper horizon of oxidized and often brown or yellowish-brown silt loam. This weathered till is very dry and structureless, loose or soft in consistency, and from 15 to 25 cm thick. Frequently this horizon displays no white salts but may show a moderate carbonate reaction. The remnants of disintegrated boulders are often present. Often occurring with a sharp break below the oxidized upper horizon is a layer of similar texture and consistency but including white salts and producing strong carbonate reaction. *Nichols* [1963, pp. 28–30] has discussed the composition and origin of a similar subsurface efflorescence of salt in a deposit of Loop glaciation age in Wright Valley; the salt in Wright Valley is predominantly halite, but includes Ca, Mg, CO_3, and SO_4. In McKelvey Valley, these visible salts may occur as narrow bands 5 to 10 cm thick, or may decrease gradually downward. The lower limit of the oxidation profile is not well defined. *Gibson* [1962] discussed similar salts, including rare nitrates and iodates in various occurrences in the Victoria Valley system.

This weathered till color and the occurrence of

the white salt horizon are variable, but carbonate reaction is always strong in the silty material, a fact probably directly related to its great age [*Ugolini and Bull*, 1965]. The depth and coloring of the oxidized layer is closely related to the proportion of fine dolerite particles rich in ferro-magnesium and hence related more to provenance than to age.

This profile described above for the older deposits is not significantly different from that in the younger deposits, but it is more consistently developed.

Discussion. The more subdued topography, the consistently greater weathering, fine texture, greater depth to ice-cemented layer, and the distinctive lithological composition that this Insel drift was deposited before the neighboring Bullseye and McKelvey moraines. Furthermore, this till was deposited by glaciers moving eastward from the inland ice plateau. It has thus been related to the earliest recognized glacial deposits and the Insel glaciation.

The greater weathering of the higher deposits of the Insel Range may be due to their more exposed position, the rapid removal or initial absence of fine material, and the longer period of exposure. Rapid area in 1968 revealed some evidence that the whole reexamination of the McKelvey Valley–Bull Pass McKelvey moraine and the drift to the east and southeast within Bull Pass might be of Insel age. However, these deposits have been tentatively correlated with the Bull drift on evidence cited farther on in this paper.

Glacial meltwater channels of southwestern Victoria Valley. In the southwestern corner of Victoria Valley, a system of wide channels has been cut in dolerite sill *b* and basement rock (Figure 7). Leading from the pass at the far eastern end of McKelvey Valley and having gradients of 2° to 5° toward Lake Vida and the Victoria Upper Valley stream, these channels show a pattern that was probably produced along the edge of a glacier retreating westward into McKelvey Valley. Consequently, the channels are probably related to the Insel glaciation. In the upper half of the slope, many of the channels are short, curved, and truncated successively upslope against one another and toward the axis of the pass. Many channels have dead ends upslope. In the lower part of the slope, the channels are resolved into several cross-contour cuts that are obscured beneath a blanket of (Bull drift) moraine near the valley bottom. The channels and 30 to 90 meters wide.
are quite uniform, often being 5 to 10 meters deep

The area occupied by these channels was overridden by glaciers flowing westward during the succeeding Bull drift episode of the Victoria glaciation. These glaciers left the till that now obscures the upper and lower ends of these channels.

Extent and History of Retreat of Insel Glaciation

The Insel glaciation is that during which the oldest glacial deposits recognized in the Victoria Valley system (Insel drift) were laid down. During this glaciation, the inland ice plateau west of the Victoria Valley system was higher than at any time since; ice flowed eastward over the bedrock thresholds into the valleys and probably through to McMurdo Sound.

The Insel glaciation was brought to a close by the lowering of the plateau ice level, the emergence of the high rock thresholds, and hence the cutting off of ice flow to the valleys [*Bull et al.*, 1962]. This sequence of events is reflected in the present conditions in valleys north and south of the Victoria Valley system.

Some details of the vertical extent of the glaciers and of the recession can be determined from the distribution and lithological composition of the remaining drift.

At some early stage, the ice covered all the Insel Range, but, as the ice plateau gradually lowered, the east end of the range (Mount Insel mesa) emerged, and the ice level in the McKelvey and Balham valleys dropped to about 1200 meters at the west end. It may have remained at this position for a long time, where its base reached several hundred meters below the dolerite mesa top into the basement complex. At this stage, the slope of the glacier surface in the McKelvey and Balham valleys was less than 1 in 60, and both its eastward flow and its cutting power were negligible.

The Webb and Victoria Upper glaciers are fed in part by local ice fields and have lower bedrock thresholds at the plateau edge. During the lowering of the inland ice, these glaciers continued to flow and deepen their valleys. Later, temperatures may have risen and local accumulation diminished, so that the local glaciers and the near-stagnant and almost separated tongues of the inland ice began to retreat more rapidly. At the eastern end of the tongue in McKelvey Valley, streams initiated the system of channels now leading from McKelvey Valley to the bottom of Victoria Valley. The formation of the large McKelvey moraine may have been initiated at this time, although its present form is ascribed to the later Bull drift episode. With a continued drop in the level of the plateau ice, all ice flow to the valleys was severed; the tongues retreated at least to the heads of McKelvey and Balham valleys and into the upper ends of Barwick and Victoria valleys.

At the end of the Insel glaciation, the valleys probably had much the same dimensions and appearance as they now have, especially in the eastern area of McKelvey and adjacent Victoria Valley.

Evidence on the inland ice plateau. The increase in elevation of the surface of the inland ice plateau to the west of the Victoria Valley system need be relatively slight to produce the 'flood' glaciations that are recorded in the Victoria Valley system and in other valleys. Such a rise is indicated by minimum ice limits shown by nunataks at the western edge of the mountains of southern Victoria Land (Figure 1). At the head of the Skelton glacier, to the south of the valley system, the Lashly Mountains, now some 305 meters above the level of the inland ice, and Portal Mountain, have been overridden by ice. Detour, Gateway, and Carapace nunataks at the head of the Mackay glacier to the north have also been overridden, although they now extend 305 to 488 meters above the ice surface [*Gunn and Warren*, 1962, pp. 53–54].

Age. There is no direct way to date the advance or retreat of the Insel glaciation in the Victoria Valley system. However, on the basis of the evidence that this was the last major advance from the inland ice plateau and one that probably reached through the valley system, the Insel glaciation is tentatively correlated with the corresponding last major episode of invasion and retreat in Wright and Taylor valleys. In Wright Valley, the correlative event is the Vanda glaciation of *Nichols* [1971] or equivalent last valley cutting episode of *Calkin et al.* [1970]. The Vanda glaciation has been assigned a minimum age of 3.7×10^6 years [*Denton et al.*, 1970; *Calkin et al.*, 1970]. The Vanda glaciation is probably equivalent to the Taylor 5 glaciation of *Denton et al.* [1970] in Taylor Valley, which must have occurred more than 2.7 to 3.5×10^6 years ago. In both valleys, the ages are based on potassium-argon dating of basaltic cones (Wright Valley) or flows (Taylor Valley) that occur in the lower part of the glacial-cut valleys.

VICTORIA GLACIATION

All the glacial deposits of the Victoria Valley system not assigned to the Insel glaciation are here assigned to the second and last glaciation recognized in the valley system, the Victoria glaciation, named after the valley where the most complete sequence of its deposits occurs.

The deposits are divided into the Bull, Vida, and Packard drifts. These show distinguishable characteristics and were exposed by glacial retreat and largely deposited during three consecutive episodes of the Victoria glaciation, each of which marked a change in glacial regimen believed to be correlative between the individual valleys.

Bull Drift and Associated Deposits

Slightly more than half of the valley floor area of the Victoria system is blanketed by highly weathered and topographically very subdued Bull drift and associated deposits which represent the initial and most extensive advance and recession of the Victoria glaciation.

The type deposits in Victoria Valley and adjacent Bull Pass consist largely of till, and there are a few small areas of lacustrine silts. Other deposits include large debris fans emplaced by mass-wasting in Bull Pass and on the adjacent floor of Wright Valley, undifferentiated talus, and extensively developed solifluction sheets that mantle and substantially reduce the areas of exposed drift. Important alluvial deposits are lacking or are indistinguishable from much younger material.

Till distribution and morphology. Till of the Bull drift (Bull till) is particularly extensive in areas bordering the Insel Range mesas. In all areas except Bull Pass, the Bull deposits are separated from the distal ends of occupied valleys by the distinctly younger Vida and Packard drifts. Till covering most of Bull Pass north of Orestes Valley has been assigned to the Bull drift on the basis of stone counts, channel orientation, and soil studies in preliminary reports [*Calkin*, 1964b] and is so assigned here (Figure 2). However, reexamination of the area by the author in 1968 suggests that the evidence is not as definitive as he originally believed.

Because of the high elevation (800 meters) and the western position of the area in the valley system, there is some chance that the Bull Pass deposits may be of Insel age.

Bull till occurs largely as subdued ground moraine and varies considerably in thickness. Accumulations, possibly tens of meters thick, occur in the depressions in the west of McKelvey and Barwick valleys, and extensive areas of thin and spotty till or free stones on highly weathered bedrock are found in the valley bottoms.

Southwest of Lake Vida, cavernously weathered projections of the dolerite sill stand as much as 5 meters above the surrounding sill rock. In other areas, ice covered during the Bull drift episode, basaltic and felsitic porphyry dikes have often been weathered out 1.5 meters above granitic bedrock.

The two most well-defined end moraines in the Victoria Valley system are apparently products of this advance. The largest, the McKelvey moraine, stretches across the eastern end of McKelvey Valley and is concave eastward. It has a broadened, inverted V-shaped cross section with the gentle face on the concave side and has an average relief of about 50 meters (Figure 8). A resistant dolerite knob above its southern end suggests that the moraine may have a bedrock core.

A smaller, probably contemporaneous end moraine forms a loop around the southern side of Bullseye Lake (Figure 4). This is more narrow and steep, sloping 5° to 10° on the ice contact (proximal) side and 15° south, its well-defined crest is uniformly about 25 meters high. Neither the Bullseye nor McKelvey moraines have been dissected by postglacial stream action. These end moraines, which have steeper (distal) sides, appear to suggest a lodgment rather than ablation type of emplacement. However, they do not represent the maximum extent of the glaciers; Bull drift in the valley system is attenuated, and terminal positions are represented only by a slight increase in boulder concentration in the till.

Other glacial margin features, including glacial meltwater channels, occur over the drift, but these features are now discontinuous and barely distinguishable as topographic forms except in Victoria and eastern Barwick valleys. Many indistinct and poorly defined concentric linear elements seen in air photographs proved to be either marginal channels in the till or slightly raised ridges of till. Such lineations are concave eastward in northern Bull Pass and McKelvey Valley. They appear to indicate glacier movement from Victoria Valley. In Barwick Valley, barely discernible marginal channels and low lateral moraines sloping westward are probably products of glacial retreat into upper Victoria Valley.

The major alteration of these deposits has been by weathering and wind action, but recent alluvial action has altered and buried both ground and end moraines in eastern Barwick Valley and on the north edge of Lake Vida. Solifluction sheets probably cover half of the original total area of deposits. Polygons are absent or only weakly active on the Bull drift.

Nature and preservation of till constituents. Most of the surficial boulders have been reduced to ground level, and many have disintegrated completely below as above the surface (Figure 17). The average frequency of upstanding boulders averages about 36 boulders per 300-meter traverse (Table 2). An early second cycle of cavernous weathering and associated granular disintegration is displayed by the upstanding boulders.

Quantitative study of surficial stones at more than 100 localities on the till of the Bull drift reveals some marked differences from valley to valley which, when related to known bedrock outcrops, suggests a pattern of glacier movements identical and complementary to that displayed by topographic lineation of moraines and channels [*Calkin*, 1964b].

For example, the easterly limit of ice invasion from tongues of the inland ice is marked by the eastern limit of granitic gneiss in the deep depression of western McKelvey Valley. Tough, well-cemented Beacon sandstone fragments are more prevalent east of the depression, where they can be traced through the main gap in the Insel Range to a source in Balham Valley.

Large percentages of the nondoleritic boulders elsewhere on the Bull drift, including the eastern parts of Barwick, Balham, and McKelvey valleys and upper Victoria Valley, resemble those rocks exposed on the northern and western walls of upper Victoria and Barwick valleys (Figure 6) and appear to be related to a former divergent extension of the Victoria Upper glacier.

Fig. 17. Typical weathered and abraded surface of Bull drift showing boulders completely and partly worn to ground level.

The western limit of the Victoria Lower glacier is marked by deep salmon pink stones on the south side of Lake Vida, which contrast sharply with the paler granites carried eastward by the Victoria Upper glacier.

The texture and weathering of the Bull till within the ice-free layer are similar to that developed on the Insel till (Table 3), but the morphology is more variable. Boulders are frequently highly disintegrated below the surface to a depth of 40 cm, particularly where the tills are silty, and polygons are absent or very poorly developed.

In the upper layers of the Bull till, significant differences occur from place to place in the texture, color, salt content, degree of oxidation and weathering, and vertical sorting. Most of these variations can be related directly to the local lithologic composition of the till. Such differences distinguish the Bull drift from the younger and much more uniform deposits.

Glacial marginal channels. A series of glacial marginal channels occurs in the far eastern McKelvey and Barwick valleys, northwestern Bull Pass, and on the southwestern side of Lake Vida (Figure 18). In the first three areas, the channels are indistinct and may exhibit little more than 1 or 2 meters relief. A regularly spaced series in Barwick Valley below Sponsors Peak may suggest a regularly intermittent glacier retreat.

The morphology of the belt of ground moraine bordering the southwest end of Lake Vida is dominated by a cross-hatched pattern of glacial meltwater channels (Figure 18). The deepest and most distinct extend directly downslope, while shallower and more irregular marginal channels closely follow the contours of the slope. The former are the buried equivalent of the bedrock channels exposed upslope (see Insel glaciation) and related to the westerly retreating McKelvey Valley glacier of the Insel glaciation.

The marginal channels slope gently from west to east and are marked by concentrations of boulders that are more numerous in the higher channels than near the lake. The channels vary in width from 5 to 200 meters and are generally less than a few meters deep. The lateral spacing of the marginal channels here is irregular.

These marginal channels probably delineate the retreating, easterly sloping terminus of the Victoria Upper glacier. The erratic spacing may indicate irregular retreat. Some anomalous inter-relationships revealed (Figure 18) may be explained by local and small readvances, or by postglacial, downslope runoff of meltwater from snow accumulating in the channels. Such patterns are often also indicative of sublateral and subglacial drainage [*Mannerfelt*, 1949; *Sissons*, 1960].

Deposits related entirely to local ice accumula-

tion. Till at the mouth of Orestes Valley and large debris fans in the same area have been assigned to the Bull drift on the basis of weathering. They may actually have been deposited in very late Bull drift to earliest Vida drift time.

Debris fans of southern Bull Pass. A large subdued fan of bouldery debris emanates from a deep bedrock notch leading from a hinging cirque valley above the southeast end of Bull Pass (Figure 19). The fan has a maximum width of more than 2.5 km and a slope of 3° to 8°. Covering the upper third of its area is a second, steep-fronted fan which, in turn, is covered in part by a third more hummocky and very much smaller fan.

The ice-free layer of the lowest fan consists of large boulders in a silty to sandy matrix that is partly stratified. More than two-thirds of the boulders have been weathered to ground level. The upper fans show successively less weathering and have sandy matrices.

The three fans have been deposited at three distinct times by flow of debris with some alluvial action. This debris was probably moved from the cirque and valley wall by the release of large amounts of meltwater, temporarily impounded between the small glacier and its terminal moraine in the cirque above. The largest and earliest flow must have been thin, perhaps associated with considerable alluvial action; it included and redistributed much of the debris left by earlier, through-valley glaciers. The less extensive second and third flows displaying steeper fronts were successively more viscous and were possibly emplaced more slowly than the first.

Pecten debris fan of Wright Valley. In Wright Valley, the scattered erosional remnants of a debris fan now stand as much as 10 meters in relief. This fan has been very deeply stream-dissected and is now largely covered by alluvium (Figure 19). It was apparently formed by the flow of meltwater charged with debris from Bull Pass. The fan debris consists of lithologic types derived from surficial deposits and bedrock. Great quantities of silt that forms part of its matrix may have been derived from thick lake deposits in the southern end of Bull Pass. The unsorted structure, alluvial gravels, and apparent irregular margins of the fan suggest that it may have been formed by a series of mudflows with associated alluvial action.

Near the axis of the Wright Valley, a meltwater

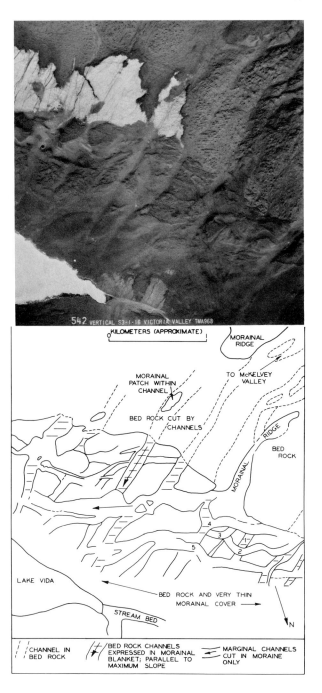

Fig. 18. Glacial meltwater channel series in Bull drift southwest of Lake Vida. Till blankets downslope channels cut in bedrock and in turn is cut by marginal channels trending obliquely to the older bedrock channels. Numbers 1 through 5 show probable sequence of channel cutting and indicate possible submarginal origin. Note channel to left of downslope arrow which makes sharp bend toward lake. This may be a subglacial chute. U.S. Navy photograph, January 27, 1962.

Fig. 19. Aerial view from approximately 6000 feet (1800 meters) looking north from Wright Valley into Bull Pass. Symbols are: *BD*, Bull drift; *br*, bedrock (Ferrar dolerite); *df*, debris fan; *c*, marginal channels; *s*, solifluction. U.S. Navy photograph, TMA-350, no. 154, F 31, January 1, 1958.

stream from Bull Pass has cut through the debris fan to expose an underlying section of gravelly glacial outwash some 4 meters thick. Within one of the gravel beds of this outwash buried by the fan, *Nichols* [1961] found a bed of pecten shells. *Bull* [1962] infers that these pecten shells were carried into the Wright Valley by invasion of ice from McMurdo Sound. They have been dated by C 14 analysis as more than 35,000 years old [*Nichols*, 1971]; Ra-U measurements suggest that they are at least 200,000 and possibly more than 800,000 years old [*Nichols*, 1971] (Wallace Broecker, personal communication, 1969).

This debris fan appears to be slightly older than the large debris fan in Bull Pass. Almost all boulders are reduced to ground level, and in a 300-meter traverse across two fan tops, only three upstanding boulder shells were encountered. Judging from the degree of weathering, the fan deposits must be of Insel or Bull drift age.

The pecten debris fan formed after the ice had retreated eastward in Wright Valley after the Pecten glaciation, which deposited the shells and produced the enclosing gravelly outwash [*Nichols*, 1961, 1966]. Therefore, taking the highly weathered condition into account, a strong argument for linking the pecten debris fan and the debris fans in southern Bull Pass to the Bull drift rather than to Insel deposits is the similarity in the pattern of glacial flow in Wright Valley and in the Victoria Valley system. *Nichols* [1966, p. 11] notes that, of the three younger glaciations marked by deposits from glaciers which moved westward up the Wright Valley, the oldest is the Pecten glaciation, typical deposits of which are the outwash sands and gravels containing pecten shells. This glaciation, he notes, followed the earlier episode (Vanda glaciation) when an outlet glacier moved eastward down Wright Valley.

It is likely that the formation of the debris fan may have been related to local melting of glaciers that reached into Bull Pass during Bull drift time; alternatively, it may be related to cirque glacier activity in Bull drift time or post-Bull–pre-Vida time.

Solifluction (undifferentiated). Sheets of debris believed to be products of solifluction extend from the foot of steep talus or bedrock slopes of greater than 30° down over inclines below as low as 3° (Figure 20). Unsorted stripes [*Washburn*, 1956, p. 837] are common, and boulders show a preferred downslope orientation of long axes. The weathering of these boulders in the solifluction sheets is nearly as great as that in the adjacent unmantled glacial deposits. Probably most of the movement occurred soon after glaciers retreated, when valley walls were over-steepened and meltwater was plentiful.

The best-developed solifluction sheets of the Victoria Valley system occur below the talus aprons in Bull Pass, and in Balham and McKelvey valleys. In these areas, occupied during the Insel glaciation and Bull drift episode and free from glaciers for many thousands of years since, solifluction sheets mask glacial drift over more than half the area of the valley floor.

The sheets of solifluction debris vary in texture, depending on the source; however, all, regardless of surficial weathering, show a moderately sorted, medium sandy matrix with less than 5% silt and clay [*Calkin*, 1964b].

History of the Bull Drift Episode

During the Victoria glaciation, the maximum volume of ice came from local ice fields and from glaciers moving west from the area of McMurdo Sound. The earliest and major advance of the Victoria glaciation took place during the Bull drift episode. Glaciers that moved into the valley system from five distal valley ends, together with local ice fields and cirque or piedmont ice, occupied nearly the whole valley bottom and much of the upland, cirque-dissected area (Figure 21). Approximately

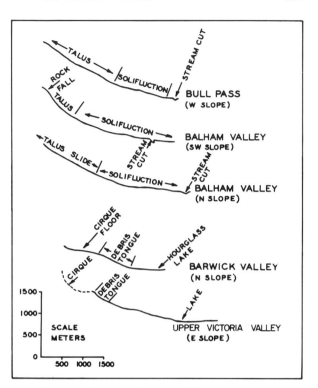

Fig. 20. Slope profiles formed by mass-wasting.

246 km² more valley floor was covered by ice than at present. Ice in the valleys attained a thickness of as much as 500 to 600 meters in the area of Lake Vida but failed to cover the Insel Range mesas by only a few tens of meters.

Advance: Victoria Lower glacier. The Victoria Lower glacier flowed westward, joined by the Clark glacier from the south. Whether the Packard glacier and adjacent cirque glaciers from the north were extended at this time is not known. The Victoria Lower glacier probably reached at least 9 km farther west than its present position and met the Victoria Upper glacier.

This westward-flowing ice was probably derived from névé fields around the 1400- to 1500-meter peaks flanking the east end of the valley and augmented by the accumulation and buildup of glacial ice in McMurdo Sound at this time.

Victoria Upper glacier. The Victoria Upper glacier with its tributaries expanded at least 15 km beyond its present position before merging with glaciers from the lower Victoria, Balham, and Barwick valleys. This expansion may have been due mainly to an increase in accumulation on the névé fields north of Sponsors Peak, or all or part of the ice may

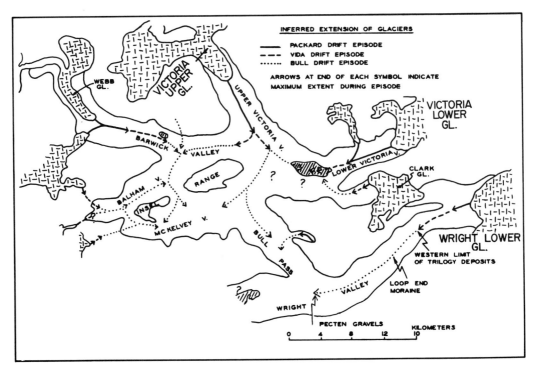

Fig. 21. Inferred extensions of glaciers in Wright Valley and the Victoria system during the Victoria glaciation.

have come from the area of the Mackay Outlet glacier to the north.

South of Sponsors Peak, the Victoria Upper tongue split into two branches, one flowing west into Barwick Valley and the southern branch flowing around the east flanks of the Insel Range and probably into the eastern end of McKelvey Valley and northern Bull Pass (Figure 21). This southern branch probably merged with the westward-flowing Victoria Lower glacier, but convincing evidence in the form of interlobate moraine or indicator erratics is lacking. Glacier flow into McKelvey Valley brought very large sandstone boulders from the upper Victoria Valley and formed the high McKelvey moraine across eastern McKelvey Valley.

At the intersection with Balham Valley, the Barwick Valley branch of the Victoria Upper glacier met the eastward flow from the expanded Webb glacier and a short, south-flowing tongue from the ice fields to the north (Figure 21). The composite glacier flowed southwestward to the area of Bullseye Lake.

Balham and McKelvey ice tongues. The inland ice plateau did not thicken as much as did the ice in the McMurdo Sound area [*Hollin*, 1962; *Bull*, 1962], so that the overflow into Balham and McKelvey valleys was small. Most of it came into Balham Valley from two ice lobes on the sides of Shapeless Mountain (Figure 4).

This ice reached through western Balham Valley to the Bullseye Lake pass area, where it merged with the westward-moving tongue from Barwick and upper Victoria valleys. Into Barwick Valley, with its lower bedrock threshold, the overflow from the plateau was greater. From this junction, a glacier tongue flowed south through the pass to the eastern edge of the depression in western McKelvey Valley (Figure 21). The glacier terminus apparently did not remain at this position long, but retreated to Bullseye Lake, where it stood to form the Bullseye moraine. At the same time, the Balham ice tongue overflowed into the western end of McKelvey Valley (Figure 21), joined with the less extensive tongue there, and apparently terminated in the deep hollow at the head of McKelvey Valley, where it may have been joined by tongues from the adjacent Olympus Range cirques.

Cirque glaciers. Many of the cirques were probably occupied by ice at this time; however, recent studies in the adjacent Wright Valley [*Calkin et al.*,

1970] and in the Taylor Valley area [*Denton et al.*, 1970] suggest that advances of the bordering alpine glaciers may have been limited by the availability of open water in the adjacent Ross Sea–McMurdo Sound. If this was the case, major advances of the alpine glaciers may have preceded or followed the major axial glacier advances of the Victoria glaciation, which were probably activated at least partly by grounded ice occupying the adjacent Ross Sea (see later section 'Sea Level Control'). Nevertheless, there is some evidence that three cirque-fed glaciers north of the west end of Lake Vida coalesced at this (Bull) time to form a piedmont, which pushed the Victoria Upper glacier to the south side of the valley. A similar piedmont exists now on the northeast side of the St. Johns Range (Figure 7).

The glacier in Orestes Valley must have reached its maximum extent at about this time or shortly after the Bull drift maximum (as was suggested above). Climatic conditions that caused the glacier to advance may have also allowed large quantities of meltwater charged with debris to form and flow southward, down over the lip of Bull Pass into Wright Valley. The resulting debris fan buried outwash and shells left earlier by the Pecten glaciation of Wright Valley.

Recession in the Victoria Valley system. As the inland ice plateau was lowered below the bedrock threshold, the supply to the McKelvey, Balham, and Barwick ice tongues was cut off and these glaciers began to stagnate. At the same time, there was a steadier and slower retreat of the Victoria Lower glacier, where nourishment diminished more gradually. The greatly extended Victoria Upper glacier, disconnected from the inland ice but still fed from its own névé basin, retreated relatively slowly from its positions in McKelvey and Bullseye Lake areas. The beds of many small lakes that formed against the retreating termini or in natural depressions nearby now comprise the only significant stratified deposits. The dearth of recognizable outwash accumulation suggests that temperatures were not high enough to produce great amounts of meltwater. This may be in agreement with findings of *Ronca* [1964] that, based on studies of thermoluminescence, frigid conditions have persisted in the McMurdo Sound area for a minimum of 210,000 years.

During the long period after the retreat of the ice tongues from McKelvey, Balham, and Barwick valleys and Bull Pass, the presence of over-steepened valley walls, higher temperatures, and perhaps wetter conditions allowed the formation of extensive solifluction sheets.

Much of the valley floor that was ice-free at the end of the Bull drift episode has not been ice-covered since. This includes most of Bull Pass and McKelvey and Balham valleys. The Webb and Victoria Lower glaciers retreated behind Lakes Vashka and Vida, respectively, but how far is not known. The Victoria Upper glacier shows no evidence of having retreated completely from the valley at that time. However, it must have retreated northwestward at least to a point some 2 km west of Lake Vida, as is evidenced by a series of end moraines and a narrow gradational zone to fresher (Vida) drift at this location.

Sea level control. In contrast with the conditions in the Insel glaciation, the Bull drift episode of the Victoria glaciation involved only a small overflow of ice from the inland plateau. However, in the Bull drift episode, a large quantity of ice invaded the Victoria Valley system from the east, which cannot be explained in terms of the present physiography of the coastal area.

Both *Hollin* [1962] and *Bull* [1962] have considered the problem. Hollin demonstrated that the horizontal extent of the grounded ice sheets of Antarctica is determined by sea level. The lowering of the sea level produced by the Pleistocene glaciations in the northern hemisphere was followed, with a slight time lag, by a seaward extension and an increase in volume of the antarctic ice sheet. Because the profile of a grounded ice sheet is approximately parabolic, horizontal expansion causes a large increase in thickness at points near the edge, while inland the increase is very much less.

Alternatively, *Wilson* [1964a] has theorized that such expansion might be caused by periodic thermal surges of the ice sheet after buildup at the ice base of normal geothermal heat to the pressure melting point. The resulting expanded ice shelves might then lower the earth's albedo enough to induce glaciation in the rest of the world. Although such surges may have taken place, proof has not yet been presented [*Hollin*, 1964]. The hypothesis set forth above will be further developed here.

Considering, for a start, the time of maximum Pleistocene glacierization, the Illinoian, the lowering of sea level may have been about 150 meters [*Donn et al.*, 1962]. This probably caused an extension northward of the eastern antarctic ice sheet of about 90 km, but the increase in thickness in the inland parts west of the ice-free valley area was only of the order of 100 meters [*Bull*, 1962]. Hence the increased flow from the plateau into the valley system was small.

A second effect of the decrease in sea level in the Ross Sea area is that the Ross ice shelf became grounded, probably as far north as 76°30'S. Ice, mainly from western Antarctica, accumulated in the area to produce the equilibrium parabolic profile of a grounded ice sheet. East of the ice-free area, this ice sheet attained a thickness of at least 1200 meters [*Bull*, 1962].

The implications of such a situation are clear. *Hollin* [1962, p. 191] says that:

> The 'dry valleys' of Victoria Land would be filled not as much by the frequently postulated "ice floods" from the interior to the west (though such may have occurred) as by the intrusion of ice lobes from the grounded shelf to the east.

The ice thicknesses during the maximum extensions of ice from McMurdo Sound into the Victoria Valley system and Wright Valley are broadly in accord with the thicknesses calculated for this grounded ice sheet during Illinoian time. However, present uncertainties of the isostatic condition of this area precludes assignment to any one period of northern hemisphere glaciation and sea level lowering. The great extension of the Victoria Upper glacier may have been a result of the backing up of the Mackay glacier (Figure 1) and diversion of a branch of this outlet glacier toward the south. However, the paucity and arrangement of drift indicates that these intrusions were not extensive enough to erode the coastal mountain valleys significantly.

The minor fluctuations in the west-flowing glaciers and the eventual deglaciation of the Victoria Valley system may be related to fluctuations and the general increase in sea level [*Hollin*, 1962, p. 190]. However, after the disappearance of the grounded ice sheet in the Ross Sea area, the controlling factors in the fluctuations are clearly local. *Bennett* [1964, p. 75] calculated that, on the basis of free-air gravity anomalies, grounded glacier ice was removed from the Ross Sea part of the continental shelf between 10,000 and 31,000 years ago. Although subject to some errors owing to simplifying assumptions of amount of postglacial sedimentation, rate of removal of grounded ice, and recovery rate of the crust, the values may be realistic. However, the values do not preclude much earlier retreats that might have controlled withdrawal of ice from Pecten and Victoria glaciation maximums.

Age. It is not possible to determine absolutely the age of the Bull drift episode. Neither the carbon 14 age of the pecten shells (>35,000 years; *Nichols*, 1961, 1966] nor the Ra-U age (>200,000 or possibly >800,000 years) reliably indicates a minimum age for the deposits of Pecten–Bull drift age. The ice advance that carried them to their approximate position may be much more recent than the 'death' of the pecten measured by the minimum dates [*Nichols*, 1971].

The earliest and most extensive invasion of Taylor Valley by ice from McMurdo Sound is the Ross Sea 4 glaciation of *Denton et al.* [1970], which probably correlates with the Pecten glaciation and the Bull drift episode. *Denton et al.* [1970] suggest that, in the Taylor Valley area, all the recognizable glacial invasions from the east (including Ross Sea 4) have occurred within the last 1.2 m.y.

Considerations of age on the basis of lowering of sea level are ambiguous. By the rough model possible at this time, a lowering of the order of 150 meters is suggested. However, this could be related to: the Early Wisconsin glaciation (30,000 to about 80,000 years B.P.) when the drop was 115 to 134 meters [*Donn et al.*, 1962]; the Illinoian glaciation (0.4 to 0.5 $\times$ 10^6 years ago [*Ericson and Wollin*, 1968]), when lowering was 137 to 159 meters [*Donn et al.*, 1962]; or the Kansan glaciation (1 to 1.4 $\times$ 10^6 years ago [*Ericson and Wollin*, 1968]), when sea level was somewhat lower.

It seems quite possible from the discussion above that the Bull drift was deposited by a major glacial advance that occurred during the Early Wisconsin time or as early as Kansan time.

Vida Drift and Associated Deposits

The Vida drift is named from Lake Vida, where till and outwash are well exposed. The drift is comprised of two major units: till, often partly sorted and sometimes including partly bedded sections; and silt, sand, and gravel outwash as fans or cut-fill deposits. Associated with the drift are debris lobes with large alluvial fans emanating from their fronts. Some solifluction deposits are present. These are largely confined to the steeper valley walls and benches above the valley floors.

The most abundant deposit is till; the alluvial deposits and products of mass-wasting are correlated to these deposits on the basis of their preservation and interfingering relations.

Distribution and morphology of the till. The till of the Vida drift is everywhere sharply differentiated from that of the Bull drift in the following ways:
1. It is thicker.
2. It shows less surface weathering and erosion.
3. It has greater general relief.
4. It displays widespread development of active polygons.
5. It is much less mantled by solifluction.
6. Its fine fraction is coarser.
7. It is marked by an abrupt change in type of deposition.
8. It is closely associated with alluvial deposits.

However, the transition to the younger deposits of the succeeding (Packard) episode is less sharp, and the degree of weathering grades toward the depositing glaciers. The dividing line is based largely on a change of type of deposition. The main localities of the Vida till are in lower and upper Victoria and Barwick valleys (Figure 2).

The till of the lower Victoria and Barwick valleys occurs entirely as ground moraine, usually some 20 meters thick with surfaces often standing several meters in relief above the much more weathered and thinner Bull drift. The moraines terminate against outwash deposits bordering Lakes Vida and Vashka. Here, ice-contact and channel scarps are still relatively well preserved. An ice-contact morainal front, 23 meters high, with a 35° slope, forms a semicircle around the western side of Lake Vashka; the Vida moraine itself terminates in lower and thinner deposits on the east side of the lake.

The Vida till of upper Victoria Valley occurs as a series of high, relatively closely spaced end moraines, some with as much as 35 meters relief; the largest of them is 250 meters wide. The boundary drawn between this and the Bull drift is based primarily on a fairly abrupt change from ground to end moraine deposits. To the west in eastern Barwick Valley, these Vida end moraines stand up in sharp contrast to the subdued ground moraine and boulder accumulations of the Bull drift.

In upper Victoria and Barwick valleys, at the contact with the Bull drift, Vida drift is also revealed by hummocky ground, broken by widespread, active, and well-developed polygons. A sharp micro-relief of up to 2 meters is developed by the polygons above Lake Vashka. Axial stream channels are cut as deeply as 12 meters into the Vida drift of Victoria and Barwick valleys. Solifluction on slopes adjoining the Vida drift has been slight or very localized since the retreat from the maximum of the Vida drift episode, and only very slight mantling of the Vida till has occurred.

Nature and preservation of till constituents. Boulder frequencies of Vida deposits average between 58 and 74 per 300-meter traverse (Table 2), compared with less than 35 on the older Bull and Insel deposits and more than 100 on most of the younger (Packard drift) deposits. Considerably less than 50% of the boulders have been worn to the ground, and of these more than 80% are medium- to coarse-grained dolerites that weather rapidly.

Vida till in the Victoria Valley system is distinguished from older drifts by the presence of a few lithologic types other than those derived from the immediate valley walls and by the frequent presence of Beacon sandstones not preserved in the Bull or Insel drifts.

The number of cavernously weathered boulders is much greater on the Vida deposits than on either the younger or older deposits (Figure 22). The effects of wind abrasion are restricted, probably because of the hummocky topography. However, well-faceted ventifacts in well-developed lag pavements are present in open areas, and wind-cut hollows, 5 to 15 cm deep around the bases of boulders, are more common than on most of the younger deposits.

The Vida till within the near-surface, ice-free zone is coarser than the older till (Table 3). Excluding deposits in Barwick Valley, whose finer texture is due to the presence of easily weathered gneiss and schist, the mean size of the material of less than 2 mm consists of coarse sand, and the silt-clay fraction makes up only from 1 to 9%. Samples from within the ice-cemented zone in Barwick and Victo-

Fig. 22. Cavernously weathered boulders on Vida ground moraine of lower Victoria Valley. Looking west toward Lake Vida.

ria valleys show nearly the same particle size. Thus surface texture is not due to weathering but probably to the original composition of the deposits.

In the Vida till, the sorting is much better, the deposits are more moist, and the ice-free zone more shallow than in the Bull till. Efflorescent salt accumulations are generally lacking. Oxidation of the near-surface material is noticeable in areas high in dolerite particles. However, in general, no clear weathering profile is distinguished. The subsurface texture and weathering characteristics of the Vida till are not sharply different from those of the younger (Packard drift) till (Table 3).

Stratified drift. Glaciofluvial deposits occupy areas of from 0.25 to 0.75 km² adjacent to the termini of moraines of the Vida drift in upper and lower Victoria and Barwick valleys. At the southeast end of Lake Vida, a gravel ice-contact alluvial fan, 30 meters thick at its origin, slopes to the surface of Lake Vida (Figure 7).

Patches of green, diatomaceous (algal) peat occur at the surface of the fan up to at least 20 meters above lake level. These patches are 2 to 5 meters in diameter, usually 10 to 20 cm thick, and are often buried 20 cm in the gravel of the fan surface. These can be either pond deposits or deposits formed during a high stand of Lake Vida. If they are pond deposits, the topography of the fan must have been greatly changed, because water cannot accumulate in these areas at present.

An alluvial terrace fan 2 meters above the present stream occurs 2 km west and approximately 16 to 25 meters above Lake Vida (Figure 7). The position, shape, and texture of the terrace fan suggest that it may have been deposited in shallow water at an extended stage of Lake Vida (Figure 7). It has strong similarities with the terraces formed by damming of meltwater of the Hobbs and Salmon glaciers (Figure 1) [*Debenham*, 1921a, pp. 75–76]. However, the terrace fan is much coarser than the outwash fan deltas now forming in the shallow margins of Victoria Upper Lake, where the largest material being deposited is only about 8 mm in diameter.

Debris lobes. Two large lobes of bouldery, moraine-like debris project from the 25° to 30° slope forming the riser of the bench northwest of Lake Vida (Figures 7 and 23). Alluvial fans several meters thick radiate from the lobe margins and reach to Lake Vida, covering some of the Bull and older Vida drift.

These lobes, from 600 to 875 meters wide, 50 to 70 meters thick, have steep-sided scalloped margins that average 33° in inclination and are unstable. They consist of angular granitic and gneissic boulders or cobbles in a matrix of sand (10% silt-clay) and pebbles all derived from the slopes above. These facts, with the regular diverging shape of the western lobe, the presence of superimposed solifluction fronts, and the position of the lobes below the dolerite bench suggest that they were emplaced by mass-wasting. Movement must have been slow, since no scar was formed, and the flow may have been initiated as a solifluction sheet. Flow was probably nourished by meltwater from retreating axial glaciers of the Bull and Vida drift episodes or by meltwater from the cirques above (Figure 7), or by both.

The degree of weathering is between that of the Bull drift and of the Vida drift, and it is possible that their formation is related to an expansion of the cirque glaciers above during the interval between Vida and Bull time. This interpretation would be in accord with evidence in Wright Valley that McMurdo Sound may have been partly free of grounded ice and a source of precipitation at this time. However, in the absence of more definite evi-

Fig. 23. Front of debris lobe at the north margin of Lake Vida (see also Figure 7). Scale shown by man at foot of slope, right center. Looking northwest.

dence of age, the major areas of the lobes are correlated with the Vida drift, since the alluvial fans at their fronts truncate and cover Bull drift adjacent to Lake Vida.

History of the Vida Drift Episode

The Vida drift episode followed the major recession of the Bull drift episode of the Victoria glaciation. This episode is distinguished by a change in the regimen of the Webb and Victoria Upper and Lower glaciers. This change involved a glacial stillstand or very slow retreat of one glacier, readvance of the others, increased deposition of till, and possibly the formation of terminal glacial lakes, at the borders of which distinctive outwash deposits were formed.

Victoria Upper, Victoria Lower, and Webb glaciers. During the Vida drift episode, the Victoria Upper and Lower glaciers and the Webb glacier extended about 10 km from their present fronts (Figure 21).

In Barwick and lower Victoria valleys, strong differences in weathering between Vida and Bull drifts and lack of terminal moraines suggest a strong glacial readvance to Lakes Vashka and Vida. In upper Victoria Valley, there was probably no important readvance, but only a pause and slow retreat with deposition of large amounts of debris in the form of closely spaced recessional moraines.

Such glacial action suggests conditions of increased local accumulation and an almost equal increase in ablation. These changes were less accentuated in Barwick Valley. Here, elevation, lower accumulation, and interdependence on a more slowly reacting inland ice sheet were important.

Mass-wasting continued on most of the steep valley walls from the preceding recession through at least the early part of the Vida drift episode. The large amounts of debris contributed by mass-wasting in the Victoria and Barwick valleys before and during the early part of this Vida drift episode, along with the vigorous regimen, enabled the three main valley glaciers to deposit thick blankets of till. In upper and lower Victoria Valley and to a lesser extent in Barwick Valley, much of the glacial debris was redeposited by outwash or sorted by wind action.

Outwash fans of Victoria Valley and high stand of Lake Vida. The Victoria Lower glacier, with its northern tributaries, including the Packard glacier, was about 2 km longer on the north side of the valley than on the south side, much as it is today (Figure 3). Some of the moraine at the terminus was washed into a lake (Lake Vida) formed in the bedrock basin between this glacier and the front of the Victoria Upper glacier. However, much of the material formed a thick outwash fan, built partly over stagnant ice and into Lake Vida.

At nearly the same time, as the Victoria Upper

glacier was retreating in a slow, pulsating fashion, a broad outwash fan began to form at the terminus and probably continued to grow through the Vida drift episode.

The coarse texture of the outwash deposits at both the east and west ends of Lake Vida, compared with similar fans forming today, suggests that they were formed in wetter and perhaps warmer conditions than exist at present. The position and shape of these fans indicate that they were probably deposited at a lake edge which stood 14 to 20 meters above the present level (Figure 7).

Balham and McKelvey ice tongues. In the western half of Barwick Valley, a thick Webb glacier was fed by overflowing inland ice, as well as by local accumulation, whereas, at the heads of Balham and McKelvey valleys, the inland ice tongues extended no lower than 1000 meters above sea level, about 2 km beyond their present ice fronts.

Weathering relations of the moraines of the Vida and Bull drift episodes suggest that only a minor pause within the general retreat may have occurred at this time. However, terminal deposits occur on slopes where solifluction has partly obscured relations with the older Bull drift.

Cirque glaciers. Most of the cirques were occupied by glaciers that were only slightly smaller than in the earlier episode, but, in the cirques opening onto McKelvey Valley from the Olympus Range, former glaciers had so destroyed the peaks and scarp that insufficient area sheltered from the wind remained for accumulation of snow. End moraines were not formed in many of the cirque valleys at this time, and it is quite possible that the major cirque glaciers advanced during the interval before or after the Vida episode when McMurdo Sound may have been partly ice-free.

Age. The carbon 14 analysis of algal peat obtained from the ice-contact alluvial fan at the east end of Lake Vida suggests that the Vida drift is more than 9700 ± 350 years old (Isotopes, Inc., I-619). However, this date is questionable. *Black and Berg* [1964] have pointed out that algae in fresh-water ponds of Victoria Land may be depleted in carbon 14, because the bicarbonates in the system are probably old recycled salts, and exchange of CO_2 is hindered by the ice cover of the ponds most of the year.

Judging by the degree of weathering, the Vida drift may correlate with the oldest Trilogy deposits in Wright Valley (a correlative advance for the Loop glaciation has not been distinguished in the Victoria Valley system). On the basis of an oversimplified relationship of increase of salt accumulation in the soils with age, and a good deal of subjective correlation within Wright Valley and also between Wright and Taylor valleys, the Trilogy (oldest) deposits may be on the order of half a million years old [*Calkin et al.*, 1970] or younger. If these very crude correlations are correct, the Vida drift episode might represent a response to Illinoian glaciation in the northern hemisphere.

Packard Drift

The most recent widespread glacial deposits in the Victoria Valley system extend 5 to 8 km beyond the fronts of the Victoria Upper and Lower glaciers and the Webb glacier. These glacial deposits are distinguished from those referred to as the Vida drift by marked differences in topographic form and preservation or, in at least one area, by an abrupt change in degree of weathering. Although the distinguishing criteria vary from valley to valley, the changes from the Vida deposits suggest a significant correlative change in glacial regimen. This drift is named after the Packard glacier in lower Victoria Valley, south of which are thick ablation deposits. Similar deposits are identifiable at the fronts of Victoria Upper and Lower glaciers and Webb glacier and also near the tongues of the inland ice in Balham and McKelvey valleys. However, near the plateau edge and the bordering cirques, the Packard drift is not so easily distinguished from the Vida drift.

General characteristics. The Packard drift displays more youthful characteristics than the Vida drift, and in fact grades to and includes deposits now being formed. In Victoria Valley, where deposits are thick at the valley margins, the ground moraine is more hummocky and marginal channels are very well defined; solifluction mantling has been slight, and in many areas lateral moraines remain unaltered, some preserving slopes greater than 30° (Figure 24). However, in lower Victoria Valley, the deposits have been mantled by wind-blown sand. Polygons are well developed in the deposits, usually to within a few meters of the glacier fronts.

Weathering of till is less than in the older deposits and extends only a few centimeters below the surface. The till, excluding that of the Victoria Lower glacier, is very bouldery, with average fre-

Fig. 24. Lateral moraine terraces of the Packard drift (Webb glacier). Glacier moved from left to right. View north across Webb Lake, Barwick Valley.

quencies in excess of 100 upstanding boulders per 300-meter traverse (Table 2); cavernous weathering and granular exfoliation have been slight in most deposits (Table 1). Wind cutting is limited, only a few moderately faceted ventifacts being found, and even the less resistant surface boulders are preserved.

The Packard till is sandier and often better sorted than the Vida till (Table 3). Weathering profiles such as those of the Vida and Bull tills are absent or are developed only locally as oxidation profiles extending for a few centimeters below accumulations of dolerite stones or where conditions have favored the formation of a subsurface salt layer.

Knob and kettle and ice-cored moraines in Barwick Valley. In the center of the valley, 5 km east of Webb glacier, is a relatively abrupt change from hummocky (Vida drift) ground moraine of a few meters relief to ridges and knob and kettle topography in (Packard) till (Figure 5). Relief is considerable in this material, greatest (30 meters) in a north-south belt halfway between Hourglass Lake and Webb Lake (*kk* in Figure 5). No break in surficial weathering is distinguished at the contact here with the lower ground moraine of the Vida drift; west of Webb Lake, a few hundred meters within the Packard drift, cavernous weathering in the boulders completely disappears, but the knob and kettle topography is continuous (Figure 25). To the south of the lake, the knob and kettle moraine is bordered sharply by a high-standing ablation moraine which appears to overlie stagnant glacial ice (Figure 5).

The knob and kettle moraine east of Webb Lake contains deep, channel-like troughs between ridges and is pockmarked by elongated kettles. The clearly displayed irregular configuration of the channels and slopes opposing that of the glacier suggests that they were formed around stagnant ice masses. No glacial ice was found within the upper meter of these deposits, and they are probably not ice-cored.

Many striated boulders of medium-grained dolerite and a few of sandstone occur both on ridge tops and channel bottoms of the knob and kettle moraine, but are rare elsewhere in the valley system.

The ice-cored ablation moraine adjacent to the lake and along the west margin of the Webb glacier preserves stagnant ice, probably of glacial origin, by virtue of its great thickness of dolerite boulders (Figure 14). These have been carried from the plateau over the Webb and Haselton icefalls and from immediate valley walls by glacier undermining and talus creep. The presence of a 900-meter-wide belt of ice-cored moraine below the Webb icefall sug-

Fig. 25. Kettles in Packard till. View northwest across Webb Lake to Webb glacier, Barwick Valley.

gests that thinning and severance from the inland ice was the principal cause of stagnation.

Material smaller than cobble size occupies considerably less than 50% of the surface area of these ice-cored moraines, and boulders larger than 1 meter occur at the surface every 1 to 3 meters. In other respects, these ice-cored moraines more than a few tens of meters from active glaciers are typical of those found along the coast of McMurdo Sound. The boulders are very angular and exhibit little or no wind-cutting; hummocky relief is augmented by particularly well-developed polygons, often with marginal furrows more than 1 meter deep (Figure 14); many closed depressions contain frozen ponds; and accumulations of mirabilite and associated salts occur in some areas where ice has recently been exposed and ponds have drained. The numerous isolated ponds within moraines of dolerite can usually be ascribed to a melting ice core, as snow often sublimes from the dark rocks long before the temperature is high enough for melting.

The isolated position of the fresh, ice-cored moraine adjacent to but higher than the material carrying more weathered boulders requires some explanation. The glacier that stagnated and is still preserved in the ice-cored moraine must also have occupied the whole valley area and covered or deposited the apparently older, lower-lying knob and kettle moraine at the southeast end of Webb Lake. One hypothesis is that, during a short readvance, the part of the glacier in the valley center was unencumbered by thick talus accumulations so that it picked up, and perhaps overrode, weathered material without depositing fresh till on retreat.

Till and washed drift in upper Victoria Valley. The boundary between the Vida and Packard drifts is not sharp. The weathering is apparently gradational from the Vida end moraines to the present ice front. The dividing line has been placed at a broad end moraine showing some 20 meters relief, which crosses the valley 8 km south of the Victoria Upper glacier. The closely spaced, hummocky sequence of recessional moraines of the Vida drift south of this moraine contrasts strongly with the channel-dissected ground moraine of the Packard drift bordering Victoria Upper Lake to the north (Figure 7).

Modified till in lower Victoria Valley. On the south side of the valley, 6 km west of the Victoria Lower glacier, an end moraine of 17 meters relief is distinguished sharply in degree of weathering from the Vida drift farther west. This moraine marks the western limit of the Packard till. Neither the plentiful Vida granite boulders nor the less resistant boulders on the valley bottom to the east show significant cavernous weathering (Table 1), and no boulders have been reduced to ground level. On the north side of the valley, the contact is not sharply defined and is represented by a slight concentration of boulders and a wide outwash fan. The deposits of the Packard drift on both sides of the valley bottom here are uniformly fresher in appearance than the deposits in the other valleys. No recessional moraines occur in this till sheet. Above the valley floor, the till is contaminated by older, cavernously weathered boulders, and on the valley floor large areas of wind-deposited sand mantle the deposit.

Stagnant ice that extends 100 to 300 meters from the front of the Victoria Lower glacier is overlain by eolian sand and drift from the active glacier terminus. The material here is gradational in weathering with that farther west. With its sandy texture, better than average sorting, and occasional bedding, it is typical of the deposits extending westward to Lake Vida.

Balham and McKelvey valleys. In the northwest fork at the head of Balham Valley (Figure 4), deposits containing fresh angular dolerite boulders are distributed 1200 meters south of the ice front and may contain a glacial ice core over this distance. At the head of McKelvey Valley, a morainal lobe of dolerite boulders terminates in a depression 1000 meters beyond the McKelvey ice tongue; it is not ice-cored.

Cirques. From the cirques, ground moraine of the Packard drift often extends only a few hundred meters beyond the present glaciers. However, most deposits, even where end moraines occur, are gradational in degree of preservation with the Vida deposits. The termini of most of the cirque glaciers consist of a section of melting stagnant ice overlain by fresh morainal debris (Figure 26).

History of the Packard Drift Episode

The last major glacial episode in the Victoria Valley system for which there is widespread depositional evidence, the Packard drift episode, began with stillstands or minor readvances of Victoria Upper and Lower glaciers and the Webb glacier. At the close of the Vida drift episode, these glaciers stretched no more than 5.5 to 8.9 km from their present positions (Figure 21) and were probably between 300 and 600 meters thick at the positions of the present fronts. The bulk of the deposits of the Packard drift episode appear to evidence a general recession from these to their present positions,

Fig. 26. Ice-cored moraine at terminus of a cirque glacier north of Lake Vida. Looking northeast.

rather than strong readvance from past Vida minimum positions followed by retreat.

The thicker lateral accumulations but very thin deposits in the valley centers, particularly in the area of the present proglacial lakes, contrast with the uniformly thicker deposits left during the Vida drift episode. For example, along the west and south margins of the Webb glacier, drift was thick enough to preserve some glacier ice to the present time. These relationships in part suggest that retreat was more sluggish and glaciers less active than during Vida drift time.

Cirques, alpine glaciers, and the tongues of the inland ice in Balham and McKelvey valleys underwent short stillstands or minor retreats followed by retreat up to 1.2 km. Subsequently the tongues in Balham and McKelvey valleys were almost completely isolated from the inland ice by glacier thinning and emergence of dolerite thresholds at the plateau edge.

During the major part of the Packard drift episode, over-steepened valley walls, frost action, and available moisture allowed the formation of frost rubble and talus, minor amounts of solifluction, alluvial fans, and a few small mudflows. However, slope movement was not so active during the Packard as during the earlier glacial episodes. This was probably a strong factor in the paucity of material deposited by the trunk glaciers, particularly during the latter half of the recession.

There is no evidence that the production of meltwater during the main part of the Packard drift episode greatly exceeded that of the present volume. Apparently, early in the episode, there was a change in stream regimen from deposition to erosion, perhaps induced in part by decrease in glacier load and deposition after the Vida drift episode. Evaporation was also strong during the major part of the episode. In western Barwick Valley, salt accumulations up to 50 cm thick were formed near the ice-cored moraine.

Cause of recession and implications of ice-cored moraines. The cause of the general recession of the episode seems to have been the reduction in local snow accumulation, and, in some cases, a reduction in ice flow fed to the valley glacier tongues from the inland ice plateau. There was probably no abrupt or strong increase in ablation due to higher temperatures for the reasons given above.

The ice is ablating each summer, and there has probably been no extended period of temperatures much higher than the present average since its deposition. The glacial ice core beneath the Barwick deposits and its apparent absence beneath deposits older than the Packard drift in the Victoria Valley system suggest that under present temperature conditions there has been insufficient time since their deposition for any buried ice to be removed. Although the higher level of Lake Vida and the formation of outwash fans and kames during the Vida drift episode can probably be accounted for under present temperature conditions, the possibility of higher temperatures during the Vida drift episode cannot be ruled out. The stillstand or minor readvance initiating the Packard drift episode in Victoria and Barwick valleys may have resulted from a temporary cooling. Further reduction in accumulation could have followed, so that glacial retreat (Packard drift episode) continued without the formation of end moraines and with generally diminishing glacial activity and deposition.

Drift that is similar in degree of weathering to Packard drift and also is frequently ice-cored occurs in the extreme eastern ends of Wright Valley and in the eastern ends of valleys opening onto McMurdo Sound to the southeast. These deposits (the Wright Lower glaciation [*Calkin et al.*, 1970] and the Ross Sea glaciation I [*Denton et al.*, 1970]) appear to be due to westward advances from grounded ice in McMurdo Sound. Although the ice sheet in McMurdo Sound at this time would not have been thick enough to have flowed directly into Victoria Valley, the minor initial advance recorded by the Webb and Victoria glaciers may have been in response to a lowering of sea level and the expansion of the inland ice into the adjacent Ross Sea. Furthermore, the subsequent decrease in local accumulation and retreat of the glaciers of the Packard drift episode would have been the local response expected with the loss of this local precipitation source.

Age. Whether some or all of the glaciers in the Victoria Valley system reacted in response to the last suggested sea-level lowering and buildup of ice in McMurdo Sound is not known. *Denton et al.* [1970] note that, on the basis of carbon 14 dates of bracketing organic debris, the possible correlative of the Packard glacier, the Ross Sea glaciation I, was initiated in McMurdo Sound more than 49,000 years B.P. and terminated before 9490 ± 140 years B.P. (sample Y-2399). *Bennett* [1964] has collected gravity data that suggest that grounded ice was removed from the Ross Sea–McMurdo Sound area between 10,000 and 31,000 years B.P.

SUMMARY AND CORRELATION OF GLACIAL GEOLOGY

The glacial geology of the Victoria Valley system is summarized in Table 4, and the major correlations presented in this paper are summarized in Table 5. It should be emphasized that correlations are presented as a basis of discussions and are speculative in nature. Furthermore, correlations given within the Wright Valley and within the McMurdo Sound–Taylor Valley area between respective glaciations of alpine, Ross Sea, and inland ice origins are still in the preliminary stage. Many differences shown between areas may be resolved by further study within individual valleys, particularly within the Wright Valley.

The correlation of glaciation within the Victoria Valley system and within Alatna Valley to the north appears simpler than that in Wright and Taylor valleys to the south. If there is a real difference in complexity and not just a lack of information, it may be due in part to the greater elevation of the

TABLE 4. Nomenclature and Summary of Glacial Geology

Period	Glacier Action	Deposits	Character of Deposits
Victoria glaciation Packard drift episode	Possible minor readvances; general slow retreat or stagnation to present positions.	Packard drift and associated deposits	Till and morainal debris, sandy and often very bouldery, some washed and sorted drift; knob and kettle areas; ice-cored moraines; little mantling by solifluction. Associated deposits include all those forming in the valley system today i.e. extensive alluvial and eolian deposits.
	Change in glacial regimen		
Vida drift episode	Probable minor readvances to, and stillstand at, 9 to 11 km beyond present positions followed by retreat.	Vida drift and associated deposits	Till and morainal debris, sandy; moderately preserved moraines; thick or broad outwash-kame deposits; associated debris lobes; minor mantling by solifluction.
	Change in glacial regimen		Marked difference in preservation
Bull drift episode	Strong westward advance of Victoria Upper and Lower glaciers extending from 12 to 20 km beyond present positions; reduced invasion from inland ice plateau; followed by stillstand and retreat. (Major advance to and retreat from terminal positions of Victoria glaciation.)	Bull drift and associated deposits	Till, usually very silty; subdued topography but two large, well-preserved end moraines; extreme cavernous weathering of boulders and bedrock; isolated areas of erratic boulders; local lake deposits; associated debris fans; extensive mantling by solifluction.
	Many glacier reversals		Marked Difference in Preservation
Insel glaciation	Strong advance of inland ice into and probably eastward through valley system followed by partial or complete retreat of glaciers.	Insel drift and associated deposits	Till, very silty; no morainal topography, few upstanding resistant boulders; extensive mantling by solifluction. Very resistant, erratic ventifacts and associated frost rubble preserved on high benches.

Alatna and Victoria valley areas and their greater isolation from the Ross Sea. The contribution of local accumulation to axial or through-valley ice movements in the Victoria Valley system or Alatna Valley is not easily distinguished from the contribution from extra-valley sources. Many other factors must be considered before glaciations or parts of glaciations in areas of southern Victoria Land that are not contiguous can be even tentatively correlated. Without recognized interglaciations, it is difficult to delimit the scale of a glaciation or to determine the extent of an advance if it is not known how far ice receded after the last advance [*Angino et al.*, 1962], or whether ice receded completely or stagnated [*Bull et al.*, 1962, p. 72]. In addition, many factors other than regional climatic character or latitude, in the ice-free areas of Antarctica, control the contemporaneity of advances, retreats, or stillstands. These additional factors include exposure, the presence or absence of bedrock thresholds at valley heads, and the position or effect of sea level.

Acknowledgments. This report represents a revision and condensation of Report 10 of The Ohio State University Institute of Polar Studies. Financial support was provided by National Science Foundation grant G-13848 to Tufts University and G-17160 to The Ohio State University Research Foundation. Logistic support of the U.S. Navy is acknowledged. Robert L. Nichols suggested the problem and Richard P. Goldthwait supervised the writing of the original report. Valuable information was obtained from Peter Webb, Graham W. Gibson, A. D. Allen, and A. T. Wilson of the Victoria University of Wellington, New Zealand. Field companion Andre Cailleux of the University of Paris and Herbert Wright, Jr., of the University of Minnesota contributed numerous ideas. George M. Haselton and Thomas C. Davis ably assisted in the field and, with Gerald Holdsworth and Henry Brecher, undertook specialized work to aid this

TABLE 5. Correlation

Mt. Gran-Alatna Valley [Calkin, 1964a]	Victoria Valley System (Calkin, this paper)	Wright Valley [Calkin et al., 1970]				McMurdo Sound and Taylor Valley [Denton et al., 1970]		
		Ross Sea Glaciations (Grounded Ross ice shelf)	Alpine Glaciations	Wright Upper Glaciations (inland ice)		Ross Sea Glaciations (Grounded Ross ice shelf)	Alpine Glaciations	Taylor Glaciations (inland ice)
'B' glaciation Episode 2	Victoria glaciation Packard drift episode	Wright Lower	1	1		9490 yr B.P. 1 >49,000 yr B.P.	1 12,200 yr B.P.*	1
				2			2 0.4 × 10^6 yr	
Episode 1?	Vida drift episode	Trilogy (oldest)	2	↑		2		2
Episode 1	Bull drift episode	Loop Pecten		↓	3 ?	3 4 1.2 × 10^6 yr		
							2.1 × 10^6 yr 3	3 1.6 × 10^6 yr 2.1 × 10^6 yr 4
			3	4 3.7 × 10^6 yr Valley-cutting episodes			3.5 × 10^6 yr	2.7 × 10^6 yr 3.5 × 10^6 yr 5, etc.
'A' glaciation	Insel glaciation							

*C 14 date after *Black and Bowser* [1969].

study. The manuscript was carefully reviewed by Kaye R. Everett, John H. Mercer, and Colin B. Bull of the Institute of Polar Studies, and Harold Borns of the University of Maine. The many discussions with Colin Bull and with George Denton of the University of Maine were invaluable.

Contribution 98 from the Institute of Polar Studies, Ohio State University.

REFERENCES

Allen, A. D., and G. W. Gibson, Outline of the geology of the Victoria Valley region, 6, Geological investigations in southern Victoria Land, Antarctica, *New Zealand J. Geol. Geophys.*, 5, 234–242, 1962.

American Commission on Stratigraphic Nomenclature, Code of stratigraphic nomenclature, *Amer. Assoc. Petrol. Geol. Bull.*, 45, 648–665, 1961.

Angino, E. E., M. D. Turner, and E. J. Zeller, Reconnaissance geology of lower Taylor Valley, Victoria Land, Antarctica, *Geol. Soc. Amer. Bull.*, 73, 1553–1561, 1962.

Bandy, O. L., and R. E. Casey, Major Late Cenozoic planktonic datum planes, Antarctica to the tropics, *Antarctic J. U.S.*, 4, 170–171, 1969.

Bennett, H. F., A gravity and magnetic survey of the Ross ice shelf area, Antarctica, *Univ. Wisc. Geophys. Polar Res. Ctr., Res. Rep. Ser. 64-3*, 97 pp., 1964.

Berg, T. E., and R. F. Black, Preliminary measurements of growth of nonsorted polygons, Victoria Land, Antarctica, in *Antarctic Soils and Soil Forming Processes, Antarctic Res. Ser. 8*, pp. 61–108, AGU, Washington, D.C., 1966.

Beskow, Gunnar, Soil freezing and frost heaving with special application to roads and railroads (translation by J. O. Osterberg, Technological Inst., Northwestern Univ., 1947), 145 pp., 1935.

Black, R. F., and T. E. Berg, Hydrothermal regimen of patterned ground, Victoria Land, Antarctica, *Intern. Assoc. Sci. Hydrol., Comm. Snow Ice, Publ. 61*, 121–127, 1963.

Black, R. F., and T. E. Berg, Glacier fluctuations recorded by patterned ground, Victoria Land, in *Antarctic Geology*, edited by R. J. Adie, pp. 107–122, Wiley, New York, 1964.

Black, R. F., and C. J. Bowser, Salts and associated phenomena of the termini of the Hobbs and Taylor glaciers, Victoria Land, Antarctica, *Comm. Snow and Ice, General Assembly of Bern, 1967*, pp. 227–238, 1969.

Bull, C., Gravity observations in the Wright Valley area, Victoria Land, Antarctica, *New Zealand J. Geol. Geophys.*, 3, 543–552, 1960.

Bull, C., Quaternary glaciations in southern Victoria Land, Antarctica, *J. Glaciol.*, 4, 240–241, 1962.

Bull, C., Climatological observations in ice-free areas of southern Victoria Land, Antarctica, in *Studies in Antarctic Meteorology, Antarctic Res. Ser., 9*, pp. 177–194, AGU, Washington, D.C., 1966.

Bull, C., B. C. McKelvey, and P. N. Webb, Quaternary glaciations in southern Victoria Land, Antarctica, *J. Glaciol.*, 4, 63–78, 1962.

Calkin, Parker, Glacial geology of the Mount Gran area, southern Victoria Land, Antarctica, *Geol. Soc. Amer. Bull.*, 75, 1031–1036, 1964a.

Calkin, Parker, Geomorphology and glacial geology of the Victoria Valley system, southern Victoria Land, Antarctica: *Ohio State Univ. Inst. Polar Stud. Rep. 10,* 66 pp., 1964b.

Calkin, P. E., R. E. Behling, and C. Bull, Glacial history of Wright Valley, southern Victoria Land, Antarctica, *Antarctic J. U.S., 5,* 22–27, 1970.

Calkin, P. E., and C. Bull, Lake Vida, Victoria Valley, Antarctica, *J. Glaciol., 6,* 833–836, 1967.

Calkin, Parker, and A. Cailleux, A quantitative study of cavernous weathering (taffonis) and its application to glacial chronology in Victoria Valley, Antarctica, *Z. Geomorphol., N.F., 6,* 317–324, 1962.

Crary, A. P., Results of United States traverses in East Anarctica, 1958–1961, *IGY Glaciol. Rep. 7,* American Geographical Society, New York, 144 pp., 1963.

Crary, A. P., Mechanism for fiord formation indicated by studies of an ice-covered inlet, *Bull. Geol. Soc. Amer., 77,* 911–929, 1966.

David, T. W. E., and R. E. Priestley, Glaciology, physiography, stratigraphy, and tectonic geology of South Victoria Land, in *Scientific Investigations, Geology, Reports: British Antarctic Exped. 1907–9,* vol. 1, William Heinemann, London, 319 pp., 1914.

Davies, W. E., Surface features of permafrost in arid areas (abstracts for Symposium on physical geography of Greenland, pp. 1–5), 19th Intern. Geol. Cong., Copenhagen, 1960.

Debenham, F., Recent and local deposits of McMurdo Sound region, in *British Antarctic ("Terra Nova") Expedition, 1910, Brit. Mus. (Nat. Hist.), Geol., 1*(3), 63–100, 1921a.

Debenham, F., The sedimentary rocks of South Victoria Land, 4a, The sandstone, etc., of the McMurdo Sound, Terra Nova Bay, and Beardmore Glacier regions, in *British Antarctic ("Terra Nova") Expedition, 1910, Brit. Mus. (Nat. Hist.), Geol., 1*(4), 104–110, 1921b.

Denton, G. H., R. L. Armstrong, and M. Stuiver, Histoire glaciaire et chronologie de la région du Détroit de McMurdo, sud de la terre Victoria, Antarctide, Note préliminaire: *Rev. Géograph. Phys. Géol. Dynamique,* [2]*11,* 265–278, 1969.

Denton, G. H., R. L. Armstrong, and M. Stuiver, Late Cenozoic glaciation in Antarctica: The record in the McMurdo Sound region, *Antarctic J. U.S., 5,* 15–21, 1970.

Donn, W. L., W. R. Farrand, and M. Ewing, Pleistocene ice volumes and sea-level lowering, *J. Geol., 70,* 206–214, 1962.

Ericson, D. B., and G. Wollin, Pleistocene climates and chronology in deep-sea sediments, *Science, 162,* 1227–1234, 1968.

Evernden, J. F., and J. R. Richards, Potassium-argon ages in eastern Australia, *J. Geol. Soc. Australia, 9*(1), 24, 1962.

Fikkan, Philip R., Granitic rocks in the dry valley region of southern Victoria Land, Antarctica, M.S. thesis, Univ. of Wyoming, Laramie, 1968.

Gibson, G. W., Geological investigations in southern Victoria Land, Antarctica, 8, Evaporite salts in the Victoria Valley region, *New Zealand J. Geol. Geophys., 5,* 361–374, 1962.

Giovinetto, M. B., The drainage systems of Antarctica, in *Antarctic Snow and Ice Studies, Antarctic Res. Ser., 2,* pp. 127–155, AGU, Washington, D.C., 1964.

Gunn, B. M., and G. Warren, Geology of Victoria Land between the Mawson and Mulock glaciers, Antarctica, *New Zealand Geol. Surv. Bull., 71,* 157 pp., 1962.

Hamilton, W., Diabase sheets of the Taylor Glacier region, Victoria Land, Antarctica, *U.S. Geol. Surv. Prof. Pap. 456-B,* 1965.

Hollin, J. T., On the glacial history of Antarctica, *J. Glaciol., 4,* 173–195, 1962.

Hollin, J. L., Origin of ice ages: An ice shelf theory for Pleistocene glaciation, *Nature, 202,* 1099–1100, 1964.

Hough, J. L., Pleistocene lithology of Antarctic ocean-bottom sediments, *J. Geol., 58,* 254–260, 1950.

Inman, D. L., Measures for describing the size distribution of sediments, *J. Sed. Petrol., 22,* 125–145, 1952.

Lachenbruch, A. H., Mechanics of thermal contraction cracks and ice-wedge polygons in permafrost, *Geol. Soc. Amer., Spec. Pap. 70,* 69 pp., 1962.

Loewe, F., The scientific observations of the Ross Sea party of the Imperial Trans-Antarctic Expedition of 1914–1917, *Ohio State Univ. Inst. Polar Stud. Rep. 5,* 43 pp., 1963.

Mannerfelt, C. M., Marginal drainage channels as indicators of the gradients of Quaternary ice caps, *Geograf. Ann., 31,* 194–199, 1949.

McDougall, I., Potassium-argon age measurements on dolerites from Antarctica and South Africa, *J. Geophys. Res., 68,* 1535–1545, 1963.

McKelvey, B. C., and P. N. Webb, Geological reconnaissance in Victoria Land, Antarctica, *Nature, 189,* 545–547, 1961.

McKelvey, B. C., and P. N. Webb, Geological investigations in southern Victoria Land, Antarctica, 3, Geology of Wright Valley, *New Zealand J. Geol. Geophys., 5,* 143–162, 1962.

Mercer, J. H., Glacier variations in the Antarctic, *Amer. Geog. Soc., Glaciol. Notes, 11,* 5–29, 1962.

Mercer, J. H., The discontinuous glacio-eustatic fall in Tertiary sea level, *Palaeogeogr., Palaeoclimatol., Palaeoecol., 5,* 77–85, 1968.

Minshew, V. H., and J. H. Mercer, Miocene volcanism and possible uplift of the Transantarctic Mountains, Antarctica, in preparation, 1971.

Mirsky, A., Reconsideration of the "Beacon" as a stratigraphic name in Antarctica, in *Antarctic Geology,* edited by R. J. Adie, pp. 364–378, Wiley, New York, 1964.

Mirsky, A., S. Treves, and P. Calkin, Stratigraphy and petrography, Mount Gran area, southern Victoria Land, Antarctica, in *Geology and Paleontology of the Antarctic, Antarctic Res. Ser., 6,* pp. 145–175, AGU, Washington, D.C., 1965.

Nichols, R. L., Geomorphology of Marguerite Bay, Palmer Peninsula, Antarctica, *Ronne Antarctic Research Exped. Tech. Rep. 12,* Office of Naval Research, Dept. of the Navy, Washington, D.C., 151 pp. 1953.

Nichols, R. L., Geomorphology of Marguerite Bay area, Palmer Peninsula, Antarctica, *Geol. Soc. Amer. Bull., 71,* 1421–1450, 1960.

Nichols, R. L., Multiple glaciation in the Wright Valley,

McMurdo Sound, Antarctica (abstract), 10th Pacific Science Congress, Honolulu, 1961.

Nichols, R. L., Geology of Lake Vanda, Wright Valley, South Victoria Land, Antarctica, in *Antarctic Research, Geophys. Monogr. 7*, pp. 47–52, AGU, Washington, D.C., 1962.

Nichols, R. L., Geologic features demonstrating aridity of McMurdo Sound area, Antarctica, *Amer. J. Sci., 261*, 20–31, 1963.

Nichols, R. L., Present status of Antarctic glacial geology, in *Antarctic Geology*, edited by R. J. Adie, pp. 123–137, Wiley, New York, 1964a.

Nichols, R. L., Four-fold check on mean annual temperature, McMurdo Sound, Antarctica, *J. Glaciol., 5*, 353–355, 1964b.

Nichols, R. L., Antarctic interglacial features, *J. Glaciol., 5*, 433–449, 1965.

Nichols, R. L., Geomorphology of Antarctica, in *Antarctic Soils and Soil Farming Processes, Antarctic Res. Ser. 8*, pp. 1–46, AGU, Washington, D.C., 1966.

Nichols, R. L., Glacial geology of the Wright Valley, McMurdo Sound, Antarctica, Proceedings of Symposium on Antarctica, American Association for the Advancement of Science, Dallas, December 1968, in press, 1971.

Opdyke, N. D., B. Glass, J. N. Hays, and J. Foster, A paleomagnetic study of Antarctic deep sea cores, *Science, 154*, 349–357, 1966.

Péwé, T. L., Quaternary glacial geology of the McMurdo Sound region, Antarctica: A progress report, *Amer. Geog. Soc., IGY Glaciol. Rep. 1*, VI-1-4, 1958.

Péwé, T. L., Sand-wedge polygons (tesselations) in the McMurdo Sound region, Antarctica: A progress report, *Amer. J. Sci., 257*, 545–552, 1959.

Péwé, T. L., Multiple glaciation in the McMurdo Sound region, Antarctica: A progress report, *J. Geol., 68*, 498–514, 1960.

Péwé, T. L., Age of moraines in Victoria Land, Antarctica, *J. Glaciol., 4*, 93–100, 1962.

Péwé, T., and R. Church, Glacier regimen in Antarctica as reflected by glacier-margin fluctuation in historic time with special reference to McMurdo Sound, *Intern. Assoc. Sci. Hydrol., Comm. Snow Ice, Publ. 58*, pp. 295–305, 1962.

Ronca, L. B., Minimum length of time of frigid conditions in Antarctica as determined by thermoluminescence, *Amer. J. Sci., 262*, 767–781, 1964.

Różycki, S. Z., Changements Pléistocènes de l'extension de inlandis en Antarctide orientale d'après l'étude des anciennes plages élevées de l'Oasis Bunger Queen's Mary Land (Pleistocene changes in the extension of the inland ice of East Antarctica, according to a study of ancient raised beaches of Bunger's Oasis, Queen Mary Land), *Biul. Peryglacjalny*, no. 10, 257–283, 1961.

Rutford, R. H., C. Craddock, and T. W. Bastien, Late Tertiary glaciation and sea-level changes in Antarctica, *Palaeogeogr., Palaeoclimatol., Palaeoecol., 5*, 15–39, 1968.

Scott, R. F., Results of the National Antarctic Expedition, 1, *Geog. J., 25*, 353–373, 1905.

Sissons, J. B., Subglacial, marginal, and other glacial drainage in the Syracuse-Oneida area, New York, *Geol. Soc. Amer. Bull., 71*, 1575–1588, 1960.

Smith, H. T. U., Anomalous erosional topography in Victoria Land, Antarctica, *Science, 141*, 941–942, 1965.

Taylor, Griffith, The physiography of the McMurdo Sound and Granite Harbour region, in *British Antarctic ("Terra Nova") Expedition, 1910–13*, Harrison and Sons, London, 246 pp., 1922.

Thomas, C. W., Late Pleistocene and Recent limits of the Ross ice shelf, *J. Geophys. Res., 65*, 1789–1792, 1960.

Ugolini, F. C., A study of pedologic processes in Antarctica: Technical report to the National Science Foundation (unpublished), 1964.

Ugolini, F. C., and C. Bull, Soil development and glacial events in Antarctica, *Quaternaria, Roma, 7*, 251–269, 1965.

Washburn, A. L., Classification of patterned ground and a review of suggested origins, *Geol. Soc. Amer. Bull., 67*, 823–866, 1956.

Wilson, A. T., Origin of ice ages: An ice shelf theory for Pleistocene glaciation, *Nature, 201*, 147–149, 1964a.

Wilson, A. T., Evidence from chemical diffusion of a climatic change in the McMurdo dry valleys 1200 years ago, *Nature, 201*, 176–177, 1964b.

Wilson, C. R., and A. P. Crary, Ice movement studies on the Skelton glacier, *J. Glaciol., 3*, 873–878, 1961.

Wright, C. S., and R. E. Priestley, Glaciology, in *British Antarctic ("Terra Nova") Expedition, 1910–13*, Harrison and Sons, London, 581 pp., 1922.

DATE DUE

MAY 1 5 1975		
AUG 1 3 1996		